CANADIAN CITIES IN TRANSITION

NEW DIRECTIONS IN THE TWENTY-FIRST CENTURY

FOURTH EDITION

Trudi Bunting
Pierre Filion
Ryan Walker

OXFORD
UNIVERSITY PRESS

OXFORD

UNIVERSITY PRESS

8 Sampson Mews, Suite 204, Don Mills, Ontario M3C 0H5
www.oupcanada.com

Oxford University Press is a department of the University of Oxford.
It furthers the University's objective of excellence in research, scholarship,
and education by publishing worldwide in

Oxford New York

Auckland Cape Town Dar es Salaam Hong Kong Karachi
Kuala Lumpur Madrid Melbourne Mexico City Nairobi
New Delhi Shanghai Taipei Toronto

With offices in

Argentina Austria Brazil Chile Czech Republic France Greece
Guatemala Hungary Italy Japan Poland Portugal Singapore
South Korea Switzerland Thailand Turkey Ukraine Vietnam

Oxford is a trade mark of Oxford University Press
in the UK and in certain other countries

Published in Canada
by Oxford University Press

Library and Archives Canada Cataloguing in Publication

Canadian cities in transition : new directions in the twenty-first century / edited by
Trudi Bunting, Pierre Filion and Ryan Walker. — 4th ed.
Includes bibliographical references and index.
ISBN 978-0-19-543125-4

1. Cities and towns—Canada—Textbooks. 2. Urbanization—Canada—Textbooks.
I. Bunting, Trudi E., 1944–   II. Filion, Pierre, 1952–   III. Walker, Ryan Christopher, 1975–

HT127.C32 2010   307.760971   C2010-901555-X

Cover image: 401 Richmond Street West roof garden in Toronto / Urbanspace Property Group

This book is printed on permanent (acid-free) paper ∞.
Printed and bound in Canada.

1 2 3 4 — 13 12 11 10

Contents

Part III   'Placing' and Planning the Twenty-First-Century City

Part IV   Towards a Sustainable, Healthy, and 'Smart' Future for Canadian Urban Communities

Publisher's Preface

Oxford University Press is proud to introduce the fourth edition of *Canadian Cities in Transition: New Directions in the Twenty-First Century*, edited by Trudi Bunting, Pierre Filion, and Ryan Walker. Designed as a core text for courses in urban geography and urban planning with a Canadian focus, this relevant and comprehensive survey of the disciplines examines urbanization from both modern and traditional perspectives. Students will glean a rich understanding of urban geography and planning in Canada, from the founding principles to the current trends shaping the discipline today. By introducing new material, this edition of *Canadian Cities in Transition* views recent and ongoing changes to our urban environment as transformational, while still maintaining the strength of the previous editions in providing authoritative selections.

Highlights of the New Edition

- **New voices.** A mix of new and established authors in the field of urban geography and planning ensures the balance of authoritative voices to interpret trends, both past and present, that are shaping Canadian cities in the twenty-first century.
- **New topic coverage.** Chapters covering Aboriginal peoples living in urban areas, environmental challenges, obesity and the auto-dependent city, and urban food systems offer students an even broader introduction to a wide range of issues.

- **Updated visuals.** Added visuals and the new 2-colour format make this edition more accessible and visually appealing. The art program has been revised to include images that help contextualize key locales and important material.
- **New pedagogical aids.**
  - The text has been organized into 4 parts, with introductions to help contextualize the material.
  - New chapter-end review questions and cross-chapter references help students to synthesize and engage with the material.
  - New appendices provide students with updated census data on metropolitan areas and information on recent trends in urban research.
  - A glossary has been added to this edition to facilitate comprehension of key concepts and terms presented in the text.
  - A new study guide and test bank provided to students online will enhance the student's learning experience and comprehension of the subject matter.
  - Foundational chapters from the third edition are provided as resource material and made available at the website for *Canadian Cities in Transition*: www.oupcanada.com/bunting4e.

Preface: A Guide to the Text

This fourth edition of *Canadian Cities in Transition: New Directions in the Twenty-First Century* is designed to serve a number of purposes. It is an introduction for university students to the Canadian urban phenomenon, presenting different facets of the city: its historical evolution, economic dynamics, environmental impacts, dependence on natural systems, urban lifestyles, cultural makeup, social structure, infrastructures, governance, planning, and appearance. The volume also is designed to assist the next generation of citizens, consumers, experts, business people, and politicians in their efforts to solve the urban problems—traffic congestion, different forms of environmental damage, crime, social segregation, governance—they are inheriting. Canadian cities are not simply a collection of problems to be solved, however, and this book helps to articulate the promise of urban Canada, where people and public space are recentred for economic, environmental, and social reasons, and where 'quality' instead of simply 'growth' becomes a unifying hallmark of urban policy. The book offers a state of the art of the knowledge on the Canadian city. The contributors review the recent literature and research on different aspects of the city, and provide their expert opinion on how to focus our examination of contemporary urban issues. Finally, the volume provides an update on urban Canada by identifying the main characteristics of the contemporary Canadian urban phenomenon, its problems, achievements, and opportunities. In this regard the text will help citizens make sense of the vast flow of information on cities circulated by the media. Because quality information is a condition for judicious decisions, knowledge of the city is vital to effective planning, private and public development, and consumer choice.

The text is situated at the confluence of different disciplines with an urban dimension: mainly geography and planning, but also economics, political science, sociology, ecology, and history. It focuses on different urban themes and draws on all the disciplines relevant to their exploration. It also considers cities belonging to all size categories as well as to different Canadian regions. Contributors who represent all parts of the country are able to highlight cross-country differences as well as similarities by drawing on examples from their own regions.

It must be stressed that this volume is very different from earlier versions of *Canadian Cities in Transition* because it heralds the new urban entity of the twenty-first century and explores aspects of cities not considered in previous editions. This fourth edition acknowledges that we are confronting a new kind of urban Canada after having been on the brink of a major transformation for many years. The picture we have put together here is one of a different city from that depicted in earlier editions. While all three previous editions of this text have been concerned with change and transition, this edition might be said to be about transformative change. The text also differs from earlier editions because it adds to the dimensions of cities considered in prior versions. Space does not allow us to repeat lessons learned about fundamental structural parameters of the city that were discussed in earlier versions of this book and that remain relevant today. To this end many of the chapters of the third edition have been placed on the Internet at www.oupcanada.com/bunting4e. These chapters deal with essential aspects of Canadian cities that are not covered in the fourth edition because of its attempt to broaden the scope of urban topics and to highlight new areas of urban inquiry brought to light by the postmodern city.

The 25 chapters of this edition are organized into four parts. The three chapters in Part 1, written by Trudi Bunting and Pierre Filion, serve an introductory role by setting the context for the inquiries that occupy the following chapters. In Chapter 1, they lay out seven universal properties common to all cities. These properties explain the existence of the urban phenomenon as well as its different manifestations over time. The next chapter provides historical background for the material contained in the book. It describes different epochs of urban development in Canada, from the origins of resource-oriented colonial settlements to the urban surge associated with industrialization, and ultimately to the present post-industrial period. Chapter 3 explores the transition and transformation theme at the centre of the book. The chapter concentrates on two periods, the 1950s and the present. It reflects on the conditions that fostered a deep and lasting urban transformation in the 1950s, and considers how the present epoch offers a context similarly suited to a radical change of direction.

The 10 chapters in Part 2, 'The Structuring Parameters of Twenty-First-Century Urbanization', look at the fundamental aspects of cities that determine their form, dynamics, and evolution. This second part also examines how cities adapt to changing societal and global contexts. In Chapter 4, Peter V. Hall demonstrates how global trends—economic, demographic, political, and cultural—reverberate on Canadian cities. He argues that Canadian cities are increasingly integrated into global networks, a source of prosperity for some and decline for others. Chapter 5, by William Rees, is about the environmental impact of cities. It pictures them as an important contributor to global environmental damage. The chapter also explores the vulnerability of cities to environmental deterioration and the need for them to deploy long-term sustainability strategies. With

Chapter 6, by Tara Vinodrai, the book begins an examination of urban economics, which will be the subject of a number of subsequent chapters. Vinodrai focuses on how cities are affected by present and recent economic trends and how they have reacted to these trends. She acknowledges the shift away from the industrial to the post-industrial city and the rising importance of innovative and knowledge sectors of the economy, but sees this trend as having both positive and negative impacts on cities. Chapter 7 also focuses on the economy of cities. Here, Tom Hutton concentrates on the locational dynamics of employment. He shows how job location within cities is a consequence of both broad economic trends and the specificities of metropolitan regions, which explains why different urban areas present different employment distribution patterns. Chapter 8, by Ivan Townshend and Ryan Walker, is about social changes affecting cities: demography, life course, and lifestyles. Among other things, the chapter highlights the effects of aging, the extension of youth, and the coexistence of numerous lifestyles, as well as the impact of these trends on the built environment and community dynamics within our cities. In Chapter 9, Heidi Hoernig and ZhiXi (Cecilia) Zhuang focus on immigration and the resulting social diversity. They chart the geography of immigration in Canada—the urban areas that especially attract immigrants and where immigrants concentrate in these cities. They also describe new urban phenomena associated with immigration, such as the emergence of 'ethnoburbs', and end with two case studies: one involving ethnic retailing, the other, places of worship. The object of Chapter 10 is social polarization. R. Alan Walks identifies a range of factors accounting for growing income polarization among households over the last decades. He also paints the urban consequences of polarization, that is, an urban social geography that is increasingly characterized by unevenness. In Chapter 11, Anthony Perl and Jeffrey Kenworthy provide a wide range of observations on the state of urban transportation in Canadian cities. This particular chapter serves to illustrate the distinctive evolution of transportation in Canadian cities as compared to other developed countries across the globe. The chapter underscores the conditions that encourage and impede public transit use, walking, and cycling in urban Canada. Chapter 12, by Ahmed Allahwala, Julie-Anne Boudreau, and Roger Keil, looks at the changes in urban policies resulting from a shift to neo-liberalism. The chapter uses the amalgamations of Toronto and Montreal to illustrate this form of contemporary urban governance. The last contribution of this part of the book, Chapter 13 by Andrejs Skaburskis and Markus Moos, examines the economics of urban land use. The chapter introduces the structuring parameters of urban land use and describes the origin and operation of urban land markets as well as their outcomes. Skaburskis and Moos end by introducing the dimensions of timing and strategy to help explain development decisions that challenge our conventional views of when, where, and how intensively land is capitalized.

Part 3 is about responses to contemporary trends and issues affecting Canadian cities, as well as specific aspects of their operation. In Chapter 14, Sean Connelly and Mark Roseland review different approaches taken to abate the environmental damage caused by urban development, discerning between ecological modernization and sustainable community development. They also investigate the efficacy of these approaches through case studies of initiatives in Surrey, BC, and Toronto. Chapter 15, by Alison Bain, deals with emerging places in Canada's inner or central cities: gentrified neighbourhoods, high-rise condominiums, and the like. It relates the conditions that have led to their development and the impact these places have on the social structure and functioning of cities. If Chapter 10 described polarization *within* urban areas, Chapter 16, by Betsy Donald and Heather Hall, zeroes in on polarization *among* urban areas. It describes the multiple challenges that declining

urban areas face and policy responses that could mitigate the consequences of decline. This issue is increasingly relevant given the present concentration of demographic and economic growth in a few large metropolitan areas and their surrounding regions. Donald and Hall convey the opportunities missed by urban decision-makers too narrowly focused on growth, rather than on qualitative development. Chapter 17, by Deborah Cowen, Amy Siciliano, and Neil Smith, deals with fear and insecurity in the city. It demonstrates how fear is constructed and leads to calls for security measures that often result in infringements of freedom. The chapter shows how behaviour and politics in the city respond to fear and insecurity. In Chapter 18, Jill Grant and Pierre Filion pursue the planning transition theme introduced in Chapter 3. Chapter 18 is about the loud call within the planning profession for a change in urban development trends. It describes and evaluates attempts at intensifying the urban environment while acknowledging the counter-effect of many new automobile-dependent urban forms such as power malls. Chapter 19 concentrates on the meaning of urban space. In this chapter, Nicholas Lynch and David Ley provide an historical perspective on urban places in Canada. They also focus on the meanings of contemporary places, which echo the increasingly globalized and polarized nature of Canadian cities. Chapter 20, by Ryan Walker and Tom Carter, is devoted to housing, the single largest user of space in Canadian cities, and its centrality to neighbourhood and urban transformation. After reviewing characteristics of the Canadian housing system, they articulate a series of social, environmental, and economic arguments pertaining to the role for public policy and communities in the twenty-first-century housing system.

Subsequent chapters in Part 3 are attuned to pressing or newly emergent problems. In response to the impending demographic shift that will accompany the aging of the 'baby boom' generation, Mark Rosenberg and Dana Wilson explore in

Chapter 21 how the aging population affects Canadian cities. Equally important, the authors address the place of youth in the twenty-first-century city. Addressing both youth and aging together, Rosenberg and Wilson offer a typology for Canadian cities as younger, older, or hanging 'in the balance', using age as the primary structuring parameter. They provide examples of difficulties that both younger and older residents encounter in urban environments and conclude by offering ideas and examples for creating an 'all' age-friendly city. Chapter 22, by Evelyn Peters, explores the presence of Aboriginal peoples in Canadian cities. It documents their concentration in certain cities and the emergence of an urban culture among Canadian Aboriginal peoples—a phenomenon likely to accelerate in the future given the differential rate of natural increase among First Nations and Métis populations in comparison with other native-born Canadians and newcomers. The chapter also considers how cities and urban services can adapt to the presence of Aboriginal peoples. Chapters 23 and 24 deal with topics that have been researched only in recent years. In Chapter 23, Jason Gilliland provides evidence on the association between automobile-dependent urban settings and levels of obesity, and explores ways of promoting active living through urban form and transportation policy. Gilliland is particularly concerned with the negative impact that current urban design has on youth, and considers the long-term habits being acquired by these future adult urban citizens. Chapter 24, by Alison Blay-Palmer, looks at the new-found interest in the geography of food, as evidenced in movements such as 'the 100-mile diet'. This chapter is about how food is procured and distributed within Canadian cities. Issues include accessibility to different forms of food outlets and the problem of 'food deserts', efforts to increase reliance on food grown nearby, and food production within cities themselves.

Part 4—Chapter 25, by Larry S. Bourne and R. Alan Walks—draws on lessons from the book to sketch possible future urban scenarios. These revolve around the themes of complexity, contradictory tendencies, and uncertainty.

Contributors

Ahmed Allahwala
Department of Geography
York University

Alison Bain
Department of Geography
York University

Alison Blay-Palmer
Geography and Environmental Studies
Wilfrid Laurier University

Julie-Anne Boudreau
INRS urbanisation culture société

Larry S. Bourne
Department of Geography and Program in
    Planning
University of Toronto

Trudi Bunting
School of Planning and Department of Geography
    and Environmental Management
University of Waterloo

Tom Carter
Department of Geography
University of Winnipeg

Sean Connelly
Centre for Sustainable Community Development
Simon Fraser University

Deborah Cowen
Department of Geography and Program in
    Planning
University of Toronto

Betsy Donald
Department of Geography
Queen's University

Pierre Filion
School of Planning
University of Waterloo

Jason Gilliland
Department of Geography
University of Western Ontario

Jill L. Grant
School of Planning
Dalhousie University

Heather M. Hall
Department of Geography
Queen's University

Peter V. Hall
Urban Studies Program
Simon Fraser University

Heidi Hoernig
Office of Research Opportunities
McGill University

Thomas Hutton
School of Community and Regional Planning
University of British Columbia

Roger Keil
Faculty of Environmental Studies
York University

Jeffrey Kenworthy
Curtin Sustainability Institute
Curtin University of Technology, Australia

Paul Langlois
Department of Geography and Program in
    Planning
University of Toronto

David Ley
Department of Geography
University of British Columbia

Nicholas Lynch
Department of Geography
University of British Columbia

Markus Moos
Department of Geography
University of British Columbia

Anthony Perl
Urban Studies Program
Simon Fraser University

Evelyn J. Peters
Johnson–Shoyama Graduate School of
    Public Policy
University of Saskatchewan

William E. Rees
School of Community and Regional Planning
University of British Columbia

Mark Roseland
School of Resource and Environmental
    Management
Simon Fraser University

Mark W. Rosenberg
Department of Geography
Queen's University

Amy Siciliano
Department of Geography and Program in
    Planning
University of Toronto

Andrejs Skaburskis
School of Urban and Regional Planning
Queen's University

Neil Smith
Department of Anthropology, Earth and
    Environmental Sciences
The City University of New York

Ivan Townshend
Department of Geography
University of Lethbridge

Tara Vinodrai
Department of Geography and Environmental
    Management
University of Waterloo

Ryan Walker
Department of Geography and Planning
University of Saskatchewan

R. Alan Walks
Department of Geography and Program in
    Planning
University of Toronto

Dana H. Wilson
Department of Geography and Program in
    Planning
University of Toronto

Cecilia Zhuang
School of Urban and Regional Planning
Ryerson University

Part I

Changing Parameters of Urban Form, Structure, and Policy

This book is premised on the probability that we are on the brink of a major transition in twenty-first-century urban form, this time, one of epochal magnitude. Our main theme for all four editions has been that cities are all about change, that 'transition' is inherent to the nature of urbanity. But the change we have spoken of in the previous three editions has been of a gradual evolutionary nature (see Chapter 3), whereas this volume purports to embrace change that is more radical and transformative.

It needs to be emphasized that the third edition (Bunting and Filion, 2006) continued in the tradition of the first two editions in its attempt to generate explanation and understanding about the form and structure of the late twentieth-century city. In the case of the first three editions each chapter included illustrative material exemplifying recent (incremental) change and transition in the Canadian city. This volume is less concerned with explanation at this point in time simply because the trends that we observe here are too new; we cannot aim for explanation simply because we do not yet know enough about ongoing changes or how these changes all come together to be expressed as urban form. Accordingly, most of the chapters in this volume (particularly in Parts 2 and 3) are oriented towards elucidating the character of what might become a twenty-first-century urban entity that evolves along a radically different path from its twentieth-century roots.

Since the fourth edition of *Canadian Cities in Transition* is primarily an exploration of the new twenty-first-century Canadian city, it is imperative that this introductory part deal theoretically with what we know about urban form and structure. We do this from an ascending historic perspective that moves us from ancient times when cities first appeared (*circa* 2000 BC) to around a decade ago at the end of the twentieth century. Setting out this historic perspective is important for three reasons: first, it demonstrates the weight of theoretical understanding about urban form and structure; second, it provides important context for understanding current styles of urban development and the marked ways in which the 'new' city departs from the old; third, in cities it is especially true that 'the past is always with us' (see in particular the capitalization property in Chapter 1) and that past development styles continue as a living legacy within the urban landscape.

Three themes—complexity, uncertainty, and contradiction—underlie materials presented throughout this volume. These overriding themes help us to understand the distinctive nature of cities and also the difficulty that decision-makers have in defining appropriate policy responses to serious urban problems. Because these themes exist at the macro level, they are introduced briefly here so the reader might keep them in mind while reading the main body of the text. All three of these themes are picked up again in greater detail in the concluding part of the volume. Since this first part is mainly designed to 'fill in the blanks' from an historical perspective, it is largely devoted to the first two themes, complexity and contradiction.

Chapter 1 introduces the seven principles or properties that we believe comprise the essence of the city. These characteristics pertain to cities of the past, present—and, as far as we are able to foresee, the future. Likewise, these properties apply to urban settlements globally. Six of these principles—production, proximity, reproduction, capitalization, place, and governance—represent patterns that we discussed in the third edition, though our discussion has been reworked here to suit this volume. The seventh property, environment, by which we mean 'biophysical' environment, is one that hindsight informs us should have been featured in the previous edition but that truly only stood out during the course of researching this volume. The point is that all urban places have distinctive environments that explain their location, site or situation and that contribute to making them 'good' places to live and do business. The additional insight that we bring today to our understanding of urban environments is that the process of urbanization itself has a huge negative impact on the biophysical environment. The call for the development of new sustainable forms of urban development and the retrofitting of urban environments to reduce ecological 'footprints' and render them more 'environment-friendly' is a late twentieth-century and early twenty-first-century awakening that underscores our third macro principle of uncertainty. This recent awakening, on the part of scholars, planners, politicians, and the residents of urban environments, also helps to explain the huge nature of the change being called for (and witnessed) at present in the relationship between cities and the biophysical environment.

Chapter 2 discusses Canada's past urban evolutionary development. The first two periods, mercantile and first agricultural settlement, spanning the years of approximately 1600–1850, are dealt with succinctly. The main impact here is witnessed as cities became established from the Atlantic to the Pacific and the urban system expanded (see Simmons and McCann, 2006; McCann and Simmons, 2001; Simmons and McCann, 2001). Beginning with the era we term the 'Great Transition' (*circa* 1850–1945) we discuss developmental patterns in greater detail and, as per the remainder of the volume, our focus is more on the internal structure of the city. The Great Transition and the following two eras, Post-War Economic Boom, 1945–1975, and Neo-liberalism/Deindustrialization, which bring us almost to the present time, are elaborated with cartographic schema to illustrate the distinctive intra-urban geographic patterns associated with each period. Above all, these serve to illustrate the importance of the seven principles of urbanity introduced in the first chapter; among these, two, production technologies and proximity, stand out most clearly.

Chapter 3, 'Transition in the City', departs from the first two chapters in that its primary concern is with change as opposed to substantive urban issues. This chapter begins by considering the role that visionary thinking plays in directing future urban growth alongside the two opposing tendencies that are reflected as urban evolutionary forces: stability and change. It also picks up from Chapter 2 and reflects on the nature of change and transition since World War II. The concluding portion of the chapter is open-ended to the extent that it considers signs on the horizon suggesting that the coming transition will be one of radical proportions.

CHAPTER 1

Fundamentals of Cities

Pierre Filion and Trudi Bunting

This volume is about understanding and deal-ing with change and transition in twenty-first-century cities. At the outset, however, we need sound knowledge of how cities operate and how they relate to broader societal trends in order to address contemporary urban issues. More often than not, past mistakes can be seen in hindsight to be the result of deficient understanding of **urban dynamics**. Yet cities resist understanding because they are such complex systems. It can be argued that, along with language, the large city is the most intricate of human creations. In both cases, complexity stems from the presence of a relatively stable structure upon which interchangeable ele-ments can be affixed in different fashions. In the case of language the structure is syntax, which supports nearly unlimited combinations of words; cities, on the other hand, owe their structure to major infrastructure networks, which provide con-nections between different assemblages of build-ings and other land uses. The degree of complexity further increases when we consider economic and value systems behind the creation of the urban built environment, the multiple ways people use this environment, and the perceptions and inter-pretations of this environment and of the activities that take place therein.

The present chapter introduces universal properties that define the city and represent uni-versal features of the urban phenomenon. While

Canadian cities have their own distinctiveness—a central theme in this book—they share many similarities with their counterparts developed at different times in different parts of the globe. This chapter, then, establishes some basic principles inherent to urban areas from the very beginning of urban settlement.

Seven properties are fundamental to under-standing the urban phenomenon regardless of time or place: *production*; *proximity*; *reproduction*; *capital-ization*; *place*; *governance*; and *environment*. These properties account for the existence of cities, their diverse configurations, and the way they operate, along with the need for specialized knowledge and interventions to deal with urban issues, and have been inherent in the city from its beginnings some 4,000 years ago in China and Mesopotamia. As expected, their manifestations vary across periods, as we will see in Chapter 2. The seven properties that we set out here provide a framework to dis-cuss problems confronting the contemporary Can-adian city and possible solutions.

Production

Throughout history the foremost raison d'être for urban settlement has been accommodation of the need for specialized production activities that could not survive in isolation in rural settings—e.g., markets, production of fine crafts, centralized governance. Indeed, if asked why they live in a city, most people will respond to the effect that 'I work there or a member of my family does.' Economic production creates jobs and brings people into the city and is thus the main reason for urban growth. Economic production, too, is the property most often associated with transformative change in **urban form** and structure. Thus, for example, Canada's first mercantile settlements were established to export 'staple' products such as fish, furs, and timber to European colonial 'mother' countries. A further impetus for urban growth was the administration of resource industries and transportation systems required for their export. These conditions were conducive to the growth of only a small number of cities, however. With agricultural development, however, centres grew to service rural areas, and then with wide-scale industrialization urban growth took off in Canada as it did in other developed countries (Innis, 1995a [1931], 1995b [1938]).

Many economic activities, of course, are aimed at the consumption needs of a city's own residents (Watkins, 1980). What most sets urban settlements apart from traditional self-reliant rural economies, however, is the historic inability of the city to satisfy all its consumption requirements. As a result, it becomes imperative that cities export so as to generate the revenues needed to acquire products that can only be obtained outside their territory. Inside the urban envelope, for example, the need for specialized activities to be close to one another and to their workers (the proximity property discussed below) rules out any possibility of devoting large surfaces to agriculture as would be needed to feed a large resident population (at least this has been the case until recent attempts at urban-based agriculture) (Christensen, 2007; Lawson, 2005). A city also must reach beyond its territory to secure other products and resources essential for sustaining its population and economic activity. This includes different forms of staples and energy and, often, water. Research shows that cities depend for their natural resources on a territory (or '**ecological footprint**'; see Rees, Chapter 5) that far exceeds the urbanized perimeter.

For a city to exist, it must be in a position to export sufficient goods and services to counterbalance its imports. But exports need not be tied to products. Capital cities, for example, export decisions and derive their monetary returns from tax revenues; likewise, in medieval times, it was not usual for cities to draw taxes, often in kind, from a hinterland to which they extended military protection. Cities that fail to export decline and may disappear altogether, as illustrated by the fate of resource communities throughout the Canadian periphery whose staple has run out. Examples include Elliot Lake, Ontario; Shefferville, Quebec; and Canso, Nova Scotia (Gill and Reed, 1997; Lorch, Johnston, and Challen, 2004; see Donald and Hall, Chapter 16).

Over recent decades, goods and resources (including labour) that cities draw from outside their territory have increasingly originated from foreign countries and continents. Canadian producers have also moved towards new international markets. This change in the reach of economic exchange is loosely referred to as globalization (see Hall, Chapter 4). In the global period, the interdependence that existed in the past between the city and a well-defined hinterland has lost much of its importance (Derudder and Witlox, 2004; Friedman, 2005; Taylor, 2001). The tendency today is for cities to transform and consume goods from around the world and sell their products on international markets. Yet, there are signs that present and future change, driven by environmental awareness and escalating energy costs, may bring about a renewed emphasis on nearness in economic exchanges (Rubin, 2009). Could economic globalization survive an eventual $200 barrel of oil? Thus, Chapter 24 discusses responses to the new emphasis on 'local' procurement of goods and services, particularly foodstuffs (e.g., 'the 100-mile diet'— see Harris, 2009; *Time*, 2006), that accompanies attempts to decrease the urban 'footprint' and so make Canadian cities more environmentally sustainable (see Blay-Palmer, Chapter 24). But it remains the case that cities have to trade with areas outside their territory, either near or afar.

Though the nature of export-based specialized economic production has changed dramatically in the last hundred years in Canadian cities, economic production still is imperative for urban growth and development. With the decline of routine production, developed countries such as Canada must rely on innovation and knowledge-intensive activities in order to compete on the world stage (the 'new' economy is sometimes spoken about as a 'knowledge' economy) (see Vinodrai, Chapter 6). This explains the key role in propelling the economy that Florida (2002, 2005) attributes to a 'creative class', as well as his emphasis on measures cities can take to attract this class. Above all, the growing importance of the service sector relative to the manufacturing sector has characterized the last decades (Daniels and Bryson, 2002). This trend has been acutely felt in both Canada and the US as the economic crisis of 2008–9 has threatened to wipe out General Motors and Chrysler, and thus a large share of the North American car industry. The transition is highly visible in the urban landscape. On the one hand, we see abandoned industrial premises (or their twenty-first-century 'makeover' as lofts), and, on the other, we are confronted with an explosion of restaurants and personal services, and places of entertainment and cultural activities (Jones, Comfort, and Hillier, 2005) (see Bain, Chapter 15). The shift towards the service economy is also felt, albeit with less intensity, within cities' export sphere. Among services Canadian cities export are financial services, engineering and development expertise, and culture.

If deindustrialization persists, Canadian cities may become focused on services and resources, with major centres being responsible for the production, administration, and export of services and resources, and smaller ones for the extraction and early-stage transformation of commodities. A renewed emphasis on staples would bring us back to early phases of Canadian economic development, even if urban distribution and forms have clearly evolved over the centuries (see Chapter 2). While their common resource orientation ties eighteenth-century fur-trading depots along the St Lawrence or the fishing communities along Canada's Atlantic coast of the eighteenth and nineteenth centuries with fast-growing Alberta oil-patch communities, their configuration obviously differs sharply.

It is important, finally, to point out that production impacts on all aspects of urban life, which is particularly notable in periods of transformative change such as the present. Thus, for example, Chapter 10 speaks to increasing socio-economic **polarization** in the Canadian city in the twenty-first century. It attributes this change primarily to a shift in the prevailing economic regime, to deindustrialization and the rise of the hourglass income distribution characteristic of the growing service sector.

Proximity

Individuals and activities have always congregated in cities to facilitate communication and minimize the cost (in terms of time, effort, and money) of interaction. If we probed reasons why people live in cities, most would place the need to be close to work at or near the top of their list of answers. Other explanations would include proximity to educational establishments, shopping opportunities, cultural activities, entertainment and recreation, family and friends, and medical facilities. People opt for urban living because of their need for frequent and repeated interaction outside the home. Likewise, businesses and institutions locate in cities so they can be close to their market, labour force, and the establishments with which they maintain linkages (in other words, to enjoy agglomerative economies). By concentrating activities and people and thus creating proximity, the city makes frequent interactions affordable in terms of cost and time. In a rural setting, in contrast, many recurring contacts that are routine in the city would involve prohibitive transportation times and/or costs due to long distances. A by-product of interaction is innovation (Hall, 1999); ease of interaction is why cities have been catalysts for change in the social, economic, technological, and cultural realms. The renewed attraction of central-city living that so distinguishes cities of the twenty-first century from the previous era is closely related to widespread recognition that the urban environment, unlike the suburbs, promotes and welcomes all kinds of low-cost and/or spontaneous interactions.

The city can be perceived as comprised of numerous overlapping markets of frequently repeated exchanges. The fact that cities are fundamentally places of economic enterprise makes daily commuting between residences and workplaces of unparalleled importance in explaining urban structure. Also important from an economic perspective are linkage networks between enterprises, which benefit from proximity, especially in the case of just-in-time delivery. Other frequent exchanges that affect the size and spatial organization of cities include the connections of retail facilities to their market and of public services to their clients. In addition, markets connect cultural and recreational activities to their public—the archetypical attraction of the city's 'bright lights'. The inherent importance of proximity has a number of highly visible consequences on urbanization. One obvious repercussion is the high cost of city land relative to land in rural areas (see Skaburskis and Moos, Chapter 13). The appeal of proximity raises demand for space in the city, which, in turn, elevates land values. Price differences between urban and rural land are largely a function of the finite supply of the former as opposed to the relative abundance of the latter. The same is true for locations within urban areas, with value varying in part according to accessibility. Land values translate

into different functional distributions and density levels. High land cost compels establishments to make judicious use of their urban land and to build intensively rather than extensively.

Proximity is a relative condition, largely determined by prevailing transportation systems and activity distribution patterns (see Perl and Kenworthy, Chapter 11). In the 1960s the urban planner Richard Meier (1962) pointed out how innovation in both communication and transportation could trigger new transformative eras of urban development. Urbanist Lewis Mumford in *The City in History* (1961) provides details that lend credence to Meier's hypothesis. Mumford points out that in the pre-industrial city, which depended on non-mechanized forms of transportation (primarily walking and horse-powered), the principle of proximity dictated that important activities be centralized; likewise, the principle of proximity meant that the outer expanse of the built-up, urbanized perimeter was largely dictated by the prevailing mode of transportation.

The car- and truck-dominated contemporary city takes on a highly decentralized form. In the decentralized or dispersed metropolis, adequate accessibility levels can be maintained over large territories so that residents and activities can consume far more land than in the past (Bottles, 1987). As rush-hour gridlock across an extensive section of Highway 401 attests, effective early twenty-first-century boundaries of the Greater Toronto Area run from Kitchener–Waterloo–Cambridge in the west to beyond Clarington in the east (a distance of 165 km). Similarly, high commuter traffic extends from the City of Toronto to points south towards Niagara Falls and north towards Orillia (a distance of 180 km). This extended Toronto region is referred to as the Greater Golden Horseshoe in recent planning documents (Ontario, 2006). But the proximity principle remains influential even in these more dispersed circumstances, as evidenced by the enduring existence of higher densities in cities than in the countryside, and at accessibility peaks within the city itself (at rail transit stations and junctions of major arterials and/or expressways). Chapter 2 in this volume can be read as a testimonial to the changing role that accessibility has played since the early twentieth century in promoting, first, centralization and high-density styles of urban development, second, dispersion and low density, and recently, a return to high residential density, at least in central parts of the metropolitan envelope.

Today, debate is ongoing as to whether our ability to substitute telecommunications for actual movement holds the potential for an even more dispersed urban form. But forecasts predicting the death of the city as we know it have proven to be wrong because they did not anticipate the impact of a changing mode of production on urban form. The 'new' urban economy, with its focus on services, entertainment, and culture, has witnessed a renewed centralization of activities and has considerably elevated the importance of face-to-face contact and spontaneous connecting. In large metropolitan regions this is reflected in the increase in inner-city living, where good access to all kinds of people and activities is within a short walk.

Reproduction

As properties of the urban phenomenon, production and reproduction are intimately tied to each other. Reproduction, as understood in the Marxian sense, centres on the conditions essential to the continued provision of an ample labour force, i.e., the literal re-production of workers. These requisites include birth and child-rearing but also other conditions that relate most directly to the well-being of family/household units within the city: health care, education, social services, family and community support facilities, immigration policies, etc. (Castells, 1977; Jessop, 2002: 47, 77).

Before disease control, engineered infrastructure, public health, food security, and general welfare improved rapidly after the Industrial

Revolution, living conditions were unhealthy in cities. Afflicted by successive epidemics, for long periods cities were unable to maintain their populations without a constant inflow from the country (Howard, 1968; Russel, 1972). Only after the introduction of water treatment and sewer systems and the advent of immunization did city living cease to be a worse threat to health than living in rural areas. Indeed, historians have traced the origins of urban planning to early efforts at alleviating adverse health effects associated with the crowding and pollution of the early industrial city (Hodge, 1998). Still, we should avoid being smug about the improved health conditions of contemporary cities. The SARS epidemic, possible deadly flu epidemics, adverse health effects associated with poor air quality and other sources of pollution, and increasing obesity due to insufficient exercise in an auto-centric culture all point to health problems associated with the contemporary city (Ali and Keil, 2006; Frumkin, Frank, and Jackson, 2004).

At the most fundamental level, Canada's low birth rate—and consequent inability to reproduce its own population—is the fundamental reason why rates of foreign immigration have soared over recent decades. Immigration policies are thus central to the reproduction property of the city, especially since, in Canada, large cities are the destination of virtually all immigrants (Hou and Bourne, 2006; Knowles, 2007; see Hoernig and Zhuang, Chapter 9). Beyond demographic growth, examples of reproduction-related urban sites include homes, schools, hospitals, and water treatment and distribution systems, as well as parks and other recreational facilities that promote health and reduce stress. A smooth operation of the reproduction system allows employers to find an abundant workforce that is healthy, qualified, and possesses a work ethic compatible with types of employment present in a given city. Reproductive activities are usually centered on the home but are increasingly supplemented with services provided by outside agencies in the public and private sectors. Today, in fact, reproduction-related consumption of both services (e.g., fast food, child care) and goods (e.g., dishwashers, microwave ovens, health and hygiene and home maintenance products) represents an important outlet for the production sector. In reality, virtually all household consumption can be seen as having a reproductive aspect. Meanwhile, resources needed for reproduction-related goods and services are derived from the production sector in the form of household expenditure and tax revenues.

A clear feminist dimension to our present understanding of reproduction concerns movement away from the traditional role of women. In the past and still to a large degree, women have assumed the major burden of reproductive work without any payment. While 'equal pay for equal work' has yet to be achieved in most sectors of the economy, the majority of Canadian women (ages 15–64) now participate in the labour force (Beaudry and Lemieux, 2000). Women's roles in both the productive and reproductive spheres have considerable impact on the way we live (see Townshend and Walker, Chapter 8) and on how essential services such as health, education, and child care are delivered. Increased participation of women in productive sectors of the economy, for example, appears to be an important driver of our consumerist lifestyles as well as of demographic stagnation. Had it not been for the massive entry of women into the job market over the last 50 years, household incomes and standards of living would have declined considerably. Labour force participation of all adult household members also can be a source of tension, as in the case of more health care being off-loaded on the home at a time when there is unlikely to be a stay-at-home caregiver to assist (Allan and Crow, 1989; Wakabayshi and Donata, 2005).

Today, we witness growing public-sector difficulties in providing essential conditions for reproduction. From the Great Depression until roughly the early 1980s, governments expanded the **welfare state** and thereby their role in the reproduction sphere. In recent decades, however, the opening of international markets has allowed producers to seek low-tax and low-wage jurisdictions offshore, thus cutting corporate-based fiscal revenues; at the same time, public resentment about high tax levels has made increasing income tax a politically unrealistic option (Campeau, 2005; Finkel, 2006; Graham, Swift, and Delaney, 2009). The result has been that governments have faced reduced spending capacity (see Hall, Chapter 4). Paradoxically, the subsequent cutbacks in public-sector reproduction services have coincided with rising expectations and demand for such services. At the very time when near-total engagement of working-age adults in the labour force makes it difficult for them to attend to the reproduction needs of society (having and rearing children) and of their dependants (old parents and sick or disabled family members), government support in these matters is either stagnant or declining.

In this same vein, the knowledge-intensive economy places a growing burden on the post-secondary education system. The response has been a gradual shift from public to private funding, evidenced by inflating community college and university fees. Yet, the persistent shift towards private funding is a major source of social inequality (see Walks, Chapter 10). Under-funding of reproduction-related public infrastructures and services can cause tragedies whose costs, in terms of suffering and monetary expense, far exceed initial savings. Two examples come to mind. In Ontario, attempts to economize on water quality inspection have been associated with the 2000 Walkerton E. coli outbreak, causing seven deaths and making 2,000 residents ill (O'Connor, 2002). More recently, in 2005, insufficient maintenance of New Orleans water retention systems resulted in billions of dollars in damages (as well as numerous casualties) in the wake of Hurricane Katrina (Brunsma, Overfelt, and Picou, 2007; Horne, 2006). The same logic pertains, in a less dramatic fashion, to the consequences of reduced services in sectors such as social services, education, and health care. For example, deferred treatment of health problems can bring about enormous suffering and productivity losses, often with cumulative costs far beyond that of the treatment.

Capitalization

The capitalization property of cities derives from their compact spatial form. Because urban land is scarce, it becomes the object of substantial capital investment so its use can be maximized. Capitalization refers to the vast resources invested to accommodate agglomerations of residents, businesses, and services. The nature of capitalization and, hence, the form cities take are largely influenced by the engineering possibilities of the time. Over the centuries, improving technologies have promoted larger city size and, until the relatively recent predilection for suburban forms, higher densities.

Once capitalized and populated, built-up urban environments become highly durable and thus contribute a considerable degree of continuity to the urban landscape, thereby becoming a factor of **path dependence** (favouring the perpetuation of existing patterns at the expense of innovation) (Pierson, 2000; see Filion and Bunting, Chapter 3). However, as technology and lifestyles tend to change faster than urban form, capitalization of urban land also engenders obsolescence. Change, especially concerning modes of production or transportation technologies, demands adjustments of the built environment to new conditions. But a city is not easily retrofitted. Typically, costs of redevelopment (on **brownfield** or **greyfield sites**) are higher than those of development on greenfield sites at the urban edge. While financial constraints can play a critical role, they are not the only impediment to altering the urban environment. One obstacle to urban environment adaptability is the symbiosis that binds patterns of behaviour to built environments. For example, high-capacity road systems encourage reliance on the automobile and the truck, and high rates of car and truck use generate a continued demand for improved and expanded roads (Noland, 2000; Parthasarathi, Levinson, and Karamalaputi, 2003). Another obstacle to changing the way the built environment is capitalized comes in the form of citizen resistance that occurs when proposals for redevelopment of previously built-up areas clash with residents' strong emotional attachment to their homes and neighbourhood. An important challenge facing planners and politicians today is how to reconcile citizens' attachment to their home 'places', where they desire to maintain the status quo (**NIMBY**), with the need for change—particularly in the face of looming environmental crisis (Inhaber, 1998).

Capitalization trends over the last decades have promoted two very different urban forms. First, across Canada most urban development still occurs in suburban-like settings. Large investments are targeted at the conditions required for suburban growth: for example, peripheral expressways, arterials, local road systems, water and sewer systems, and public services such as schools and hospitals. At the same time, a sizable private industry is dedicated to suburban residential and commercial development, which caters to an enduring strong demand for suburban living. Interest groups that presently oppose measures to contain peripheral urban development mostly stem from these industries. Meanwhile, following a downtown office boom that began in the 1960s and lasted until the late 1980s, inner cities of large metropolitan regions are experiencing residential intensification, largely in the form of high-rise condominiums (Filion, 2000).

We cannot underestimate the importance on the economy of investments, both from the private and the public sectors, targeted at the urban environment. The home is indeed the main asset for a majority of households. The presence of a speculative dimension to urban-related investments, which mirrors the dynamics of the stock market, is thus not surprising. This is especially the case since space is limited within metropolitan regions and the development process can take years. Such conditions can result in an imbalance between vigorous demand and lagging supply, which translates into escalating property prices. Recent events demonstrate the adverse impacts on the economy of the bursting of a speculative property bubble. As property values inflated, many households used their rising equity as collateral for growing consumer debt, which left them exposed when house values tumbled in 2008 and 2009. The situation has been worse in the US than in Canada, where measures to encourage homeownership, and thus fuel property values, have been less aggressive (Immergluck, 2009; Zandi, 2009).

Place

Sense of place is the least tangible among urban properties discussed in this chapter. It does not mean, however, that this is any less important. Indeed, a renewed sensitivity to place most distinguishes current styles of urban development from earlier, modern (Fordist) growth. Enhanced interest in place is consistent with a shift in economic priority from the city as a centre of industrial production to the city as primarily an agglomeration of services, including culture and entertainment. Urbanites always attach meaning to space, whether it conforms to the intent of developers or not. Recently, however, more attention on the part of developers has gone to the messages conveyed by place, either in an effort to reap financial benefits or simply to enhance the well-being of users (see Lynch and Ley, Chapter 19).

The noted geographer Yi-Fu Tuan (1974) coined the term **topophilia** to denote the personal identity with, and love of, a place. Thus, whereas 'space' relations in cities are mostly about objective attributes of proximity and access, 'place' is all about subjective attachment. Design-oriented professionals—such as Jane Jacobs (1961), Jon Lang (1994), planner Kevin Lynch (1964, 1984), and architect Christopher Alexander (1979; Alexander, Ishikawa, and Silverstein, 1977)—believe that fundamental 'place' principles can provide guidelines to good urban form. They argue that, applied to urban development, such principles will lead to higher quality of life (the topophilia factor) as well as more efficient use of urban space. Scholars such as Relph (1976, 1987) use the term 'sense of place' in a related fashion to speak about subjective and emotional feelings associated with different parts of the urban environment. Relph argues that in modern times the perpetuation of monotonous landscapes in the suburbs and the lack of concern about imaging and good urban design left a vacuum in the urban entity. Others have been concerned that most people who live in car-oriented cities are missing out by virtue of being 'detached' from their surroundings, and that the quality of individual and collective life, as well as the quality of the physical environment, has deteriorated as a result (e.g., Kunstler, 1993, 1996).

The diminishment of topophilic places as intrinsic urban attributes is considered characteristic of the industrial city. Meanwhile the elevation, indeed celebration, of positive features of strong places is one of the foremost differences distinguishing the postmodern city from the modern, industrial city (Ellin, 1999; Lance, Dixon, and Gillham, 2008; see Lynch and Ley, Chapter 19). Nonetheless, even in the largest of Canadian cities, detachment from landscape remains largely characteristic of the suburban environments where most current residents of Canadian cities spend most of their time. In suburban settings, place attachment is mostly associated with the home and the neighbourhood. Other places are considered to be, at best, highly standardized (such as shopping malls and **power centres**) or, at worst, 'junkscapes' (as in the case of haphazard car-oriented retail strips).

Place, then, is the attribute of urbanity that engenders strong connections, positive or negative, between urban dwellers and the locales that surround them. Specific characteristics of place that are favoured will vary in terms of the particular style or symbolic meaning conveyed through urban developments at any point in time. Radical or transformative change, as occurred during the Industrial Revolution, can cause a shift in the extent to which topophilia is elevated as an urban property. Sense of place is thus highly fluid. Former industrial areas provide a good example of this fluidity—these were avoided by most urbanites until their rebirth as loft developments towards the end of the twentieth century (Zukin, 1982). If the industrial transformation led to a relative demise of principles of good urban design, we seem to stand today at another crossroads of transformative shift in urban development wherein the 'selling' or marketing of place has increasingly become an intended goal of municipal land-use policies. Aspects of cities that are highlighted for publicity purposes, such as economic development or tourism, reflect the evolution in the types of places that are valued. If smokestacks were considered to be iconic of progressive cities at the turn of the last century, in the 1960s city promotions featured sleek modernist office towers, thus reflecting a shift in the economy from a predominance of blue- to white-collar employment. The emphasis is now on cultural and festival places, heritage buildings and neighbourhoods, and well-designed public spaces.

In sum, place is the intangible that makes some locales feel good while others do not and that invites or repels visitors. Place interfaces with all the other properties discussed here. Under conditions of the 'new' economy, the manipulation of place properties has become a marketing device used to attract global interest and bring in outside investment (see Zukin, 1991). Thus, place features can be seen as a lure for the creative class, and are associated with the benefits of economic

development believed to derive from this class (Florida, 2002, 2005).

Governance

As understood here, **governance** consists of administrative structures and political processes aimed at generating policies suited to the specific circumstances confronting cities.[1] The proximity and capitalization properties central to urban as compared to non-urban settlements require distinct management measures for the urban community (e.g., Booth and Jouve, 2005; Lightbody, 2006). Proximity requires collective control and co-operation between nearby neighbours over communal space. The smooth functioning of cities relies on shared infrastructures (transportation, communication, electricity, water mains, and sewers) and services (e.g., policing and garbage collection) and on a battery of legal measures (property rights, payment for shared facilities, bylaws) intended to assure the orderly cohabitation of a wide variety of land uses. Haphazard development decisions can plunge a city into a state of chaos. For example, without planning controls incompatible land uses such as noisy and polluting industry and high-traffic generators could locate in residential areas. Likewise, new developments could proceed without heeding infrastructural capacity, thus provoking all sorts of bottlenecks. A pure laissez-faire approach is clearly not suited to the city.

Various types of administrative arrangements have developed to provide urban infrastructures, services, and controls. These administrations have been local or regional or have been lodged in senior governments, as is the case with provincial ministries of municipal affairs or federal housing programs. Issues of governance generally belong to the public sector, but some urban management responsibilities can be vested in community-based or private-sector organizations. Over time, as cities grew, as buildings became bigger and required more infrastructure (roads, water, sewage), as reliance on mechanized forms of transportation (particularly the automobile) increased, and as the public demanded more and better services, administrations responded by becoming larger and more complex.

The need for interventions specific to the urban context has spawned disciplines and university programs that generate and impart the knowledge essential to these interventions. Relevant disciplines include urban planning, urban geography, urban sociology, urban economics, and subfields within civil engineering.

Today, coincident with the growing realization of the importance of local governance in matters such as environmental protection, economic development, the equitable provision of services, and the promotion of health and quality of life, we face the harsh reality that municipal administrations are confronting severe financial restrictions (Donald, 2002; Lightman and Irving, 1991). Even more than higher levels of government, local municipalities have been dealt huge budget cuts. In the absence of a reliable funding stream from these higher levels, municipalities presently rely primarily on limited tax revenues. Yet, at the same time, municipalities have incurred increased costs associated with the added responsibilities they have assumed. This has led to the launching of a movement of local administrations calling for a 'new deal for cities'. Reacting to demands from cities, the federal government has allocated a substantial share of its economic stimulus program to municipal infrastructure projects, both to repair existing infrastructures and to build new ones, including new roads and public transit services (Canada, 2009: 142–6).

Environment

It is a truism to say that to survive in the long term cities must respect their natural environment. In recent years, human life everywhere is threatened by an environment where soil, air, and water quality is severely degraded. As a consequence, increased dangers to health arise from by-products of our industrial and consumption processes. In the past, cities were generally situated at favourable locales; the immediate environment would usually be chosen for reasons of 'site' (e.g., Quebec City's easily defendable position or Halifax's deep and sheltered harbour) or 'situation' (good connectivity, by water in the case of Montreal or by rail in the case of western Canadian cities—Winnipeg is a prime example). Usually, however, the regional environment had not been thought of in active terms.

As demonstrated in Chapter 5, even today cities tend to ignore to a large extent the environmental damage they inflict by exporting their pollution. Thus, for example, by sending its garbage to a Michigan landfill, Toronto does not have to deal directly with the consequences of its consumption. In a similar vein, the Montreal sewer system discharges its partially treated effluents downstream from Montreal Island in the St Lawrence River and Victoria still pumps raw sewage in the Strait of Juan de Fuca. The core argument of Chapter 5 is that cities have an environmental footprint that exceeds manifold their built area (Rees, 2008; Wackernagel and Rees, 1996). With time, environmental awareness has extended its scope from the very local to the global (Carr, 2005; Krooth, 2009). Prior to the twentieth century, the well-off classes would leave cities during times when disease spread most quickly; the urban residences of the wealthy would also command sites that were deemed to be most attractive and relatively 'risk-free' according to the dictates of the time. In the 1950s the nascent environmental movement targeted local consequences of pollution, e.g., high bacteria counts preventing swimming and air pollution caused by specific close-by industries (Crenson, 1971). Later, as air pollution worsened due mostly to rising automobile use, environmental awareness became metropolitan in scale. Residents of metropolitan regions soon became conscious of the fact that retreating to distant leafy suburbs offered little relief from many forms of air pollution. Today, environmental awareness is decidedly global. There is increasing realization that with a world population nearing seven billion, an increasing proportion of which is moving towards a consumerist lifestyle, the effects of human behaviour on the entire biosphere are inevitable (Friedman, 2008). Recognition of global environmental impacts was prompted by the depletion of the ozone layer, which led to the signing of the Montreal Protocol in 1987 to control the emission of chlorofluorocarbons—ozone-damaging compounds found in aerosol containers and refrigeration equipment. But above all, global environmental awareness is driven by concern over climate change. And as hubs of industrial production and consumption containing 50 per cent of the world population, cities are major contributors to global warming. It is thus inevitable that any attempt to control greenhouse gas emissions will have major urban repercussions.

In his book *Collapse: How Societies Choose to Fail or Succeed*, Jared Diamond has documented the disappearance of civilizations over the ages due to circumstances affecting their immediate natural environment (Diamond, 2005; see also Whyte, 2008). In some instances, climate change (then mostly the outcome of natural circumstances) caused droughts; in other cases, communities and entire civilizations carelessly depleted the natural resources on which they depended. The lesson from Diamond's book is that the survival of any human group, and of humanity itself, depends on its ability to achieve sustainable forms of development in harmony with the natural environment. Otherwise, it is doomed. The same goes for cities, except for the fact that they have long been successful in exporting, and thus overlooking, their environmental harm. But with the increasingly global impact of human activity, it is becoming difficult for cities to ignore their environmental effects. The environmentally related collapse of individual civilizations can now be transposed to a worldwide scale. It is now the fate of human civilization in its entirety (and thus the global urban system) that is at stake. Scenarios of doom in a warmer planet proliferate, with stark implications for cities: the flooding of low-lying coastal urban areas; unprecedented heat waves that will take a heavy toll on vulnerable urban populations; and violent conflicts over remaining water sources and fertile areas as deserts expand (Dyer, 2008; Schwartz and Randall, 2003; Smil, 2008). To be sure, cities have engaged in initiatives to alleviate their environmental impact (Register, 2006; Satterthwaite, 1999; see Connelly and Roseland, Chapter 14). Yet, while these measures have had positive results at local and regional levels (cleaner air and water, soil decontamination, preservation of natural areas), it remains to be seen if their scope will be sufficient to abate global trends such as climate change (Marcotullio and McGranahan, 2007).

Conclusion

In an effort to conceptualize the urban phenomenon, this chapter has considered its essentials. Seven properties—production, proximity, reproduction, capitalization, place, governance, and environment—are present in all cities across history. These properties help to explain the reasons for the existence of cities and identify the main principles that drive their functioning. While the chapter has been about shared features of cities, the remainder of the book is about ramifications of the urban phenomenon. The book investigates diverse facets of cities and differences in how these facets manifest themselves according to their place within a metropolitan region and the position of a city in Canada's urban system. It also looks at how different aspects of society are mirrored in cities. Chapter 1 was about common characteristics of urban areas; the remainder of the book is about their multiple dimensions.

Table 1.1 Urban Properties and Their Effect on the Contemporary City

Properties	Definition	Manifestation in the Contemporary City
Production	Need for cities to produce goods and services for their own residents and to be exported beyond their territories to assure the purchase of the goods and services that cannot be procured within their territory. Production attracts people to cities.	Links between cities and their hinterland are replaced by economic interconnections between cities across the world. Transition from an industrialized to a service economy along with a renewed emphasis on resources.
Proximity	Cities are made of numerous overlapping markets of frequently repeated exchanges, with a predominant structuring role taken by the labour market. Proximity makes these exchanges possible; otherwise they would be ruled out by excessive travel time and cost.	Reliance on the car has greatly extended spatial range whereby repeated exchanges can be carried out. However, decentralization tendencies are in part countered by the stress placed on culture, entertainment, and, generally, face-to-face contacts by new economic tendencies.
Reproduction	Reproduction refers to the different conditions needed for the availability of a labour force that is well suited to the needs of the production sector of an urban area. A narrow definition of reproduction relates to the replacement of generations and the presence of conditions needed to maintain health. A broader definition includes education and much of household consumption, including even entertainment.	With the vast majority of working-age adults in the labour force, Canada faces below-reproduction birth rates and households have difficulties in providing reproduction-related services to their members. The problem is that increasing demand for state reproduction-related services happens at a time when public-sector willingness and capacity to intervene is limited by insufficient resources.
Capitalization	Refers to all investments in the built environment of cities, as well as to this built environment itself. The capitalization property of cities derives from its dense urban environment. Capitalization is a factor of stability and durability for cities, and can be an obstacle to the implementation of innovations.	Over recent decades capitalization in cities has taken two forms: suburban development and inner-city intensification. Urban capitalization is conducive to speculation as demonstrated in the recent property bubble and associated adverse economic consequences.
Place	Place is about feelings, either positive or negative, associated with different locales in the urban environment. It refers to subjective reactions to these aspects of the city. Efforts are made by different professions involved in urban development to associate positive meaning with their projects. The types of urban places that are most valued vary over time.	Renewed attention given to place characteristics coincides with the growing importance of services and leisure in the post-industrial city. Quality places are seen as a way of attracting the creative class, which has the potential to propel the economy of an urban area. Meanwhile, standard and poor-quality places are still being produced, especially in the suburban retailing sector.

(continued)

Table 1.1 (*continued*)

Properties	Definition	Manifestation in the Contemporary City
Governance	Cities require interventions that are suited to their reality, and thus specialized forms of administrations to formulate and deliver these interventions. They also rely on the knowledge that is essential to these interventions. Need to deal with issues related to concentration of activities and urban infrastructures.	Expansion of administrations with responsibility for urban interventions in response to growing demand for such interventions. This expansion is followed by cutbacks in tight budgetary circumstances. Under pressure from the municipal lobby, the federal government has directed an important share of its economic stimulus budget to urban infrastructures.
Environment	Historically, to survive cities had to respect their environment. Cities that did not do so were unable to draw natural resources essential to their survival and vanished over time.	Cities are able to draw resources from ever-longer distances. They are thus less dependent on their immediate environment. At the same time, environmental awareness becomes global, and concern about different planetary impacts of cities (especially on global warming) is on the rise.

Review Questions

1. In your opinion, do the seven properties described in the chapter cover all aspects of the urban phenomenon? If not, which would you add?

2. Which properties, in your estimation, are the most important in the present context? How is this different from previous periods?

Note

1. For a fuller discussion of municipal governance, see Andrew Sancton, 'City Politics: Municipalities and Multi-Level Governance', Chapter 17 in Bunting and Filion (2006), at: <www.oupcanada.com/bunting4e>.

References

Alexander, C. 1979. *The Timeless Way of Building*. New York: Oxford University Press.

———, S. Ishikawa, and M. Silverstein. 1977. *A Pattern Language: Towns, Buildings, Construction*. New York: Oxford University Press.

Allan, G., and G. Crow, eds. 1989. *Home and Family: Creating the Domestic Sphere*. Basingstoke: Macmillan.

Ali, S.H., and R. Keil. 2006. 'Global cities and the spread of infectious disease: The case of severe acute respiratory syndrome (SARS) in Toronto, Canada', *Urban Studies* 43: 491–509.

Beaudry, P., and T. Lemieux. 2000. *Evolution of the Female Labour Force Participation Rate in Canada, 1976–1994: A Cohort Analysis*. Hull, Que.: Human Resources Development Canada, Applied Research Branch.

Booth, P., and B. Jouve, eds. 2005. *Metropolitan Democracies: Transformations of the State and Urban Policy in Canada, France and Great Britain*. Aldershot, Hants: Ashgate.

Bottles, S. 1987. *Los Angeles and the Automobile: The Making of a Modern City*. Berkeley: University of California Press.

Brunsma, D.L., D. Overfelt, and J.S. Picou. 2007. *The Sociology of Katrina: Perspective on a Modern Catastrophe*. Lanham, Md: Rowman & Littlefield.

Campeau, G. 2005. *From UI to EI: Waging War Again on the Welfare State*. Vancouver: University of British Columbia Press.

Canada. 2009. *Canada's Economic Action Plan: Budget 2009*. Ottawa: Department of Finance Canada. At: <www.budget.gv.ca/2009/pdf/budget-planbudgetaire-eng.pdf>.

Carr, M. 2005. *Bioregionalism and Civil Society: Democratic Challenges to Corporate Globalism.* Vancouver: University of British Columbia Press.

Castells, M. 1977. *The Urban Question: A Marxist Approach.* London: Edward Arnold.

Christensen, R. 2007. 'SPIN-Farming: Advancing urban agriculture from pipe dream to populist movement', *Sustainability: Science, Practice and Policy* 3, 2: 57–60.

Crenson, M.A. 1971. *The Un-politics of Air Pollution: A Study of Non-decision-making in the Cities.* Baltimore: Johns Hopkins University Press.

Daniels, P.N., and J.R. Bryson. 2002. 'Manufacturing services and servicing manufacturing: Knowledge-based cities and changing forms of production', *Urban Studies* 39: 977–91.

Derudder, B, and F. Witlox. 2004. 'Assessing central places in a global age: On the networked localization strategies of advanced producer services', *Journal of Retailing and Consumer Services* 11: 171–80.

Donald, B. 2002. 'Spinning Toronto's golden age: The making of a "city that worked"', *Environment and Planning A* 34: 2127–54.

Dyer, G. 2008. *Climate Wars.* Toronto: Random House Canada.

Ellin, N. 1999. *Postmodern Urbanism*, rev. edn. Princeton, NJ: Princeton University Press.

Filion, P. 2000. 'Balancing concentration and dispersion? Public policy and urban structure in Toronto', *Environment and Planning C, Government and Policy* 18: 163–89.

Finkel, A. 2006. *Social Policy and Practice in Canada: A History.* Waterloo, Ont.: Wilfrid Laurier University Press.

Florida, R.I. 2002. *The Rise of the Creative Class: And How It's Transforming Work, Leisure, Community and Everyday Life.* New York: Basic Books.

———. 2005. *Cities and the Creative Class.* New York: Routledge.

Friedman, T.L. 2005. *The World Is Flat: A Brief History of the Twenty-First Century.* New York: Farrar, Straus and Giroux.

———. 2008. *Hot, Flat, and Crowded: Why the World Needs a Green Revolution, and How We Can Renew Our Global Future.* London: Allen Lane.

Frumkin, H., L.D. Frank, and R. Jackson. 2004. *The Public Health Impacts of Sprawl.* Washington: Island Press.

Gill, A.M., and M.G. Reed. 1997. 'The reimaging of a Canadian resource town: Postproductivism in a North American context', *Applied Geographic Studies* 1: 129–47.

Graham, J.R., K. Swift, and R. Delaney. 2009. *Canadian Social Policy: An Introduction.* Toronto: Pearson Prentice-Hall.

Hall, P. 1999. *Cities in Civilization: Culture, Innovation, and Urban Order.* London: Phoenix.

Harris, E. 2009. 'Neoliberal subjectivities or a politics of the possible? Reading for difference in alternative food networks', *Area* 41: 55–63.

Hodge, G. 1998. *Planning Canadian Communities: An Introduction to the Principles, Practice, and Participants.* Toronto: ITP Nelson.

Horne, J. 2006. *Breach of Faith: Hurricane Katrina and the Near Death of a Great American City.* New York: Random House.

Hou, F., and L.S. Bourne. 2006. 'The migration–immigration link in Canada's gateway cities: A comparative study of Toronto, Montreal, and Vancouver', *Environment and Planning A* 38: 1505–25.

Howard, S. 1968. *Medieval Cities.* New York: Braziller.

Immergluck, D. 2009. *Foreclosed: High-Risk Lending, Deregulation, and the Undermining of America's Mortgage Market.* Ithaca, NY: Cornell University Press.

Inhaber, H. 1998. *Slaying the NIMBY Dragon.* New Brunswick, NJ: Transaction.

Innis, H.A. 1995a [1931]. 'Transportation as a factor in Canadian economic history', in D. Drache, ed., *Staples, Markets and Cultural Change: Selected Essays, Harold A. Innis.* Montreal and Kingston: McGill-Queen's University Press.

———. 1995b [1938]. 'The penetrative process of the price system on new world states', in D. Drache, ed., *Staples, Markets and Cultural Change: Selected Essays, Harold A. Innis.* Montreal and Kingston: McGill-Queen's University Press.

Jacobs, J. 1961. *The Death and Life of Great American Cities.* New York: Random House.

Jessop, B. 2002. *The Future of the Capitalist State.* Cambridge: Polity Press.

Jones, P., D. Comfort, and D. Hillier. 2005. 'Regeneration through culture', *Geography Review* 18, 4: 21–3.

Knowles, V. 2007. *Strangers at Our Gates: Canadian Immigration and Immigration Policy, 1540–2007.* Toronto: Dundurn.

Krooth, R. 2009. *Gaia and the Fate of Midas: Wrenching Planet Earth.* Lanham, Md: University Press of America.

Kunstler, J.H. 1993. *The Geography of Nowhere: The Rise and Decline of America's Man-Made Landscape.* New York: Simon & Shuster.

———. 1996. *Home from Nowhere: Remaking Our Everyday World for the Twenty-First Century.* New York: Simon & Shuster.

Lance, J.B., D. Dixon, and O. Graham. 2008. *Urban Design for an Urban Century: Placemaking for People.* New York: Wiley.

Lang, J. 1994. *Urban Design: The American Experience*. New York: Van Nostrand Reinhold.

Lawson, L.J. 2005. *City Bountiful: A Century of Community Gardening in America*. Berkeley: University of California Press.

Lightbody, J.M.A. 2006. *City Politics, Canada*. Peterborough, Ont.: Broadview Press.

Lightman, E., and A. Irving. 1991. 'Restructuring Canada's welfare state', *Journal of Social Policy* 20: 65–86.

Lorch, B., M. Johnson, and D. Challen. 2004. 'Views of community sustainability after a mine closure: A case study of Manitouwadge, Ontario', *Environments* 32: 15–29.

Lynch, K. 1964. *The Image of the City*. Cambridge, Mass.: MIT Press.

———. 1984. *Good City Form*. Cambridge, Mass.: MIT Press.

Marcotullio, P., and G. McGranahan, eds. 2007. *Scaling Urban Environmental Challenges—From Local to Global and Back*. London: Earthscan.

Meier, R.L. 1962. *A Communications Theory of Urban Growth*. Cambridge, Mass.: MIT Press.

Mumford, L. 1961. *The City in History: Its Origins, Its Transformations, Its Prospects*. New York: Harcourt, Brace and World.

Noland, R.B. 2000. 'Relationship between highway capacity and induced vehicle travel', *Transportation Research Part A, Policy and Practice* 35: 47–72.

O'Connor, D.R. 2002. *Report on the Walkerton Inquiry: The Event of May 2000 and Related Issues*. Toronto: Queen's Printer for Ontario (Parts 1 and 2). At: <www.attorney general.jus.gov.on.ca/english/about/pubs/walkerton/part1/>.

Ontario. 2006. *Growth Plan for the Greater Golden Horseshoe*. Toronto: Government of Ontario, Ministry of Public Infrastructure.

Parthasarathi, P., D.M. Levinson, and R. Karamalaputi. 2003. 'Induced demand: A microscopic perspective', *Urban Studies* 40: 1335–51.

Pierson, P. 2000. 'Increasing returns, path dependence, and the study of politics', *American Political Science Review* 94: 251–67.

Rees, W.E. 2008. 'Human nature, eco-footprints and environmental injustice', *Local Environment: The International Journal of Justice and Sustainability* 13: 685–701.

Register, R. 2006. *EcoCities: Rebuilding Cities in Balance with Nature*, rev. edn. Gabriola Island, BC: New Society Publishers.

Relph, E. 1976. *Place and Placelessness*. London: Pion.

———. 1987. *The Modern Urban Landscape*. Baltimore: Johns Hopkins University Press.

Rubin, J. 2009. *Why Your World Is About to Get a Whole Lot Smaller: Oil and the End of Globalization*. New York: Random House.

Russel, J.C. 1972. *Medieval Regions and Their Cities*. Newton Abbot: David and Charles Press.

Satterthwaite, D., ed. 1999. *Sustainable Cities*. London: Earthscan.

Schwartz, P., and D. Randall. 2003. *An Abrupt Climate Change Scenario and Its Implications for United States National Security*. A report commissioned by the US Defense Department. Washington.

Smil, V. 2008. *Global Catastrophes and Trends: The Next 50 Years*. Cambridge, Mass.: MIT Press.

Taylor, P.J. 2001. 'Urban hinterworlds: Geographies of corporate service provision under conditions of contemporary globalization', *Geography* 86: 51–60.

Time. 2006. 'The lure of the 100-mile diet', 11 June. At: <www.time.com/time/magazine/article/0,9171,1200783,00.html>.

Tuan, Y. 1974. 'Space and place: A humanistic perspective', *Progress in Geography* 6: 233–46.

Wackernagel, M., and W.E. Rees. 1996. *Our Ecological Footprint: Reducing Human Impact on the Earth*. Gabriola Island, BC: New Society Publishers.

Wakabayshi, C., and K.M. Donata. 2005. 'The consequences of caregiving: Effects on women's employment and earnings', *Population Research and Policy Review* 24: 467–88.

Watkins, A.J. 1980. *The Practice of Urban Economics*. Beverly Hills, Calif.: Sage.

Whyte, I. 2008. *World Without End? Environmental Disaster and the Collapse of Empires*. London: I.B. Tauris.

Zandi, M. 2009. *Financial Shock: A 360° Look at the Subprime Mortgage Implosion, and How to Avoid the Next Financial Crisis*. Upper Saddle River, NJ: FT Press.

Zukin, S. 1982. *Loft Living: Culture and Capital in Urban Change*. Baltimore: Johns Hopkins University Press.

———. 1991. *Landscapes of Power: From Detroit to Disney World*. Berkeley: University of California Press.

CHAPTER 2

Epochs of Canadian Urban Development

Trudi Bunting and Pierre Filion

The chapter examines distinct epochs of urban development, each characterized by relative stability and delineated by radical shifts. In this approach, we draw from the seminal work of John Borchert (1967), who described transformative change as 'epochal and open-ended', epochal because it launches a new distinctive period and open-ended because the implications of such change are seldom predictable or understood at the outset. In a similar vein, referring directly to Canadian urban development, McCann and Simmons (2001: 88) remark that: 'The urbanization process is often shaped by major transitions at critical moments that leave a distinctive mark on individual regions, as well as on the system as a whole.' Our interpretations of the different epochs thus benefit from the time that separates us from the historical content of this chapter.

Historical narratives demonstrate how technological, economic, value and demographic changes unfolding both at a national and international scale reverberate on urban development. Our discussion concentrates both on the urban system (the size and economic role of the different urban areas across the country) and on urban structure (the form of urban areas and the distribution of their population and activities). This chapter emphasizes the openness, from the outset of the colonial era to the present, of the Canadian economy to internal trade, along with the lasting importance

of resource extraction and harvesting. The successive phases of Canadian economic development have left their mark on the urban system. The east–west resource export patterns have determined the nature of Canadian transportation networks and the location of urban centres, industrialization has been favourable to western Quebec and southern Ontario centres, and the present importance of the service sector and immigration benefit the largest metropolitan regions across the country. Meanwhile, the evolution of urban structure highlights the influence of technology, urban size, and social change across all historic periods.

We identify five epochs of urban development: (1) the mercantile era, 1600–1800, marked by colonial governance and based on resource extraction; (2) agricultural settlements, 1800–50, characterized by a first wave of export-oriented agricultural production; (3) the so-called 'Great Transitions', 1850–1945, when manufacturing took hold and expanded in an increasingly dominant heartland; (4) the post-World War II Fordist and Keynesian economic boom, 1945–75, defined by a considerable growth of the industrial and service sector, as well as of government interventions; and (5) the neo-liberal turn, from 1975 until now, defined by deindustrialization and growing reliance on services in a climate of constrained government.[1] The possibility that we may now be entering a sixth urban epoch is investigated in Chapter 3. In this chapter each epoch is analyzed similarly. We sketch major economic, social, and political trends unfolding at a national and international scale. Then, we explore how these trends have shaped the Canadian urban system with emphasis on intra-urban structure covered at the end of each sub-section.[2]

The Mercantile Era, 1600–1800

During the mercantile era, Canada developed in a typically colonial fashion. The economy was oriented towards the provision of staples to the home country, and any economic development that might have competed with the ruling country was intentionally deterred. Like the economy, the political system was truncated by the home country's power in major decisions. Furs were the primary export of Canada over this period; their gathering and shipment relied on natural waterways entailing occasional portages.

Canada's population remained low throughout the period, distinguishing it from many other colonial enterprises. First Nations populations were decimated by European diseases (Cook, 1998; Houston and Houston, 2000). In the context of the vastness of the Canadian territory, the population remained small even when it reached 340,000 at the end of this period. Settlements were scattered on the sea coasts of Nova Scotia, Prince Edward Island, New Brunswick, and along the St John River in New Brunswick. Settlements were continuous on both sides of St Lawrence River, with a concentration on the Montreal plain and a more dispersed pattern on the north shore of Lake Ontario and Lake Erie and along the Grand and Thames rivers (Harris and Wood, 1987). With 5,300,000 inhabitants, the US was already at the turn of the nineteenth century a much more populous country than Canada (Haines and Steckel, 2000). The demographic distinction between the two countries was reflected in the size of their cities. In 1800, the population of New York City was 60,515, followed by Philadelphia, Baltimore,

and Boston with, respectively, 41,220, 26,514, and 24,937 residents. In contrast, the populations of the two largest Canadian cities of the time, Quebec City and Montreal, were 8,000 and 6,000 (Harris and Wood, 1987).

According to 'staples theory', developed by economic historian Harold Innis (1995 [1931]), the demographic, economic, and political evolution of Canada was a function of the nature of the resources it exported. As the prominent export changed, so did the nature of national development. The theory explains the limited demographic growth of Canada over this mercantile era by the fact that pelts were a relatively valuable and low-volume commodity, which required limited shipping capacity (Innis, 1956). As a result, ships were small and had little space available for immigrants when they returned to Canada. Staples theory also accounts for the size and distribution of settlements over this period. The fur trade resulted in a smattering of posts and a few urban centres developed at transhipment points. From the largest among these centres fur expeditions were organized and financed. The main centres played the role of mercantile depots for the collection and short-term storage of raw materials (Vance, 1970). The fledgling urban system thus reflected needs related to the exploitation and export of the resource most in demand in the mother countries, first France and then Great Britain (Sitwell and Seifried, 1984). The period also saw the origins of subsistence agriculture.

The **urban form** of the small Canadian settlements of the time was pedestrian-oriented and in most cases adhered to gridiron road plans. Quebec City was a partial exception to the rule likely because of its precipitous relief (Dechêne, 1987). The grid pattern adopted by so many New World towns constituted a departure from European cities of the time, with their tortuous road layouts (a legacy of the Middle Ages when towns also were pedestrian-oriented) with a few exceptionally broad avenues linking symbolic locales.

Vieux-Québec and Vieux-Montréal each provide portraits of the structural properties of mercantile cities. While most of their buildings that remain today date from the early nineteenth century, some were erected earlier and road patterns certainly were laid during the mercantile era.

Agricultural Settlements, 1800–50

This period was marked by much higher rates of immigration and a burgeoning population. By 1851 the population of the settled parts of Canada was 2,436,000. Meanwhile, its largest cities—Montreal, Quebec City, and Toronto—had populations of 58,000, 42,000, and 30,000, respectively (Urquhart and Buckley, 1983). The rapid rise in the Canadian population between 1800 and 1851 brought it to about one-tenth of the US population, which was 23,191,876 in 1850, a proportion that has remained relatively constant.

Many factors lay behind this accelerated growth. Natural increase was high, especially in Lower Canada (now Quebec), and circumstances in Europe, such as the Irish famine and the late effects of enclosure in the Scottish Highlands, favoured emigration and the settlement of agricultural areas across the Maritimes and Quebec, but especially in Ontario (Miller, 1985). Staples theory provides its own perspective on the demographic surge of this epoch. By the turn of the nineteenth century, lumber became a dominant export, and demand was heightened during the shipbuilding frenzy that accompanied the Napoleonic wars. The relationship between timber trade and rural expansion was dual. On the one hand, lumber was a by-product of the clearing of land for agriculture, so when more lumber was produced, more land became available for agriculture. On the other hand, in contrast to pelts, lumber was a bulky, low-value commodity requiring commodious shipboard space, so there was plentiful room for immigrants on ships returning from Europe (Innis,

1995 [1931]). During this period of agricultural expansion, the amount and quality of the land that was cleared encouraged grain export to Europe, demand for which expanded after the repeal of the British Corn Laws in 1846 (Schonhardt-Bailey, 2006; Semmel, 1970).

Growing rural settlement and agricultural production had a profound effect on the urban system. As the population and wealth of agricultural areas increased, market and service centres developed to serve the rural population. And the more populated and well-to-do the surrounding rural population, the more these urban centres grew. Early Kingston was a larger city than Toronto, but it was in a low-yield agricultural zone, where the Canadian Shield meets the Appalachians. Toronto rapidly overtook Kingston largely by virtue of its location in a rich agricultural belt.

Staples theory also provides a perspective on the impact agricultural exports had on transportation systems and political institutions. Before the advent of railways, canals were needed to carry grain to ports, so the period witnessed active canal-building, especially in Upper Canada (Ontario), where much of export-oriented agriculture was found. But the digging of canals caused the swelling of public debt in Upper Canada. The Act of Union, which unified Upper and Lower Canada (Quebec) in 1841, can, in part, be interpreted as an attempt to provide more financial stability to Upper Canada by merging it with relatively debt-free Lower Canada—a political upshot of dependence on the export of staples.

As their populations and economic activities grew, urban settlements took more space and made more intensive use of the built areas within their perimeters. The centre of the largest cities offered a mixture of stores, warehouses, and houses. In some cases, there was also a nascent industrial sector, such as shipbuilding in Quebec City's lower town. Development was relatively compact because most journeys were made on foot. Only the rich could afford to live on the outskirts and commute in their horse-drawn buggies. But density was limited by the absence of elevators and the high cost of building upward using the technologies of the time.

Great Transitions, 1850–1945

This period from about the mid-nineteenth century to the mid-twentieth century was one of major transitions: the development of a railway network, the substantial expansion of the staples economy with the settlement of the prairies and, above all, the formation of an industrial heartland. The introduction of mechanized transportation along with accelerated urban growth propelled by industrialization resulted in profound transformations of urban structures.

Responsiveness to international demand for grain, as well as a determination to prevent US encroachment on the territory of British North America, promoted farming on the prairies and the region's ultimate transformation into the 'breadbasket of the world'. The settlement of the prairies and the export of grain required a railway network (grain elevators next to train stations became emblematic of the prairies). From this perspective, Confederation can be interpreted essentially as a political arrangement serving (among other things) to pool the resources necessary to build transcontinental train lines (Waite, 1967). Table 2.1 traces the growth over this period of the three main Prairies urban centres: Winnipeg, Calgary, and Edmonton.

Compared to the US, Canada was slow to industrialize. While US industry developed in the first half of the nineteenth century, causing population from rural Canada (especially Quebec) to emigrate to mushrooming New England manufacturing towns, Canada industrialized in the latter part of the century (Naylor, 2006; Smucker, 1980). Industrial development in Canada was stimulated by the National Policy, enacted in 1879 by the John A. Macdonald government, which introduced tariffs on manufactured products from the US. Resulting industrialization was dominated by US branch plants locating in Canada to circumvent the newly imposed tariffs (Forster, 1986). For the most part, industries congregated along the Quebec City–Windsor axis, where a relatively dense rural population provided a labour force as well as a market for manufactured products. The Southern Ontario portion of the axis enjoyed the further advantage of proximity to the US Great Lakes industrial belt. The Quebec City–Windsor corridor became known as the **heartland** of Canada (sometimes also referred to as 'Mainstreet'), a region that enjoyed rapid demographic and economic growth thanks to its industrial activity (Yeates, 1975). The industrialization of Canada rendered staples theory less effective in explaining its economic, demographic, and political evolution. Henceforth, the gap between the industrial heartland and the resources **hinterland** became a major source of difference in regional economic fortunes (McCann and Gunn, 1998). Also included in the hinterland were Maritimes vestiges of early Canadian industrialization. Either way, the geographic development of Canada's central parts could no longer be explained simply by the staples model.

By 1941 the population of Canada had reached 11,506,655, with two metropolitan regions around the million mark: Montreal slightly above and Toronto slightly below (see Table 2.1). Moreover, with the introduction of streetcars and commuter trains, there was also unprecedented outward areal expansion in these two metros. What we today know as the inner city (urban areas built before 1946, according to many definitions) includes sectors developed over this and the two previous periods. (But few Canadian urban centres predate 1850, and when they do, their pre-1850 area was minuscule compared to the size of the sectors developed between 1850 and 1945.) Spatially, the cities of this era consisted of: the central business district (CBD); a 'zone of transition' with a mixed and changing land-use pattern (Preston

Table 2.1 Population of Canada and Canadian Urban Areas 1851–2006

	1861	1871	1881	1891	1901	1911	1921	1931
Canada	3,229,633	3,689,257	4,324,810	4,833,239	5,371,315	7,206,643	8,787,949	10,376,786
Halifax	49,021	56,963	67,917	71,358	74,662	80,257	97,228	100,204
Quebec	79,002	79,306	82,724	82,593	90,941	104,554	124,627	170,915
Montreal	118,015	144,049	193,171	277,525	360,838	554,761	724,205	1,003,868
Ottawa–Hull	59,665	75,989	92,994	115,342	139,734	167,716	203,387	233,910
Toronto	104,495	115,979	153,113	245,101	272,663	444,234	647,665	856,955
Winnipeg		5,157	19,297	40,367	65,346	172,126	229,084	283,828
Calgary					11,358	73,178	112,689	140,624
Edmonton					18,491	58,703	95,334	126,835
Vancouver			7,939	41,507	53,641	183,108	256,579	379,858

	1941	1951	1961	1971*	1981	1991	2001	2006
Canada	11,506,655	14,009,429	18,238,247	21,568,310	24,343,180	28,031,000	31,111,000	31,612,571
Halifax	122,656	162,217	225,723	222,650	277,991	320,501	359,183	372,858
Quebec	202,882	252,890	331,307	480,405	576,075	645,550	682,757	715,515
Montreal	1,138,431	1,358,075	1,358,075	2,743,230	2,828,349	3,127,242	3,426,350	3,635,571
Ottawa–Hull**	273,708	334,829	482,043	602,510	717,980	920,857	1,063,664	1,130,761
Toronto	951,549	1,176,622	1,733,108	2,628,250	2,998,945	3,893,046	4,682,897	5,113,149
Winnipeg	295,342	330,130	475,989	540,265	584,842	652,354	671,274	694,668
Calgary	146,990	195,352	317,989	403,320	592,743	754,033	951,395	1,079,310
Edmonton	149,193	226,199	410,679	495,700	657,057	839,924	937,845	1,034,945
Vancouver	449,376	649,238	907,531	1,082,350	1,268,183	1,602,502	1,986,965	2,116,581

*Major changes of boundaries due to the introduction of the census metropolitan area concept.

**The Ottawa–Hull CMA was renamed Ottawa–Gatineau following the 2002 amalgamation of five municipalities in the Outouais region of Quebec.

Sources: Censuses of Canada, 1861–2006.

and Griffin, 1966); factory belts along waterways and railways; and relatively high-density, though not high-rise, residential neighbourhoods, most of them segregated along lines of income and ethnicity (Park and Burgess, 1969 [1925]). Figure 2.1 presents a schematic model of the spatial structure characteristic of large Canadian cities of the time. The reader will note a clear zonal patterning of land uses with a more wedge-shaped alignment along major transit lines and/or significant topographic features—e.g., ravines and river valleys—as well as a clear distinction between 'city' and commuter or 'suburban' zones. It must be emphasized that heavy industrial production was the main force pushing urban growth over this period, though administrative activities played an important role in larger centres and resource extraction supported the economies of many hinterland communities.

Available time (the workweek included Saturday), monetary constraints, and general reliance on transit and walking (in a period when the car was the conveyance of the rich) limited journey length. Accordingly, buildings were closely packed together, residential density was high, distance between home and work was short (except for a small contingent of well-heeled suburban commuters), and shopping facilities were found throughout the city, at walking distance from virtually every house (Adams, 1970; Colby, 1933; Jacobs, 1961). Commercial streets catered to most retail needs with the exception of high-order goods found in the CBD. The high density of the built environment also made it easy to walk to schools, medical services, places of worship, and other facilities. Transit was composed primarily of fixed-line rail services (streetcars in addition to commuter trains to larger cities). Their linearity left a profound mark on the inner city's urban structure, propelling outward growth along their lines (Hoyt, 1939; Warner, 1962). The CBD enjoyed unchallenged accessibility potential with virtually all transit lines radiating outward from this central point.

Post-World War II: Fordist and Keynesian Economic Boom, 1945–75

The seeds for the **Fordist-Keynesian** period were sown in the pain of the Great Depression, when the first social programs were launched (Blake, 2009; Moscovitch and Albert, 1987). But it is over the 1945–75 period of rapid economic and demographic expansion that major changes took place. It is easy to understand why, after the hardship of the Great Depression and World War II, Canadian households readily adopted family and consumerist values, leading to the simultaneous occurrence of the baby boom and of an extended period of consumer goods accumulation. Above all, sustained prosperity over this period was propelled by concurrent productivity improvements and robust consumer demand. Several mechanisms, considered as part of the 'Fordist regime of accumulation', balanced the accelerated growth. The term 'Fordist' refers to two innovations introduced by Henry Ford early in the twentieth century: (1) standardized assembly line production resulting in massive productivity improvements, and (2) the payment of salaries to blue-collar workers that were high enough to allow them to purchase the goods they produced—e.g., cars. 'Regime of accumulation' connotes political and para-political processes that co-ordinated the production and consumption spheres in a context of rising productivity, mostly mechanisms that fuel consumption: trade unions, government redistribution programs, public sector stimulation of household expenditure, etc. (Aglietta, 1979; Jessop and Sum, 2006). Governments, inspired by the economic theories of John Maynard Keynes, readily used public expenditure to stimulate the economy (Frazer, 1994; Hamouda and Smithin, 1988; Jones, 2008; also see Vinodrai, Chapter 6).

The heartland–hinterland perspective best captures the geographical dimension of economic

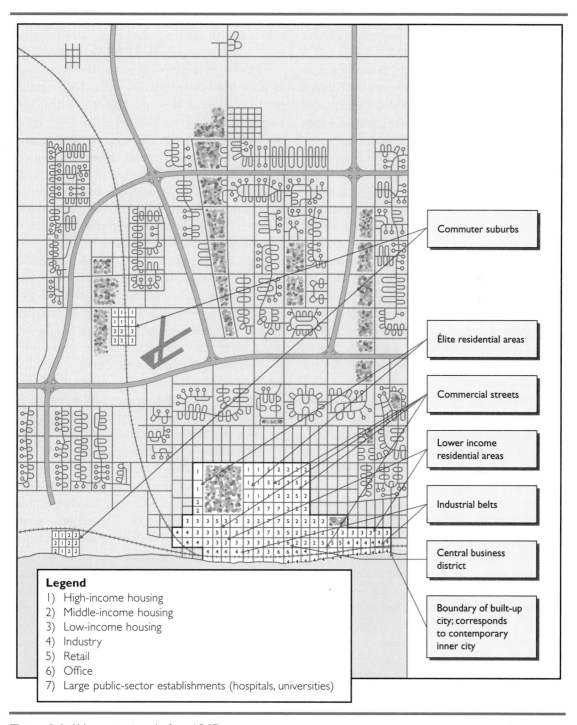

Commuter suburbs

Élite residential areas

Commercial streets

Lower income
residential areas

Industrial belts

Central business
district

Boundary of built-up
city; corresponds
to contemporary
inner city

Legend
1) High-income housing
2) Middle-income housing
3) Low-income housing
4) Industry
5) Retail
6) Office
7) Large public-sector establishments (hospitals, universities)

Figure 2.1 Urban structure before 1945

trends during these years (McCann and Gunn, 1998). Demographic and economic growth in the heartland exceeded that of the nation as a whole (Yeates, 1998: 115). Meanwhile, consistent with the interventionist ideology of the time, the federal government launched economic development programs intended to disperse industrialization—then seen as the primary source of economic growth—to the benefit of the hinterland and other less prosperous parts of the heartland (Coffey and Polèse, 1987). The success of these measures was at best lukewarm (Matthews, 1983). For example, subsidies did play a role in the decentralization of car assembly installations to Quebec, New Brunswick, and Nova Scotia, but by the early 2000s, car assembly again was confined to southern Ontario. Throughout this period, then, high rates of growth accrued to the Windsor–Quebec City corridor. Meanwhile, the fortunes of communities in 'hinterland' parts of the country rose and fell largely in accordance with international demand for staples alongside fluctuations in international money markets.

Urban development was both an outcome of, and contributor to, the economic conditions of the time. The car-oriented suburban model, developed for the most part in the 1950s, epitomized urban development over the period (Sewell, 2009; also see Perl and Kenworthy, Chapter 11). Widespread homeownership and increased indoor and outdoor space for families distinguished the new suburb from the inner city (Miron, 1988; Spurr, 1976). Well-paid blue-collar and public-sector workers swelled the ranks of the middle class so that a much higher proportion of families could afford their own homes. Meanwhile, the production of relatively large homes on relatively large lots (by comparison to earlier residential consumption standards) generated an unprecedented demand for mass-produced goods to fill them (furniture, appliances) and to access them (cars). Still, there was residential diversity in the Canadian suburb. For example, in Toronto's Don Mills, prototypical

of the 1960s and 1970s post-war suburb, approximately half of the residential units were found in apartment buildings (Sewell, 1993: 79–96).

Lifestyles, characterized as 'modern', were conformist: they were consumer-oriented and centred on the nuclear family (Clark, 1966; Dobriner, 1958; Hamel, 1993; Harvey, 1989; Hayden, 2003; Muller, 1981; Popenoe, 1977). Increased car ownership and use were inextricably linked to the rise in residential space and single-family home ownership. Facilitated by prosperity, auto ownership became the norm. The proliferation of low-density environments made automobile travel a necessity rather than a matter of choice or indulgence. In turn, this abrupt change in travel mode, along with the prevailing taste for low-density as well as new expectations regarding green space provision, profoundly transformed spatial arrangements as greatly increased accessibility translated into heightened land consumption needed to accommodate autos.

Governments intervened directly in urban development, building educational establishments, hospitals, subsidized housing, and, most notably, roads and expressways (Fishman, 1987; Harris, 2004; Rowe, 1991). The federal government actually stimulated suburban development through a mortgage subsidy/guarantee program for new single-family homes (Dennis and Fish, 1972). Planning regulations also served to standardize the main features of suburbs: large mono-functional zones, super-blocks bounded by arterial roads, accommodation of the car, and abundance of green space.

It needs to be emphasized that the predominant land use in early suburbs was residential, thus suburbs were often spoken of as 'bedroom communities' (Mackenzie and Rose, 1983; Mazy and Less, 1983; Reisman, 1958). But with previously centralized functions relocating to newly developed areas, suburbs soon came to host a wide range of functions. Figure 2.2 identifies some of the more salient properties of suburban areas developed in the post-war period. This model illustrates many

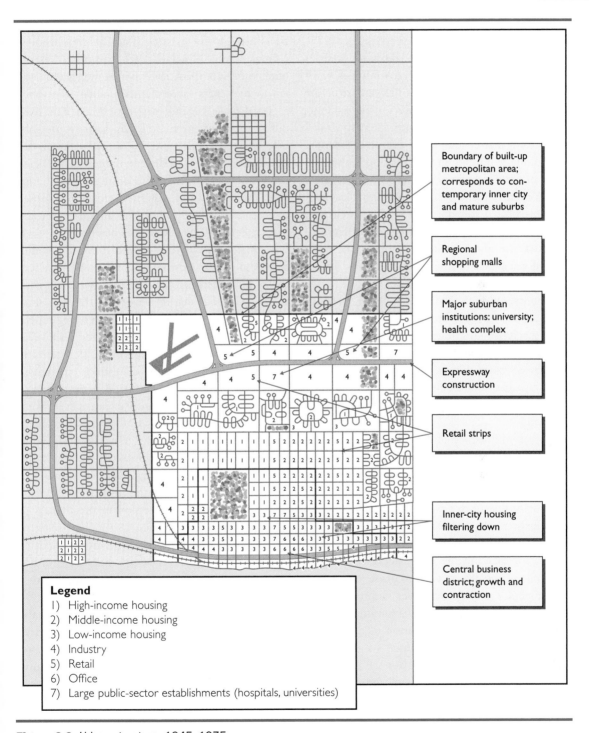

Boundary of built-up metropolitan area; corresponds to contemporary inner city and mature suburbs

Regional shopping malls

Major suburban institutions: university; health complex

Expressway construction

Retail strips

Inner-city housing filtering down

Central business district; growth and contraction

Legend
1) High-income housing
2) Middle-income housing
3) Low-income housing
4) Industry
5) Retail
6) Office
7) Large public-sector establishments (hospitals, universities)

Figure 2.2 Urban structure, 1945–1975

of the principles of suburban-style urban development first identified by Harris and Ullman (1945), e.g., the separation of unlike activities such as housing and industry, and/or proximity of compatible facilities such as heavy-duty infrastructure for industry or parks and schools for single-family residential uses. Land value and density gradients became flatter and more evenly distributed throughout suburban space, creating numerous points of equivalent importance. As a function of greatly improved auto accessibility, high accessibility points located at major expressway interchanges and arterial intersections became sites for regional malls that accommodated the largest concentrations of retail establishments, with the exception of CBDs in very large metropolitan regions (Jones, 1991; Jones and Simmons, 1990). Other locations offering good accessibility along, off, or close to high-capacity roads attracted industrial and business parks, small retail malls, and self-standing retail and service establishments (fast-food outlets, car dealerships, gas stations, and so on). Space–hungry activities such as university campuses also opted for suburban locations.

In yet another fashion, suburbs distinguished themselves from the inner city by adopting an inwardly focused system of curvilinear streets within super-blocks. Ties with older parts of the city and, for that matter, with other suburban areas were maintained via a limited number of high-capacity expressways and arterial roads. Generalized use of cars and the construction of arterial road and expressway networks made it possible for households to reach activities within a greatly enhanced perimeter, thus broadening available choices. However, the combination of the dispersion of activities and lower density of population was damaging to public transit (Bottles, 1987; Cervero, 1986, 1989; Miller and Shalaby, 2003; Perl and Pucher, 1995).

Quality of life in inner-city neighbourhoods suffered considerably from efforts to improve suburban commuters' accessibility to the CBD. Most dramatic was the impact of expressway construction through pre-existing older neighbourhoods (Nowlan and Nowlan, 1970). Another consequence of suburban development was a dramatic fall in the CBD's share of metropolitan sales once regional malls became well-established. The exodus of much of the middle class towards the suburbs also triggered a 'filtering down' of inner-city housing (that is, a decline in households' socio-economic status, conversions of single-family units into multi-family accommodation, and subsequent physical deterioration of an aging housing stock). Perceived decline in older housing stock was a major incentive for **urban renewal** schemes that demolished existing structures to make way for public housing based on growing concern for social justice. More frequently, though, 'slum clearance' was directed at private redevelopment (Birch, 1971; Bourne, 1967; Dennis and Fish, 1972; Hoover and Vernon, 1962; Miron, 1993; Smith, 1964). It must be stressed that 'inner-city blight' was much more characteristic of American than of Canadian cities. Many Canadian inner cities were fortunate in that the period of rapid suburbanization coincided with a huge wave of post-World War II immigration. This initiated a cycle of do-it-yourself private home improvement/renovation in older inner-city neighbourhoods that would subsequently find fuller expression in the process of **gentrification** (Ley, 1996).

Over the 1945–75 period, downtown redevelopment was encouraged by massive public-sector investment in road widening and, in larger agglomerations, in public transit. The erection of highly symbolic public buildings, such as the Toronto City Hall and Montreal's Place des Arts, was intended to improve the image of the core. Redevelopment was promoted by liberal zoning regulations authorizing high-rise structures over much of the inner city (Bourne, 1967; Leo, 1994; Sewell, 1993). In larger agglomerations, 1960–75 was a time of feverish office employment growth in the CBD that produced much altered cityscapes profiled by new towers (Code, 1983; Gad, 1991; Lortie, 2004). Core area retailing also was transformed. In virtually every Canadian downtown, malls reproducing conditions offered by their suburban counterparts were erected (Filion and Hammond, 2009; Frieden and Sagalyn, 1989; Paumier, 1988; Redstone, 1976). Thanks to such development, many CBDs were able to maintain, or even increase, their absolute retail activity level, at least for a time. Everywhere, however, massive suburban retail developments caused a serious drop in downtowns' relative importance within the metropolitan-wide retail marketplace.

Neo-Liberalism, Deindustrialization, and the Shift to the Service Sector

A number of events converged in the 1970s and 1980s to challenge the Fordist-Keynesian arrangements that dominated the previous period (Amin, 1994; Bonefeld and Holloway, 1991). Many identify the oil crisis of 1974 and the stagflation (the combination of economic stagnation and inflation) experienced over this decade and the early part of the following decade as major factors accounting for the unravelling of the Fordist regime of accumulation (Meade, 1982; Olson, 1982). Ballooning sums dedicated to energy expenditures reduced industrial investment and thus productivity increments. Concurrently, doubt was cast over the economic stimulation capacity of government spending as it increasingly appeared that these expenditures fuelled inflation rather than fostering genuine growth. Fordist-Keynesian arrangements were also perturbed by international trade liberalization (the Free Trade Agreement with the US and the subsequent North American Free Trade Agreement, which included Mexico as a trading partner); equally important were multilateral and bilateral agreements involving a wide range of developed and developing countries (see Friedman,

2005; also Hall, Chapter 4). Trade liberalization translated into more and cheaper products. For example, without international trade, computers would have remained a luxury item. The impact on Canadian industry was damaging, however. A succession of sectors became victims of international competition—textile and garment manufacturing, toys, electronics, appliances—and, at the time of writing in 2009, large segments of the car industry, the jewel in the Canadian industrial crown, were at risk of collapse. Until recently, the loss of industrial employment was more than made up by an expansion of the service sector. However, there is an important distinction between industrial and service-sector jobs. While industrial jobs, especially when unionized, tended to pay wages that allowed access to middle-class lifestyles, service-sector employment is highly polarized: some jobs pay high salaries but most service-sector wages are at or below subsistence levels (Canadian Centre for Policy Alternatives, 2007; McCarty, Poole, and Rosenthal, 2006; Yalnizyan, 2000).

All these circumstances favoured an ideological and political shift from **Keynesianism** to neo-liberalism (Biven, 1989; Frazer, 1994: Friedman, 1984). **Neo-liberalism** entailed a retreat in government economic intervention, a trimming of social programs, and enhanced reliance on private enterprise and market mechanisms leading to privatization of public enterprises. While neo-liberalism is most directly associated with Margaret Thatcher, Ronald Reagan, and, in Canada, Ontario Premier Mike Harris, the neo-liberal turn was felt virtually everywhere (Keil, 2002; see Allahwala, Boudreau, and Keil, Chapter 12). Many conditions contributed to a change of attitude on the part of governments. The inability to control inflation largely discredited Keynesianism. Redistribution programs were seen as inefficient, resentment towards taxation rose, and with firms moving to low-wage and low-taxation jurisdictions, governments felt compelled to concentrate on economic competitiveness at the expense of social programs

(Hackworth, 2006). In Canada, the retrenchment of state intervention was further abetted by measures to contain and, from the mid-1990s, eliminate deficits.

In Canada, as in most countries, neo-liberalism has curtailed but not eliminated government intervention (Murphy, 1999). While less generous than 30 years ago, social and regional development programs are still in place. In the same vein, there have been limitations to deregulation. For example, the urban planning apparatus is intact. And in recent years, governments have spent colossal sums to stimulate the economy in order to reduce the impact of the recession—a typical Keynesian reaction to a crisis that has been blamed, especially in the US, on the deregulation of financial markets.

Along with deindustrialization, a shift to the service sector, and neo-liberalism, two other factors have influenced the evolution of the urban system since 1975. One factor, the importance of which has increased lately, is a heightened international demand for natural resources. The rapid development of the economy of emerging countries, such as China and India, will assure ever-growing demand for resources in the future. Therefore, while economic globalization is having adverse effects on the industrial economy of the heartland, it is having opposite impacts on portions of the hinterland harvesting or extracting commodities in high demand on world markets. (On the impact of resource depletion on hinterland urban centres, see Donald and Hall, Chapter 16.) But demography is probably the factor with the most impact on the present urban system. Post-1975 demographic trends are profoundly different from those of the preceding period. The baby boom that lasted from the late 1940s to the early 1960s has given way to a 'baby bust', with birth rates below replacement since the early 1970s. To compensate, Canada has raised immigration levels (Bercuson and Carment, 2008; Frideres, Burstein and Biles, 2008; Knowles, 2007).

The joint effect of deindustrialization and high demand for commodities calls into question the validity of the heartland–hinterland perspectives in the current context. It appears that the new divide is between large metropolitan regions, irrespective of whether they are located in the heartland or hinterland, and the remainder of the country. Figure 2.3 illustrates this situation by showing that between 1971 and 2006 the hinterland has either grown much faster than the heartland or at approximately the same rate. The figure also illustrates the higher growth rate of census metropolitan areas (CMAs) with over one million population in 2006—Toronto, Montreal, Vancouver, Ottawa–Gatineau, Calgary, and Edmonton—than that of the 33 other urban areas with populations exceeding 60,000 in 1971.[3] The populations and economies of the six large urban areas have grown faster than that of the country as a whole for two reasons. First, they are much more appealing than are smaller centres to service-sector activities with a propulsive effect on the economy—business services, research and development, tourism, arts and culture (see Vinodrai, Chapter 6). Second, they are major attractors of immigrants by virtue of their diversified economy and ethnic communities (Hou and Bourne, 2004; see Hoernig and Zhuang, Chapter 9). In the near future, with the retirement and passing away of the baby boom generation, only urban areas that are magnets for immigrants will be able to expand or even maintain their economic activity and population. Signs already foreshadowing the demographic decline of small urban centres include: high average age, low proportion of people of working age, weak presence of immigrants. Moncton, Saint John, Saguenay, Quebec City, Sherbrooke, Trois-Rivières, Belleville, Peterborough, North Bay, Sudbury, Sault Ste Marie, and Thunder Bay fall within this category. If current economic and demographic trends persist, the number of declining urban areas will grow (see Donald and Hall, Chapter 16).

Turning to the urban structure dimension of Canadian cities, despite the enduring symbolic preeminence of the CBD in the largest agglomerations,

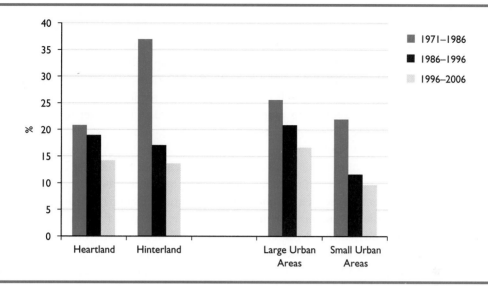

Figure 2.3 Population change, 1971–2006: heartland and hinterland, large urban areas and small urban areas (%)

Source: Censuses of Canada, 1971–2006.

Canadian metropolitan areas followed North American development trends resulting in the presence of most economic activity and population in suburban-like environments (Baldassare, 1986; Bourne, 1989; Bunting and Filion, 1999; Knox, 1994; Muller, 1981). Moreover, during this period, dispersed styles of suburbanization have resulted in a rate of outward physical expansion that far exceeds demographic growth. Even cities that saw little population increase experienced considerable peripheral suburban development, as demonstrated by the Winnipeg experience (Leo, 2006). Thus, in most mid-size and small urban areas what is left of the dense inner-city environment and its reliance on walking and transit systems has become a subsidiary urban form within agglomerations dominated by suburban, car-oriented land uses (Evenden and Walker, 1993; Kling, Olin, and Poster, 1991; Knox, 1994). This said, it must also be noted that recently some cities, such as Vancouver and Toronto, have enjoyed substantial residential development in their central parts

(Berelowitz, 2005; Punter, 2003). At this time, however, it is difficult to assess the extent to which this form of residential redevelopment represents a significant counter-tendency to suburban growth.

Today, any expressway exit constitutes a high-accessibility site for development, and any location within a 15-minute drive from such an exit offers adequate accessibility for housing (a situation called the 'exit-ramp economy'). This slackening of accessibility constraints produces the complex patchwork of land use that characterizes contemporary suburban areas. Specialized suburban zones of different sizes are juxtaposed in apparently random fashion, which makes it more difficult to grasp the structural parameters of the contemporary metropolis than those of earlier urban forms (Filion and Bunting, 1990; Filion, Bunting, and Curtis, 1996). Postmodern urban theorists have characterized the apparently random patterning of different land uses and social communities as 'keno capitalism', likening the outcome to the roll of dice on a game board (Dear and Flusty, 1998). Another result of this transportation/land-use dynamic is a further extension of access to rural zones surrounding metropolitan regions. The movement of such activities as employment and shopping to the outer edge of the built perimeter encourages residential developments in even more remote rural areas, which the new outer edge brings within commuting range (Bourne et al., 2003; Coppack, Russwurm, and Bryant, 1988; Joseph, Keddie, and Smit, 1988).

As shown in Figure 2.4, a number of innovations distinguish the current from the preceding period of suburban development (see Grant and Filion, Chapter 18). With the decentralization of offices, the post-1975 period witnessed the creation of suburban business (or office) parks, and more recently, the emergence of so-called suburban downtowns that combine office and retail concentrations with civic centres, comparable in many respects to those in the US called 'edge cities' (Cervero, 1989; Garreau, 1991; Kling, Olin, and Poster, 1991; Stanback, 1991). On the retail scene, the appearance of big-box stores and **power centres** transformed the shopping centre hierarchy (local, community, regional) inherited from the previous period (Berry, 1963; Brown, 1992; Dawson and Lord, 1985; Guy, 1994; Howe and Rabiega, 1992; Jones and Doucet, 2000; Jones and Simmons, 1990). Meanwhile, although the density of suburban employment and retail space is either steady or declining, that of residential areas is on the rise. Homes generally are larger than they were in the previous period, but sit on comparatively smaller lots. Nonetheless, auto-based configurations of residential developments continue. One major difference between the two generations of suburban residential areas concerns the nature of green space. Whereas suburban green space in the previous period came mostly in the form of large building lots and playgrounds, more space is now allocated to the preservation of natural features, such as woods, creeks, and their riparian zones and marshes/ponds (Hough, 2004; Manuel, 2003; see Connelly and Roseland, Chapter 14).

Another suburban innovation over the post-1975 period is the introduction of the New Urbanism model of development, which attempts to reproduce appealing aspects of late nineteenth- and early twentieth-century forms of urbanization: a main street, reduced car-orientation, vernacular architectural styles, picket fences, etc. (Duany, Plater-Zyberk, and Speck, 2000). However, this urban formula has not been as successful as originally anticipated. Few attempts have been made in Canada to create full-fledged New Urbanism neighbourhoods. Examples of such projects include McKenzie Towne in Calgary and, in the Toronto area, Cornell in Markham, Oak Park in Oakville, and Montgomery Village in Orangeville. In a number of cases, projects begun as New Urbanism communities reverted to conventional suburban models; others have failed in their attempt to create a commercial main street. Most importantly, there is no evidence that this form of development is able to induce, as intended, a marked increase in walking and transit use (Grant, 2006; see Grant and Filion, Chapter 18).

The image of the suburb as solidly middle class economically no longer conforms to reality, if it ever did. In large metropolitan regions, the gentrification of the inner city has resulted in a filtering down of some parts of the suburbs that were built during the previous period. These areas are undergoing profound transformations; as single-detached homes are 'duplexed' or transformed into rooming houses, density is increasing, schools add portables, and public transit use is on the rise. Meanwhile, other older suburbs, which enjoy accessibility advantages and/or exceptional urban or natural amenities, filter up as their original housing stock gives way to mega-houses. More recent suburbs still maintain a middle-class character and their residents tend to be ideologically and politically right-of-centre (Walks, 2004). With rising immigration, however, suburbs have become much more ethnically mixed than they were in the previous period.

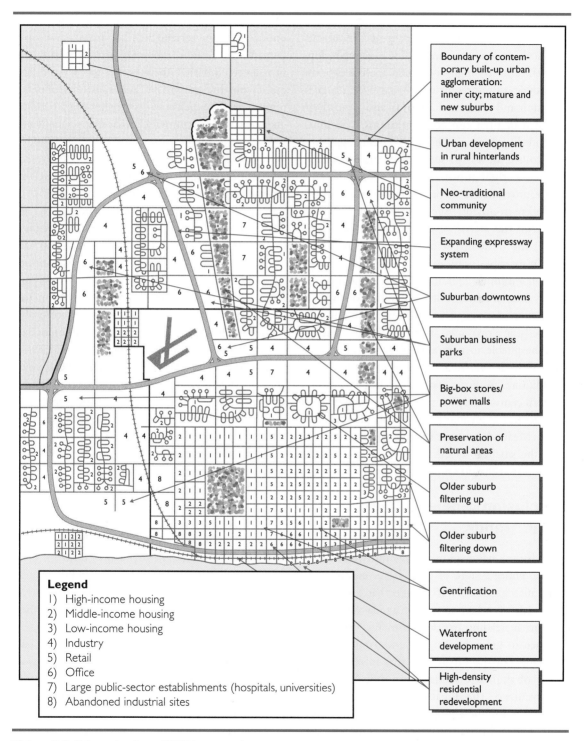

Boundary of contemporary built-up urban agglomeration: inner city; mature and new suburbs

Urban development in rural hinterlands

Neo-traditional community

Expanding expressway system

Suburban downtowns

Suburban business parks

Big-box stores/power malls

Preservation of natural areas

Older suburb filtering up

Older suburb filtering down

Gentrification

Waterfront development

High-density residential redevelopment

Legend
1) High-income housing
2) Middle-income housing
3) Low-income housing
4) Industry
5) Retail
6) Office
7) Large public-sector establishments (hospitals, universities)
8) Abandoned industrial sites

Figure 2.4 Urban structure, 1975 to present

In large metropolitan regions, the previous period was a time of downtown and inner-city large-scale commercial and residential redevelopment (i.e., mostly high-rise rental apartments); the focus during the present epoch has been on gentrification and residential intensification, mostly in the form of high-rise condominiums. The inner city and the core have become a highly appealing place to live for a large segment of the population sensitive to their urban amenities: entertainment, culture, proximity to workplaces, and walking-hospitable environments (see Bain, Chapter 15). An issue with the arrival of new residents in central areas is the difficult coexistence of different land uses, for example, noisy nightclubs close to expensive condominium buildings. There is also a cry to modify arterial roads, built a generation or two ago to accommodate large traffic flows between suburbs and the downtown, in order to render them more hospitable to new and growing populations living in their immediate environs (Hume, 2009). In addition, the disappearance of affordable housing from central areas has pushed low-income residents to less accessible parts of the metropolitan region. Meanwhile, other metropolitan areas, especially the smaller ones, have yet to experience a renaissance in the downtown and inner city.

Conclusion: Societal and Urban Trends

This chapter has explored how individual cities and the entire system of cities have responded to economic changes. We have seen that the historic settlement of urban areas in Canada was driven by shifts in leading exports: from furs to lumber and then to grain. In fact, the enduring east–west orientation of the urban system is in large part a legacy of the early staples economy. As Canada industrialized, the location of manufacturing became the foremost propellant of urban development. The industrial phase of development was favourable primarily to the heartland, that is, to the

Quebec City–Windsor corridor. The most recent epoch is characterized by deindustrialization, a demand for certain resources, and high immigration levels. Together these contemporary tendencies favour the growth of the largest metropolitan regions at the expense of the smaller ones. Gradual changes in the circumstances that shaped the urban system are thus accounted for by staples theory, the heartland–hinterland dichotomy, and present efforts at connecting neo-liberalism, a growing service sector, and rising immigration.

Intra-urban structure, too, is influenced by broad economic trends. Thus, by virtue of their rapid growth over the last decades, Toronto and Vancouver are well provided with contemporary suburban-style developments as well as with large-scale condominium redevelopment in the core. But we have seen that urban structure also is affected by values. The widespread adherence to the modernist credo during the 1950s and 1960s translated into accelerated suburban development at the expense of the inner city. More recently, aspirations of a large proportion of the population for a more urbane lifestyle have fostered a return of residents to central parts of large urban areas.

The changing impact transportation has had on urban systems and urban structure is noteworthy. While the railway system largely determined the location and expansion of western Canada's urban centres, subsequent technologies mostly adapted to the existing urban hierarchy and, indeed, reinforced it. The expressway network connected large urban areas; airlines located their hubs in major centres; and these infrastructures favoured the ongoing expansion of the urban centres they were servicing. In contrast to the inter-urban systems, all new modes of transportation have had a profoundly transformative effect on intra-urban structure, as evidenced by the different configurations of walking-, streetcar-, and automobile-oriented urban environments.

Adaptation to coming societal circumstances likely will be difficult at the level of both the individual urban entity and the urban system. More than in the past, the urban system will be divided between winners and losers, that is, large metropolitan regions experiencing rapid growth alongside smaller urban areas confronted with demographic and economic decline. Governments, therefore, will need to deal simultaneously with costs associated with accelerated urban expansion and those required to react to steep decline. At the same time, urban structure will be forced to respond to tensions raised by the disjunction between prevailing urban configurations and dynamics (in particular a generalized dependence on the car) and societal changes on the horizon, particularly as regards decreasing oil supplies (and consequent price escalation), rising awareness of global warming (and the eventual adoption of stringent greenhouse gas controls), and an overall change of values regarding consumption and urban living. Chapter 3, which explores the possibility of an eventual transition in urban development patterns in the face of these changing circumstances, examines anticipated tensions between forces of inertia and of change.

Review Questions

1. How does the Canadian urban system reflect Canadian history?
2. What are the relations between trends affecting society as a whole and urban development patterns?
3. What are the emerging features of contemporary Canadian urbanization? Do these foreshadow different forms of urban development?

Notes

1. The years defining the five epochs are approximate.
2. For a fuller discussion of the evolution of the Canadian urban system, see Jim Simmons and Larry McCann, 'The Canadian Urban System: Growth and Transition', Chapter 3 in Bunting and Filion (2006), at: <www.oupcanada.com/bunting4e>.

3. Note that in Figure 2.3 the populations of CMAs adjacent to the Toronto, Montreal, and Vancouver metropolitan regions were included with those of the central three regions.

References

Adams, J.S. 1970. 'Residential structure of Midwestern cities', *Annals, Association of American Geographers* 60: 37–62.

Aglietta, M. 1979. *A Theory of Capitalist Regulation: The U.S. Experience*. London: New Left Books.

Amin, A., ed. 1994. *Post-Fordism: A Reader*. Oxford: Blackwell.

Baldassare, M. 1986. *Trouble in Paradise: The Suburban Transformation in America*. New York: Columbia University Press.

Bercuson, D.J., and D. Carment, eds. 2008. *The World in Canada: Diaspora, Demography, and Domestic Policy*. Montreal and Kingston: McGill-Queen's University Press.

Berelowitz, L. 2005. *Dream City: Vancouver and the Global Imagination*. Vancouver: Douglas & McIntyre.

Berry, B.J.L. 1963. *Commercial Structure and Commercial Blight: Retail Patterns and Process*. Chicago: University of Chicago, Department of Geography Research Paper No. 85.

Birch, D. 1971. 'Toward a stage model of urban growth', *Journal of the American Institute of Planners* 37: 78–87.

Biven, W.C. 1989. *Who Killed John Maynard Keynes? Conflicts in the Evolution of Economic Policy*. Homewood, Ill.: Dow Jones–Irwin.

Blake, R.B. 2009. *From Rights to Needs: A History of Family Allowances in Canada, 1929–1992*. Vancouver: University of British Columbia Press.

Bonefeld, W., and J. Holloway, eds. 1991. *Post-Fordism and Social Form: A Marxist Debate on the Post-Fordist State*. Houndmills, Basingstoke: Macmillan.

Borchert, J. 1967. 'American metropolitan evolution', *Geographical Review* 59: 301–32.

Bottles, S. 1987. *Los Angeles and the Automobile: The Making of a Modern City*. Berkeley: University of California Press.

Bourne, L.S. 1967. *Private Redevelopment of the Central City*. Chicago: University of Chicago, Department of Geography Research Paper No. 112.

———. 1989 'Are new urban forms emerging? Empirical tests for Canadian urban areas', *Canadian Geographer* 33: 312–28.

——— and D. Ley, eds. 1993. *The Changing Social Geography of Canadian Cities*. Montreal and Kingston: McGill-Queen's University Press.

———, M. Bunce, L. Taylor, N. Luka, and J. Maurer. 2003. 'Contested ground: The dynamics of peri-urban growth in the Toronto region', *Canadian Journal of Regional Science* 26: 251–70.

Brown, S. 1992. *Retail Location: A Micro-Scale Perspective*. Aldershot: Ashgate.

Bunting, T., and P. Filion. 1999. 'Dispersed city form in Canada: A Kitchener CMA case study', *Canadian Geographer* 43: 268–84.

——— and ———, eds. 2006. *Canadian Cities in Transition: Local Through Global Perspectives*, 3rd edn. Toronto: Oxford University Press.

Canadian Centre for Policy Alternatives. 2007. *What Can Governments Do About Canada's Growing Gap? Canadian Attitudes toward Income Inequality*. Toronto: CCPA.

Cevero, R. 1986. 'Urban transit in Canada: Integration and innovation at its best', *Transportation Quarterly* 40: 293–316.

———. 1989. *America's Suburban Centers: The Land-Use Transportation Link*. London: Unwin-Hyman.

Clark, S.D. 1966. *The Suburban Society*. Toronto: University of Toronto Press.

Code, W.R. 1983. 'The strength of the centre: Downtown offices and metropolitan decentralization in Toronto', *Environment and Planning A* 15: 1361–80.

Coffey, W.J., and M. Polèse, eds. 1987. *Still Living Together: Recent Trends and Future Directions in Canadian Regional Development*. Montreal: Institute for Research on Public Policy.

Colby, C. 1933. 'Centripetal and centrifugal forces in urban geography', *Annals, Association of American Geographers* 23: 1–20.

Cook, N.D. 1998. *Born to Die: Disease and New World Conquest, 1492–1650*. Cambridge: Cambridge University Press.

Coppack, P., L.H. Russwurm, and C.R. Bryant, eds. 1988. *Essays on Canadian Urban Process and Form III: The Urban Field*. Waterloo, Ont.: University of Waterloo, Department of Geography Publication Series No. 30.

Dawson, J., and J. Lord. 1985. *Shopping Centre Development: Policies and Prospects*. London: Croom Helm.

Dear, M.J., and S. Flusty, eds. 2002. *The Spaces of Postmodernity: Readings in Human Geography*. Oxford: Blackwell.

Dechêne, L. 1987. 'The town of Quebec, 18th century', in R.C. Harris and G.J. Matthews, eds, *Historical Atlas of Canada: Volume 1, From the Beginning to 1800*. Toronto: University of Toronto Press (Plate 50).

Dennis, M., and S. Fish. 1972. *Programs in Search of a Policy*. Toronto: Hakkert.

Dobriner, W.M., ed.1958. *The Suburban Community*. New York: Putnam's.

Duany, A., E. Plater-Zyberk, and J. Speck. 2000. *Suburban Nation: The Rise of Sprawl and the Decline of the American Dream*. New York: North Point Press.

Evenden, L.J., and G. Walker. 1993. 'From periphery to centre: The changing geography of the suburbs', in L.S. Bourne and D. Ley, eds, *The Changing Social Geography of Canadian Cities*. Montreal and Kingston: McGill–Queen's University Press.

Filion, P. Forthcoming. 'Growth and decline in the Canadian urban system: The impact of emerging economic, policy and demographic trends', *Geojournal*.

——— and T. Bunting. 1990. *Affordability of Housing*. Ottawa: Minister of Supply and Services, Statistics Canada Catalogue 93–130.

———, ———, and K. Curtis. 1996. *The Dynamics of the Dispersed City: Geographic and Planning Perspectives*. Waterloo, Ont.: University of Waterloo, Department of Geography Publication Series No. 47.

——— and K. Hammond. 2009. 'When planning fails: Downtown malls in mid-size cities', *Canadian Journal of Urban Research* 17, 2: 1–27.

Fishman, R. 1987. *Bourgeois Utopias: The Rise and Fall of Suburbia*. New York: Basic Books.

Forster, J.J.B. 1986. *A Conjunction of Interests: Business, Politics, and Tariffs, 1825–1879*. Toronto: University of Toronto Press.

Frazer, W.J. 1994. *The Legacy of Keynes and Friedman: Economic Analysis, Money, and Ideology*. Westport, Conn.: Praeger.

Frideres, J., M. Burstein, and J. Biles, eds. 2008. *Immigration and Integration in Canada in the Twenty-First Century*. Montreal and Kingston: McGill–Queen's University Press.

Frieden, B.J., and L. Sagalyn.1989. *Downtown, Inc.: How America Rebuilds Cities*. Cambridge, Mass.: MIT Press.

Friedman, M. 1984. *Politics and Tyranny: Lessons in the Pursuit of Freedom*. San Francisco: Pacific Institute for Public Policy Research.

Friedman, T.L. 2005. *The World Is Flat: A Brief History of the Twenty-First Century*. New York: Farrar, Straus, and Giroux.

Gad, G. 1991. 'Office location', in T. Bunting and P. Filion, eds, *Canadian Cities in Transition*. Toronto: Oxford University Press.

Garreau, J. 1991. *Edge City: Life on New Frontier*. New York: Doubleday.

Grant, J. 2006. *Planning the Good Community: New Urbanism in Theory and Practice*. London: Routledge.

Guy, C. 1994. *The Retail Development Process: Location, Property and Planning*. London: Routledge.

Hackworth, J. 2006. *The Neoliberal City: Governance, Ideology, and Development in American Urbanism*. Ithaca, NY: Cornell University Press.

Haines, M.R., and R.H. Steckel. 2000. *A Population History of North America*. Cambridge: Cambridge University Press.

Hamel, P. 1993. 'Modernity and postmodernity: The crisis of urban planning', *Canadian Journal of Urban Research* 2: 16–29.

Hamouda, O.F., and J.N. Smithin. 1988. *Keynes and Public Policy after Fifty Years*. New York: New York University Press.

Harris, R. 2004. *Creeping Conformity: How Canada Became Suburban, 1900–1960*. Toronto: University of Toronto Press.

Harris, R.C., and D. Wood. 1987. 'Eastern Canada in 1800', in R.C. Harris and G.J. Matthews, eds, *Historical Atlas of Canada: Volume 1, From the Beginning to 1800*. Toronto: University of Toronto Press (Plate 68).

Harvey, D. 1989. *The Condition of Postmodernity*. Oxford: Blackwell.

Hayden, D. 2003. *Building Suburbia: Green Fields and Urban Growth, 1820–2000*. New York: Pantheon Books.

Hoover, E.M., and R. Vernon. 1962. *Anatomy of a Metropolis*. New York: Doubleday Anchor.

Hou, F., and L. Bourne. 2004. *Population Movement into and out of Canada's Immigrant Gateway Cities: A Comparative Study of Toronto, Montreal and Vancouver*. Ottawa: Statistics Canada, Business and Labour Market Division, Research Paper Series No. 229.

Hough, M. 2004. *Cities and Natural Processes: A Basis for Sustainability*. London: Routledge.

Houston, C.S., and S. Houston. 2000. 'The first smallpox epidemic on the Canadian Plains: In the fur-traders' words', *Canadian Journal of Infectious Disease* 11: 112–15.

Howe, D., and W. Rabiega. 1992. 'Beyond strips and centres: The ideal commercial form', *Journal of the American Planning Association* 58: 213–19.

Hoyt, H. 1939. *The Structure and Growth of Residential Neighborhoods in American Cities*. Washington: Federal Housing Administration.

Hume, C. 2009. 'Fear and cycling on Jarvis: Wherein our columnist puts aside his terror to jump on a bike and brave a mean street', *Toronto Star*, 30 May.

Innis, H.A. 1956. *The Fur Trade in Canada: An Introduction to Canadian Economic History*. Toronto: University of Toronto Press.

———. 1995 [1931]. 'Transportation as a factor in Canadian economic history', in D. Drache, ed., *Staples, Markets and Cultural Change: Selected Essays, Harold A. Innis*. Montreal and Kingston: McGill-Queen's University Press.

Jacobs, J. 1961. *The Death and Life of Great American Cities*. New York: Random House.

Jessop, B., and N.-L. Sum. 2006. *Beyond the Regulation Approach: Putting Capitalist Economies in their Place*. Cheltenham: Edward Elgar.

Jones, J.P. 2008. *Keynes's Vision: Why the Great Depression Did Not Return*. London: Routledge.

Jones, K. 1991. 'The urban retail landscape', in Bunting and Filion (2006).

——— and M. Doucet. 2000. 'Big-box retailing and the urban retail structure: The case of the Toronto area', *Journal of Retailing and Consumer Services* 7: 233–47.

——— and J. Simmons. 1990. *The Retail Environment*. London: Routledge.

Joseph, A., P. Keddie, and B. Smit. 1988. 'Unraveling the population turnaround', *Canadian Geographer* 32: 17–31.

Keil, R. 2002. '"Common-sense" liberalism: Progressive conservative urbanism in Toronto, Canada', *Antipode* 34: 541–601.

Kling, R., S. Olin, and M. Poster. eds. 1991. *Postsuburban California: The Transformation of Orange County since World War II*. Berkeley: University of California Press.

Knowles, V. 2007. *Strangers at Our Gates: Canadian Immigration and Immigration Policy, 1540–2007*. Toronto: Dundurn.

Knox, P. 1994. *Urbanization: An Introduction to Urban Geography*. Englewood Cliffs, NJ: Prentice-Hall.

Leo, C. 1994. 'The urban economy and the power of the local state: The politics of planning in Edmonton and Vancouver', in F. Frisken, ed., *The Changing Canadian Metropolis: A Public Policy Perspective*. Berkeley and Toronto: Institute of Governmental Studies Press, University of California Press and Canadian Urban Institute.

——— and K. Anderson. 2006. 'Being realistic about urban growth', in Bunting and Filion (2006). At: <www.oupcanada.com/bunting4e>.

Ley, D. 1996. *The New Middle Class and the Remaking of the Central City*. Oxford: Oxford University Press.

Lortie, A. 2004. *The 60s: Montreal Thinks Big*. Montreal: Canadian Centre for Architecture.

McCann, L., and A. Gunn, eds. 1998. *Heartland–Hinterland: A Regional Geography of Canada*, 3rd edn. Toronto: Prentice Hall.

——— and J. Simmons. 2001. 'The core–periphery structure of Canada's urban system', in Bunting and Filion (2001).

McCart, N.M., K.T. Polle, and H. Rosenthal. 2006. *Polarized America: The Dance of Ideology and Unequal Riches*. Cambridge, Mass.: MIT Press.

Mackenzie, S., and D. Rose. 1983. 'Industrial change, the domestic economy and home life', in A.S. Duncan and R. Hudson, eds, *Redundant Spaces in Cities and Regions?* London: Academic Press.

Manuel, P.M. 2003. 'Cultural perceptions of small urban wetlands: Cases from the Halifax Regional Municipality, Nova Scotia, Canada', *Wetlands* 23: 921–40.

Matthews, R. 1983. *The Creation of Regional Dependency.* Toronto: University of Toronto Press.

Mazy, M.E., and D.R. Lee. 1983. *Her Space Her Place.* Washington: Association of American Geographers.

Meade, J.E. 1982. *Stagflation.* London: Allen and Unwin.

Miller, E.J., and A. Shalaby. 2003. 'Evolution of personal travel in Toronto area and policy implication', *Journal of Urban Planning and Development* 129: 1–26.

Miller, K.A. 1985. *Emigrants and Exiles: Ireland and the Irish Exodus in North America.* New York: Oxford University Press.

Miron, J. 1988. *Housing in Postwar Canada: Demographic Change, Household Formation and Housing Markets.* Montreal and Kingston: McGill-Queen's University Press.

———, ed. 1993. *House, Homes, and Community: Progress in Housing Canadians, 1945–1986.* Montreal and Kingston: McGill-Queen's University Press.

Moscovitch, A., and J. Albert, eds. 1987. *The 'Benevolent' State: The Growth of Welfare in Canada.* Toronto: Garamond Press.

Muller, P. 1981. *Contemporary Suburban America.* Englewood Cliffs, NJ: Prentice-Hall.

Murphy, B. 1999. *The Ugly Canadian: The Rise and Fall of a Caring Society.* Winnipeg: J. Gordon Shillingford.

Naylor, R.T. 2006. *The History of Canadian Business, 1867–1914.* Montreal and Kingston: McGill-Queen's University Press.

Nowlan, D., and N. Nowlan. 1970. *The Bad Trip: The Untold Story of Spadina Highway.* Toronto: Anansi Press.

Olson, M. 1982. *The Rise and Decline of Nations: Economic Growth, Stagflation, and Social Rigidities.* New Haven: Yale University Press.

Park, R., and E.W. Burgess, eds. 1969 [1925]. *The City.* Chicago: University of Chicago Press.

Paumier, C.B. 1988. *Designing Successful Downtowns.* Washington: Urban Land Institute.

Perl, A., and J. Pucher. 1995. 'Transit in trouble? The policy challenge posed by Canada's changing mobility', *Canadian Public Policy* 21: 261–83.

Popenoe, D. 1977. *The Suburban Environment.* Chicago: University of Chicago Press.

Preston, R.E., and D.W. Griffin. 1966. 'A restatement of the zone of transition concept', *Annals, Association of American Geographers* 56: 339–50.

Punter, J. 2003. *The Vancouver Achievement: Urban Planning and Design.* Vancouver: University of British Columbia Press.

Redstone, L. 1976. *The New Downtowns: Rebuilding Business Districts.* New York: McGraw-Hill.

Reisman, D. 1958. 'The suburban sadness', in W.M. Dobriner, ed., *The Suburban Community.* New York: Putnam's.

Rowe, P.G. 1991. *Making a Middle Landscape.* Cambridge, Mass.: MIT Press.

Schonhardt-Bailey, C. 2006. *From the Corn Laws to Free Trade: Interests, Ideas, and Institutions in Historical Perspectives.* Cambridge, Mass.: MIT Press.

Semmel, B. 1970. *The Rise of Free Trade Imperialism: Classical Political Economy, the Empire of Free Trade and Imperialism 1750–1850.* Cambridge: Cambridge University Press.

Sewell, J. 1993. *The Shape of the City: Toronto Struggles with Modern Planning.* Toronto: University of Toronto Press.

———. 2009. *The Shape of the Suburbs: Understanding Toronto's Sprawl.* Toronto: University of Toronto Press.

Sitwell, O.F.G., and N.R.M. Seifried. 1984. *The Regional Structure of the Canadian Economy.* Toronto: Methuen.

Smith, W. 1964. *Filtering and Neighborhood Change.* Berkeley: University of California, Center for Real State and Urban Economics, Report No. 24.

Smucker, J. 1980. *Industrialization in Canada.* Scarborough, Ont.: Prentice-Hall.

Spurr, P. 1976. *Land and Urban Development: A Preliminary Study.* Toronto: James Lorimer.

Stanback, T.M. 1991. *The New Suburbanization: Challenge to the Central City.* Boulder, Colo.: Westview Press.

Urquhart, M.C., and K.A.H. Buckley. 1983. *Historical Statistics of Canada*, 2nd edn. Ottawa: Ministry of Supply and Services, Canada, Series A2-14.

Vance, J.E. 1970. *The Merchant's World: The Geography of Wholesaling.* Englewood Cliffs, NJ: Prentice-Hall.

Waite, P.B. 1967. *The Life and Times of Confederation. 1864–1867: Politics, Newspapers and the Union of British North America*, 2nd edn. Toronto: University of Toronto Press.

Walks, R.A. 2004. 'Place of residence, party preferences, and political attitudes in Canadian cities and suburbs', *Journal of Urban Affairs* 26: 269–95.

Warner, B.J. 1962. *Streetcar Suburbs: The Process of Growth in Boston.* Cambridge, Mass.: Harvard University Press.

Yalnizyan, A. 2000. *Canada's Great Divide: The Politics of the Growing Gap between Rich and Poor in the 1990s.* Toronto: Centre for Social Justice.

Yeates, M. 1975. *Main Street: Windsor to Quebec City.* Toronto: Macmillan.

———. 1998. 'The industrial heartland: Its changing role and internal structure', in L. McCann and A. Gunn, eds, *Heartland–Hinterland: A Regional Geography of Canada*, 3rd edn. Toronto: Prentice-Hall.

CHAPTER 3

Transition in the City

Pierre Filion and Trudi Bunting

*C*anadian Cities in Transition emphasizes change. Most chapters in the book deal with urban change in one form or another, and this one explores the concept of how change unfolds in cities. Change is perceived as the outcome of tensions between opposing categories of forces, the first of which promotes change while the second favours stability. We will demonstrate that cities are inherently resistant to drastic change due to the durability of their built environment and their modes of transportation that determine patterns of behaviour and the distribution of activities. At the same time, however, they do experience both incremental change (which takes place mostly *within* prevailing urban forms and transportation systems) and, more infrequently, radical transformations, which entail deep modifications of form and systems. Our perspective on urban change interprets past urban transitions and explores the possibility of a radical shift arising in the way cities function and grow. While avoiding specific predictions, our approach nonetheless makes it possible to define a range of future developments.

Visions of the City vs Urban Complexity

Any avid reader of the urban literature could be pardoned for taking the impression that the city is constantly undergoing deep transformations, one major change after another. Much of this literature stresses change at the expense of stability. After all, one will attract much less attention with a business-as-usual scenario than with either the proposal of an innovative model of urban development or the discovery of a new trend. Urban visionaries have played a major role in this literature, but their models have faced difficulties in implementation and the new trajectories they have detected generally have failed to do justice to the comprehensive urban reality.

Urban planning has been profoundly influenced by reformers who advanced new urban development formulas intended to address downsides of the industrial city.[1] The two reformers whose influence was strongest during the twentieth century are Ebenezer Howard and Le Corbusier (Fishman, 1977). The objective of Howard was to provide an alternative to the excessive density, unsanitary conditions, and high costs of living of the early twentieth-century industrial city. His Garden City concept entailed the creation of cities where population density would be considerably lower, green space much more abundant, and population would be kept to a manageable size with between 32,000 and 58,000 residents. Each Garden City would be surrounded by a green belt; but Howard envisioned networks of such cities eventually accommodating much of Great Britain's population and employment (Howard, 2003 [1902]).

Le Corbusier, who was born in Switzerland but spent his professional life in Paris, had a different perspective on the city. Like Howard, Le Corbusier was driven by concern about unsanitary conditions in the cities of the early part of the twentieth century, but his solution to that problem distinguishes him from Howard. For Le Corbusier technology made it possible to provide a sanitary urban environment while preserving the high-density characteristic of the European cities of the time. His formula, **Tower in the Park**, consisted of high-rise buildings erected at a distance from each other to allow for ample green space at ground level (Le Corbusier, 1973 [1933]).

The visions of both Howard and Le Corbusier were integrated in some measure in Canadian urban development. For example, the suburbs of Toronto offer a combination of low-density housing on curvilinear streets, recalling the Garden City model, and pockets of high-rise residential buildings configured in a Tower in the Park fashion. But the implementation of these visions in the Canadian urban landscape did not live up to expectations. The sprawling Canadian suburb is a far cry from the Garden City with its limited size and green belt, and the Tower in the Park formula was implemented on a much smaller scale than anticipated by Le Corbusier. Moreover, the green space of such developments usually was poorly used and in the worst cases became crime-ridden.

More recently, the architect Andres Duany has played a role that recalls the two visionaries, even if his ambition is more modest. Duany's focus is mostly on problems inherent in suburban development since World War II, in particular heavy dependence on the automobile, traffic congestion, exaggerated land-use specialization, and the blandness of suburban environments (Duany, Plater-Zyberk, and Speek, 2001). Duany's New Urbanism model calls for a return to the layout and appearance of the traditional small town, with houses built close to the sidewalk, a gridiron layout, and a main commercial street. The model de-emphasizes the presence of the car, in large part by relegating garages to back lanes. However, New Urbanism communities have remained a niche market, hardly making a dent in the popularity of conventional suburban subdivisions. New Urbanism, therefore, has had a negligible impact on the development of urban areas. What is more, New Urbanism communities do not actually function as originally intended. Reliance on the automobile typically remains as high as it is in other suburban areas and rarely have New Urbanism suburban communities been able to create viable pedestrian-friendly commercial streets (Grant, 2006; Talen, 2000).

Urban change is emphasized also by postmodernist scholars attempting to identify new urban trajectories. Dear and Flusty (2002), for example, describe how space within the postmodern metropolis has become hyper-specialized, causing disconnection between some urban sectors. More recently, urban studies have been dominated by the creative class perspective put forth by Richard Florida (2002, 2005, 2008). According to this view, economic development has become reliant on a creative class whose members are a source of entrepreneurship and of attraction for sectors of the economy with high growth and innovation potential. Among other things, the creative class approach calls for a profound revision of economic development policies. In the past these policies mostly involved incentives intended to attract large manufacturing facilities; today, creative class proponents advocate the fashioning of a tolerant, culturally rich, and diversified urban environment apt to appeal to members of the creative class (see Vinodrai, Chapter 6).

All the above approaches, whether proposing or interpreting urban change, can be criticized for their reductive view of the urban phenomenon, that is, for failing to do justice to the complexity of the city. Their blinkered focus on only one aspect of the city can be held responsible for the failure of their visions to materialize as intended. The

Garden City perspective overlooked the appeal of the diversity of activities provided by large centres; the Tower in the Park model often failed to generate community ties; and New Urbanism did not account for journey patterns in car-oriented environments. The more contemporary post-modernist and creative class perspectives are flawed by their incapacity to account for the multiple facets of cities' social and economic realities, in particular, the circumstances perpetuating conventional economic and urban development patterns.

The most compelling critique of one-sided approaches to the city and reliance on simple solutions came from Jane Jacobs. In the early 1960s, she convincingly demonstrated that traditional inner-city neighbourhoods (at that time such neighbourhoods often were targeted for **urban renewal**) encouraged the formation of intense community ties—in sharp contrast to the social desolation of some projects inspired by the Tower in the Park model. She particularly praised the intricate nature of the traditional inner-city commercial street, with its unplanned mix of stores at ground level and residential units on upper floors. She described the intense social interactions that took place on these streets and their inherent safety, thanks to the presence of 'eyes on the street' (Jacobs, 1961). Unlike the proponents of urban change, Jane Jacobs celebrated the complexity of the city—especially its expression in inner-city streets and neighbourhoods—and defended the physical variety and multi-functionality of traditional urban environments against attempts at destroying them to make way for the highly standardized urban patterns typical of the time. Twice she was instrumental in saving mixed-use neighbourhoods slated to make way for expressways.

In this chapter we attempt to correct the literature bias in favour of change by identifying factors both of transformation and of stability, and by demonstrating how they account for occasional radical urban transitions followed by long periods of relative stability. A note of warning, however:

we have emphasized in this section the visions and perspectives of urban experts but many other factors shape the evolution of cities, including the economy, technology, political systems, and the preferences of consumers. Only when visions of the city are consistent with these other factors will they influence development.

The Issue of Change

Let us imagine that you return to your childhood neighbourhood after a 15-year absence. You would expect to notice profound changes: after all, 15 years comprise a significant part of one's life. There are different people as the neighbourhood now hosts a new crop of young families. Perhaps there has been a modification in the ethnic mix. Trees have grown, and some houses have been renovated while others have simply been repainted a different colour. In the retail area, which was a typical suburban strip mall, many stores have changed, but overall the strip mall caters to about the same needs as before. You are surprised that things have not changed as much as you expected. The houses are the same even if they look a little different. While you don't recognize most of the people, the sounds of the neighbourhood—cars, children playing—are just as you remember them, as are its daily rhythms with people leaving the home and returning at the same time as before. You go to the school and find that apart from some rearrangement—the use of some school space for a community centre and a preschool program—all is the same. And then you realize that 15 years may have been a long time for you but is a brief episode in the life of a city.

In this section, we examine circumstances that either propel or limit the transformation of cities, which will help us chart possible urban outcomes. But before focusing on the urban phenomenon, we briefly consider forces of transformation and of inertia operating at the scale of society. Nothing is set in society (on approaches to societal change, see Boudon, 1986; Fischer-Kowalski and Haberl, 2007; Janos, 1986). For example, even the best programs in the best universities are the objects of constant revision. The same applies to successful government programs and to the organizational structures of thriving corporations. The ubiquity of change in human affairs is illustrated by the fashion and design industries, where transformations are not the outcome of any ameliorative purpose but rather the reflection of desire for stimulation and for creativity.

Aspects of society that naturally evolve are thereby drivers of change. Such is notably the case of *technological* advancements. One need think only of the enormous social impact of the progression of electronic communication over the past hundred years: from the wireless telegraph, to the radio, television, cellular telephone, and Internet. The outcome has been a much expanded access to information and entertainment, whereby communities of proximity have been replaced in some cases by more distant communities of interest (Krug, 2005; Lin and Atkin, 2007). Shifting *values* also have had profound societal impacts in heightened individualism and acceptance or tolerance towards a wider spectrum of lifestyles. Evidently, the *economy* also has played a profound role in the transformation of society. Collective wealth has increased over the last decades, albeit unevenly and at the cost of a higher participation rate in the labour force (Amin, 1995; Levin-Waldman, 2005). All the same, levels of consumption for goods and services have risen over this period. *Demography* is another factor of change. Since World War II, the maturing of the baby boom generation has dominated the demographic scene, forcing institutional and economic adjustments along the way (Foot with Stoffman, 1996). Moreover, as a result of the more recent 'baby bust', the coming decades will be characterized by population growth only in those regions that attract immigration (see Bunting and Filion, Chapter 2; Donald and Hall, Chapter 16).

In the presence of these factors of change, social organizations and economic processes demonstrate remarkable durability. The concept of **path dependence** has been invoked to account for their relative stability (Boas, 2007; Brian, 1994; Pierson, 2000). According to this view, considerable advantages are to be gained by maintaining rather than replacing existing systems and processes: they generally yield increasing returns over time; vested interests have developed around them; habits have formed; and, of course, the 'devil you know' syndrome has its effects.

Clearly, we are in the presence of two opposing tendencies. One promotes change while the other fosters stability. In fact, the differing perspectives on social phenomena can be distributed along a continuum according to the emphasis they place on either stability or change. Thus, when we consider, for example, economic norms one end would be occupied by neo-classical economics, which centre on the attainment and preservation of conditions of equilibrium, and the other by Marxist perspectives, which focus on the need for far-reaching transformation.

The tension between stability and change pertains equally to cities. The factors of societal change we listed above all have specific urban impacts. Regarding the effect of *technology*, we have demonstrated in Chapter 2 how successive modes of transportation have shaped different urban forms (the compact downtown-focused, rail-transit urban form vs the low-density, dispersed, automobile-oriented city). As *values* became more diversified so did preferences for certain urban forms and their attendant lifestyles, which has led over the last decades to a **gentrification** of certain inner-city neighbourhoods, high-rise condominium living, the transformation of some industrial premises into lofts, paralleled by ongoing low-density suburban development. A generally positive performance of the *economy* has resulted in larger homes, greater accessibility to cars, and the expansion of the service sector (Miron, 1998). Fluctuations in the respective proportion of single-detached and multiple housing starts have largely been driven by specific *demographic* demand, the passage of the baby boom through successive life stages (Foot with Stoffman, 1996). Now, in large metropolitan regions, immigration is having a major impact on housing development.

There is another, often overlooked, factor of change affecting cities: urban growth itself. We all know that there are important size-related differences between urban areas (Bessey, 2002). Sudbury has little traffic congestion in comparison with Toronto, but public transportation in Sudbury is much less developed, even accounting for difference in the sizes of the two metropolitan regions. Likewise, there are fewer restaurants and less cultural activity in Sudbury than in Toronto. If we compare Toronto to the New York metropolitan region, it is now New York that offers the more developed public transportation system and the wider array of restaurants and cultural offerings. So, as an urban area grows, it changes many of the ways it functions. Presently, planners hold the view that population growth forces Toronto to raise its density and reliance on public transportation lest it become mired in ever-worsening traffic congestion (Metrolinx, 2008).

Concurrently, change in the city is moderated by stabilizing forces. In fact, urban areas provide one of the best examples of path dependence, for they are subjected not only to stabilizing effects felt throughout society, but are affected equally by intrinsic factors of inertia. Indeed, radical change tends to be offset by the difficulty of overcoming the inertia of urban form and dynamics. We understand **urban form** to consist of the built structure of the city, such as buildings and roads, and **urban dynamics** to comprise behaviours that take place within these city forms. Of course, an adaptation between form and dynamics is constantly underway, but this adaptation tends to be of a mutual nature and in itself contrives mostly to add to the stabilizing influences. To modify large expanses

of the built environment is difficult because so much capital has been sunk into it. Moreover, a great complex of sentimental feeling and of economic thought becomes attached to the existing built environment, which explains the intensity of NIMBY (not in my back yard) reactions against attempts at transforming it (Curic and Bunting, 2006; Schively, 2007). The distribution of activities in the built environment and uneven accessibility to parts of that environment together shape behaviour in the city, although not without occasional cognitive dissonance, which occurs when people act in ways that contradict their values (Cooper, 2007). A typical case of cognitive dissonance arises when someone with deep environmental values is forced to drive long distances to reach work and other activities because of dispersed activity patterns and an absence of adequate public transportation—a situation characteristic of most suburban environments. One aspect of the interrelation between urban form and dynamics concerns the relation between land use and journey patterns, themselves shaped by transportation systems. Accessibility peaks generally attract activities that benefit from such locations (workplaces, retailing, services, institutions), but configurations vary according to the dominant mode of transportation (high and low density, respectively, in transit- and car-oriented areas) (Boarnet and Crane, 2001; Frank, Kavage, and Appleyard, 2007; Millward and Xue, 2007; van de Coevering and Shawnen, 2006).

Each perspective—the one that emphasizes change and the one that highlights the role of factors of stability—takes a useful view on the evolution of cities. The visit to the childhood neighbourhood has shown that many aspects have changed but that the fundamentals (built environment, forms of transportation, daily rhythms) have remained largely the same, which leads us to an examination of the relationship between stability and change in the urban context. Change can be of an incremental nature. When change does not challenge predominant urban forms and dynamics (or, in this instance, organizations and processes), the incremental alterations fit within the paradigm called path dependence. Transformations to the childhood neighbourhood fall within this category of change.

Alternatively, change can be more radical, involving the transformation of urban fundamentals. Change pressures, even incremental ones, can reach a tipping point that will subvert prevailing urban form and dynamics (Gladwell, 2002). The path dependence perspective cannot account for this second category of change, because of a profound transformation of prior channels of path dependence, such as the built environment, distributions of activities, and transportation systems. If your grandparents were to return to their childhood neighbourhood, chances are that it would bear the mark of deep urban transitions (such as the one that took place in the 1950s and is explored later in the chapter). They might not recognize the neighbourhood since so many buildings have either been torn down or replaced and its streetcar orientation has made way for full reliance on the automobile, causing considerable space to be occupied by parking lots. Perhaps the very function of the neighbourhood has changed, resulting in an absence of children.

Past Urban Transitions

Chapter 2 has described transitions in Canadian cities, driven by the passage from a mercantile to an industrial and then to a post-industrial economy, as well as by shifts in modes of transportation. But the transition that most importantly determined the form and dynamics of our contemporary cities took place over one decade, the 1950s. Conditions were ripe for a major urban transition. Little development had taken place over the previous 20 years due to the Great Depression and World War II, and when urban growth resumed after the war it began with an extension of earlier urban patterns. However, the 1930s and 1940s had witnessed the formulation of urban innovations—with the concept of the expressway having greatest impact on the future of cities (Lewis, 1997). Sluggish urban development in these decades precluded the implementation of such innovations.

The 1950s rode a wave of sustained prosperity, driven in large part by government investment in infrastructures and housing, and by pent-up demands for consumption and innovation. The economy was propelled by explosive demands for automobiles and for homes, many of which were single-family houses built in new suburban subdivisions. The automobile orientation of new housing developments in turn became a catalyst for urban innovations over the 1950s, establishing a cycle that continues to the present. This scenario stimulated all sorts of automobile-oriented facilities, including the strip mall, the outdoor and then indoor shopping mall, the motel, the drive-in restaurant, and the drive-in cinema (Hardwick, 2004). At the same time, single-family zones were planned to discourage through traffic in order to improve security for children and to provide a quiet residential environment (Wang and Smith, 1997). From the late 1950s, intricate patterns of curvilinear streets and cul-de-sacs and crescents increasingly typified newly developed low-density residential areas. The adaptation to the car persisted over subsequent decades, with garages expanding to accommodate two and then three vehicles, and with retail formulas catering to ever-larger, automobile-based catchment areas, culminating over the 1990s and 2000s with big-box stores, power malls, and gigantic supermarkets (Evans-Cowley, 2006).

It would be difficult to overestimate the impact of the urban transformation that was set in motion in the 1950s. Before this decade, urban areas were focused predominantly on their central business district, which encompassed virtually all their high-order services and much of their office employment. Land values were largely determined by accessibility to the CBD. In addition, public transportation accounted for a majority of journeys until the precipitous fall in patronage during the 1950s. Henceforth, automobile-based accessibility patterns would result in much increased per-capita land consumption, principally in the form of green space. All land uses were adapted to the car, which meant ubiquitous parking spaces. The creation of automobile-oriented environments was particularly painful in older parts of the city. Tree-lined avenues were widened to make room for automobiles, buildings were sacrificed to create parking lots, and worst of all, large swatches were cut through the urban texture to make way for expressways (see Bunting and Filion, Chapter 2).

With the generalization of car use new urban areas adopted a dispersed configuration, where, in sharp contrast with earlier concentrated urban patterns, retailing, services, institutions, and employment were scattered throughout the numerous sites enjoying high automobile accessibility (Filion, Bunting, and Warriner, 1999; Lang, 2003). Consequently, activity in the downtown experienced relative decline in large metropolitan regions and an absolute regression in all other urban areas. Moreover, the car made it possible to travel easily from one large specialized zone to another, hence the highly functionally segregated land-use patterns that characterize the suburb.

There were counter-tendencies to the patterns of development launched in the 1950s. When time came to find a residence of their own, some of the better-paid professional people, some of them children of the first settlers of post-war suburbs, decided to locate in the inner city of large metropolitan regions, thus launching gentrification (Lees, Slater, and Wyly, 2008). In these urban regions, gentrification was paralleled with residential intensification in and around the downtown and near public transit stations. Residents of these developments post a much lower than average reliance on the automobile (McLeod, 1999). Still, for all its visibility and media coverage, gentrification affected a much smaller proportion of the population than did ongoing suburbanization.

The extent and nature of the urban transition that was launched during the 1950s and that soon affected all parts of urban areas were unexpected. No one in 1945 could have predicted the deep urban transformations that were about to unfold; no glaring signs of change were visible at this time. To be sure, Utopian visions of the car-dominated city could be found: for example, Frank Lloyd Wright's Broadacre City; but they failed to comprehend the full implications of a total adaptation of urban areas to the car (Fishman, 1977). One of the best known among these Utopias was an animated model labelled Futurama, which was the centrepiece of the General Motors pavilion at the 1939–40 New York World's Fair. It portrayed the city of the future, 20 years hence (in 1960). Not surprisingly, considering the sponsor, the city was fully adapted to the car and was criss-crossed by a dense expressway network (Leinberger, 2008). In light of what we now know about the relation between land use and reliance on the automobile, the urban landscape of Futurama was unrealistic. The model depicted a city without a central business district but with many clusters of high-rise buildings along expressways. Absent were large-surface parking lots and, perhaps most significantly, no traffic congestion was evident despite universal reliance on the automobile and the presence of high-density developments.

Difficulties in anticipating the consequences of the urban form and dynamics that evolved in the 1950s are illustrated by ill-fated adaptive strategies. One particularly unsuccessful strategy was the attempt over the 1950s and 1960s by interests associated with downtowns of mid-size urban areas to adapt their sector to the automobile in order to resist suburban competition. Downtown merchants, who were especially vulnerable to emerging suburban shopping malls, spearheaded this effort. The logic seemed unassailable: since there still were more shops in downtowns than in individual suburban malls, improved automobile accessibility would assure the retail supremacy of the downtown. The outcome of improved parking did not meet expectations. Despite the widening of roads and new provisions for parking in downtown areas, suburban locations with their expressway access, wide arterials, and ample, free, surface parking maintained their car accessibility advantage. To make matters worse, accommodation to the car caused the disappearance of many downtown buildings to make way for surface parking, thus reducing the amount and diversity of their activities and hampering their pedestrian connectivity (Jakle and Sculle, 2004). Had they maintained their distinctiveness relative to the suburb, downtowns could have targeted markets looking for a more vibrant urban environment than could be found in the suburbs, along with alternatives to the standardized products and services available in suburban malls. As it happened, rather than rescuing mid-size city downtowns the automobile accommodation strategy precipitated their decline and compromised future attempts at **revitalization** (Filion et al., 2004). In large metropolitan regions circumstances were more favourable to downtowns thanks to much higher public transportation patronage, large concentrations of employment and residents, and the presence of tourists.

In an attempt to understand what triggered such a momentous and long-lasting urban change, we end this section by identifying different factors that drove the 1950s transition. The decade was marked by an unusual alignment of factors favourable to a radical transformation of the city. After the hardship of the Great Depression and World War II and in a period when modernity was highly valued at the expense of tradition, there was widespread willingness to adopt a new lifestyle. As it turned out, this lifestyle required high levels of consumption, stimulated by the emerging urban form and dynamics. Lower density provided households with more space to accumulate durable goods and the ownership of a car became a necessity for a growing proportion of the population. The relationship between the transformation

of cities and economic prosperity is reciprocal. If prosperity led to an adaptation of urban forms to a consumerist lifestyle and provided public-sector resources needed for new infrastructure systems, the argument also can be made that rising consumption and public investments associated with the 1950s urban transition fuelled economic growth. Another factor of change, which contributes to the durability of the urban form and dynamics emanating from the 1950s, is the mutual adaptation of travel patterns, increasingly reliant on the car, and land use. Finally, we cannot exclude the role of numerous agents with an interest in the then emerging form of urban growth: the asphalt lobby (the car industry, traffic engineers, road builders), the development industry, and anti-intensification NIMBY organizations.

The Coming Transition?

Signs of change are gathering on the horizon. The feeling grows that present development patterns cannot continue. These patterns are criticized for their adverse environmental impacts such as their contribution to global warming and, recently, the health consequences of air pollution along arterial roads and expressways (Jerrett et al., 2009). Critiques also target the impossibility of walking to most destinations and a consequent rise in obesity, the poor appearance of much post-war development, and the economic loss and aggravation caused by worsening traffic congestion (Downs, 1992; Kay, 1997). The costs of creating and maintaining present urban environments represent another issue with current forms of urban growth (Burchell, Downs, and McCann, 2005).

Planning documents long have called for a reorientation of urban development so as to elevate density, achieve mixed use, and diminish dependence on the automobile. Over recent decades, these types of urban change were associated with two broad perspectives. First, sustainable development concentrated on the need to address the adverse environmental consequences of urbanization; smart growth followed with a focus on the environmental, quality-of-life, and expenditure downsides of prevailing urban development (Barnett, 2007; Roseland, 1992; Tomalty, Alexander, and Grammenos, 2005). New urban development models have been proposed repeatedly over the last decades, but with relatively modest effects on the dominant urban development trajectory.

This time, however, things may be different. Factors of urban change may be aligned, as they were in the 1950s, and their impact may be sufficient to shift patterns of development inherited from this decade. Along with widespread reaction against environmental, financial, and quality-of-life issues related to automobile-oriented urban forms and dynamics, an increasing segment of the population aspires to a more urban lifestyle and is willing to live in high densities (Berelowitz, 2005). This group is made of young adults, empty nesters, and recent immigrants who are accustomed to high-density living. Other households are compelled for financial reasons to opt for higher-density living than they would wish. Many households cannot afford single-family homes and a continuous rise in energy prices (a distinct possibility in the near future) that would place long-distance car commuting beyond the financial capacity of growing numbers of individuals. Yet another factor of change looms, the effect of which remains highly uncertain at this time. We have seen that the Great Depression played a role in the urban transition of the 1950s. Could history be repeating itself with the present recession? At the time of writing, the Obama administration intends to use the economic recovery package to reduce oil dependency in the US and to promote green energy. In the words of Rahm Emmanuel, President Obama's chief of staff, 'You never want a serious crisis to go to waste. What I mean by that is that it's an opportunity to do things you could not do before' (Leonhardt, 2009: 25). Will such measures have an impact on urban development? Will Canada follow the lead of the US in this matter?

Presently, draft or approved plans for Toronto, Montreal, and Vancouver propose momentous changes in urban development patterns. The development models they advance comprise much-heightened reliance on vastly expanded public transit networks, as well as on walking and cycling; further downtown growth and multi-functionality; systems of sub-centres that combine multi-use, high density, and public transit access; and limits on outward metropolitan expansion (CMM, 2008; Metro Vancouver, 2008; Ontario, 2006). Also, the City of Toronto plans to intensify its car-oriented arterials in order to transform them into public transit- and pedestrian-hospitable environments (Toronto, 2002). Toronto and Vancouver already feature elements of the multi-nodal systems, the outcome of prior planning initiatives (Filion, 2007; Hutton, 2004; Punter, 2003). If fully implemented, these proposals would have the potential of abating the effect of decades of sprawling auto-oriented suburban development by channelling an important share of new development into a public transit-oriented, high-density, multi-nodal urban structure.

Of course, all of this hinges on the capacity to induce a new urban transition. Obstacles are daunting, stemming from the path dependence generated by multiple features of the automobile-dependent suburban model. The prevailing relationship between specialized and low-density land use and car dependence is entrenched. Important and co-ordinated public transit investments and land-use interventions will be needed to alter this relationship (Forkenbrock, 2002). A related difficulty is the vast amount of capital sunk in the urban form. An urban transition will require large expenditures to develop parallel urban forms that would be associated with alternative dynamics. If the recession is long-lasting, will it be possible to raise these important sums in the foreseeable future? It may also be difficult to overcome attachment for extant urban forms. While they loathe traffic congestion and the homely retail strip, suburban residents generally are fond of their homes and its surroundings. They value the low density of the suburbs as well as the open areas and greenness of their residential properties. Perhaps what many people oppose is not so much the suburban form and lifestyle as their downsides. Finally, habits tied to the prevailing urban environment would need to be changed. These habits die hard for many individuals reliant on the car, for it is highly convenient in a dispersed environment, barring traffic congestion.

We turn now to explore differences between the contemporary transitional context and the one that prevailed 60 years ago. First, the present clamour for another urban development model is not accompanied by a consensus on alternative urban lifestyles. We have seen that, over the last decades, growing proportions of households have demonstrated a preference for core and inner-city living, where they enjoy public transit and walking access to a wide range of activities and, thus, reduced dependence on the car. Such a residential choice is in tune with the increased place taken by the consumption of services, which require good accessibility, relative to that of goods, which require residential space. The preference of these households was responsible for a wave of residential intensification and rising housing prices in the core and inner city of some large metropolitan regions. Yet, no present urban form matches the broad appeal suburbs enjoyed in the 1950s. Residential preferences are more fragmented today than they were 60 years ago.

Another difference between the two periods concerns the likelihood of the large investments needed to create a new system of infrastructure apt to contribute to the reshaping of urban form and dynamics. Massive investments would be required to counteract 60 years of well-funded automobile-oriented development (White, 2003). The previous transition took place at a time of accelerated economic growth that allowed governments to invest massively in urban infrastructure. Furthermore, public expenditure in urban matters was seen through the mid-twentieth-century Keynesian lens as a stimulus to the overall economy. Comparatively, economic growth has been sluggish for almost 20 years, as has governments' spending capacity. Since the 1980s, cross-Canada government expenditure on infrastructure was further eroded by the rise of the neo-liberal ideology and from the 1990s by deficit elimination measures. In these circumstances, not only did infrastructure expansion fail to keep pace with growth, but even the maintenance of existing infrastructure was underfunded. Because of their high cost, public transit projects have been disproportionately hampered by tight public-sector budgets. It is difficult to predict the extent to which reliance on public–private partnerships could provide alternative funding, as has been the case for some North American toll highway developments, due to an absence of public transit profit potential.

Among uncertainties about factors that could play a role in the coming transition is the influence of actors promoting urban change. Environmental groups have long dominated movements insisting on a different type of urban development. Until recently, however, calls for urban change were opposed by powerful economic interests with a stake in prevailing development patterns. Of late, economic voices favourable to intensification and public transit development have been heard, notably from developers engaged in urban intensification. But clearly, the elephant in the room regarding possible factors of change is the present economic recession. We have seen that recovery measures may augur an urban transition. But we cannot affirm that economic stimulation instruments will have the desired impact on the economy as a whole and/or on cities. A protracted recession would freeze urban development, leaving little opportunity for change—a situation that would recall the conditions experienced by cities in the 1930s.

To explore possible futures we sketch three potential scenarios. We do not attempt a

Table 3.1 The 1950s Urban Transition and the Possible Coming Transition

1950s Urban Transition		
Conditions for Change	**The Transition**	**Urban Outcomes**
–Little urban development over the 1930s and 1940s –Urban innovations happen over these decades, but are not implemented until the 1950s –Wave of prosperity over the 1950s and 1960s; flush government infrastructure budgets –Pent-up demand for consumption –Explosive demand for automobiles and homes –Height of modernism; readiness to adopt a new lifestyle breaking from tradition	–Transformation of urban form and dynamics: from transit-oriented centralized urban area to auto-oriented dispersed pattern with low density and specialized land uses –Auto-oriented innovations: shopping malls, the suburban configurations (super blocks, curvilinear streets) –Reciprocal relation between urban transition and economic growth	–Deepening of the 1950s transition over the next 50 years: urban areas become increasingly car-oriented, rising automobile use and ongoing adaptation of land use –More car-oriented innovations: big-box stores, power malls –Some counter-tendencies, gentrification and intensification, but overall dispersion remains the dominant trend
Possible Coming Transition		
Conditions for Change	**The Transition**	**Urban Outcomes**
–Growing dissatisfaction with present forms of urban development due to environmental impacts, health consequences, deterioration of quality of life, high expenditures –Planning documents advance alternative models of development related to sustainable development and smart growth, but wide gap between scale of proposals and of urban change –Aspiration on the part of many for a more urban lifestyle –Urban transformation associated with economic stimulation measures, which would reflect concern for the environment and energy conservation **Conditions Countering Change** –Path dependence associated with prevailing relationship between land use and transportation, capital sunk in the built environment, attachment to present urban forms, convenience of car-oriented lifestyle –Lack of consensus over a common lifestyle, in contrast with the attraction of the suburb in the 1950s –Economic conditions unfavourable to large investments in urban infrastructures, uncertainty about the effects of the recession	Models proposed involve: –Downtown intensification and enhanced multi-functionality –Sub-centres that are multi-functional, high-density, and public transit-oriented –Transit and high-density avenues –Public transit expansion and improvement of walking and cycling conditions –Control of outward urban expansion	Three possible outcomes: –More people and activities live or operate in a high-density public transit- and walking-oriented environment, but auto-oriented dispersed pattern remains in place –Path dependence dominates forces of change, thus, the persistence of development models inherited from the 1950s –Protracted recession delays urban development, the maintenance of the development status quo, or transition

Futurama-like vision of the Canadian city in 2030, but consider ranges of possibilities in light of the above discussions. The first scenario projects the tension between path dependence and rising forces calling for an urban transition. According to this scenario there will indeed be a transition, but it will be partial. As is presently the case in Canada's largest metropolitan regions, two urban forms and dynamics will be superimposed: one that is car-oriented and the other that is adapted to public transit and walking (Filion, 2000). The effect of the transition will be not so much to challenge the prominence of an automobile-dependent urban form and dynamics, as to expand the proportion of people and activities living and operating within a transit- and walking-oriented urban realm. Under present circumstances this may well be the most we can expect in terms of urban transition.

The second scenario mirrors a victory of path dependence over forces of change, whereby present forces that determine the form urban development takes will persist, largely unaffected by calls for a transition. The gap between planning proposals and the urban reality that has characterized the past decades will endure. The city 20 years from now will be similar to what it is today, notwithstanding the effects of expansion in some cases and shrinkage in others (see Bunting and Filion, Chapter 2; Donald and Hall, Chapter 16).

The third scenario anticipates a stalling of urban development. From its perspective, little if any construction will happen due to a deep and long-lasting recession, which will preclude both ongoing development tendencies and transition. In the third scenario the possibility of both an urban transition and business-as-usual is delayed, as was the case during the Great Depression.

Conclusion

We have discussed urban transition, cautioning that the urban literature exaggerates transition at the expense of the ongoing reproduction of the defining features of the contemporary city. A more balanced perspective on urban change focuses on a tension between path dependence and forces that propel change. This tension was illustrated during the 1950s, a decade of radical urban change, and in the period since then, which witnessed incremental rather than radical transformations. We concluded by comparing present circumstances to those of the 1950s to determine if we are at the cusp of another radical urban transition. Present powerful forces of change are confronted with a more deeply entrenched path dependence associated with existing urban conditions than was the case 60 years ago. It is likely, therefore, that despite loud calls for change, the outcome will be more restrained developmental change than in the previous urban transition.

Review Questions

1. What factors favour urban stability and what factors encourage urban change? Which factors will likely have the most impact in the future?
2. What circumstances triggered the deep urban transformation of the 1950s? Are present conditions likely to launch a transition of a comparable scale (but different orientation)?

Note

1. For a detailed account of the history of urban planning in Canada, see Jill Grant, 'Shaped by Planning: The Canadian City Through Time', Chapter 18 in Bunting and Filion (2006), at: <www.oupcanada.com/bunting4e>.

References

Amin, A., ed. 1995. *Post-Fordism: A Reader*. Oxford: Blackwell.

Barnett, J. 2007. *Smart Growth in a Changing World*. Chicago: American Planning Association.

Berelowitz, L. 2005. *Dream City: Vancouver and the Global Imagination*. Vancouver: Douglas & McIntyre.

Bessey, K.M. 2002. 'Structure and dynamics in an urban landscape: Toward a multiscale view', *Ecosystems* 5: 360–75.

Bhaskar, R.A. 2008 [1975]. *A Realist Theory of Science*. Milton Park, Oxon: Routledge.

Boarnet, M.G., and R. Crane. 2001. *Travel by Design: The Influence of Urban Form on Travel*. Oxford: Oxford University Press.

Boas, T.C. 2007. 'Conceptualizing continuity and change: The composite-standard model of path dependence', *Journal of Theoretical Politics* 19: 33–54.

Boudon, R. 1986. *Theories of Social Change: A Critical Approach*. Cambridge: Polity Press.

Brian, A.W. 1994. *Increasing Returns and Path Dependence in the Economy*. Ann Arbor: University of Michigan Press.

Burchell, R., A. Downs, and B. McCann. 2005. *Sprawl Costs: Economic Impacts of Unchecked Development*. Washington: Island Press.

Communauté métropolitaine de Montréal (CMM). 2005. *Cap sur le monde: Pour une région métropolitaine de Montréal—Projet de schéma métropolitain d'aménagement et de développement*. Montreal: CMM.

Cooper, J. 2007. *Cognitive Dissonance: Fifty Years of a Classic Theory*. Los Angeles: Sage.

Curic, T.T., and T. Bunting. 2006. 'Does compatible mean same as? Lessons learned from the residential intensification of surplus hydro lands in four older suburban neighbourhoods in the City of Toronto', *Canadian Journal of Urban Research* 15: 202–24.

Dear, M.J., and S. Flusty, eds. 2002. *The Spaces of Postmodernity: Readings in Human Geography*. Oxford: Blackwell.

Downs, A. 1992. *Stuck in Traffic: Coping with Peak-Hour Traffic Congestion*. Washington: Island Press.

Duany, A., E. Plater-Zyberk, and J. Speck. 2001. *Suburban Nation: The Rise of Sprawl and the Decline of the American Dream*. New York: North Point Press.

Evans-Cowley, J. 2006. *Meeting the Big Box Challenge: Planning, Design, and Regulatory Strategies*. Chicago: American Planning Association.

Filion, P. 2000. 'Balancing concentration and dispersion? Public policy and urban structure in Toronto', *Environment and Planning C, Government and Policy* 18: 163–89.

———. 2007. *The Urban Growth Centre Strategy in the Greater Golden Horseshoe: Lessons from Downtowns, Nodes and Corridors*. Toronto: Neptis Foundation.

———, T. Bunting, and K. Warriner. 1999. 'The entrenchment of urban dispersion: Residential preferences and location patterns in the dispersed city', *Urban Studies* 36: 1317–47.

———, H. Hoernig, T. Bunting, and G. Sands. 2004. 'The successful few: Healthy downtowns of small

metropolitan regions', *Journal of the American Planning Association* 70: 328–43.

Fischer-Kowalski, M, and H. Haberl, eds. 2007. *Socioecological Transitions and Global Change: Trajectories of Social Metabolism and Land Use*. Cheltenham: Edward Elgar.

Fishman, R. 1977. *Urban Utopias in the Twentieth Century: Ebenezer Howard, Frank Lloyd Wright, and le Corbusier*. New York: Basic Books.

Florida, R.L. 2002. *The Rise of the Creative Class: And How It's Transforming Work, Leisure, Community and Everyday Life*. New York: Basic Books.

———. 2005. *Cities and the Creative Class*. New York: Routledge.

———. 2008. *Who's Your City? How the Creative Economy Is Making Where to Live the Most Important Decision of Your Life*. New York: Basic Books.

Foot, D.K., with D. Stoffman. 1996. *Boom, Bust and Echo: Profiting from the Demographic Shift in the New Millennium*. Toronto: Macfarlane Walter and Ross.

Forkenbrock, D.J. 2002. 'Transportation investments and urban form', *Transportation Research Record* 1805: 153–60.

Frank, L.D., S. Kavage, and S. Appleyard. 2007. 'The urban form and climate change gamble', *Planning* 73: 18–23.

Gladwell, M. 2002. *The Tipping Point: How Little Things Can Make a Big Difference*. Boston: Little, Brown.

Grant, J. 2006. *Planning the Good Community: New Urbanism in Theory and Practice*. London: Routledge.

Hardwick, M.J. 2004. *Mall Maker: Victor Gruen, Architect of an American Dream*. Philadelphia: University of Pennsylvania Press.

Howard, E. 2003 [1902]. *To-morrow: A Peaceful Path to Real Reform*. London: Routledge.

Hutton, T.A. 2004. 'Post-industrialism, post-modernism, and the reproduction of Vancouver's central area: Retheorizing the twenty-first century city', *Urban Studies* 41: 1953–82.

Jacobs, J. 1961. *The Death and Life of Great American Cities*. New York: Random House.

Jakle, J.A., and K.A. Sculle. 2004. *Lots of Parking: Land Use in a Car Culture*. Charlottesville: University of Virginia Press.

Janos, A.C. 1986. *Politics and Paradigms: Changing Theories of Change in Social Science*. Stanford, Calif.: Stanford University Press.

Jerret, M., et al. 2009. 'A cohort study of traffic-related air pollution and mortality in Toronto, Canada', *Environmental Health Perspectives*. At: <www.ehponline.org/members/2009/11533/11533.pdf>.

Kay, J.H. 1997. *Asphalt Nation: How the Automobile Took Over America and How We Can Take It Back*. New York: Crown.

Krug, G. 2005. *Communication, Technology and Cultural Change*. London: Sage.

Kunstler, J. 1993. *The Geography of Nowhere: The Rise and Decline of America's Man-made Landscape*. New York: Simon & Schuster.

Lang, R.E. 2003. *Edgeless Cities: Exploring the Elusive Metropolis*. Washington: Brookings Institution Press.

Le Corbusier. 1973 [1933]. *The Athens Charter*. New York: Grossman.

Lees, L., T. Slater, and E.K. Wyly. 2008. *Gentrification*. New York: Routledge.

Leinberger, C.B. 2008. *The Option of Urbanism: Investing in a New American Dream*. Washington: Island Press.

Leonhardt, D. 2009. 'The big fix', *New York Times Magazine*, 1 Feb., 22–9, 48, 50–1.

Levin-Waldman, O.M. 2005. *The Political Economy of the Living Wage: A Study of Four Cities*. Armonk, NY: M.E. Sharpe.

Lewis, T. 1997. *Divided Highways: Building the Interstate Highways, Transforming American Life*. New York: Penguin Putnam.

Lin, L.A., and D.J. Atkin, eds. 2007. *Communication Technology and Social Change: Theory and Implications*. Mahwah, NJ: Lawrence Erlbaum Associates.

McLeod, M. 1999. *Auto Passenger Travel and Auto Occupancy in the Greater Toronto Area: 1996 Results and Recent Trends*. Toronto: Data Management Group, Joint Program in Transportation, University of Toronto.

Marshall, A. 2000. *How Cities Work: Suburbs, Sprawl, and the Roads Not Taken*. Austin: University of Texas Press.

Metrolinx. 2008. *The Big Move: Transforming Transportation in the Greater Toronto and Hamilton Area*. Toronto: Metrolinx.

Metro Vancouver. 2008. *Our Livable Region: Metro Vancouver's Growth Strategy (Preliminary Draft)*. Burnaby, BC: Metro Vancouver.

Millward, H., and G. Xue. 2007. 'Local urban form measures related to land-use and development period: A case study for Halifax, Nova Scotia', *Canadian Journal of Urban Research* 16: 53–72.

Miron, J. 1988. *Housing in Postwar Canada: Demographic Change, Household Formation, and Housing Demand*. Montreal and Kingston: McGill-Queen's University Press.

Ontario, Ministry of Public Infrastructure Renewal. 2006. *Growth Plan for the Greater Golden Horseshoe*. Toronto: Government of Ontario.

Peck, J. 2005. 'Struggling with the creative class', *International Journal of Urban and Regional Research* 29: 740–70.

Pierson, P. 2000. 'Increasing returns, path dependence, and the study of politics', *American Political Science Review* 94: 251–67.

Punter, J. 2003. *The Vancouver Achievement: Urban Planning and Design*. Vancouver: University of British Columbia Press.

Roseland, M. 1992. *Towards Sustainable Communities: A Resource Book for Municipal and Local Governments*. Ottawa: National Round Table on the Environment and the Economy.

Schively, C. 2007. 'Understanding the NIMBY and LULU phenomena: Reassessing our knowledge base and informing future research', *Journal of Planning Literature* 21: 255–66.

Talen, E. 2000. 'New urbanism and the culture of criticism', *Urban Geography* 21: 318–41.

Tomalty, R., D. Alexander, and F. Grammenos. 2005. *Smart Growth in Canada: Implementation of a Planning Concept*. Ottawa: Canada Mortgage and Housing Corporation.

Toronto, City of. 2002. *Toronto Official Plan*. Toronto: City of Toronto.

Van de Coevering, P., and T. Schwanen. 2006. 'Re-evaluating the impact of urban form on travel patterns in Europe and North America', *Transport Policy* 13: 229–39.

Wang, S., and P.J. Smith. 1997. 'In quest of "forgiving" environment: Residential planning and pedestrian safety in Edmonton, Canada', *Planning Perspectives* 43: 225–50.

White, R. 2003. *Urban Infrastructure and Urban Growth in the Toronto Region: 1950s to the 1990s*. Toronto: Neptis Foundation.

Zukin, S. 1982. *Loft Living: Culture and Capital in Urban Change*. Baltimore: Johns Hopkins University Press.

Part II

The Structuring Parameters of Twenty-First-Century Urbanization

In this section we attempt to capture what we suspect may prove to be the structuring parameters of the twenty-first-century metropolis. Undoubtedly, there will be factors we have overlooked or have yet to recognize as significant to moulding today's city. Likewise, there may be features included that have only a secondary impact on current and future urban development trends. Either way, the following section includes 10 chapters, each considering how general forces of change have resonated within the built environment over recent decades.

Many of the structuring parameters identified throughout the book are new to this edition, and we would argue pertain to the postmodern, as distinct from the modern, city. Features of globalism, environmental sustainability, and the new service economy were not recognized as relevant to city structure 40 or 50 years ago, in the period we write about in Chapter 2 occurring at the height of the modern 'Fordist' boom. Though widely recognized and even talked about by the average person on the street today, globalism in its many guises—international trade, immigration, heightened international political and institutional collaboration, imported consumer goods and services, etc.—was a less significant force on the urban landscape before 1975. Likewise, neither recognized nor understood in your grandparents' days were the many negative impacts that city building and day-to-day routines (e.g., the impact of auto and truck travel on fossil fuel depletion and global climate change) exert on the urban biophysical environment. Twenty-five years ago, terms like 'ecological footprint' or 'urban metabolism' would probably not have made much sense within either scholarly or policy-oriented circles.

Fifty years ago, our modern cities and city systems were understood to be propelled by the forces of industrialism. Indeed, in the 1990s at least one scholar who wrote about impending deindustrialization foresaw 'the end of work' as forces of automation and economic globalization devalued manual and routine labour in developed countries (Rifkin, 1995). As it turned out, technological, cultural, and economic changes did not so much bring an end to work as they profoundly transformed it. For example, virtually costless telecommunications have renewed the importance of face-to-face contacts for creative enterprises. A generation or more ago, art and culture were interpreted as singularly high-order services found to be localized in one or more entertainment districts near the centre of the city. At that time scholars and policy-makers could hardly be blamed for viewing arts and culture as something of an urban 'frill'. Only with hindsight can one understand that it would be most unrealistic for 'modern' urban thinkers to have anticipated that, by the twenty-first century, the creative sector would come to be understood as a driving force of economic growth and an essential tool in the kitbag of local economic developers everywhere.

As we move through Part 2, with Chapter 7, 'Economic Change in Canadian Cities: Locational Dynamics of Employment', we begin to examine parameters that have long been identified as important structuring features for cities and that have been clearly articulated in previous eras of urban development. Distinctive features of these long-recognized parameters have shifted markedly, however, in the twenty-first-century city and so today have a significantly different impact on city structure than in the past. Taking employment as an example, we witness the demise of not just inner-city but also much suburban industry. Likewise, the high-order producer services sector has expanded significantly, especially for business in the FIRE (finance, insurance, real estate) sector, a distinctive marker of the new economy in the core of our largest metropolitan areas. Those who understand historic patterns of urban form will quickly grasp that location of employment has a significant effect on where different classes of residents/workers choose to live and subsequently on how they travel to work. Less obvious perhaps is the role that interaction between any one urban parameter, such as employment location, and other urban parameters might play—e.g., the effect that the new creative economy has had on the overall regeneration of the central city.

Yet another major parameter of urban structure has long been identified in terms of social status. Geographers in the past have most often thought about patterns of urban social division in terms of a general integrative model that has been widely researched under the rubric of the 'factorial' or 'social ecology' of cities.[1] In the modern city important differentiation of urban social space was found to be associated with three interrelated dimensions: socio-economic status, stage in the life cycle, and ethnicity. Today these same dimensions continue to give rise to important distinctions in both social space and activities or lifestyle, but urban social space has become much more complicated. Thus, for example, rather than a life-cycle continuum, as was characteristic of the 'modern' city where most households were of the nuclear family type, we now find entirely new arrangements, such as 'blended families', alternative household types, and the growing predominance of single-person households. In the twenty-first-century city, each of the three dimensions of intra-urban social space—life cycle and demography, economic status, and ethno-cultural diversity—has taken on sufficient differentiation to act as a somewhat independent dimension. Where once one model of social space could suffice, today's city requires much more multi-faceted analyses.

Another structuring feature of urban form is witnessed in the apparent inseparable dependence of urban form on prevailing modes of transportation. Thus, whether speaking of cities of the present or ancient past, it seems axiomatic to state that land-use patterns, along with other important parameters such as land value, will be a function of movement patterns and, conversely, that movement and transportation also becomes a function of land use and urban form. Chapter 3 described how the newer suburban parts of the modern Fordist city have evolved through an intrinsic relationship between low-density suburban form and automotive dependency. Likewise, the predominant territorial portion of current cities, i.e., the area that includes the inner and outer suburbs, perpetuates the low-density-auto-dependent dynamic, so explaining our inability to accommodate apparently inexhaustible demands for more expressway lanes on metropolitan highways. In Canada's largest city this insatiable demand for freeway space has actually led to the construction of a public–private partnership toll road, since fully privatized, that is reasonably expensive to use but nonetheless usually runs at or near capacity during weekday rush hours. Either way, though, the most dramatic expression of what is new in the contemporary urban transportation scene is neither the privatized expressway nor gridlock, but the overpowering recognition that the entrenched relationship between exclusively zoned, low-density suburbs and automobile dependency is dysfunctional to the point of engendering irreversible environmental damage with anticipated catastrophic consequences.

The unprecedented level of uncertainty about the urban future that we now face is in large part a product of this dysfunctional but entrenched relationship between urban form and auto dependency. Thus, the call today across many disciplines and professions is for retrofitting Canada's cities, redesigning them to support primarily transit, cycling, and pedestrian movement—no small feat, though Perl and Kenworthy (Chapter 11) suggest that, by international standards, Canada is already reasonably well poised to pursue the transit planning alternative. Most certainly it is clear that current, restructured styles of neo-liberal governance (Chapter 12) will be hard-pressed, if even willing, to provide the resources needed to bring about reversal of this land-use transportation dynamic, involving, as it must, policy that will find disfavour with large segments of the urban population.

At the end of Part 2, we come to urban land values. As the final parameter introduced here, land values provide a good example of the complex interrelationships that exist within and between structuring forces in urban areas. We see, for example, that urban land values are both an expression of the other critical forces, such as demography and globalization and, at the same time, a shaper of urban structure in their own right, determining what type of activities can afford which locations.

Note

1. For spatial analysis of Canadian cities and Canada's urban social mosaic, see Robert A. Murdie and Carlos Teixeira, 'Urban Social Space', Chapter 9 in Bunting and Filion (2006), at: <www.oupcanada.com/bunting4e>.

Reference

Rifkin, J. 1995. *The End of Work: The Decline of the Global Labor Force and the Dawn of the Post-Market Era.* New York: G.P. Putnam's Sons.

CHAPTER 4

The Global Imperative

Peter V. Hall

Introduction

From New Westminster, once the capital of British Columbia but now a suburb of Greater Vancouver, one can watch log booms containing trees felled elsewhere in the province being pulled up the Fraser River. After being processed at one of the dwindling number of urban riverfront mills, they return downriver in vast open containers as woodchips, ready for processing into chipboard, pulp, and paper at both remote and local sites. Other wood products are placed on specialized railcars or are taken to the port for export. On the surface these flows appear timeless and unchanging; they are the tangible manifestations of the imprint of Canadian staples on the city.

We know, however, that this imprint has changed, and continues to change. In other parts of Greater Vancouver, most notably in the False Creek area adjacent to the downtown, industrial land uses related to the staples economy have given way to condos and live/work spaces. Here, flows of wood products have been replaced by flows of real estate investment, immigrants, and digitally delivered services, resulting in a dramatically new form of urban (re)development (Hutton, 2008; Olds, 2001). Even the flow of material goods has changed dramatically in the past decades. Today, fewer milled products and more wood pellets and raw logs (Parfitt, 2007) are going east as well as south.

Across the Fraser River from New Westminster, in the edge cities of Surrey and Delta, the infrastructure of a new system of flow is taking shape. Most tangible is the South Fraser Perimeter Road, under construction since early 2009 and intended to expedite the truck drayage of ocean shipping containers between Deltaport and the CN rail yard in north Surrey, and from there to distribution centres across North America. Funded by both provincial and federal governments, the Perimeter Road is a centrepiece in the Asia-Pacific Gateway and Corridor Initiative designed to 'contribute significantly to Canada's competitiveness in the rapidly changing world of global commerce' (Transport Canada, 2007).

At the core of the globalization perspective is the idea that the global flows that connect cities to other places around the world play an important role in shaping those urban places. These flows include products, goods and services, people, finance and investment, ideas, and policies—even diseases and illegal drugs. Global flows are not a new thing. Every city in Canada was formed in relation to some other far-flung location. Yet, flows alone do not have the power to re-shape cities; globalization also involves new sets of social, political, and economic practices, relations, and discourses. Thus, the increasingly global origin and destination of the increased flows, plus their complex network organization, scale and diversity, and the ways people in their cities and regions have responded to them, give the globalization perspective its analytical edge.

This chapter endeavours to capture the core of the globalization perspective while also reflecting on some of the debates that have swirled about it. In what follows we will outline some of the dimensions of the heightened global flows and the ways that scholars have theorized about their impact on the city. The globalization perspective is not a single, unified approach to understanding contemporary cities, nor is it without its critics who assert that local dynamics are of equal if not greater importance. These critics assert that despite all the increased flow, indeed partly because of it, some local, city-regional dynamics have gained heightened importance. This is one of the ironies of globalization; when many things are more mobile, some immobile people and places might be bypassed and ignored, while other immobile practices and assets may become more valuable.

Globalization operates unevenly across space; cities do have, and will continue to have, very different experiences. The accelerated flows are structured at various scales, and most do not operate at a truly global scale. Instead, there are important continental (i.e., North American), regional (i.e., Cascadian or Midwestern), metropolitan (i.e., Greater Golden Horseshoe), and trans-oceanic (i.e., Pacific Rim) dimensions to globalization. It makes a difference that Windsor interacts more with Detroit, while Vancouver interacts more with Hong Kong. Even within a city-region, the effects of flow are spatially uneven: the core areas of greater Vancouver are more susceptible to flows of people and property investment, while some suburbs are more heavily influenced by the flow of goods.

Today, it is more important to pay attention to the limitations of the globalization perspective. This chapter was written in the midst of what may be the most serious crisis to face economic globalization, with global economic growth forecast to be negative for the first time since the Great Depression (IMF, 2009), and global trade forecast to decline even more sharply. World Trade Organization statistics show that in the period 1950 to 2007, the year-on-year growth in the volume of merchandise exports out-stripped growth in the total volume of merchandise production in all but six years (WTO, 2008). In '2000, Canada exported 45.6% of what it made and imported 39.8% of what it bought' (Yalnizyan, 2009: 15). Although both import and export proportions did fall back to about a third in 2008, the comparable proportions were under one-fifth in 1960. However, in Vancouver, after more than a decade of barely interrupted growth, the number of ocean shipping containers handled by the local port declined sharply in early 2009. Just as trade flows grew more rapidly during the heyday of globalization, we have every reason to think that the contraction in international flows may exceed the economic contraction overall, as national governments reassert themselves as domestic spending engines and as polities rethink their engagement with globalism (Saul, 2005). It remains to be seen, however, whether this contraction will continue, and whether the practices and discourses of globalization will retreat as sharply as the flows themselves.

Cities and Flows

You can see patterns of flow similar to those on the Fraser River and the new perimeter road in and around all Canadian cities. Indeed, the roads and highways leading from Toronto, Montreal, Calgary, and Vancouver each follow a familiar transect. Departing from the central city one passes suburbs, agricultural lands, and market towns before reaching such places as Sarnia, Trois-Rivières, Drumheller, and Hope once known for the natural resources they contained or to which they gave access. The landscape transitions that you can observe from your vehicle tell one kind story about the forces shaping our urban agglomerations. It is an unequal, if co-dependent, landscape knit together by flows of people, money, food, water, energy, and—particularly in the Canadian context—flows of natural resources (cf. Jacobs, 1985; McCann and Gunn, 1998). Some flows can easily be seen on the highway: passenger vehicles, school buses, delivery vans, and logging trucks. Evidence of flow also can be found in connective infrastructure; the highway itself as well as the canalized rivers, railroads, and power lines that run alongside, and in nodes such as office parks, gas stations, and shopping malls. Some of the flows are not easily observed: natural gas pipelines are buried in underground infrastructure, while exchanges of knowledge, finance, and sharing are subsumed in banking, family, community, and professional social networks.

We can understand quite a lot about the social and economic functioning of our cities by examining these city–metropolitan and regional–hinterland flows, both the daily movements within and around a given city or contiguous metropolitan area, and the less frequent ones between the city and its regional hinterland. Flows are evidence of interaction, of valued relationships between people who are not in constant close proximity. For example, commuting patterns speak volumes about how we organize and separate our work and home lives, just as the log booms that arrive in one place and that leave as wood chips tell us something about the structure of the forestry industry. Very little about these flows is random; often they are quite stable and durable. An individual commuter might follow the same well-trodden path to work for years. No one builds a pulp mill without first ensuring a continuous availability of feedstock. So flows can help us understand what is happening at the places where they originate and end, and about places in-between. Supporting all these flows are massive investments in infrastructure that, once sunk into the ground, shape and direct future interactions and relationships (Harvey, 1982). Flows are not passive; they exert an influence by virtue of their constant repetition.

It is thus a remarkable and important thing when flows change; it suggests that something profound has shifted at their origin or their destination, in the nature of their interactions, in the infrastructure and networks that connect them, or in some combination of these. Back on the highway leading from any Canadian city, it is clear that some flows, and some of the infrastructure that supports them, are not exclusively or even primarily related to the daily functioning of the city, the metropolitan area, or its regional hinterland. Instead, we see evidence of global flows that cross one or more international boundaries. Sometimes these global flows rely on the same infrastructure that organizes the more local flows, but they may also have their own new and more specialized network infrastructure (Graham and Marvin, 2001). For example, commuters will share the South Fraser Perimeter Road with trucks moving ocean shipping containers for a small, but vitally important portion, of their long journey between remote locations of production and consumption.

Airports have expanded to handle more and larger airplanes flying longer distances, but once passengers leave the terminal, they often share the same roads and transit systems as local residents. A distinctive social and economic landscape has emerged that reflects the flows through the airport—more immigrants live in the suburbs and more national distribution headquarters of global firms have located in the industrial parks near major airports. The core areas of the major cities also reflect a new set of flows of people, finance, and urban planning ideas; Canadian downtowns increasingly display less of a national character, becoming more like the cities elsewhere that they both model and inform.

More Flow, Differently Organized

Observations such as these lie at the heart of what we might call the 'globalization perspective' that associates globalization with the 'accelerated circulation of people, commodities, capital, money, identities and image through global space' (Brenner, 1999: 431). At the core of the globalization perspective is the idea that connections and flows between places make a difference to the places they connect. At least four characteristics of contemporary flows differentiate them from more conventional city–metropolitan and regional–hinterland flows. First, flows have changed qualitatively in regard to novelty and diversity in what is carried and in their more complex, network-based organization. Second, flows have increased quantitatively: more goods are moving through our cities. Third, the origins and destinations have changed: it is not that national, provincial, and regional flows have ceased; rather, international flows have become relatively more important. Many of the flows moving through Canadian cities are increasingly international as opposed to local, regional, or national in origin or destination. Fourth, responses to flows—reactive and proactive, resistant and embracing—have become more intense and influential in shaping cities.

Consumer Products and Services

While consumer products with global cachet have been around for centuries, some of globalization's most iconic images are provided by products that are now consciously marketed to global consumers. Of course, McDonald's, Coca-Cola, Canada Dry, Blackberry, Microsoft, and any number of consumer electronics companies spring to mind. Yet, global consumer products are not only for the mass market; truly global brands are marketed in the luxury automobile, watch and jewelry, clothing, and bottled water markets. Even some consumer services have gone transnational and global, including H&R Block, Manulife, and Marriott, as well as Internet- and call centre-mediated services provided by such firms as Expedia, Amazon, and MasterCard. Despite their global brand-image, when we scratch the surface we find often that many of these products and services are not truly global; they are heavily customized to meet local cultural and market preferences. Either way, global consumer products and services exert a powerful set of images that reshape local consumption patterns.

Intermediate Goods and Producer Services

Behind all these consumer brands are complex flows of **intermediate goods** and **producer services**. Intermediate goods include the raw and bulk materials, parts and components that are assembled to create consumer products. These components increasingly are organized in supply chain systems or production networks (Dicken, 2003). The implications of this reorganization for places in southern Ontario that are integrated into the North American automobile industry have been profound. When this industry was booming, pressure was constant for improvements in road and border crossing infrastructure to speed the movement of goods within this regional, cross-border production network. These networks are exceedingly complex, and apparently are constantly changing. For example, as automobile assemblers have merged to become dominated by a smaller number of large, global players (e.g., Toyota, GM, VW) and have created common platforms, the tiers of suppliers also have become more hierarchical. First-tier auto component suppliers (e.g., Magna, Denso, Bosch) are now as big as some of the assemblers they supply. In fact, the Canadian supply firm, Magna, recently sought (unsuccessfully) to purchase GM's European automobile assembler, Opel. Beneath the assemblers and first-tier suppliers are hundreds of small-component suppliers.

These production networks represent new and more complex organizational forms and they entail new ways of organizing goods shipment, most notably, in the intermodal ocean shipping container (see Levinson, 2006). The word 'intermodal' means that the standard container can be moved from one transport mode to another with unprecedented speed and efficiency. The implications for urban development of this transport revolution have been enormous. Some ports have become much larger, serving more extensive hinterlands, and new forms of urban development have appeared, such as the intermodal yard and the suburban warehouse estate (Hesse, 2008; McCalla et al., 2001). In other former port locations, vacant waterfront industrial land blights the surrounding neighbourhoods until redevelopment dollars are attracted.

Producer services also have been restructured on a global basis—accounting firms such as Price-WaterhouseCoopers, advertising firms like Ogilvie & Mather, and banks such as HSBC and Citibank. These firms mobilize flows of finance, information, and highly skilled workers along the networks established by advanced telecommunications and airlines, and are concentrated in just a handful of cities, primarily London, New York, Hong Kong, Tokyo, and Paris (Taylor, 2005); Toronto is the only Canadian city that ranks high in global producer services.

People

The flow of people across international borders has always been a central feature of urban development in Canada, but the shift in the 1960s to an immigration policy that favoured human capital (rather than national origin), and subsequent increases in the annual targets, profoundly changed the impact of immigrants on Canadian cities. According to the 2006 census, one in five people living in Canada were immigrants or non-permanent residents. Hence, the nature and destination of these international movements of people has huge consequences for the shape of the Canadian space-economy. Immigrants settle overwhelmingly in the gateway cities of Montreal, Toronto, and Vancouver (Hiebert, 2000; see also Hoernig and Zhuang, Chapter 9). Although there were some indications in the 2006 census of a small shift in the settlement of immigrants from the Big Three to other metropolitan areas, immigration remains an overwhelmingly large-city phenomenon in Canada.

Table 4.1 shows the percentage of immigrants or non-permanent residents in each province and in the largest city (census metropolitan area or census agglomeration) in each province. Note that the largest city in each province has a higher percentage of immigrants than the province as a whole, with the exception of Moncton, New Brunswick. Almost half (47 per cent) of those living in Toronto, and two-fifths of those living in Vancouver, are immigrants. Meanwhile, we have seen increasing efforts by provincial- and municipal-level governments to develop their own immigration policies (e.g., provincial nominees) and immigrant settlement programs (e.g., the Toronto Regional Immigrant Employment Council; see Lo, 2008).

Immigrants are important not only because of their numbers and selective location, but also because of how they profoundly transform places where they settle. Scholars examining cities and regions in the United States have noted that immigrants play an important role in transnational business networks that (re)shape development trajectories at both 'ends' of the network (see Lo and Wang, 2007; Saxenian, 2006). Immigrants to Canada once clustered in inner-city ethnic enclaves but today settle across metropolitan areas in a more dispersed way (see Hiebert, 2000; Luk, 2007; Ooka, 2007), resulting in urban development patterns that Li (1998) has termed 'ethnoburbs'. But, movements of people under globalization are not only one-way permanent immigration moves. Canada is a stopping place for **transnational** migrants who may move on to new countries or return to their point of origin.

Finance

Financial deregulation has come to be regarded as one of the key factors underpinning economic globalization (Dicken, 2003). Changes in banking and foreign exchange regulations made it easier to move money across national borders; much of

Table 4.1 Immigrants Concentrate in the Largest Cities

Province Largest City (CMA/CA)	2006 Population	Per Cent Immigrants and Non-Permanent Residents
Newfoundland and Labrador	500,605	1.9%
St John's	179,270	3.5%
Nova Scotia	903,090	5.4%
Halifax	369,455	8.0%
Prince Edward Island	134,205	3.8%
Charlottetown	57,715	4.8%
New Brunswick	719,650	4.0%
Moncton	124,055	3.7%
Quebec	7,435,900	12.1%
Montreal	3,588,520	21.8%
Ontario	12,028,895	29.2%
Toronto	5,072,075	47.2%
Manitoba	1,133,515	14.0%
Winnipeg	686,040	18.5%
Saskatchewan	953,850	5.5%
Saskatoon	230,855	8.7%
Alberta	3,256,355	17.0%
Calgary	1,070,295	24.7%
British Columbia	4,074,385	28.7%
Vancouver	2,097,965	41.5%
Canada	**31,241,030**	**20.7%**

Source: Author's analysis of Community Profiles, Statistics Canada.

which has found its way into investments in property developments that have transformed cities around the world. Flows of finance and investment are not blind to the particularities of place; patterns of property investment are highly selective and uneven. When real estate investors enter a new city, it takes time for them to learn about the local market, assemble land, and secure planning permission; often, they rely on existing social relationships and local partners. Edgington (1996), for example, shows that between 1985 and 1993 Japanese-based real estate investors focused on tourism facilities in Vancouver and surrounding areas frequented by Japanese tourists. Over a similar time span, Olds (2001) traced investment by Hong Kong-based investors in Vancouver's residential real estate. He argues that even vast and fungible financial flows do not simply land in the city to transform its built environment. Rather, these flows are mediated and reshaped by the complex interaction between local and external actors.

Information and Ideas

Much attention has been paid to the role of computing and telecommunications technologies, which allow the rapid transfer of information across the globe (see Castells, 1996). The rapid movement of some—but not all—kinds of information from one place to another certainly has facilitated the emergence of global production alongside, in some cases, actual service networks, and withdrawal of some production activities from cities in the developed world. Information technologies arguably contribute to relatively more service, cultural, and multi-media production in cities. More rapid and cheaper flow of digital information supports other flows of goods, via order and shipment tracking, people, via electronic messaging and cheaper phone calls home, and money, via electronic transfers.

Beyond the transfer of information in the form of digital data, globalization is associated with more rapid flows of ideas. Often, these ideas are directly relevant to urban development, involving linkages between cities through city twinning, networking, and even global associations of local governments (see Borja and Castells, 1997: ch. 2). Scholars have begun to pay more attention to the messy process of transferring policies from one jurisdiction to another. For example, Peck and Theodore (2001) have written about the transfer of welfare-to-work policies, while Ward (2006) explored the dispersion of **business improvement areas**. In his study of Vancouver's drug policy, McCann (2008: 2) uses the term 'urban policy mobilities' to describe the way 'in which policy knowledge and policy models move from city to city'. Rejecting any implication that local actors are passive recipients of global ideas, he argues that the 'how of policy-making in global relational context entails a range of locally based actors, from politicians to policy professionals to activists and drug users, developing connections with experts from elsewhere and with related flows of knowledge in order to operationalize a new strategy to fight drug related harm' (15).

'Nasty' Flows

It comes as no surprise that some undesirable flows—such as pathogens, illicit drugs, weapons, organized crime, human trafficking, and pollutants—are transported via the networks discussed above. These flows also disrespect national borders, perhaps in surprising ways. For example, since 2003 the City of Toronto has been exporting its solid waste to dumps in Michigan, allowing more consumption in Toronto but evoking strong opposition from US-based communities. In their edited volume on the SARS outbreak in cities such as Toronto, Hong Kong, and Singapore, Ali and Keil (2008: 12) point out that 'the SARS virus represented another flow type that connected global cities. The spread of SARS in this manner therefore underscores the fact that today infectious diseases cannot simply be considered a public health issue that is exclusively confined to the developing world or pegged to a particular spatial scale (whether it be the local, regional, or national).' Likewise, terrorists have sometimes sought out global cities precisely because they are well-connected places from which media images resound.

The Effects of Flow: Theorizing the Global City

Thinking about globalization's impacts on the city via a list of flows is only a starting point. Having defined globalization as increased flow, we are immediately confronted by more complex questions: Which flows, if any, are more important than others? How do the various flows combine in particular places? Do they influence all cities in similar ways, or is their impact mediated by local forces? What are the actors and forces that underlie the flows? And, most importantly, what are the effects, both positive and negative, of increased flow? While there is no one comprehensive explanation, a number of theorists have attempted to understand the combined effects on the city of these global flows.

The concept of the **world city** or **global city** focuses on the role of cities in the organization of international economic activity. In their original statement of the 'world city hypothesis', Friedmann and Wolff (1982: 309) argued that cities are nodal points in all those flows described above. Furthermore, some cities occupy a more dominant position than others: 'at the apex of this hierarchy are found a small number of massive urban regions that we shall call world cities. Tightly interconnected with each other through decision-making and finance, they constitute a worldwide system of control over production and market expansion.' In the cities that are at the top of this global hierarchy, we expect to find concentrations of command-and-control functions, and advanced financial and other business services.

In giving prominence to the interconnections between dominant places, Friedmann and Wolff were influenced by established theories that emphasized the importance of flows between places tracing back to Sir Peter Hall's observations about a class of world cities, rising out of their respective national urban hierarchies and containing the headquarters or centres of international political, cultural, and economic power (Hall, 1966). Another line of influence was world systems theory (see Wallerstein, 1984), which analyzed the global dimensions of resource flows, dividing nations into an industrialized and democratic core, and a resource-exporting and underdeveloped periphery. Also, Friedmann and Wolff were influenced by the notion of the 'new international division of labour', which emphasizes how international corporations separate their functions geographically, placing routine functions in low-cost locations, while keeping more specialized decision-making and control functions in core metropolitan areas (see Massey, 1984).

Friedmann's (1986) follow-up article identified a world city hierarchy, based on large cities' relative concentration of functions such as high finance, transnational corporate headquarters, and international institutions alongside rapid growth of population and of locally based business services,

manufacturing and, transport facilities. Toronto is the only Canadian city mentioned by Friedmann; he identifies it as a secondary core city, alongside such cities as Miami, San Francisco, Madrid, and Sydney, but behind such primary cities as London, Paris, New York, Chicago, and Tokyo. This pattern is replicated in much of the subsequent research on 'world city rankings'; for example, in Taylor's (2005) taxonomy of leading cities in globalization, Toronto is identified as an 'incipient global city', alongside Boston, Chicago, and Madrid, but behind London and New York (the 'leading duo') and Tokyo, Los Angeles, Paris, and San Francisco.

More important than the rankings are Friedmann's seven 'theses' about how cities are affected by their place in, and relationship to, the world city hierarchy. A recurring theme is that world cities are characterized by a great degree of internal **polarization**. While '(w)orld cities are major sites for the concentration and accumulation of international capital' (1986: 73), they are 'characterized by a dichotomized labour force: on the one hand, a high percentage of professionals specialized in control functions and, on the other, a vast army of low-skilled workers engaged in manufacturing, personal services and the hotel, tourist and entertainment industries that cater to the privileged classes for whose sake the world city primarily exists' (1986: 73). Furthermore, '(w)orld cities are points of destination for large numbers of both domestic and/or international migrants' (1986: 75). These immigrants have unmet social needs, so '[w]orld city growth generates social costs at rates that tend to exceed the fiscal capacity of the state' (1986: 77).

Other researchers, most notably Sassen (1991), have picked up these themes, chronicling the rising inequality in most global cities. One challenge in interpreting these arguments in Canada is that Toronto is the only Canadian city with a claim to global status. However, other theories that look beyond the iconic global cities also predict rising polarization in the most globally connected places. Castells (1996), for example, has argued that the global economy is organized through a series of information networks. The fortunes of places are increasingly determined by their connections to these networks and while all places potentially connect to these networks, many may be bypassed. A central element of Castells's ideas about the impact of global flows is unevenness, in both spatial and social domains. In the social domain, Castells supports much of what Friedmann and Sassen argue, namely that the global economy is associated with highly unequal labour markets. He also argues that the spatial organization of cities is likely to be highly uneven because, while some parts connect to global networks, other parts of the city are actively excluded from them (see Borja and Castells, 1997: ch. 2).

The argument that polarization is most severe in the largest and most globally connected places resonates in the Canadian urban system. Among the major Canadian cities a general increase in income inequality is evidenced. Comparing family income in the 1980s with the 2000s in Toronto, Montreal, and Vancouver, Heisz (2006) finds that in each city, inequality rose as higher-income families experienced higher rates of income growth than low-income families. A key factor driving this inequality is the fact that, since the 1980s, immigrant labour market incomes have lagged behind those of non-immigrants. Hall and Khan (2008) show that the immigrant earnings penalty is greatest in the largest Canadian cities where there are the most immigrants.

In Canada, too, increasing income inequalities are exacerbated by rapidly rising rents and/or house prices in the largest cities, which means increasing housing-affordability challenges (Bunting, Walks, and Filion, 2004; Moore and Skaburskis, 2004). In 2008, at the height of the boom in the oil patch, the president and CEO of the Calgary Chamber of Commerce commented that '(t)he current price of housing relative to incomes poses a significant challenge to Calgary businesses

seeking to attract and retain employees—the number one issue confronting Chamber members' (Calgary Chamber of Commerce, 2008). Overall, there are indications that the trends in social inequality and housing unaffordability are combining to create some especially deprived neighbourhoods in the most global of Canadian cities. For example, writing about Toronto, Heisz (2006: 11) notes that the 'rising income gap between high- and low-income families was mirrored by a rising gap between high- and low-income neighbourhoods.' Walks and Bourne (2006) examine whether **ghettos**, neighbourhoods that combine racial or ethnic segregation with concentrated poverty, are emerging in Canadian cities. While they find some evidence of emerging ghettos in Toronto, they show that increased income inequality has not (yet?) led to the formation of ghettos in Canadian cities. Instead, they conclude that 'the confluence of increasing income inequality and the particular geography of housing in each given place, including that of tenure, form and price, are more important in determining overall patterns of segregation' (Walks and Bourne, 2006: 295).

This last observation highlights one of the key challenges facing the 'world city hypothesis', namely, how to relate outcomes in specific places to the general increase in global flow. What are we to make of observations that flows of immigrants and rising labour market inequality do not necessarily lead to the predicted outcomes? One way of dealing with this problem is to recognize the increasing complexity and fragmentation of the numerous flows that combine in unpredictable and often temporary ways under globalization. Appadurai (1996) for example writes of the 'disjunctive order' of the new global cultural economy (32). Instead of trying to relate globalization to a small number of potential outcomes, he developed an overlapping typology of different '-scapes' that are created by global flows. These are (1) ethnoscapes involving the movement of people, (2) mediascapes involving the creation and circulation of images, (3) technoscapes involving mobile technology and information, (4) financescapes involving global financial capital, and (5) ideoscapes involving political ideals and ideologies. Appadurai further argues that understanding the impact of global flows on a given city requires understanding the ways in which diverse flows come together in different combinations and interact uniquely with particular localities. Likewise, Norcliffe (2001: 27) also argues that 'there is not one narrative but many regional and local narratives that describe Canada's global connections, all of which are interrelated, with each one having greatest resonance for certain groups, in specific regions, at particular times.' The implication is that comprehending the complexity of globalization, means thinking locally as well as globally.

Reasserting the Local

The insistence on the importance of local dynamics despite increased global flows echoes older debates about the relative importance of those factors which make a place or region distinctive, and those factors or processes which connect them. One such debate involved two famous American economists, Douglas North and Charles Tiebout in discussion about the origins of regional economic growth and development. Drawing on the ideas, amongst others, of Canadian Harold Innis (1930) about the society-shaping role of export staples, North (1955) argued that regional economies in the new world all developed in relation to some far-flung place. After initially exporting only a small number of resource-based products, economies would then diversify and urbanize. Tiebout's (1956) response was to insist that there is nothing automatic in the relationship between exports and regional growth; local conditions influence the success of the exporting enterprise, but only some places are able to capture and reinvest the value created by those exports. So regional development depends on exports; but exports are themselves dependent on regional qualities. This 'chicken and egg' debate continues in a variety of forums, with most scholars accepting the perspective that both local and global dimensions are important (see Lipietz, 1993).

A further complication in this debate is that what we call 'local' (or, more precisely, 'not global') today may have altered by tomorrow. Cox (1997) and others point out that spatial scales are not static, and that under globalization various groups may try to redefine them in order to advance their agenda. Returning to Vancouver, a local-centred perspective would respond to the globalization perspective by observing that the flow of containerized goods by itself did not create the Asia-Pacific Gateway, and its constituent warehouses, terminals, and new highways. Rather, these physical investments in the built environment are the result of a lot of hard work—both to assemble the finance and overcome community opposition—by those whose interests are served by more goods movement. For example, the Greater Vancouver Gateway Council, a coalition of port-related industries, lobbied for several years to attract government support for the infrastructure now under construction. These interests themselves were strengthened by changes in trade regulations and the rise of global production systems that stimulated more trade-supporting infrastructure. They also were strengthened by alliances with regional public-sector agencies and provincial governments across western Canada. In the process, Vancouver-based actors engaged others outside of the Lower Mainland, thus creating new scales of activity.

How much weight should we place on local versus global dynamics? It probably makes a difference which flows we are analyzing and against which specific urban developments. For example, in trying to explain why dynamic economic growth has remained concentrated in just a few regional clusters, scholars have focused on the factors that promote learning, innovation, and risk-taking—conditions not easily replicated that are the outcomes of distinctive local social relationships (see Storper, 1997). Canadian scholars have drawn on these ideas to understand clusters ranging from aerospace in Montreal to wireless telecommunications in Waterloo (Wolfe and Gertler, 2004). Although clusters are all different, each has developed through a combination of external sources of knowledge alongside local mechanisms of knowledge sharing and deployment.

However, this kind of theory is not as good at providing an account of whole cities and the multiplicity of flows and local dynamics that shape them. One type of argument made against the idea that there is something distinctive about 'the global city' is the 'ordinary cities' perspective (see Amin and Graham, 1997; Robinson, 2006). Developing out of criticisms of globalization perspectives that tend to focus on one or two attributes that make cities more or less competitive with other cities as sites for innovation or investment, Amin and Graham argue against concentrating 'exclusively on one element of urban life and city development "culture", social polarization, housing, industrial districts, politics, transport, governance, property development, planning and so on. Thus the very essence of the city—the concentration of diverse relational intersections between and within such activities and elements—tends to be lost. Oscillations between dire predictions of urban doom and optimistic portrayals of an urban renaissance serve further to confuse' (Amin and Graham, 1997: 411). Similarly, Robinson (2006), whose work is informed by cities in both developed and developing countries, insists that actual cities are diverse and heterogeneous. Rather than cities reflecting the impact of just a handful of global forces, what actually happens in a particular place is the result of the 'unique' assembly of local and global social, institutional, cultural, and economic webs.

Echoes of this perspective can be found in recent writing about Canadian cities. Examining the case of Hong Kong immigrants in Vancouver, Mitchell argues that 'certain kinds of flows . . .

are central to neoliberal state formations but also are deeply disruptive to the national liberal, social, and political narratives as they have developed and become embedded through time in the crusty layers of urban social life' (2004: 3). The implications of this line of argument are that globalization impacts the city, not only through a process by which new, extra-local and de-territorialized flows displace existing social relations. Global flows, such as those embodied in new immigrants with their own ideas, investments, and practices, also re-constitute spatial arrangements by working through existing and newly emerging social practices and politics.

Global and Local: Caveat and Conclusion

While Mitchell's case study points to the disturbing reactions of some Vancouverites to new immigrants, her conclusion is hopeful in the sense that it leaves the outcome of globalization open-ended. Thus, from cities in the Prairie provinces of Canada, so often portrayed in pessimistic terms, comes Silver's (2008) optimistic analysis of inner-city development in Winnipeg and Saskatoon. Silver relates the problems of concentrated poverty, especially among Aboriginal residents, to wider global and regional economic restructuring processes, but argues that a distinctive form of development is emerging in response:

> It is a unique form of development that has emerged largely spontaneously from the harsh realities of urban poverty in inner-city neighbourhoods, and that has been driven for the most part by inner-city residents themselves. Out of this process there has been created in Winnipeg's inner city, and is emerging in Saskatoon's core neighbourhoods, an 'infrastructure' of community-based organizations with a particular way of working, guided by a distinctive and commonly-held philosophy. This infrastructure holds considerable promise

for resolving the complex and now deeply entrenched problems arising from spatially concentrated racialized poverty, if it can continue to be patiently nurtured, and if it can be linked to an expanded and revised role for the state. (Silver, 2008: 1)

In conclusion, it seems sensible to accept the fundamental assertion of the globalization perspective that the flows which connect cities to other places around the world play an important role in shaping those urban places. At the same time, it is important to remember that local dynamics—both the unique constellation of flows in particular places and the way actors in those localities shape, understand, and respond to those flows—are vital to understanding actual patterns of urban development. Focusing on the processes of global–local interaction is essential to good analysis and policy-making.

We also need to be aware that globalization, or at least some of the most significant global flows, has its limits in this time of economic recession, climate change, and questioning of the carbon-based economy. Canadian philosopher John Ralston Saul's 2005 book, *The Collapse of Globalism and the Reinvention of the World*, seems especially prescient. Saul noted an emerging locally and nationally based countermovement to economic globalization and the neo-liberal spending policies, deregulation, and trade liberalization that have accompanied it. The potential demise of this set of economic policies and practices will surely change the relationship between cities and the global economy; but it will not change the fact that all Canadian cities have been, and will continue to be, shaped through their interaction with other, more-or less-remote places.

Review Questions

1. Identify major forces of globalization that have impacted Canadian cities.

2. How do forces of globalism interact with more regional or local forces?

3. Give an example of a global force that has impacted most Canadian cities; show how the impact is different in different places.

References

Ali, S.H., and R. Keil, eds. 2008. *Networked Disease: Emerging Infections in the Global City*. Malden, Mass.: Blackwell.

Amin, A., and S. Graham. 1997. 'The ordinary city', *Transactions of the Institute of British Geographers* 22: 411–29.

Appadurai, A. 1996. *Modernity at Large: Cultural Dimensions of Globalization*. Minneapolis: University of Minnesota Press.

Borja, J., and M. Castells. 1997. *Local and Global: Management of Cities in the Information Age*. London: Earthscan.

Brenner, N. 1999. 'Globalization as re-territorialization: The re-scaling of urban governance in the European Union', *Urban Studies* 36, 3: 431–51.

Bunting, T., A.R. Walks, and P. Filion. 2004. 'The uneven geography of housing affordability stress in Canadian metropolitan areas', *Housing Studies* 19, 3: 361–93.

Calgary Chamber of Commerce. 2008. 'Chamber research report recommends solutions to Calgary's long-term housing affordability challenge', news release, 1 May. At: <www.calgarychamber.com/resources/docs/May 1-Chamber Releases Major Calgary Housing Market Study.pdf>.

Castells, M. 1996. *The Rise of the Network Society*. Cambridge, Mass.: Blackwell.

Cox, K.R. ed. 1997. *Spaces of Globalization: Reasserting the Power of the Local*. New York: Guildford.

Dicken, P. 2003. *Global Shift: Reshaping the Global Economic Map in the 21st Century*, 4th edn. New York: Guildford.

Edgington, D.W. 1996. 'Japanese real estate investment in Canadian cities and regions, 1985–1993', *Canadian Geographer* 40, 4: 292–305.

Fong, E., and C. Luk, eds. 2007. *Chinese Ethnic Businesses: Global and Local Perspectives*. London and New York: Routledge.

Friedmann, J. 1986. 'The world city hypothesis', *Development and Change* 17, 1: 69–84.

——— and G. Wolff. 1982. 'World city formation: An agenda for research and action', *International Journal of Urban and Regional Research* 6, 3: 309–44.

Graham, S., and S. Marvin. 2001. *Splintering Urbanism: Networked Infrastructures, Technological Mobilities and the Urban Condition*. New York: Routledge.

Hall, P.G. 1966. *The World Cities*. New York: McGraw-Hill.

Hall, P.V., and A. Khan. 2008. 'Differences in hi-tech and native-born immigrant wages and earnings across Canadian cities', *Canadian Geographer* 52, 3: 271–90.

Harvey, D. 1982. *The Limits to Capital*. New York: Oxford University Press.

Heisz, A. 2006. *Canada's Global Cities: Socio-economic Conditions in Montreal, Toronto and Vancouver*. Ottawa: Statistics Canada.

Hesse, M. 2008. *The City as Terminal: The Urban Context of Logistics and Freight Transport*. Aldershot: Ashgate.

Hiebert, D. 2000. *The Social Geography of Immigration and Urbanization in Canada: A Review and Interpretation*. Research on Immigration and Integration in the Metropolis, Working Paper #00–12.

Hutton, T. 2008. *The New Economy of the Inner City: Restructuring, Regeneration, and Dislocation in the 21st Century Metropolis*. New York: Routledge.

Innis, H. 1930. *The Fur Trade in Canada*. Toronto: University of Toronto Press.

International Monetary Fund (IMF). 2009. 'Global economy contracts, with slow recovery next year', *IMF Survey Magazine*, 22 Apr.

Jacobs, J. 1985. *Cities and the Wealth of Nations*, reprint edn. New York: Vintage Books.

Levinson, M. 2006. *The Box: How the Shipping Container Made the World Smaller and the World Economy Bigger*. Princeton, NJ: Princeton University Press.

Li, W. 1998. 'Los Angeles's Chinese ethnoburb: From ethnic service center to global economy outpost', *Urban Geography* 19: 502–17.

Lipietz, A. 1993. 'The local and the global: Regional individuality or inter-regionalism?', *Transactions of the Institute of British Geographers* 18: 8–18.

Lo, L. 2008. 'DiverCity Toronto: Canada's premier gateway city', in M. Price and L. Benton-Short, eds, *Migrants to the Metropolis: The Rise of Immigrant Gateway Cities*. Syracuse, NY: Syracuse University Press.

——— and S. Wang. 2007. 'The new Chinese business sector in Toronto: A spatial and structural anatomy of medium-sized and large firms', in Fong and Luk (2007).

Luk, C. 2007. 'The global–local nexus and ethnic business location', in Fong and Luk (2007).

Massey, D. 1984. *Spatial Division of Labour: Social Structures and the Geography of Production*. London: Methuen.

McCalla R.J., et al. 2001. 'Intermodal freight terminals: Locality and industrial linkages', *Canadian Geographer* 45, 3: 404–14.

McCann, E.J. 2008. 'Expertise, truth, and urban policy mobilities: Global circuits of knowledge in the development of Vancouver, Canada's "four pillar" drug strategy', *Environment and Planning A* 40: 885–904.

McCann, L.D., and A.M. Gunn. 1998. *Heartland and Hinterland: A Regional Geography of Canada*. Scarborough, Ont.: Prentice-Hall Canada.

Mitchell, K. 2004. *Crossing the Neoliberal Line: Pacific Rim Migration and the Metropolis*. Philadelphia: Temple University Press.

Moore, E., and A. Skaburskis. 2004. 'Canada's increasing housing affordability burdens', *Housing Studies* 19, 3: 395–413.

Norcliffe, G. 2001. 'Canada in a global economy', *Canadian Geographer* 45, 1: 14–30.

North, D. 1955. 'Location theory and regional economic growth', *Journal of Political Economy* 63, 3: 243–58.

Olds, K. 2001. *Globalization and Urban Change: Capital, Culture and Pacific Rim Mega-projects*. Oxford: Oxford University Press.

Ooka, E. 2007. 'Going to malls, being Chinese? Ethnic identity among Chinese youths in Toronto's ethnic economy', in Fong and Luk (2007).

Parfitt, B. 2007. *Wood Waste and Log Exports on the BC Coast*. Canadian Centre for Policy Alternatives—BC Office.

Peck, J., and N. Theodore. 2001. 'Exporting workfare/importing welfare-to-work: Exploring the politics of Third Way policy transfer', *Political Geography* 20, 427–60.

Robinson, J. 2006. *Ordinary Cities: Between Modernity and Development*. New York: Routledge.

Saul, J.R. 2005. *The Collapse of Globalism: And the Reinvention of the World*. Toronto: Viking Canada.

Sassen, S. 1991. *The Global City: New York, London, Tokyo*. Princeton, NJ: Princeton University Press.

Saxenian, A. 2006 *The New Argonauts: Regional Advantage in a Global Economy*. Cambridge, Mass.: Harvard University Press.

Silver, J. 2008. *The Inner Cities of Saskatoon and Winnipeg: A New and Distinctive Form of Development*. Ottawa: Canadian Centre for Policy Alternatives. At: <www.policyalternatives.ca/documents/Manitoba_Pubs/2008/Inner_Cities_of_Saskatoon_and_Winnipeg.pdf>.

Storper, M. 1997. *The Regional World: Territorial Development in a Global Economy*. New York: Guildford.

Taylor, P. 2005. 'Leading world cities: Empirical evaluations of urban nodes in multiple networks', *Urban Studies* 42, 9: 1593–1608.

Tiebout, C. 1956. 'Exports and regional economic growth', *Journal of Political Economy* 64, 2: 160–4.

Transport Canada. 2007. 'Canada's new government announces projects for the Asia-Pacific Gateway and

Corridor Initiative', press release. Ottawa: Government of Canada. At: <www.tc.gc.ca/mediaroom/releases/nat/2007/07-gc016e.htm#bg>.

Walks, R.A., and L.S. Bourne. 2006. 'Ghettos in Canada's cities? Racial segregation, ethnic enclaves and poverty concentration in Canadian urban areas', *Canadian Geographer* 50, 3: 273–97.

Wallerstein, I. 1984. *The Politics of the World Economy*. Cambridge: Cambridge University Press.

Ward, K. 2006. '"Policies in motion", urban management and state restructuring: The trans-local expansion of Business Improvement Districts', *International Journal of Urban and Regional Research* 30: 54–75.

Wolfe, D., and M. Gertler. 2004. 'Clusters from the inside and out: Local dynamics and global linkages', *Urban Studies* 41, 5–6: 1071–93.

World Trade Organization (WTO). 2008. International trade statistics 2008, author's analysis of Appendix Table A1a. At: <www.wto.org/english/res_e/statis_e/its2008_e/its08_appendix_e.htm>.

Yalnizyan, A. 2009. *Exposed: Revealing Truths about Canada's Recession*. Ottawa: Canadian Centre for Policy Alternatives.

CHAPTER 5

Getting Serious about Urban Sustainability: Eco-Footprints and the Vulnerability of Twenty-First-Century Cities

William E. Rees

At the heart of this assessment is a stark warning. Human activity is putting such a strain on the natural functions of the Earth that the ability of the planet's ecosystems to sustain future generations can no longer be taken for granted. (MEA, 2005: 5)

Introduction: Framing the Analysis

This chapter is concerned with the long-term sustainability of cities. My starting premise is that because of accelerating global ecological change, cities everywhere are facing unprecedented challenges to their functional integrity and even survival. Unprecedented challenges require unprecedented solutions. In keeping with this reality, I depart from most urban scholarship, which assumes a humanities and social science perspective. Instead, I approach 'the urban question' from a mainly biophysical point of view. Accelerating global change makes clear that society will not be able to assure the sustainability of cities without a much fuller understanding of cities as ecological entities subject to biophysical laws.

With this in mind, the chapter begins with a brief consideration of the organic origins of cities and subsequent evolution of cities. Permanent settlements became possible as a result of technology-induced changes in the ecological 'niche' of humans 10 millennia ago. However, the subsequent alienation of urban techno-industrial society from nature has produced modern cities that are not only incomplete as human ecosystems but that exist in essentially hostile relationship to the natural ecosystems that sustain them.

The next section uses ecological footprint analysis to illustrate the ecological load imposed on the natural world by people and to estimate the de facto surface area of the earth occupied *ecologically* by the inhabitants of modern cities to sustain their material lifestyles. Pay attention, urban planners! The eco-footprints of typical cities are hundreds of times larger than their political or built-up areas. In any functionally meaningful sense, doesn't this 'hinterland' area constitute urban land as much as does a parking lot within the city limits?

I then consider the increasing vulnerability of modern cities to global ecological change. Urbanization represents the greatest mass migration of people ever. More people will be added to the world's cities in the first three or four decades of the twenty-first century, mostly through inmigration, than had accumulated on the entire planet by 1930. But urbanization implicitly assumes climate stability, reliable supplies of vital resources, and geopolitical calm. Just how secure will the world's six billion urbanites be when cities are besieged by climate change, rising sea levels, energy and food shortages, and resultant violent conflicts by mid- to late century?

The final section examines the ecological leverage that cities can exercise in society's general quest for sustainability. What can cities do to decrease their eco-footprints and enhance their own survival prospects? How might rethinking the 'city-as-ecosystem' help humanity to live sustainably within the carrying capacity the earth?

Setting the Ecological Stage

While few people think of them as such, cities are biophysical entities. The fundamentally *organic* nature of cities is underscored by the fact that permanent settlements are actually a product of a change in human ecological circumstances. 'The city' is an emergent phenomenon made possible by people's adoption of agriculture ten millennia ago. Humanity's slowly developing ability to produce regular food surpluses triggered a truly 'autocatalytic process—one that catalyses itself in a positive feedback cycle, going faster and faster once it has started' (Diamond, 1997). More food made higher population densities possible, enabled large permanent settlements with the specialized skills and inventiveness this implies, and shortened the time-spacing between children. This, in turn, enabled the higher populations to produce still more people, which increased both the demand for food and the technical and organizational capacity to produce it.

The first small, more-or-less permanent human settlements appeared for these reasons barely 9,000 years ago and another 1,500 years passed before the first definable cities, with socially stratified societies and marked division of labour, emerged in southwest Asia (today's Middle East) around 5500 BC. In short, while we tend today to take the existence of cities for granted, cities actually have a remarkably short history. They have been part of human reality for merely 1.5 per cent of the time since 'modern' humans—*Homo sapiens*—stumbled onto the world stage about 500,000 years ago.

But there is more to this story than 'surplus food leads to urban civilization'. The shift from the nomadic hunter-gatherer lifestyle to a more agriculture-based, settlement-centred way of life represents a major transformation of human ecological reality and may well constitute the most critical branch-point to date in the evolution of *Homo sapiens*. First, consider that with large-scale agriculture, people switched from merely taking what wild nature had to offer, to manipulating entire landscapes in order to redirect as much as possible of nature's productivity to strictly human ends. In this way, humans became the most significant 'patch-disturbance' species on the earth (Rees, 2000). Indeed, agriculture and agriculture-induced urbanization constitute a great leap forward in an accelerating process that has gradually seen humans become the most important geological force changing the face of the planet.

Second—and, regrettably, given the enormous ecological impacts of industrial cities—the very process of urbanization insulates city-dwellers from the negative consequences of human ecological dysfunction. Initially, the migration of people to cities distances them physically from the ecosystems that support them and, even more important, from the direct negative consequences of subsequent landscape degradation and from having to acknowledge that resource drain. The separation of people's lives and livelihoods *from* the land diminishes urbanites' sense of felt connectedness *to* the land. In short, humanity's apparent abandonment of the countryside is critically reshaping billions of people's spatial relationships and psychological sensitivities to nature. Thus, doubly blinded, many urbanites, particularly in high-income developed countries, remain blissfully unaware that they remain ecological actors and of the growing threat their consumer lifestyles pose to distant ecosystems on which they remain utterly dependent.

One effect of this alienating process is that city-dwellers generally don't think of 'the city' in ecological terms. Even urban scholars have only recently acknowledged and begun to study the human ecological dimensions of urbanization and cities. Most discussions of urbanization still view the process mainly as a demographic or economic phenomenon made possible by the intensification of agriculture, increased resource productivity, and improvements in communications and transportation technology. Cities are perceived as concentrations of people; areas dominated by the built environment; places of intense social interaction; the seats of government; hotbeds of political conflict; the nexus of national transportation and communication systems; and as the engines of national economic growth—but rarely as a biological phenomenon. Some observers actually—and falsely—interpret urbanization as evidence that humanity is *transcending* nature, that the human economy is 'decoupling' from 'the environment'.

Modern humans' failure to appreciate themselves as ecological beings reflects a deep cognitive bias. Over the past 300 years, our evolving techno-scientific paradigm has erected a self-serving perceptual barrier between humanity and the rest of the natural world. Indeed, this so-called 'Cartesian dualism' is a defining characteristic of industrial society that strongly reinforces the physical and psycho-separation of urban humans from their roots in nature.

This chapter is intended to address this perceptual gap. A major purpose is to show that, while urbanization represents a dramatic shift in urbanites' spatial/psychological relationships to the land, *there is no corresponding change in eco-functional relationships*. Indeed, far from reducing people's dependence on productive ecosystems, urbanization generally implies an increase in our per capita **ecological footprints**. From this perspective, urbanization and the modern city remain bio-ecological phenomena fully explicable only in light of human evolutionary history and fundamental biophysical laws. Failure to understand ourselves and our cities as ecological agents will doom our quest for global sustainability and exposes an increasingly vulnerable global urban civilization to the spectre of collapse.

Cities and the Human Ecosystem

By now almost everyone interested in cities is familiar with the term **urban ecosystem**. Prominent urban analysts have long recognized that the city could be conceived as an ecosystem (e.g., Douglas, 1981) and today there is even a scientific journal called *Urban Ecosystems*. Nevertheless, the concept itself remains ambiguous. For example, a majority of the papers in *Urban Ecosystems* focus on the impacts of urbanization on plants and animals or on remnant 'natural' ecosystems within the city. This shows that most natural scientists who study 'urban ecosystems' cast the city as a somewhat unnatural habitat for *other* species.[1] To ecologists, the 'urban ecosystem' consists of the assemblage of non-human species in the city and the purpose of their inquiries is to determine how these species have adapted to the structural and chemical characteristics of the 'built environment' (Rees, 2003). Remarkably, humans are excluded from the analysis except as their actions affect these other species. This conception of urban ecosystems is a clear reflection of the Cartesian dualism that separates people from nature in the human mind.

On the other hand, those who do acknowledge humans as the major ecological actors in the city err if they see 'the city' per se as the modern human ecosystem. To qualify as a complete human ecosystem, a city would have to contain a sufficient complement of producer organisms (green plants), macro-consumers (animals, including humans), micro-consumers (bacteria and fungi), and abiotic factors to support its human population indefinitely. Any complete ecosystem consists of a self-organizing, self-sustaining assemblage of living species existing in complementary relationship with each other and the physical environment. Ecosystems are energized by the unidirectional cascade of solar energy and maintained in perpetuity by the continuous recycling of chemical nutrients.

Clearly, from this perspective no modern city qualifies as a functionally complete human ecosystem. Some essential defining parts are missing altogether (e.g., virtually the entire relevant producer complex) and others (micro-consumers) are insufficiently abundant for functional integrity. As significantly, the spatial separation of people from the rest of their supportive ecosystems (e.g., agricultural and forest lands) inhibits the on-site application of organic matter and the recycling of phosphorus, nitrogen, and other nutrients contained in human wastes. In effect, urbanization transforms local, integrated, cyclical ecological production systems into global, horizontally disintegrated, unidirectional, throughput systems (Rees, 1997). Ironically, the resultant continuous 'leakage' of nutrients from farmland in shipments of food to cities threatens to undermine organic agriculture even as it gains ground in the urban marketplace.

In a crude but useful metaphor, the city might be described as a livestock feedlot (Rees, 2003). Like cities, feedlots are densely populated almost entirely by a single macro-consumer species—for example, cattle (or pigs, or chickens, which today are raised using even more constrained industrial methods). However, the grain fields that produce the feed for feedlot animals may be located hundreds of kilometres from the feedlot itself. Also missing are adequate populations of micro-consuming decomposers. Having separated the functionally inseparable, industrial feedlots short-circuit even the possibility of within-system decomposition and nutrient recycling. As a result, vast quantities of manure containing vital nutrients rarely are re-deposited on range or cropland for nutrient recycling, but rather are disposed of inappropriately, contaminating soils and surface and subsurface waters at a distance and over large areas.[2]

Of course, cities are more ecologically complex than feedlots. However, in structural terms cities are to their human inhabitants what feedlots are to cattle. The largest and functionally most important components of urbanites' ecosystems—the assemblage of producer organisms that feed them and provide them with oxygen, most of the micro-consumers that complete their nutrient cycles and the various sub-systems that perform myriad other vital life-support functions—all are found in rural 'environments' increasingly scattered all over the planet. Also, like feedlots, cities generate enormous quantities of waste that cannot be assimilated within the city, making cities the major source of pollution of the global commons. In short, cities are nodes of intense resource consumption and waste generation entirely dependent for their survival on the productive and assimilative capacities of ecosystems increasingly located at great distances from the cities themselves. In both ecological and spatial terms, 'the city' constitutes only a fraction of the total urban-centred human ecosystem.

The Human Ecological Footprint

The next question is: just how big is the human ecosystem? How much of the productive capacity of the ecosphere do humans need to sustain urban industrial society? One way to approach this question is through ecological footprint analysis (Rees, 1992; Wackernagel and Rees, 1996). A variant and extension of energy and material flows assessment (Haberl et al., 2004), eco-footprint analysis (EFA) starts from a comprehensive inventory of the annualized energy and material flows required to support any specified population—an individual, a city, a nation or the entire world. We also quantify the flows of certain critical wastes generated in this production/consumption process, particularly carbon dioxide (the carbon footprint). Eco-footprinting is further based on the fact that many of these material and energy flows can be converted into a corresponding area of productive land and water ecosystems. Thus, we formally define the ecological footprint of a specified population as:

> The area of land and water ecosystems required, on a continuous basis, to produce the resources that the population consumes and to assimilate the wastes that the population produces, wherever on Earth the relevant land/water is located (Rees 2001, 2006).[3]

The area of a population's theoretical eco-footprint (EF) depends on four factors: the population size, the average material standard of living, the productivity of the land/water base (whether local or 'imported' in trade goods), and the efficiency of resource harvesting, processing, and use. Regardless of the relative importance of these factors and how they interact, every population has an ecological footprint, and the productive land and water area captured by eco-footprint analysis (EFA) represents much of the 'natural capital' (the productive natural resource base or *biocapacity*) required to meet the study population's consumptive and assimilative demand.[4] Furthermore, one can interpret the eco-footprint in thermodynamic terms as the area of natural 'solar collector' needed to regenerate the biomass and chemical energy equivalents of the useful resources and fossil energy consumed and dissipated by the study population.

It is important to recognize that population EFs constitute mutually exclusive appropriations of nature. The biocapacity used by one population is not available for use by another. True, the grain grown in a particular region may wind up in export shipments to several countries, but the total area of cropland involved is the sum of the areas required by the individual populations. In the final analysis, *all human populations are competing for the productive capacity (biocapacity) of the earth*.

Table 5.1 shows the equivalence-adjusted[5] per capita EFs and domestic biocapacities for a selection of countries from among the richest to among the poorest using 2005 data from the *Living Planet Report 2008* (WWF, 2008). Note the vastly larger load imposed on the ecosphere by wealthy, mainly urban consumers compared to that imposed by mainly rural peasants. The citizens of wasteful, high-income countries like the US and Canada have average EFs of six to 10 global hectares (gha), or up to 20 times larger than the EFs of the citizens of the world's poorest countries, such as Bangladesh and Malawi. European countries and Japan typically have per capita EFs in the four to six gha range. China is fairly representative of the emerging economies which show rapidly growing EFs of 1.5 to three gha. These data reflect the growing global income gap: the richest 20 per cent of the human family spend more than 75 per cent of world income; the poorest 20 per cent subsist on just 1.5 per cent (UNDP, 2007).

The final column of Table 5.1 shows each country's 'overshoot factor'. This is a simple ratio of the national average eco-footprint compared to per capita domestic biocapacity. Countries with overshoot factors larger than one impose a greater burden on the ecosphere than can be supported by their domestic ecosystems. That is, these countries are at least partially dependent on trade and on exploitation of the global commons to maintain their current lifestyles (i.e., average per capita consumption levels). The Netherlands, for example, uses almost four times as much productive land/water outside its borders as is found within the country. Japan's demand for biocapacity is eight times its domestic supply. Such countries are running 'ecological deficits' with the rest of the world.

A few countries with overshoot ratios less than one are living within their 'natural incomes' and thus seem to have ecological surpluses. They only seem to have surpluses because the extra biocapacity in most cases is being traded away to cover the ecological deficits of other countries. The agricultural, forestry, and fisheries surpluses of Canada, for example, serve a large export market. Trade, therefore, contributes proportionately to the

Table 5.1 The Eco-Footprints and Biocapacities of Selected Nations

Country	Per Capita Eco-Footprint (global ha)	Per Capita Domestic Biocapacity (gha)	Overshoot Factor
World	2.7	2.1	1.3
United States	9.4	4.9	1.9
Australia	7.8	15.4	0.5
Canada	7.1	20.0	0.4
Greece	5.9	1.7	3.5
United Kingdom	5.3	1.6	3.3
France	4.9	3.0	1.6
Japan	4.9	0.6	8.2
Germany	4.2	1.9	2.2
Netherlands	4.0	1.1	3.6
Hungary	3.5	2.8	1.3
Mexico	3.4	3.3	1.0
Malaysia	2.4	2.7	0.9
Brazil	2.4	7.3	0.3
China	2.1	0.9	2.3
Thailand	2.1	0.8	2.6
Peru	1.6	4.0	0.4
Ethiopia	1.4	1.0	1.4
Nigeria	1.3	1.0	1.3
Indonesia	0.9	1.4	0.6
India	0.9	0.4	2.3
Bangladesh	0.6	0.3	2.0
Malawi	0.5	0.5	1.0

Source: WWF (2008).

ongoing degradation of the nation's soils, forests, and fish stocks (Kissinger and Rees, 2009).

Ominously, the world as a whole is in a state of overshoot (Table 5.1). Human demand exceeds the earth's regenerative capacity by about 30 per cent. We are living, in part, by depleting and dissipating as waste the enormous stocks of potentially renewable natural capital (fish, forests, soils, etc.) that have accumulated in ecosystems over millions of years.

The Global Reach of Cities

Cities, of course, are virtually all ecological deficit. Urban populations are almost totally dependent on rural people, ecosystems, and life-support processes increasingly scattered all over the planet (Girardet, 2004; Newman and Jennings, 2008; Rees, 1992, 2003). In some respects, this relationship is a two-way, mutualistic one—rural areas benefit from urban markets, the products of urban factories, urban-based services, technology transfers from urban areas, etc. However, while rural populations have survived historically without cities the ecological dependence of urbanites on 'the **hinterland**' is absolute. *There can be no urban sustainability without rural sustainability* even if the 'rural' for any particular city is scattered all over the planet. Understanding the nature of rural–urban interdependence is essential to understanding the total human ecosystem and to understanding **urban sustainability** (see Filion and Bunting, Chapter 1).

In theory, estimating the eco-footprint of a city is not different from estimating that of an entire country. In practice, however, matters are more complicated, hindered by the lack of local data. No statistical or planning agencies monitor the flow of biophysically significant goods and services across municipal boundaries. While some urban EF studies do attempt to compile local data, others use 'quick and dirty' extrapolation from national eco-footprint estimates, sometimes with adjustments for local conditions, income differences, etc. (e.g., FCM, 2005). This method produces more accurate city footprint numbers for highly urbanized, high-income countries than for less-urbanized, poorer countries.

So, just how great is a typical modern city's debt to the global countryside? Despite methodological and data-quality differences, urban eco-footprint studies invariably show that the EFs of typical modern high-income cities exceed their geographic or political areas by two to three orders of magnitude. For example:

- Based on locally adjusted per capita EF estimates, the people of Toronto and Vancouver 'occupy' an ecosystem area outside their municipal boundaries 292 and 390 times larger, respectively, than the cities themselves (FCM, 2005). Even the lower-density metropolitan areas of these cities have EFs 57 times bigger than the respective urban regions (Table 5.2). The citizens of Toronto and Vancouver might want to contemplate the implications of this growing extra-territorial dependence as they sprawl out over Canada's most productive farmland in an era of global change. Where will they turn when they can no longer import essential foods from distant elsewheres?

- Under varying management assumptions of cities' ability to cope with regional waste management issues, Folke et al. (1997) estimated that the 29 largest cities of the Baltic region require, for resources and certain categories of waste assimilation, an area of forest, agricultural, marine, and wetland ecosystems 565–1,130 times larger than the area of the cities themselves.

- With a population of 33 million and a per capita EF of about 4.9 gha, metropolitan Tokyo's total eco-footprint is 161,700,000 gha. However, the entire domestic biocapacity of Japan is only about 76,860,000 gha. In short, Tokyo, with only 26 per cent of Japan's population, lives on an area of productive ecosystems 2.1 times larger than the nation's entire terrestrial biocapacity.[6] Clearly, if Japan were required by changing global circumstances to subsist on its domestic biocapacity, the country would have difficulty supporting even the population of its capital city.

- Warren-Rhodes and Koenig (2001) estimated that Hong Kong, with almost seven million people, has a total eco-footprint of 332,150 km^2 to 478,300 km^2 (5.0–7.2 ha per capita) (the range reflects two estimates of carbon sink land requirements).

Hong Kong's eco-footprint is at least 303 times the total land area of the Hong Kong Special Administrative Region (1,097 km²) and 3,020 times the built-up area of the city (110 km²).

These data show clearly that, in material terms, 'sustainable city' is an oxymoron (Rees, 1997). Modern cities are urban black holes sweeping up the productivity of a vastly larger and increasingly global resource hinterland and spewing an equivalent quantity of waste back into it. They are compact nodes of consumption living parasitically on the productivity and assimilative capacity of a vastly larger 'undeveloped' area, portions of which may be thousands of kilometres from the built-up area at the centre.

While some have interpreted the consumptive and polluting power of cities as an anti-urban argument, it is nothing of the sort. Cities do have enormous ecological footprints; however, as we shall see, cities actually offer several advantages over more dispersed settlement patterns in the quest for sustainability.

Eco-footprinting also suggests several other paradoxes about current perceptions of cities. For example, why is the lifeless asphalt of the mall parking lot considered to be urban land, while cropland vital to the survival of the city is not? What does it mean for urban planning if 99.5 per cent of the de facto urban (eco)system lies outside the municipal boundaries, out of sight and beyond the control of those it supports? Perhaps we should redefine what we mean by urban (eco)system.

Finally, eco-footprinting underscores a material reality that is all but ignored in the sustainability literature—no individual, no city, and no country can achieve sustainability if the system of which it is a part is unsustainable. Vancouver, Toronto, or Montreal might become exemplars of sustainable urban design and lifestyles, but if the global system of which they are a part remains on an unsustainable path, then even our model cities would be taken down by, for example, severe climate change, depleted resources, and resultant geopolitical instability. Given such interdependence, the best any sub-global system can attain independently is a state of quasi-sustainability. 'Quasi-sustainable' describes that level of economic activity and energy/material consumption per capita that, if extended to the entire system, would result in global sustainability (Rees, 2009). In 2009, quasi-sustainability implies a per capita eco-footprint of about 2.1 gha (2.1 gha represents an equitable per capita share of global biocapacity). Since Canadians' average eco-footprint is 7.1 gha per capita, we would have to reduce consumption by 70 per cent to meet the quasi-sustainability standard!

Table 5.2 The Eco-Footprints of Toronto and Vancouver (global average hectares)

City or Region	Population (2006)	Per Capita Eco-Footprint (gha)	Area (hectares)	Total Eco-Footprint (gha)	Ratio of EF to Actual Area
Vancouver	578,041	7.71	11,400	4,456,696	390
Metro Vancouver	2,116,580	7.71	278,736	16,318,832	57
Toronto	2,503,281	7.36	63,000	18,424,148	292
Greater Toronto	5,555,912	7.36	712,500	40,891,512	57

The Vulnerability of Modern Cities

Increasing global interdependence obviously has enormous implications for the security of urban populations in an era of global change. (For globalization processes, see Hall, Chapter 4.) Cities have grown so large and have such enormous eco-footprints not because size necessarily confers great advantage but simply because they could—historically, globalization and trade have assured the abundant supplies and uninterrupted flows of the energy and other material resources required to grow the modern metropolis. But this raises an increasingly awkward question in an era of global change: just how secure is any city of millions, or even a relative 'town' of 100,000, if resource scarcity, shifting climate, or geopolitical unrest threatens to cut it off from vital sources of supply? There are several interrelated reasons to believe this is not an idle question. For example:

1. Reliable food supplies should be of increasing concern to urbanizing populations (Kissinger and Rees, 2009). Global grain production is levelling off. Yet, just to keep pace with UN medium population growth projections, agricultural output will have to increase by over 50 per cent by 2050 and improving the diets of malnourished people would push this towards 100 per cent.[7] Achieving increases of this magnitude may be difficult. By 1990, 562 million hectares (38 per cent) of the world's roughly 1.5 billion hectares of cropland had become significantly eroded or otherwise degraded; 300 million hectares (21 per cent) of cultivated land—enough to feed almost all of Europe—has been lost to production, and we are losing five to seven million hectares annually (FAO, 2000; SDIS, 2004). Depending on the climate and agricultural practices, topsoil is being 'dissipated' 16 to 300 times as fast as it is regenerated. So far, the impact has been

masked because we have managed to substitute fossil fuel for depleted soils and landscape degradation—but that capacity for substitution may be about to change.

2. Cities are very much the product of abundant, cheap fossil fuel. Indeed, no other resource has changed the structure of economies, the nature of technologies, the balance of geopolitics, and the quality of human life as much as petroleum (Duncan and Youngquist, 1999). Fossil fuels, especially oil, currently supply about 85 per cent of humanity's total energy demand and are essential for transportation, space and water heating, and the generation of electricity. Oil is also a major factor in the green revolution. Mechanization, diesel-powered irrigation, the capacity to double-crop, and agro-chemicals (fertilizers and pesticides) made from oil and natural gas account for 79–96 per cent of the increased yields of wheat, rice, and maize production since 1967 (Cassman, 1999; Conforti and Giampietro, 1997). For all these reasons, some analysts argue that the peaking of global petroleum extraction (see Figure 5.1) anticipated by 2010 represents a singular event in modern history and poses a greater challenge to geopolitical stability and urban security than any other factor (Campbell, 1999; Duncan and Youngquist, 1999; Laherrere, 2003). In addition to shrinking the supply and exploding the price of food, 'peak oil' could have an enormous impact on urban transportation, **urban form** and structure, and the future size of cities.

3. Other analysts see climate change as the greatest threat to urban civilization. Even modest shifts in weather patterns could disrupt historic water availability and distribution, thus undermining both agricultural production and urban water supplies. By some accounts, more severe climate change could bring the world to the edge of anarchy (e.g., CSIS, 2007; Schwartz and Randall, 2003). In *The Age of Consequences*, Washington's Center for

Strategic and International Studies suggests that human-induced climate change driven by burning fossil fuels could end peaceful global integration as various nations contract inwardly to conserve what they need—or expand outwardly to *take* what they need—for survival. In the event of 'severe climate change', corresponding to an average increase in global temperature of 2.6°C by 2040 (now deemed to be increasingly likely), major non-linear changes in biophysical systems will give rise to major non-linear socio-political events. Shifting climate will force internal and cross-border migrations as people abandon areas where food and water are scarce. People will flee rising seas and areas devastated by increasingly frequent droughts, floods, and severe storms. Dramatic increases in migration combined with food, energy, and water shortages

will impose great pressure on the internal cohesion of nations. War is likely and nuclear war is possible (CSIS, 2007).

Such dismal scenarios seem increasingly likely in light of findings of recent climate studies (e.g., Anderson and Bows, 2008; Hansen et al., 2008). Current loose 'targets' for controlling climate change include stabilization of carbon dioxide at 350 parts per million by volume (it is currently 380 ppmv) and maintaining temperature increases below 2°C. However, according to Anderson and Bows (2008), 'an optimistic interpretation of the current framing of climate change implies that stabilization [of green house gases] much below 650 ppmv CO_2e [carbon dioxide equivalents] is improbable.' This is partly because, in order to stabilize at 650 ppmv CO_2e, the majority of OECD nations will soon have to begin decarbonizing at rates in excess of 6 per

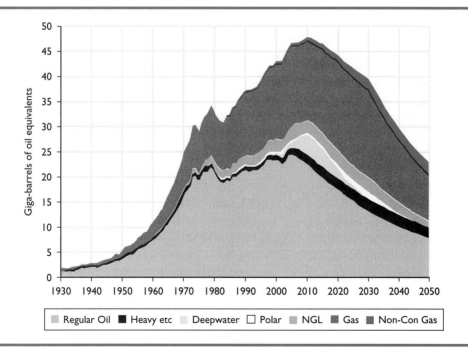

Figure 5.1 Oil and gas production profiles, 2007 base case

Source: ASPO (2008).

cent per year, which would likely require *a planned economic recession* (Anderson and Bows, 2008). We should note that atmospheric GHG concentrations of 650 ppmv CO^2e imply a catastrophic 4°C increase in mean global temperature, compared to the 'mere' 2.6°C increase assumed in CSIS's already horrific 'severe climate change' scenario. Models suggest that 4°C warming would be sufficient to convert much of the US, Southern Europe, China, India, Africa, and South America into uninhabitable wastelands (see Vince, 2009), displacing billions of people and jeopardizing prospects for maintaining any form of global civilization.

The good news in all this is that determined action to address climate change could help avoid the peak oil problem and vice versa. For example, if the world were to take the action necessary to reduce CO^2 emissions by 6 per cent per year to avoid the worst of climate change, the drop in demand for oil would keep pace with or exceed the anticipated decline in extraction rate.

Towards the 'One Planet' City

Industrialized world reductions in material consumption, energy use, and environmental degradation of over 90 per cent will be required by 2040 to meet the needs of a growing world population fairly within the planet's ecological means. (BCSD, 1993)

This is a world in overshoot, yet both population and per capita consumption are increasing and material expectations are rising all over the world. This is a fundamentally unsustainable situation—to raise just the present world population sustainably to North American material standards would require the biocapacity of four additional earth-like planets (Rees, 2006). The *really* inconvenient truth is that, to achieve sustainability, global energy and material throughput must decrease, not grow.

Techno-industrial society is a self-proclaimed, science-based society but we are not acting

consistently with our best science. The risks associated with anticipated global change certainly demand major restructuring and may well require a planned economic contraction. For sustainability with greater equity, wealthy countries will have to free up the ecological space necessary for needed growth in the developing world (Rees, 2008). High-income nations should therefore work to limit the energy and material throughput required to sustain urban life (Lenzen, Dey, and Barney, 2004; Newman and Jennings, 2008; Rees, 2009). To achieve one-planet living, Canadians should be taking steps *now* to reduce their eco-footprints by 70 per cent from 7.4 gha to their 'equitable earth-share' of 2.1 gha per capita; similarly, Americans need to trim their eco-footprints by 78 per cent (see Table 5.1).[8]

Clearly, achieving such targets will require a dramatic shift in prevailing economic beliefs, values, and most particularly in consumer lifestyles—as much as 60–70 per cent of the material flows through cities are attributable to personal consumption. Fortunately, the technology is available to make this a relatively painless transition (Weizsäcker, Lovins, and Lovins, 1995). Moreover, 'managing without growth' is both economically possible and could well *improve* quality of life (see Victor, 2008).

Regrettably, there is scant evidence that the necessary cultural shift is underway. Certainly no national government, the United Nations, or any other official international organizations have begun openly to contemplate the implications for humanity if the science is correct, let alone articulate in public the kind of policy responses the science evokes. Despite repeated warnings that staying our present course spells catastrophe for billions of people (MEA, 2005; UCS, 1992), the modern world remains mired in a swamp of cognitive dissonance and collective denial (Rees, 2009).

Consequently, mainstream responses to our ecological conundrum to date do not address the fundamental problem but seem designed, instead, to reproduce the status quo by other means. Such 'innovations' as hybrid cars, green buildings, smart growth, the New Urbanism, green consumerism, and even much of the eco-cities movement all assume that we can achieve sustainability through technological innovation and greater material and economic efficiency. This is a conceptual error—historically, efficiency has actually *increased* consumption by, for example, raising incomes and lowering prices. With more money chasing cheaper goods and services, throughput rises. In effect, improved efficiency simply makes industrial growth-bound society more efficiently unsustainable.

The Urban Sustainability Multiplier

> The climate crisis won't be solved by changing light bulbs and inflating your tires more, planting a tree and driving a little less. It's going to require a truly fundamental shift in how we build our cities and live in them. (Register, 2009)

Getting serious about urban sustainability obviously requires more determined action than society has yet been willing to contemplate. Fortunately, the very factors that make wealthy cities weigh so heavily on the ecosphere—the concentration of people and the localized intensity of energy/material consumption and waste generation—also give them considerable economic and technical leverage in shrinking their eco-footprints.

To enable cities to take full advantage of this leverage, provincial and municipal governments must create the land-use legislation and zoning bylaws that urban planners need to eliminate sprawl and consolidate and densify existing built-up areas. Compact cities have the potential to be vastly less energy- and material-intensive than today's sprawling suburban cities. The economies of scale and the agglomeration economies associated with high-density settlements confer a substantial 'urban sustainability multiplier' on cities (Rees, 1999). For example:

- lower biophysical and economic costs per capita of providing piped treated water, sewer systems, waste collection, and most other forms of infrastructure and public amenities;
- a greater range of options for material recycling, reuse, remanufacturing, and a concentration of the specialized skills and enterprises needed to make these things happen;
- reduced per capita demand for occupied land;
- greater possibilities for electricity co-generation, district heating/cooling, and the use of waste process heat from industry or power plants, to reduce the per capita use of fossil fuel for water and space-heating;
- more opportunities for co-housing, car-sharing, and other co-operative relationships that have lower capital requirements (consumption) per household and individual;
- more ways greatly to reduce the (mostly fossil) energy consumption by motor vehicles through walking, cycling, and public transit;
- more 'social contagion', facilitating the spread of more-nearly sustainable lifestyle choices (e.g., 'voluntary simplicity');
- the potential to implement the principles of low throughput 'industrial ecology' (i.e., the ideal of closed-circuit industrial parks in which the waste energy or materials of some firms are essential feedstocks for others).

Walker and Rees (1997) show that the increased density and consequent energy and material savings associated with condos and high-rise apartments, compared to single-family houses, can reduce that part of the per capita urban ecological footprint associated with housing type and related transportation needs by about 40 per cent.[9]

As noted, however, efficiency gains alone will not enable society to achieve 'one-planet living'. Sustainability and security demand that cities everywhere become less consumption-driven and more materially self-reliant. Indeed, cities may be forced down this unfamiliar path by either or both the rising cost of oil-based transportation or/and the rapid phase-out of fossil fuels needed to avoid severe climate change (target: at least 80 per cent decarbonization by 2050). Certainly, there is no place for the fossil-fueled automobile—or any cars at all—in the eco-cities of the future (Register, 2009). For these reasons, urban designers must begin now to rethink cities so they function as complete ecosystems. This is the ultimate form of bio-mimicry.

Bio-mimicry at the city level requires the re-localization of many ecological and economic functions. The least vulnerable and most resilient urban ecosystem might be a new form of regional eco-city state (or bioregion) in which a densely built-up core is surrounded by its essential ecosystems (see Connelly and Roseland, Chapter 14; Grant and Filion, Chapter 18).[10] The central idea is to consolidate as much as possible of the city's productive hinterland in close proximity to its consumptive urban core. In effect, without preventing essential trade, this would internalize the widely scattered external eco-footprints of our present cities into more compact and manageable city-centred regions that could function as complete human ecosystems. Such a transformed home place, 'rather than being merely the site of consumption, [would], through its very design, produce some of its own food and energy, as well as become the locus of work for its residents' (Van der Ryn and Calthorpe, 1986). Eco-city states would be less a burden on, and more a contributor to, the life-support functions of the ecosphere than contemporary cities.

Most importantly, the bioregional city would reconnect urban populations both physically and psychologically to 'the land'. Because inhabitants would be more directly dependent on local ecosystems, they would have a powerful incentive—currently absent—to manage their land and water resources sustainably in the face of global change. (Ideally, political control over the productive land and resource base of the consolidated region would pass from the provincial to new eco-city state governments.) Less reliant on imports, their populations would be partially insulated from climate vagaries, resource shortages, and distant violent conflicts. Note that if the entire world were organized into a system of self-reliant bioregions, managed to conserve adequate per capita stocks of natural capital, the aggregate effect would be global sustainability.

Canada is one of the few countries with sufficient space and resources for many of its cities to be readily reorganized along bioregional lines. Conversely, for many large cities, particularly wealthy cities in small countries, reorganization into bioregions is no longer an option. Remember Tokyo with its eco-footprint spanning two Japans? The best such cities can do in the face of global change is to reduce their material demands as much as possible and hope their national governments can negotiate reliable supplies of vital resources from areas that still have surplus capacity. It may not be possible for all cities to be self-sustaining, but in a crowded world, every city has an obligation to contribute to global sustainability by taking appropriate action within its own boundaries (McGranahan and Satterthwaite, 2003).

Of course, 'appropriate action' will vary greatly among cities. This discussion has focused mainly on the sustainability implications of global change for wealthy cities. We also acknowledge that many millions of people live under deplorable conditions in the barrios, slums, and squatter settlements of Third World cities. The greatest environmental problems facing these people are inadequate diets and the lack of potable water and sanitary sewage. Unlike the already wealthy, the world's urban poor would benefit from rising incomes and greater consumption. The world community should be assisting them to *increase* their eco-footprints. In this light, an appropriate strategy for rich and poor cities alike would be to manage gross consumption so that average per capita eco-footprints converge from above and below toward 2.1 gha, each person's equitable share of global biocapacity.[11]

Epilogue

Historically, environmental concerns about cities were confined to the local public health effects of air, water, and land pollution that preoccupy impoverished cities today. However, the four-fold rise in human numbers and the order of magnitude increase in economic activity during the twentieth century undermined basic life-support systems and raised ecological concerns to the global scale.

Despite increasing costs and risks, the world community remains addicted to material growth. Thus, many studies of urban metabolism seem concerned mainly with alternative technologies and better resource management that would enable cities to maintain their growth trajectories (e.g., Brunner, 2007). Kennedy, Cuddihy, and Engel-Yan (2007) acknowledge that energy, material, and water throughput is increasing with modernization, even on a per capita basis; therefore, we should expect greater loss of ecosystem function and biodiversity, and, *if necessary*, urban policy-makers might consider strategies to slow exploitation. However, the authors convey no particular sense of urgency. Similarly, Decker et al. (2000) recognize the accelerating degradation of earth systems, but imply a fairly smooth succession to the point where 'modern megacities will . . . begin to climax when global fossil fuel reserves are exhausted and global water and food resources are maximally utilized'. Few academic studies acknowledge the possibility of implosion or urge the kind of dramatic response to prevent it that now seems justified by global change science.

This chapter attempts to fill the gap. I argue that we are witnessing the dissipative destruction of essential ecosystems and vital life-support functions and that the process is accelerating with population growth and rising material expectations. Given the increasing probability of severe climate change, resource shortages, large-scale population displacements, and resultant geopolitical chaos, the world is justified in taking decisive action now. The 'transition' to sustainability will be anything but smooth and predictable if there is further delay—urban civilization itself is at stake.

Environmental scientists are sometimes dismissed as purveyors of gloom and doom but, if the science is correct, it is our present compass-setting that leads towards the abyss. It need not continue in that direction. Humans are an intelligent species uniquely capable of forward planning. Properly warned, we should be able to manage our way out of the crisis.

As a first step, society must acknowledge that (un)sustainability represents both gross market failure and the greatest collective problem humankind has ever faced. All of us are part of the same ecosystemic hierarchy in which all sub-systems (e.g., cities and nation-states) are dependent on the operational integrity of the whole (the ecosphere itself) for survival. For perhaps the first time in history, individual self-interest now coincides with humanity's collective interest on the level of basic survival.

In these circumstances, government intervention in the economy for the common good is both necessary and justifiable. Indeed, achieving global sustainability will require a concerted *inter*governmental program including global population control, the development of more satisfying but less material-intense lifestyles, and reshaping our cities in the image of natural ecosystems. Rich countries, at least, will have to abandon the idea of continuous economic growth and learn to share the economic and ecological output of the planet more equitably. The immediate material goal of this endeavour should be to reduce gross consumption and waste production to match the regenerative capacity of the ecosphere. The ultimate objective is to create a positive future for all, characterized by ecological stability, greater social equity, and enhanced economic security. Recognition that no person, city, or nation can be sustainable on its own should be an incentive for collective action.

The transition to an ecologically sustainable urban society is a 100-year project and there is no excuse not to begin immediately. Certainly, the problems of both developed and developing world cities are well documented and many partial solutions have been proposed (e.g., Marcotullio and McGranahan, 2007; Martine et al., 2008; Satterthwaite, 1999). Some of the best and most accessible handbooks for urban sustainability are explicitly based on treating cities as true ecosystems (Register, 2006; Newman and Jennings, 2008). The only question is whether our growth-addicted global culture is capable of acknowledging and responding to the challenge in time. *Homo sapiens* are facing their greatest test of whether they are truly a rational species. Humanity can continue to thrive, but only if urban civilization adapts purposefully to living within the carrying capacity of the earth.

Review Questions

1. Why are cities not functioning as true ecosystems?

2. What are the implications of the ecological-footprint perspective in terms of the global environmental impacts of cities, and in terms of the environmental vulnerability of cities?

3. What can be done to reduce the ecological footprint of cities?

4. What are the main environmental threats on the horizon for cities?

Notes

1. One paper in the March 2009 issue of the journal even struggles with the question of whether 'the ecosystem concept [is] relevant when humans are part of the system' (Pickett and Grove, 2009). The remaining eight articles focus on other aspects of non-human ecology.

2. Since livestock feedlots are a sub-system of the human urban industrial system, it is not surprising that they are similar to cities in eco-structure.

3. For full details of the method, including inclusions, exceptions, and limitations, see Rees (2003, 2006), WWF (2008), and various links at <www.footprint network.org/en/index.php/GFN/>.

4. EFA obviously does not capture the entire human impact on earth, only those dimensions for which the ecosphere has regenerative capacity. For example, various wastes such as ozone-depleting chemicals or the toxic chemical residues accumulating in our food chain cannot be converted into a corresponding ecosystem area.

5. To enable fair comparisons among countries, the national EF and biocapacity data in Table 5.1 are presented in terms of global hectares, i.e., the equivalent area of ecosystems of global average productivity.

6. The area of Japan is only about 37,770,000 ha but Japan's terrestrial ecosystems are considerably more productive than the world average, which increases the country's biocapacity to almost 77,000,000 gha.

7. This situation is complicated by the diversion of grain, especially maize, to biofuel production.

8. Cuba, South Africa, and Thailand are countries with per capita eco-footprints in the sustainable range.

9. Many North Americans fear density, but the City of Vancouver illustrates that excellent urban design can actually draw families from the suburbs to live in high-density urban communities. In 2008, the city actually had to restrict further high-rise development in its high-amenity downtown core because it was displacing office development.

10. For a history and philosophy of the bioregional movement, see Carr (2005). Also see Bain, Chapter 15 and Lynch and Ley, Chapter 19.

11. Note that as the human population increases and productive ecosystems are degraded, this 'equitable earth-share' will decline.

References

Anderson, K., and A. Bows. 2008. 'Reframing the climate change challenge in light of post-2000 emission trends', *Philosophical Transactions of the Royal Society A* 266: 3863–82. At: <http://rsta.royalsocietypublishing.org/content/366/1882/3863.abstract>. (June 2009)

Association for the Study of Peak Oil and Gas (ASPO). 2008. *Newsletter* No. 91.

Brunner, P.H. 2007. 'Reshaping urban metabolism', *Journal of Industrial Ecology* 11, 2: 11–13.

Business Council for Sustainable Development (BCSD). 1993. *Getting Eco-Efficient*. Report of the BCSD First Antwerp Eco-Efficiency Workshop, Nov. 1993. Geneva: BCSD.

Campbell, C.C. 1999. *The Imminent Peak of World Oil Production*. At: <www.hubbertpeak.com/campbell/commons.htm>. (May 2009)

Carr, M. 2005. *Bioregionalism and Civil Society: Democratic Challenges to Corporate Globalism*. Vancouver: University of British Columbia Press.

Cassman, K.G. 1999. 'Ecological intensification of cereal production systems: Yield potential, soil quality, and precision agriculture', *Proceedings of the National Academy of Science* 96: 5952–9.

Center for Strategic and International Studies (CSIS). 2007. *The Age of Consequences: The Foreign Policy and National Security Implications of Climate Change*. Washington: CSIS. At: <www.csis.org/media/csis/pubs/071105_ageof consequences.pdf>. (May 2009)

Conforti, P., and M. Giampietro. 1997. 'Fossil energy use in agriculture: An international comparison', *Agriculture, Ecosystems and Environment* 65: 231–43.

Decker, E.H., S. Elliott, F.A. Smith, D.R. Blake, and F.S. Rowland. 2000. 'Energy and material flow through the urban ecosystem', *Annual Review, Energy and Environment* 25: 685–740.

Diamond, J. 1997. *Guns, Germs, and Steel: The Fates of Human Societies*. New York: Norton.

Douglas, I. 1981. 'The city as an ecosystem', *Progress in Physical Geography* 5: 315–67.

Duncan, R.C., and W. Youngquist. 1999. 'Encircling the peak of world oil production', *Natural Resources Research* 8: 219–32.

Federation of Canadian Municipalities (FCM). 2005. *Ecological Footprints of Canadian Municipalities and Regions*. Edmonton: Report for the Federation of Canadian Municipalities prepared by Anielski Management. At: <www.anielski.com/Documents/EFA%20Report%20 FINAL%20Feb%202.pdf>. (May 2009)

Folke, C., A. Jansson, J. Larsson, and R. Costanza. 1997. 'Ecosystem appropriation by cities', *Ambio* 26: 167–72.

Food and Agriculture Organization (FAO). 2000. *Land Resource Potential and Constraints at Regional and Country Levels*. Rome: Land and Water Development Division, Food and Agriculture Organization of the United Nations.

Girardet, H. 2004. 'The Metabolism of Cities', in S. Wheeler and T. Beatley, eds, *Sustainable Urban Development Reader*. London: Routledge.

Haberl, H., M. Fischer-Kowalski, J. Krausmann, H. Weisz, and V. Winiwarter. 2004. 'Progress toward sustainability? What the conceptual framework of material and energy flow accounting (MEFA) can offer', *Land Use Policy* 21: 199–213.

Hansen, J., M. Sato, P. Kharecha, D. Beerling, R. Berner, V. Masson-Delmotte, M. Pagani, M. Raymo, D.L. Royer, and J.C. Zachos. 2008. 'Target atmospheric CO_2: Where should humanity aim?', *The Open Atmospheric Science Journal* 2: 217–31.

Kennedy, C., J. Cuddihy, and J. Engel-Yan. 2007. 'The changing metabolism of cities', *Journal of Industrial Ecology* 11, 2: 43–59.

Kissinger, M., and W.E. Rees. 2009. 'Footprints on the prairies: Degradation and sustainability of Canadian agricultural land in a globalizing world', *Ecological Economics* 68: 2309–15.

Laherrere J. 2003. 'Forecast of Oil and Gas Supply to 2050', paper presented to Petrotech 2003, New Delhi. At: <www.hubbertpeak.com/laherrere/Petrotech 090103.pdf>. (May 2009)

Lenzen, M., C. Dey, and F. Barney. 2004. 'Energy requirements of Sydney households', *Ecological Economics* 49: 375–99.

McGranahan, G., and D. Satterthwaite. 2003 'Urban centres: An assessment of urban sustainability', *Annual Review of Environmental Resources* 28: 243–74.

Marcotullio, P., and G. McGranahan, eds. 2007. *Scaling Urban Environmental Challenges—From Local to Global and Back*. London: Earthscan.

Martine, G., G. McGranahan, M. Montgomery, and R. Fernández-Castilla, eds. 2008. *The New Global Frontier—Urbanization, Poverty and Environment in the 21st Century*. London: Earthscan.

Millennium Ecosystem Assessment (MEA). 2005. 'Living beyond our means: Natural assets and human well-being (Statement from the Board)', *Millennium Ecosystem Assessment*. At: <www.millenniumassessment.org/documents/document.429.aspx.pdf>. (May 2009)

Newman, P., and I. Jennings. 2008. *Cities as Sustainable Ecosystems*. Washington: Island Press.

Pickett, S.T.A., and J.M. Grove. 2009. 'Urban ecosystems: What would Tansley do?', *Urban Ecosystems* 12: 1–8.

Rees, W.E. 1992. 'Ecological footprints and appropriated carrying capacity: What urban economics leaves out', *Environment and Urbanization* 4: 120–30.

———. 1997. 'Is "Sustainable City" an oxymoron?' *Local Environment* 2: 303–10.

————. 1999. 'The built environment and the ecosphere: A global perspective', *Building Research and Information* 27: 206–20.

————. 2000. 'Patch disturbance, eco-footprints, and biological integrity: Revisiting the Limits to Growth', in D. Pimentel, L. Westra, and R. Noss, eds, *Ecological Integrity: Integrating Environment, Conservation and Health*. Washington: Island Press.

————. 2001. 'Ecological footprint, concept of', in Simon Levin, ed., *Encyclopedia of Biodiversity* 2: 229–44. San Diego, Calif.: Academic Press.

————. 2003. 'Understanding urban ecosystems: An ecological economics perspective', in A. Berkowitz et al., eds, *Understanding Urban Ecosystems*. New York: Springer-Verlag.

————. 2006. 'Ecological footprints and biocapacity: Essential elements in sustainability assessment', in J. Dewulf and H. Van Langenhove, eds, *Renewables-Based Technology: Sustainability Assessment*. Chichester, UK: John Wiley and Sons.

————. 2008. 'Human nature, eco-footprints and environmental injustice', *Local Environment: The International Journal of Justice and Sustainability* 13: 685–701.

————. 2009. 'The ecological crisis and self-delusion: Implications for the building sector', *Building Research and Information* 37: 300–11.

Register, R. 2006. *EcoCities: Rebuilding Cities in Balance with Nature*, rev. edn. Gabriola Island, BC: New Society. See also: 'An interview with Richard Register, author of *Ecocities: Building Cities in Balance with Nature*'. At: <www.sustainablecityblog.com/2009/03/richard-register-interview/>. (June 2009)

————. 2009. 'Cities can save the earth', *Foreign Policy in Focus*, 12 May. At: <www.fpif.org/fpiftxt/6113>.

Satterthwaite, D., ed. 1999. *Sustainable Cities*. London: Earthscan.

Schwartz, P., and D. Randall. 2003. *An Abrupt Climate Change Scenario and Its Implications for United States National Security*. Washington: A report commissioned by the US Defense Department.

Sustainable Development Information Service (SDIS). 2004. *Disappearing Land: Soil Degradation*. Washington: SDIS, Global Trends, World Resources Institute.

Union of Concerned Scientists (UCS). 1992. *World Scientists' Warning to Humanity*. At: <www.ucsusa.org/about/1992-world-scientists.html>. (May 2009)

United Nations Development Program (UNDP). 2007. *Human Development Report 2007*. New York: UN.

Van der Ryn, S., and P. Calthorpe. 1986. *Sustainable Communities: A New Synthesis for Cities and Towns*. San Francisco: Sierra Club Books.

Victor, P. 2008. *Managing without Growth: Slower by Design, Not Disaster*. Cheltenham, UK: Edward Elgar.

Vince, G. 2009. 'Surviving in a warmer world', *New Scientist* 201, 2697: 29–33.

Wackernagel, M., and W.E. Rees. 1996. *Our Ecological Footprint: Reducing Human Impact on the Earth*. Gabriola Island, BC: New Society.

Walker, L., and W.E. Rees. 1997. 'Urban density and ecological footprints: An analysis of Canadian households', in M. Roseland, ed., *Ecocity Dimensions*. Gabriola Island, BC: New Society.

Warren-Rhodes, K., and A. Koenig 2001. 'Ecosystem appropriation by Hong Kong and its implications for sustainable development', *Ecological Economics* 39: 347–59.

Weizsäcker, E.V., L.H. Lovins, and A.B. Lovins. 1997 [1995]. *Factor Four: Doubling Wealth—Having Resource Use: A Report to the Club of Rome*. London: Earthscan/James and James.

World Wildlife Fund (WWF). 2008. *Living Planet Report 2008*. Gland, Switzerland: World Wide Fund for Nature.

CHAPTER 6

The Dynamics of Economic Change in Canadian Cities: Innovation, Culture, and the Emergence of a Knowledge-Based Economy

Tara Vinodrai

Cities—despite their enduring materiality—are dynamic and ever-changing, responding to and being shaped by economic, social, political, and cultural forces. While there is no question that cities are important social, cultural, and political spaces, it is their key role as *economic spaces* that is the focus of this chapter. While cities have always been important economic spaces, the precise nature of their economic roles and activities has evolved over time. In the past three decades, cities in advanced capitalist societies have undergone significant deindustrialization and economic restructuring marked by the growth of a (highly variegated) service sector, the increasing importance of a highly educated (and sometimes highly mobile) workforce valued for their embodied knowledge and capacity to innovate, as well as a decline in 'blue collar' jobs. These economic changes have often been accompanied by labour market disorganization, characterized by declining private-sector unionization and the transformation of employment relations, including the increase of flexible, part-time, temporary, and precarious work. It is common for scholars, policy-makers, and the popular and business press to refer to the **new economy** or the **knowledge-based economy**, or—more controversially—the 'creative economy'; these terms become shorthand to describe the contemporary economy. In this chapter, I examine how these changes have been experienced in Canadian cities.

I begin by briefly exploring the context in which the economic landscapes of Canadian cities have changed over the past three decades. The chapter proceeds to document some key socio-economic transformations in the Canadian economy since the 1970s. The third section more closely examines three key dynamics that characterize the economic transitions in Canadian cities: the changing landscape of manufacturing; the transformation of urban economies towards more innovation-based and knowledge-intensive forms of economic activity; and the ascendance of creative and cultural activities. However, these changes have not affected all Canadian cities in the same way. Thus, while the broader shift from a natural resource-based or manufacturing-based economy to one based on tertiary and quaternary-order service activities generally holds true across Canadian cities, the contours of this change are path-dependent (that is, determined by historical contexts) and vary from city to city based on their unique local histories, institutions, politics, economic development strategies, and regional contexts. Therefore, we pay attention to how different Canadian cities have developed and adapted to the conditions of the new economy.

Beyond Fordism: Towards a Post-Industrial, Knowledge-Based Economy

Before looking at the specific changes that have occurred in Canadian cities, it is important to understand how advanced capitalist societies have evolved over the past several decades. There is general agreement that the nature of production and its associated labour market structures, employment relations, and forms of **governance** have evolved away from those dominating the post-World War II period, characterized as **Fordism** (Amin, 1994; Bryson and Henry, 2005; Storper and Scott, 1992). In the narrowest sense, Fordism is associated with a set of production practices pioneered by Henry Ford and applied to the automotive manufacturing industry in the early twentieth century. These practices included a detailed division of labour resulting in simplified, deskilled tasks requiring minimal job training; a high degree of labour control; highly standardized and routinized manufacturing; and dedicated machinery organized in an assembly line model allowing for mass production. However, the ascendance and widespread adoption of this organizational form rested not only on the production practices themselves but on an accompanying set of institutional and organizational practices that enabled their success. Thus, 'Fordism' is understood more broadly by social scientists as an economy-wide set of institutional architectures that sustained production and promoted economic growth in North America and Europe from the end of World War II until the early 1970s. In other words, what the French Regulation School refers to as the 'regime of accumulation' (in this case, Fordist mass production) requires a complementary 'mode of regulation' (Boyer, 2005; Elam, 1994; Jenson, 1990). In this case, most advanced, capitalist countries found an appropriate regulatory apparatus in the form of the Keynesian **welfare state** articulated through

strong national state intervention, regulation and control, and social welfare provision, although the contours of this model varied across nation states (Boyer, 2005; Tickell and Peck, 1992). Jenson (1990) has argued that Canada's particular form of Fordism, what she refers to as 'permeable Fordism', was produced through a set of historical processes and negotiated between government, business, and the organized labour movement beginning in the interwar period; it was designed to deliver higher wages, full employment, minimal labour unrest, as well as a set of social programs to support social reproduction. Alongside a leading role for a redistributive welfare state during the post-war period, other institutional forms supported the reproduction and sustainability of the broader economic system. However, in the early 1970s, it became apparent that these particular institutional arrangements were increasingly strained across advanced, capitalist economies as profit levels declined, global competition intensified, and broader economic restructuring resulting from rising oil prices placed the system in crisis (Harvey, 1990).

In seeking to characterize the subsequent developments of the capitalist system, scholars have referred to the coming of post-industrial (Bell, 1973) or risk society (Beck, 2000), **post-Fordism** or after-Fordism (see Amin, 1994), an era of flexible specialization (Piore and Sabel, 1984) or flexible accumulation (Harvey, 1990), and—most recently—an emerging form of cognitive-cultural capitalism (Scott, 2007).[1] Despite differences among social scientists in terms of the language and theoretical apparatus used to describe the nature of contemporary capitalism, there is widespread agreement that the economy has changed in both quantitative and qualitative terms over the past several decades, particularly in the types of industries leading growth and in the forms of work associated with this paradigm. The most striking shift has been the tremendous growth of service-based industries, which can be attributed to the introduction of new technologies allowing for

significant automation, the outsourcing and relocation of production-oriented activities to offshore
locations, and changing patterns of consumption
including demand for improved and differentiated
services, as well as differentiated goods that require
more creative and service-based inputs in their
production and distribution. Scott (2007: 1466)
notes that:

> much of productive activity today involves
> digital technologies and flexible organization
> sustaining the expansion of sectors that thrive
> on innovation, product diversity and the pro-
> vision of personalized services. . . . Labor pro-
> cesses have come to depend more and more
> on intellectual and affective human assets (at
> both high and low levels of remuneration), and
> are increasingly less focused on bluntly routin-
> ized mental or manual forms of work.

He identifies technology-intensive manufacturing;
services; fashion-oriented, neo-artisanal production; and cultural-products industries as the key
drivers of growth and innovation in the contemporary economy.

The question remains: what role do cities
play in this new economy? Observers identify two
competing trends that raise questions about the
continued vitality of cities in an age of globalization. On the one hand, transportation and communication technologies have allowed for shifts in
terms of the locational choices of firms, industries,
and—increasingly—workers. In this view, economic activities are no longer tied to particular
locations; capital and labour are increasingly foot-
loose—able to relocate to any location. This has
led to pronouncements of the 'death of distance',
the 'end of geography', and the emergence of a 'flat
world' (Cairncross, 1997; Friedman, 2005). On the
other hand, scholars have argued that place—and
especially cities—has become even more important in the contemporary economy (Morgan, 2004).
Despite technological advances in transportation

and communications, the advent of the Internet,
and—more recently—the rise of Web 2.0 tools
that reduce the costs and barriers to interaction
over space and time, many of the industries critical
to producing the content, tools, and infrastructure
of this new economy are agglomerated in particular *urban* locations. In other words, economic activity—particularly the forms favoured by the new
economy—remains spatially concentrated in cities.

It is this spatial agglomeration—what Michael
Porter (1998) calls 'clustering'—that enables firms
involved in similar activities to draw on shared
local resources and collective infrastructure, reduce
transaction costs, and gain other efficiencies. Firms
benefit from local access to deep pools of specialized, skilled labour; proximity to suppliers, service
providers, and other related and supporting industries; and closeness to sophisticated customers who
provide demand-side impetus for developing and
improving products and services, although this latter characteristic remains contested (Wolfe and
Gertler, 2004). In addition to these traditional
agglomeration economies or externalities, firms
benefit from knowledge spillovers, the ability to
monitor their competition, and access to local institutional supports. In other words, agglomeration
enables learning, knowledge flows, co-operation
and competition (Maskell and Malmberg, 1999;
Wolfe and Gertler, 2004). Of course, firms are
connected to other places through complex divisions of labour, supply and commodity chains, and
markets and firms access knowledge through both
local and non-local partners and networks (Bathelt,
Malmberg, and Maskell, 2004; Gertler, 2008). However, there are compelling reasons why firms often
remain agglomerated in particular cities. Most fundamental amongst these—from an economic perspective—is the growing importance of knowledge
and learning to innovation and the creation of economic value in the contemporary era.

Cities provide many of the necessary conditions to support the innovation, learning, and
knowledge generation and circulation processes

integral to the development of dynamic clusters or specialized agglomerations of activity. Moreover, scholars increasingly recognize that learning and innovation—critical for firms' competitiveness—are inherently *social* processes that rely on interaction between different economic actors such as firms, industry, and professional associations, universities, government, and private research and development laboratories, technology transfer offices, unions, venture capitalists, and other intermediaries (Gertler, 2001; Wolfe and Bramwell, 2008). While these interactions can be in person (e.g., meetings, conferences, trade shows, other events) or electronically mediated (e.g., email, videoconferencing, various Web 2.0 platforms, virtual environments like Second Life), there is agreement that the primarily *tacit* nature of a lot of knowledge demands that some of the most important interactions occur face-to-face in order for understanding, learning, and the sharing of practice and ideas to take place (Gertler, 2003, 2008; Storper and Venables, 2004). Cities provide an ideal environment where such interactions are easily made possible and are more likely to occur since codes, norms and values are shared among actors.

Human capital, particularly highly skilled labour—referred to as knowledge workers, talent, the 'creative class' (Florida, 2002), or the 'new class' (Gouldner, 1979)—should not be overlooked as an important embodied input into the innovation process. These workers often provide specialized skills, creativity, ideas, and know-how to knowledge-intensive production and the innovation process. Again, cities have been increasingly described as the preferred location for these workers. An emerging view most clearly articulated in Richard Florida's (2002) controversial work on the 'creative class' suggests that highly skilled workers are attracted to cities, but not just any cities; those cities that have high-quality social environments, low barriers to entry, and are diverse and open to differences will be particularly attractive (for critiques, see Markusen, 2006; Peck, 2005; Storper

and Scott, 2009). One significant implication arising from this work is the argument that economic development should shift focus away from creating a favourable climate for business towards one favourable to attracting people, specifically the 'creative class'.

Finally, one important characteristic of the contemporary economy is that competition increasingly rests on intangibles (such as symbolic, design, and aesthetic content and inputs) rather than only on tangible assets in the production of goods and services (Lash and Urry, 1994; Scott, 2001). Thus, one group of industries are assuming a prominent role: cultural products industries, which include activities related to film and television production, music recording, book publishing, video game production, live theatre and other entertainment, and so on. Studies consistently demonstrate that these industries have high levels of creative content, are engaged in a constant search for novelty, and are often susceptible to rapid shifts in consumer demands, necessitating higher rates of innovation and easy access to information about changing tastes and styles (Power and Scott, 2004; Scott, 2004). Therefore, the agglomeration economies described above are of paramount importance; moreover, these creative and cultural activities tend to agglomerate in *major urban centres* for reasons outlined below (Pratt, 1997; Scott, 2001). First, firms need constant access to cutting-edge knowledge and pools of highly skilled labour. Second, workers benefit from being able to develop their careers while living in 'cool', diverse urban environments and neighbourhoods, although this can lead to polarizing, **gentrification** dynamics (Ley, 1996; Zukin, 1982; see Walks, Chapter 10; Bain, Chapter 15). Third, these urban environments are said to provide inspiration (Lloyd 2006) and, in turn, elements of place become embedded in outputs (Molotch, 2002; Rantisi, 2004). Fourth, this type of work often involves flexible organizational forms such as project-based work, which expose workers (despite their professional status) to high levels of individual risk through self-employment, contract-based work, and freelancing (Ekinsmyth, 2002; Grabher, 2002; Vinodrai, 2006). Thus, the nature of the industries and of work itself requires well-developed (local) social and knowledge networks to tune into the necessary background 'noise' about jobs, projects, styles, and leading-edge developments (Grabher, 2004)—all of which are most readily available in large, diverse cities.

The above discussion foregrounds how broader economic changes (re)shape urban economies and labour markets and highlights how cities have become even more important sites of economic activity in the contemporary era. The remainder of this chapter documents how Canada's urban economies have changed since the crisis of Fordism and the rise of a knowledge-based economy, marked by the dominance of new forms of work, industrial activity, and technologies.

The Rise of Canada's Post-Industrial Economy?

To put an empirical face to this far more abstract discussion, we examine how Canada's economic landscape has changed since the early 1970s. Specifically, we focus on the significant shifts in the relative importance of different industries, as well as changes in the characteristics of the Canadian labour force. To emphasize the shift in the types of industries that have gained prominence, we distinguish between goods- and services-producing industries. Goods-producing industries include agriculture, forestry, fishing, mining, oil and gas, utilities, construction, and manufacturing. Services-producing industries include retail, business, personal care, health care, education and public administration. As is evident from this list, the services sector is incredibly diverse, entailing many types of work and activities (Bryson, Danielson, and Warf, 2004; Harrington and Daniels, 2006; Marshall and Wood, 1992). Moreover, the cultural

products industries identified earlier cannot be easily classified as either goods or services in the traditional sense (Scott, 2001, 2007). Figure 6.1 shows the relative contributions of the goods- and services-producing sectors to Canadian employment between 1976 and 2008. It demonstrates that the growth of the Canadian economy has rested on the services-producing industries. In 1976, goods-producing industries accounted for 34.5 per cent of Canadian employment; by 2008, this had declined to 23.4 per cent. However, employment in goods-producing industries did not decline in absolute terms, suggesting that this remains an important sector. Table 6.1 examines this dynamic more closely by comparing employment gains and losses across 16 sectors of the Canadian economy between 1976 and 2008. Over 90 per cent of the employment gains were in the services-producing sector and the highest annual growth rates were in the professional, scientific, and technical services (5.0 per cent) and the business, building, and other support services (4.6 per cent); followed by accommodation and food services (3.0 per cent); health care and social assistance (2.8 per cent); and information, culture and recreation industries (2.5 per cent). Agriculture was the only industry to experience an overall decline in employment. Manufacturing, a key part of the Canadian economy during the post-war period, had a very low growth rate (0.2 per cent). However, these broad shifts in industrial structure have not been experienced evenly across the Canadian urban system, as we shall see later in this chapter.

Alongside shifts in the industrial structure of the Canadian economy have been parallel changes in the characteristics of the workforce. Figure 6.2 reveals that the proportion of the Canadian workforce with a bachelors degree or higher almost tripled between 1976 and 2007, increasing from 7.2 per cent to 19.4 per cent; levels of human capital (as measured by formal education levels) in Canada have increased steadily since the mid-1970s. During this same period, the gap in

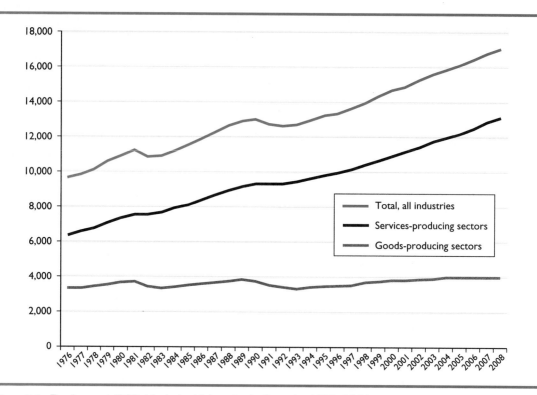

Figure 6.1 Employment (000s) by industrial sector in Canada, 1976–2008
Source: Statistics Canada, *Labour Force Survey Historical Review, 1976–2008* (author's calculations).

labour force participation rates among men and women has narrowed considerably (Figure 6.3). In 1976, there was a 32 per cent difference between the participation rates of males (77.7 per cent) and females (45.7 per cent). By 2007, there was only a 10 per cent difference in the participation rates of males (72.7 per cent) and females (62.7 per cent), marking a convergence in the participation rates of men and women and a significant feminization of the workforce in the post-industrial era. Figure 6.4 demonstrates that job tenure (the average length of time that a worker holds a particular job) has fluctuated over the past several decades. While easy interpretation is confounded by factors such as the aging of the workforce (reflected in rising average job tenure from the mid-1970s to the mid-1990s) and the arrival of new entrants into the workforce

(i.e., increased participation by women, new immigrants), there are some telling signs that these fluctuations are not only related to socio-demographic shifts. The flattening of average job tenure in the late 1980s and early 1990s reflects the economic recession that resulted in many workers being laid off and losing their jobs. The steady decrease in job tenure beginning in the mid-1990s reflects the more flexible and precarious nature of work in emerging industries, as well as the restructuring of work in existing industries to include the greater prevalence of temporary, contract, and part-time work, as well as higher levels of self-employment (Benner, 2002; Christopherson, 2002; Vosko, 2006).

Finally, Table 6.2 shows how Canada's occupational structure has changed between 1971 and 2006. While the overall labour force grew at a rate

Table 6.1 Employment (000s) by Industry in Canada, 1976–2008

	1976	2008	Change	% Change	Annual Growth Rate
Goods-producing sector	*3,371*	*4,021*	*650*	*19.3*	*0.6%*
Agriculture	464	327	(137)	−29.5	−1.1%
Forestry, fishing, mining, oil and gas	255	340	85	33.5	0.9%
Utilities	110	152	42	38.3	1.0%
Construction	682	1,232	551	80.8	1.9%
Manufacturing	1,861	1,970	109	5.9	0.2%
Services-producing sector	*6,377*	*13,105*	*6,728*	*105.5*	*2.3%*
Trade	1,572	2,679	1,107	70.5	1.7%
Transportation and warehousing	563	858	295	52.3	1.3%
Finance, insurance, real estate, and leasing	526	1,075	549	104.4	2.3%
Professional, scientific, and technical services	253	1,200	947	375.1	5.0%
Business, building, and other support services	161	687	525	325.3	4.6%
Educational services	677	1,193	516	76.3	1.8%
Health care and social assistance	794	1,903	1,110	139.8	2.8%
Information, culture, and recreation	347	760	413	119.1	2.5%
Accommodation and food services	413	1,074	660	159.7	3.0%
Other services	427	751	324	75.9	1.8%
Public administration	645	926	281	43.6	1.1%
Total, all industries	*9,748*	*17,126*		*75.7*	*1.8%*

Source: Statistics Canada, *Labour Force Survey Historical Review, 1976–2008* (author's calculations).

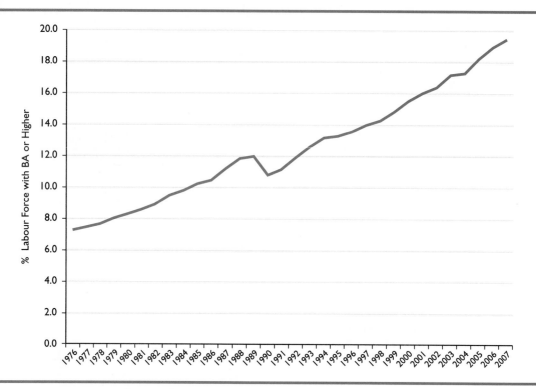

Figure 6.2 Proportion of labour force with a university degree or higher, 1976–2007

Note: The small discrepancy in the levels of university degree holders in the pre-1990 and post-1990 data can be attributed to the introduction of new data classification standards in 1990 for the Labour Force Survey.

Source: Statistics Canada, *Labour Force Survey Historical Review, 1976–2007* (author's calculations).

of 2 per cent annually over this period, there is variation among occupational groups. Growth rates were higher than the national average in management (4.4 per cent), professional (4.1 per cent), medicine and health (3.2 per cent), business and finance (2.3 per cent), and sales and services (2.4 per cent) occupations. Barring the latter category, these occupational groups are the building blocks of what Richard Florida (2002) describes as the 'creative class' and provide empirical confirmation of the increasing professionalization of the workforce over the past several decades. It is worth noting that while sales and services occupations did not grow as fast as the occupations typically identified as 'knowledge-intensive' or

'creativity-intensive', employment in this category (which includes caregiving, hairdressing, cleaning, groundskeeping, and security) grew the most in absolute terms and accounted for just over one-quarter of the employment gains between 1971 and 2006. Work in these areas, while essential to the (re)production of cities, typically offers lower wages, is highly feminized, requires fewer credentials, and exposes workers to higher levels of job insecurity (Cervenan, 2009; Vosko, 2006). This is a worrying trend that reveals the paradox that the very economic activities associated with the knowledge-based economy may also lead to rising inequality and social exclusion (Gertler, 2001; Scott, 2007; Storper and Scott, 2009).

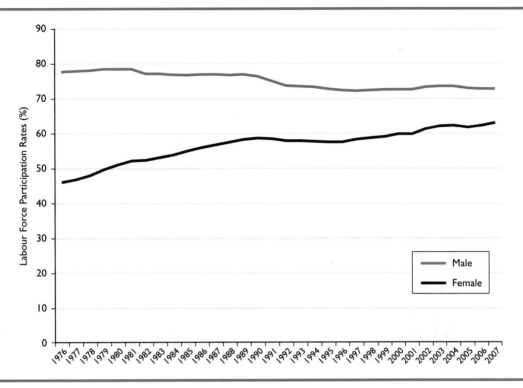

Figure 6.3 Participation rates in the Canadian labour force by gender, 1976–2007

Source: Statistics Canada, *Labour Force Survey Historical Review, 1976–2007* (author's calculations).

The occupation groups that experienced decline were primary (–13.3 per cent) and processing and machining (–5.1 per cent) occupations, which are most closely associated with the primary and manufacturing industries, reconfirming the trend towards deindustrialization.

Towards a Knowledge-Based Economy? The Economic Transformation of Canadian Cities

We must understand economic change in Canadian cities in the context of these broader shifts in the Canadian economy. In this section, we examine how deindustrialization and the growth of the knowledge-based economy have

occurred unevenly across Canada's urban system (on resulting urban decline in some cases, see Donald and Hall, Chapter 16). We focus on three significant changes that epitomize this economic transition. First, given the wrenching deindustrialization experienced by many Canadian cities, we examine the evolving manufacturing landscape, paying attention to Ontario's cities where these changes have been most acute. Second, given the increasing emphasis on innovation and knowledge to the growth of cities, we explore the economic development paths taken by different Canadian cities in responding to these changes. Finally, given the increased attention being paid to culture and creativity in cities, we examine the cultural industries, paying special attention to Canada's three largest urban centres where these activities are more

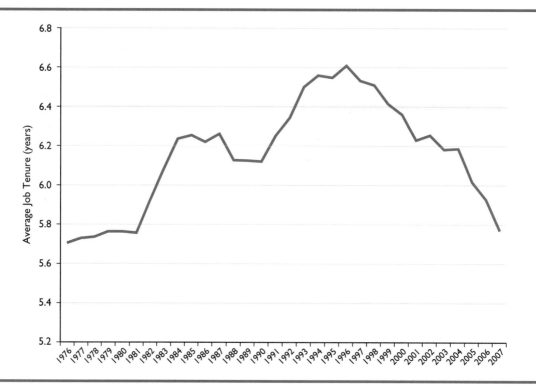

Figure 6.4 Average job tenure in years of the Canadian labour force, 1976–2007

Source: Statistics Canada, *Labour Force Survey Historical Review, 1976–2007* (author's calculations).

prominent and well developed compared to other Canadian cities.

Evolving Manufacturing Landscapes

While the above discussion highlighted the decline in the relative importance of manufacturing to the Canadian economy, this has not affected all Canadian cities in the same way. Table 6.3 shows that, in 1971, manufacturing accounted for more than a third of employment in Brantford (42.7 per cent), Kitchener (41.5 per cent), Oshawa (41.5 per cent), Hamilton (36.8 per cent), Windsor (35.6 per cent), and St Catharines–Niagara (35.5 per cent). By 2006, these percentages had dropped to 21.6 per cent in Brantford, 23.1 per cent in Kitchener, 15.5 per cent in Oshawa, 16.2 per cent in Hamilton, 24.5 per cent in Windsor, and 13.8

per cent in St Catharines–Niagara. These dramatic changes in the relative importance of manufacturing reflect large employment losses primarily experienced in southern Ontario and Quebec, the historical manufacturing centres of Canada. For example, between 1971 and 2006, manufacturing employment declined by 13,700 in St Catharines–Niagara and 14,800 in Hamilton. This can be attributed to significant restructuring in the dominant industries in these cities: automotive and steel (Warrian and Mulhern, 2009). A similar portrait of the dramatic reconstitution of Ontario's cities emerges when we look at the occupations most closely associated with the production side of manufacturing: processing and machining occupations (considered the core of blue-collar work). Table 6.4 shows that, between 1971 and 2006, processing and

Table 6.2 **Employment (000s) by Occupation in Canada, 1971–2006**

	1971	2006	Change	% Change	Annual Growth Rate
Managerial occupations	372	1,632	1,259	338.4	4.4%
Professional occupations	767	3,025	2,258	294.6	4.1%
Medicine and health-related occupations	327	950	624	191.0	3.2%
Business, finance, and clerical occupations	1,374	3,025	1,652	120.3	2.3%
Sales and service occupations	1,786	4,038	2,252	126.1	2.4%
Primary occupations	666	648	–17	–2.6	–0.1%
Processing and machining occupations	1,210	993	–217	–18.0	–0.6%
Construction, trades, transport-equipment operating occupations	2,127	2,550	424	19.9	0.5%
Total, all occupations	8,627	16,861	8,234	95.4	2.0%

Source: Statistics Canada, *Canadian Census of Population, 1971 and 2006* (author's calculations).

machining occupations declined in their relative importance across Canadian cities, yet these shifts were felt unevenly across Canadian cities. Perhaps most striking are the differences between Montreal and Toronto. Toronto experienced a slight absolute increase in the number of processing and machining jobs (13,075) even though, by 2006, workers in these occupations accounted for a smaller proportion of employment, declining from 14.6 per cent to 7.1 per cent. However, Montreal experienced a sharp absolute decline in this type of employment (62,940) leading to a decline from 15.7 per cent to 5.5 per cent in the proportion of the labour force accounted for by processing and machining occupations. This reflects a significant restructuring of Montreal's labour market beginning in the 1970s. Elsewhere, places such as St Catharines–Niagara and Hamilton saw employment in processing and machining occupations decrease by 15,045 and 18,580 respectively.

However, manufacturing remains important in several Canadian cities, and—in some cases—deindustrialization has not been an important dynamic of local economic change. For example, western Canadian cities like Vancouver and Calgary were never major manufacturing sites and their economies are more connected to extractive and natural resource-based industries. And even though cities like Toronto and Montreal have witnessed significant changes in their manufacturing base, manufacturing industries still account for 14.4 per cent of employment in both cities, including manufacturing in areas such as biotechnology, pharmaceuticals, medical equipment and assistive technologies (MAT), and other life sciences. The life sciences provide an interesting case of the evolution of manufacturing, including the strong links between high technology manufacturing and related services. Toronto and Montreal both have firms engaged in a range of life science activities,

Table 6.3 Manufacturing Employment in Canadian Cities, 1971–2006

	1971		2006		1971–2006	
	Local Employment	% Local	Local Employment	% Local	Local Employment	% Local
Toronto	315,570	27.3	394,882	14.4	79,312	(12.9)
Montreal	276,770	28.2	274,382	14.4	(2,388)	(13.9)
Vancouver	78,750	18.0	109,550	9.6	30,800	(8.4)
Calgary	19,830	12.0	52,334	8.1	32,504	(3.9)
Ottawa–Gatineau	20,670	8.6	35,304	5.7	14,634	(2.9)
Edmonton	24,965	12.3	52,925	8.8	27,960	(3.5)
Quebec City	21,910	13.7	38,127	9.7	16,217	(4.0)
Winnipeg	44,425	19.6	44,733	11.8	308	(7.7)
Hamilton	73,255	36.8	58,478	16.2	(14,777)	(20.7)
Kitchener	41,360	41.5	58,338	23.1	16,978	(18.4)
London	28,670	23.7	38,070	15.5	9,400	(8.1)
Halifax	7,885	8.9	12,858	6.2	4,973	(2.7)
St Catharines–Niagara	41,355	35.5	27,636	13.8	(13,719)	(21.7)
Victoria	7,135	9.5	9,131	5.1	1,996	(4.4)
Oshawa	19,375	41.5	27,501	15.5	8,126	(26.1)
Windsor	34,820	35.6	39,692	24.5	4,872	(11.1)
Saskatoon	5,230	10.4	12,078	9.3	6,848	(1.1)
Regina	5,815	10.1	6,434	5.9	619	(4.2)
Sherbrooke	6,890	23.7	16,676	17.5	9,786	(6.2)
St John's	3,335	7.6	5,220	5.5	1,885	(2.2)
Sudbury	8,030	14.3	5,172	6.5	(2,858)	(7.7)
Kingston	4,490	13.1	4,717	6.0	227	(7.0)
Guelph	8,245	31.6	17,296	23.9	9,051	(7.7)
Trois-Rivières	6,890	23.7	11,303	16.5	4,413	(7.2)
Brantford	14,245	42.7	14,348	21.6	103	(21.0)
Saint John	6,935	17.7	5,657	9.0	(1,278)	(8.7)
Thunder Bay	7,430	17.1	5,242	8.4	(2,188)	(8.7)
Peterborough	8,285	32.5	7,006	11.8	(1,279)	(20.7)
Sarnia	9,465	30.9	7,068	15.7	(2,397)	(15.3)
Sault Ste Marie	11,235	36.8	5,104	13.1	(6,131)	(23.7)
Canada	1,593,595	19.6	2,108,784	12.6	515,189	(7.0)

Source: Statistics Canada, *Canadian Census of Population, 1971 and 2006* (author's calculations).

Table 6.4 Employment in Processing and Machining Occupations in Canadian Cities, 1971–2006

	1971		2006		1971–2006	
	Local Employment	% Local	Local Employment	% Local	Local Employment	% Local
Toronto	182,255	14.6	195,330	7.1	13,075	(7.6)
Montreal	169,310	15.7	106,370	5.5	(62,940)	(10.1)
Vancouver	59,000	12.5	45,840	4.0	(13,160)	(8.5)
Calgary	15,505	8.7	19,870	3.0	4,365	(5.7)
Ottawa–Gatineau	16,580	6.4	9,630	1.5	(6,950)	(4.8)
Edmonton	22,150	10.1	18,615	3.1	(3,535)	(7.0)
Quebec City	17,105	9.6	13,050	3.3	(4,055)	(6.3)
Winnipeg	32,670	13.4	21,200	5.6	(11,470)	(7.9)
Hamilton	42,335	20.0	23,755	6.5	(18,580)	(13.4)
Kitchener	27,565	26.0	29,620	11.6	2,055	(14.4)
London	17,780	13.7	19,980	8.1	2,200	(5.6)
Halifax	6,705	7.1	4,420	2.1	(2,285)	(5.0)
St Catharines–Niagara	27,620	22.1	12,575	6.2	(15,045)	(15.9)
Victoria	6,875	8.5	2,930	1.6	(3,945)	(6.9)
Oshawa	11,870	24.1	14,320	8.0	2,450	(16.1)
Windsor	23,920	22.8	21,330	13.1	(2,590)	(9.7)
Saskatoon	4,560	8.6	5,185	4.0	625	(4.6)
Regina	4,655	7.6	2,260	2.0	(2,395)	(5.5)
Sherbrooke	4,865	15.1	7,615	7.9	2,750	(7.2)
St John's	3,585	7.5	1,965	2.1	(1,620)	(5.4)
Sudbury	7,845	12.6	1,720	2.2	(6,125)	(10.5)
Kingston	3,545	9.5	1,930	2.4	(1,615)	(7.0)
Guelph	5,515	19.6	9,350	12.8	3,835	(6.8)
Trois-Rivières	6,850	20.1	5,105	7.4	(1,745)	(12.7)
Brantford	8,980	25.1	7,535	11.3	(1,445)	(13.8)
Saint John	4,875	11.6	1,900	3.0	(2,975)	(8.5)
Thunder Bay	6,440	13.8	2,175	3.5	(4,265)	(10.4)
Peterborough	4,775	17.6	3,560	5.9	(1,215)	(11.6)
Sarnia	4,860	14.9	2,680	5.9	(2,180)	(9.0)
Sault Ste Marie	6,540	20.0	1,695	4.3	(4,845)	(15.7)
Canada	*1,210,025*	*14.0*	*992,765*	*5.9*	*(217,260)*	*(8.1)*

Source: Statistics Canada, *Canadian Census of Population, 1971 and 2006* (author's calculations).

although Montreal is more specialized in pharmaceuticals whereas Toronto has a more diverse mix of activities. In each city, the life sciences industries are embedded in a broader regional and national innovation system and benefit from an array of institutional supports (see Gertler and Vinodrai, 2009; Lowe and Gertler, 2005, 2009). National strategies and policies such as the 1983 National Biotechnology Strategy, tax incentives to stimulate R&D expenditures (particularly the Scientific Research and Economic Development tax credits), and—in Montreal—the creation of National Research Council (NRC) laboratories focused on biotechnology were critical to the industry's development. Different provincial policies related to public health care also contribute to each city's development trajectory. Quebec's provincial drug plan provided a 15-year exclusive approval guarantee for brand-name drugs, even if generic alternatives were available. However, since the 1970s, Ontario has required patients to purchase cheaper, generic versions of drugs when available. Consequently, Montreal has a high proportion of brand-name drug producers whereas generic producers are more prominent in Toronto. At the local level, publicly funded research occurring in the universities and teaching hospitals has created opportunities for entrepreneurship and innovation. Toronto has a long history of medical research and innovation at the University of Toronto and elsewhere in the city, including the founding of Dow Pharmaceuticals in the 1880s, the establishment of the Connaught Laboratories at the University of Toronto in 1914, the discovery of insulin in the early twentieth century, as well as other medical breakthroughs. Moreover, Lowe and Gertler (2005: 26) suggest that Toronto's economic diversity, including 'a wide range of sophisticated service industries, including finance and professional/**producer services**, [and] a strong manufacturing base in industries such as automotive, food products, electronics, specialized machinery and aerospace', has provided opportunities for cross-sectoral

knowledge exchange and convergence between different types of technologies leading to the diversity of highly innovative life science activities.

In work on southern Ontario's automotive manufacturing cluster (including vehicle assembly and parts manufacturing), Rutherford and Holmes (2008a, 2008b) note that the sector has been one of Ontario's strongest economic sectors, accounting for nearly 20 per cent of the provincial manufacturing gross domestic product and almost 150,000 jobs, mostly in southern Ontario cities.[2] In 2006, manufacturing still accounted for almost one-quarter of employment in Kitchener (23.1 per cent) and Windsor (24.5 per cent), where automotive manufacturing remains a key source of employment. However, Kitchener's manufacturing base is more diverse. For instance, Research In Motion (RIM), known for its popular Blackberry, has manufacturing facilities in the city-region, as do several large, industrial-scale food manufacturers, including Schneider Foods and Dare Foods Limited. In addition to these larger employers, a wide range of smaller companies engaged in information and communications technology (ICT), machinery, furniture, and other manufacturing continue to operate. With the understanding that Canadian manufacturing faces challenges, particularly during recessionary periods, business leaders, universities, and government in Waterloo Region have made an effort to sustain local manufacturing activity by establishing the Manufacturing Innovation Network (MIN) (Nelles and Wolfe, 2009). The MIN is a regional initiative intended to encourage manufacturers to adopt new business practices and technologies, become more innovative, and solve common problems through sharing information and expertise with the end goal of helping local manufacturers remain competitive in the global economy.

Overall, it is clear that the transformation and restructuring of the manufacturing industries is far more complex than what first meets the eye. There has been a significant decline in the relative importance of the manufacturing industries and an even more acute decline in the proportion of the labour force in production-oriented occupations. Yet, there remain pockets of high-technology and traditional manufacturing. This raises two related points regarding deindustrialization and the nature of manufacturing activity itself. First, even though employment in the manufacturing sector declined in general, the workers most affected are those in production-oriented jobs. Deindustrialization involves both the transformation of work and sectoral shifts in employment. Second, manufacturing comprises a diverse range of sub-sectors including automotive and automotive parts, steel, office furniture, machinery, food, and other goods which are affected by different dynamics related to innovation, technological change, levels of unionization, and market demand. Some forms of manufacturing remain prominent in the Canadian landscape and some Canadian cities still have a relatively large manufacturing presence.

Innovation and the Knowledge-Based Economy

In the previous section, our discussion highlighted the significant decline of the manufacturing base and its associated forms of work, as well as the place-specific evolution of this transition given the specific histories and geographies of particular Canadian cities. Likewise, the unfolding of the knowledge-based economy has been experienced differently across Canadian cities. As noted earlier, there has been a widespread professionalization of the workforce, leading to the growth of occupations that require higher levels of formal education and specialized knowledge such as scientists, engineers, lawyers, architects, writers, designers, and other artistic, cultural, social science, and technical occupations. Table 6.5 shows the growth of this occupational group across 30 Canadian cities between 1971 and 2006. Professionals accounted for less than 10 per cent of the labour force in 1971 but almost doubled in their relative importance by

2006, accounting for 17.9 per cent of the workforce. Again, although this group of occupations grew in both absolute and relative terms across all of these cities, this did not occur evenly and there is a distinct relationship to city size (see Beckstead and Brown, 2006; Wolfe and Bramwell, 2008). Canada's largest cities (Toronto, Montreal, Vancouver, Calgary, Ottawa) experienced double-digit increases in the proportion of the labour force accounted for by professionals, as did several smaller regional centres such as St John's, Victoria, and Quebec City. Beckstead and Vinodrai (2003) have shown that growth in these occupations has been experienced across almost all industrial sectors, including typically 'low-tech' sectors, such as mining and other natural resource-based sectors. Further work by Beckstead and Gellatly (2006) confirms that mining and oil and gas extraction industries, as well as the manufacturing industries, rank quite highly in terms of employing scientists and engineering. In other words, even traditional industries are increasingly relying on knowledge-based inputs.

Sudbury provides an interesting case of the shift towards **knowledge-intensive economic activity**. As a northern Ontario city dominated by the mining sector, including large multinational companies such as Inco (owned by the Brazilian company, Vale) and Falconbridge (owned by the Swiss company, Xstrata), the city has been strongly influenced by the restructuring of the mining industry. Between 1971 and 2006, the proportion of employment accounted for by Sudbury's primary industries (e.g., mining, forestry, and other extractive and natural resource-based activities) declined from 26.6 per cent to 7.8 per cent. This reflects the ongoing layoffs and outsourcing that occurred from the mid-1970s onward among local mining firms (Warrian and Mulhern, 2009). While such downsizing has unquestionably had negative effects on Sudbury and its surrounding communities, at the same time, this has led to the emergence of a small, but technology-intensive,

Table 6.5 Employment in Professional Occupations in Canadian Cities, 1971–2006

	1971		2006		1971–2006	
	Local Employment	% Local	Local Employment	% Local	Local Employment	% Local
Toronto	122,055	9.8	562,550	20.4	440,495	10.6
Montreal	106,520	9.9	403,440	21.0	296,920	11.1
Vancouver	41,600	8.8	231,515	20.1	189,915	11.3
Calgary	20,035	11.3	140,010	21.4	119,975	10.2
Ottawa–Gatineau	36,905	14.2	175,110	27.9	138,205	13.7
Edmonton	22,650	10.3	107,580	17.7	84,930	7.4
Quebec City	19,480	11.0	88,785	22.4	69,305	11.5
Winnipeg	22,700	9.3	69,595	18.3	46,895	8.9
Hamilton	18,645	8.8	61,985	17.0	43,340	8.2
Kitchener	9,105	8.6	45,155	17.7	36,050	9.2
London	12,320	9.5	43,170	17.5	30,850	8.0
Halifax	9,945	10.5	42,170	20.1	32,225	9.6
St Catharines–Niagara	10,140	8.1	27,850	13.8	17,710	5.7
Victoria	7,450	9.2	37,735	20.8	30,285	11.6
Oshawa	3,950	8.0	29,500	16.5	25,550	8.5
Windsor	8,255	7.9	25,260	15.5	17,005	7.7
Saskatoon	6,125	11.5	24,055	18.4	17,930	6.9
Regina	6,080	9.9	21,090	19.1	15,010	9.2
Sherbrooke	4,000	12.4	18,535	19.2	14,535	6.8
St John's	4,540	9.5	20,005	20.9	15,465	11.4
Sudbury	5,390	8.7	13,375	16.7	7,985	8.1
Kingston	4,990	13.3	17,020	21.5	12,030	8.2
Guelph	3,315	11.8	14,875	20.4	11,560	8.6
Trois-Rivières	3,330	9.7	12,660	18.3	9,330	8.6
Brantford	2,475	6.9	8,460	12.7	5,985	5.8
Saint John	3,215	7.6	10,365	16.4	7,150	8.8
Thunder Bay	3,850	8.3	10,785	17.1	6,935	8.9
Peterborough	2,980	11.0	10,735	17.9	7,755	6.9
Sarnia	3,735	11.5	7,185	15.8	3,450	4.3
Sault Ste Marie	2,895	8.9	7,130	18.2	4,235	9.3
Canada	766,550	8.9	3,024,560	17.9	2,258,010	9.1

Source: Statistics Canada, Canadian Census of Population, 1971 and 2006 (author's calculations).

cluster of mining supply and service companies. The layoffs of highly skilled workers from the large mining companies in the city-region resulted in the emergence of a new set of firms that drew on the existing local skill base and provided services to the large mining companies. Moreover, Sudbury's high-technology mining services sector is increasingly supported through the emergence of several intermediary organizations such as the Northern Centre for Advanced Technology (NORCAT), the Centre for Excellence in Mining Innovation (CEMI), the Sudbury-and-area Mining Supply and Service Association (SAMSSA), and the Ontario Mining Industry Cluster Council (OMICC); these organizations are actively promoting the sector's development as a means of sparking broader economic development in the city and the wider region (Hall and Donald, 2009). In addition, there is a growing role for Laurentian University, which increasingly is viewed as a key supplier of highly skilled labour and provider of leading-edge research (Warrian and Mulhern, 2009).

Parallel stories of how the pre-existing economic base of cities shape and condition their subsequent reinvention and transformation emerge elsewhere in Canada. Turning to Hamilton, where the steel industry has left a heavy imprint on the city's economy, the city's transition to knowledge-intensive activities has been shaped by its existing industrial base and through the activities of the local, publicly funded university. One focus of Hamilton's shift towards a more diversified, service- and knowledge-based economy has related to health care and health sciences. While a significant part of Hamilton's growing reputation in health is due to McMaster University's School of Medicine, which offers a distinctive, experientially based form of medical education, one of the primary reasons for health services emerging in Hamilton can be attributed to the pre-existing automotive and steel industries. As Warrian (2009: 18) notes, 'Hamilton citizens also tend to have generous medical benefits from unions and the

government which means that specialized medical services—such as designed orthotics—are often paid for. Union health benefits have provided the financial base and market for these emergent health services firms.' In other words, the presence of the Canadian Auto Workers and United Steel Workers unions in Hamilton was critical in driving demand for and financing Hamilton's emerging health services sector, which has produced world-class innovations.

Elsewhere, the knowledge-based economies of Canadian cities have similarly been built on long-term public investments. The emergence of Ottawa's ICT cluster specializing in photonics and telecommunications is partly the result of a long-standing investment in publicly funded government laboratories, as well as a product of particular historical events and regulatory decisions such as the founding of a key firm, Bell Northern Research (BNR), which anchored the ICT industry in Ottawa beginning in the 1950s. As Lucas, Sands, and Wolfe (2009: 194) write:

> the original decision by Northern Electric in the late-1950s to establish a research facility in the region was made after a judicial decision in the US cut off its ready access to patents from the Western Electric Co. Its purchase of a substantial tract of land on the outskirts of Ottawa as the future home of Bell Northern Research (BNR) was largely because the concentration of federal government laboratories in the nation's capital created a steady stream of industrial engineers, researchers, and managers moving into the region.

BNR (and a subsequent failed subsidiary) were the training ground for several entrepreneurs who left the company to start their own businesses in Ottawa, leading to the development of a critical mass of firms and workers specialized in ICT and related activities. Moreover, several organizations have emerged that are dedicated to promoting networking among firms to encourage firm learning, knowledge sharing, and the solving of common problems. This local associative activity, alongside support from local and provincial government programs, the presence of national research laboratories, and the research activity of local universities encourage the ongoing development and growth of the industry within the city.

The Cultural Economies of Canadian Cities

As noted earlier, the creative and cultural industries are increasingly important to urban economies. Figure 6.5 shows the growth of employment in the cultural industries in Toronto, Montreal, and Vancouver between 1987 and 2007; these three cities account for almost half of Canadian employment in the cultural industries. Over this 20-year period, the average annual growth rates of the cultural industries in Toronto (2.1 per cent), Montreal (3.0 per cent), and Vancouver (3.6 per cent) significantly outpaced the overall growth rates in each city. While these industries experienced employment growth throughout the 20-year period, Figure 6.5 demonstrates accelerated growth beginning in the mid-1990s, which may be attributed to the increased adoption and widespread use of the Internet as a platform for new media applications (Britton, Tremblay, and Smith, 2009). However, the trajectories in each of these three cities have been shaped by their own local histories, institutions, and policies. In other words, although all three of Canada's largest urban centres have a large concentration of cultural industries, we should not simply assume their character or development paths to be

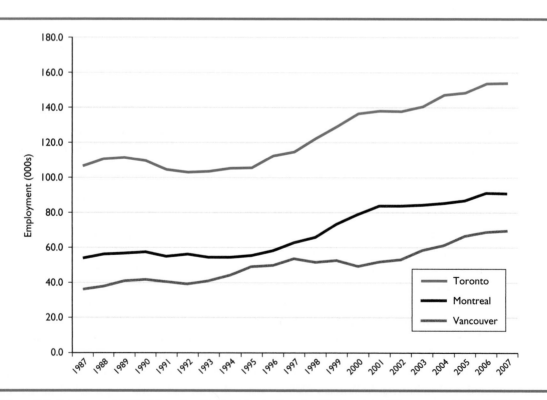

Figure 6.5 Employment (000s) in cultural industries in Toronto, Montreal, and Vancouver, 1987–2007

Source: Statistics Canada, *Labour Force Survey, 1987–2007* (custom tabulations; author's calculations).

identical, given different pre-existing conditions and institutions.

For example, Vancouver's development trajectory is unique compared to many Canadian and US cities. As Barnes and Hutton (2009) note:

> to use the Canadian economic historian Harold Innis's (1930) term, Vancouver first developed as a 'local metropole' within British Columbia's staples economy and based on the extraction, processing and export of natural resources, most prominently in forestry, fishing and mining. Vancouver's role was primarily a control and distribution centre for staples goods, and only secondarily, a processing site.

Yet, the new economy has taken hold in Vancouver and the city is now viewed as an appealing location for firms and workers involved in a range of creativity- and culture-oriented endeavours. Certainly, the quality of life factors identified by Florida (2002) related to lifestyle amenities figure heavily in accounts of why workers and firms are attracted to Vancouver (Britton, Tremblay, and Smith, 2009). However, this only partially explains the growth of Vancouver's creative and cultural industries. Vancouver has emerged recently as a growing hot bed for video game production and other new media applications. Companies like Electronic Arts—one of the largest video game developers in the world—have a significant presence in the city and other gaming companies are also located there (ibid.). But, as discussed elsewhere in this chapter, there is a longer history that explains Vancouver's particular development. First, Vancouver's new media industry draws on resources that were already present in the local economy, including the film and television industry, which provided the initial talent base for this emerging industry. Second, Vancouver's video game industry can be traced back to at least the early 1980s with the founding of Distinctive Software, which was acquired by Electronic Arts in the

early 1990s. Similar to Vancouver's other cultural industries (such as film and television), entertainment-oriented new media firms in Vancouver generally are focused on international markets and are tied to other places through international capital flows (Barnes and Hutton, 2009; Britton, Tremblay, and Smith, 2009). Finally, several unique public and quasi-public institutions provide support to Vancouver's new media industry, such as the University of British Columbia's Media and Graphics Interdisciplinary Centre (MAGIC), which connects researchers with other local players to enable new product development; New Media BC, a dedicated umbrella industry association, brings together various institutions, organizations, and other interested parties to promote and develop the industry.

In Toronto, the municipal government has identified the cultural sector as one of several key clusters that are critical to its economic development trajectory. A recent study found that employment in Toronto's cultural industries grew at a faster rate than did the cultural industries in several other major North American cities, including San Francisco, Los Angeles, and Chicago (Gertler et al., 2006; see also Vinodrai and Gertler, 2007). Yet, such growth rests on several important territorial assets and institutional supports. Toronto's cultural industries have been nurtured by public funding for the arts and culture at the local, provincial, and national levels. Moreover, a diverse set of public and private organizations support the development of these industries. For example, Toronto's film industry is supported by several city-level initiatives including: (1) the City of Toronto's Film and Television office, which provides logistical and regulatory support; (2) a Film Commissioner and Film Board that create strategies and policies that promote and support Toronto's film and television industry; and (3) Filmport, Canada's largest film and media production complex located along the city's waterfront. In addition to these city-supported initiatives, Toronto's film industry has benefited from federal institutions such as the National Film Board (NFB) and the Canadian Broadcasting Corporation (CBC), as well as provincial programs like the Ontario Film and Television Tax Credit (Gertler et al., 2006). Toronto has also become a destination for film consumption, hosting the Toronto International Film Festival (TIFF), with its planned home at the Bell Lightbox in Toronto's entertainment district, as well as other events, such as the Canadian Film Centre's Worldwide Short Film Festival (WSFF). In addition to film and television, there is a critical mass of fashion, design, art, architecture, music, new media, and other cultural activity (Britton, Tremblay, and Smith, 2009; Leslie and Brail, 2008; Vinodrai, 2009; Vinodrai and Gertler, 2007). Publicly funded institutions of higher education, such as Sheridan College, Seneca College, the Ontario College of Art and Design, Ryerson University, and George Brown College, offer programs in animation, design, and other culture-related programs that develop the local talent so critical to the survival of these industries. For new media and advertising in particular, the presence of local demand generated by other sectors, especially financial and business services, has helped to buoy these industries.

Conclusion

This chapter has traced how the economic landscapes of Canadian cities have profoundly changed since the crisis and decline of Fordism and its associated production and employment relations, regulatory mechanisms, and other institutions. On the one hand, the occupational composition of Canadian cities reflects an increasing demand for knowledge, creativity, and formal qualifications. On the other hand, the sectoral composition of Canadian cities has slowly transformed to one where services-producing industries (including higher-order, knowledge-intensive services and lower-order services) have eclipsed goods-producing industries in terms of their relative

importance to urban economies. We examined the three key features of this economic transition. First, Canada's manufacturing base has declined and transformed over the past several decades. Second, there has been a shift towards knowledge-based and innovation-oriented activity. Third, creative and cultural economic activities have become increasingly important, particularly in large cities. However, as we saw throughout the chapter, economic transitions unfold slowly and urban economies evolve over time—not divorced from either broader macroeconomic, political, social, and cultural conditions or their particular local contexts. The transition towards a post-industrial economy has been experienced differentially across Canadian cities and their economic development trajectories are path-dependent. In other words, the specific industrial histories, individual characteristics, and territorial assets of city-regions matter to their economic development. Furthermore, it is clear that public investments at the local, provincial, and national levels have been important in shaping these outcomes.

On a final note, the emphasis on innovation, and on knowledge-based economic development and growth, points to some troubling questions about the nature of work and the fortunes of individuals, communities, and cities. Scholars have raised questions about the polarizing dynamics related to the emergence of a knowledge-based, creative economy, often associated with an increasingly deepening divide between professional, managerial, technical, and scientific workers and workers occupying lower-wage, routinized service work jobs (Donald and Morrow, 2003; Gertler, 2001; Scott, 2007; Storper and Scott, 2009). Moreover, some of these same tendencies exist within the industries associated with the emerging knowledge-based economy itself, with significant implications for the urban labour markets in which they are situated. These issues remain challenging questions for scholars and policy-makers to understand and address. As students and scholars interested in contemporary urban and economic change, it is our task to understand the ongoing economic transitions in Canadian cities and the implications for individuals, communities, and cities.

Review Questions

1. What economic transitions are Canadian cities presently undergoing?
2. How can cities promote economic change? How can they be made to embrace economic performance?
3. What are the consequences of de-industrialization on Canadian cities?
4. How does the rise of the knowledge-based economy affect cities?

Notes

1. Debates continue about the particular expressions or 'varieties' of capitalism that may exist across different nation-states and how this (re)produces different 'neo-liberal' governance regimes at a variety of scales (Hall and Soskice, 2001; Peck and Theodore, 2007); a detailed discussion is beyond the scope of this chapter.
2. The North American automotive industry is currently restructuring and faces several challenges, including General Motors entering into bankruptcy protection and receiving unprecedented financial backing from the Canadian and US governments. This raises questions about the future of the Canadian automotive industry, particularly in southern Ontario.

References

Amin, A., ed. 1994. *Post-Fordism: A Reader*. Cambridge, Mass.: Blackwell.

Barnes, T., and T. Hutton. 2009. 'Situating the new economy: Contingencies of regeneration and dislocation in Vancouver's inner city', *Urban Studies* 46: 1247–69.

Bathelt, H., A. Malmberg, and A. Maskell. 2004. 'Clusters and knowledge: Local buzz, global pipelines and the process of knowledge creation', *Progress in Human Geography* 28: 31–56.

Beck, U. 2000. *The Brave New World of Work*. Cambridge: Cambridge University Press.

Beckstead, D., and M.W. Brown. 2006. *Innovation Capabilities: Science and Engineering Employment in Canadian and US Cities*. Ottawa: Analytical Studies Branch, Statistics Canada, Canadian Economy in Transition Series. Catalogue no. 11–622–MIE2006012.

——— and G. Gellatly. 2006. *Innovation Capabilities: Science and Engineering Employment in Canada and the United States*. Ottawa: Analytical Studies Branch, Statistics Canada, Canadian Economy in Transition Series. Catalogue no. 11–622–MIE2006011.

——— and T. Vinodrai. 2003. *Dimensions of Occupational Change in Canada's Knowledge Economy, 1971–1996*. Ottawa: Analytical Studies Branch, Statistics Canada, Canadian Economy in Transition Series. Catalogue no. 11–622–MIE2003004.

Bell, D. 1973. *The Coming of Post-Industrial Society: A Venture in Social Forecasting*. New York: Basic Books.

Benner, C. 2002. *Work in the New Economy: Flexible Labor Markets in Silicon Valley*. Malden, Mass.: Blackwell.

Boyer, R. 2005. 'How and why capitalisms differ', *Economy and Society* 34: 509–57.

Britton, J.N.H., D.-G. Tremblay, and R. Smith. 2009. 'Contrasts in clustering: The example of Canadian new media', *European Planning Studies* 17: 281–301.

Bryson, J., and N. Henry. 2005. 'The global production system: From Fordism to post-Fordism', in P. Daniels, M. Bradshaw, D. Shaw, and J. Sidaway, eds, *An Introduction to Human Geography: Issues for the 21st Century*. Harlow: Pearson.

Bryson, J.R., P.W. Danielson, and B. Warf. 2004. *Service Worlds: People, Organizations, Technologies*. New York: Routledge.

Cairncross, F. 1997. *The Death of Distance: How the Communications Revolution Is Changing Our Lives*. Cambridge, Mass.: Harvard Business Press.

Cervenan, A. 2009. *Service Class Prosperity in Ontario*. Toronto: University of Toronto, Martin Prosperity Institute (Ontario in the Creative Age Working Paper Series).

Christopherson, S. 2002. 'Project work in context: Regulatory change and the new geography of media', *Environment and Planning A* 34: 2003–15.

Donald, B., and D. Morrow. 2003. *Competing for Talent: Implications for Social and Cultural Policy in Canadian City-regions*. Gatineau, Que.: Department of Canadian Heritage (Strategic Research and Analysis, Strategic Planning and Policy Coordination).

Ekinsmyth, C. 2002. 'Project organization, embeddedness, and risk in magazine publishing', *Regional Studies* 36: 229–44.

Elam, M. 1994. 'Puzzling out the post-Fordist debate: Technology, markets and institutions', in Amin (1994).

Florida, R. 2002. *The Rise of the Creative Class*. New York: Basic Books.

Friedman, T. 2005. *The World Is Flat: A Brief History of the Twenty-First Century*. New York: Farrar, Straus, and Giroux.

Gertler, M.S. 2001. 'Urban economy and society in Canada: Flows of people, capital and ideas', *ISUMA: The Canadian Journal of Policy Research* 2, 3: 119–30.

———. 2003. 'Tacit knowledge and the economic geography of context, or the undefinable tacitness of being (there)', *Journal of Economic Geography* 3: 75–99.

———. 2008. 'Buzz without being there? Communities of practice in context', in A. Amin and J. Roberts, eds, *Community, Economic Creativity and Organization*. Oxford: Oxford University Press.

——— and T. Vinodrai. 2009. 'Life sciences and regional innovation: One path or many?', *European Planning Studies* 17: 235–61.

——— et al. 2006. *Imagine a Toronto: Strategies for a Creative City*. Toronto: University of Toronto, Munk Centre for International Studies.

Gouldner, A. 1979. *The Future of Intellectuals and the Rise of the New Class*. New York: Seabury.

Grabher, G. 2002. 'The project ecology of advertising: Talents, tasks, and teams', *Regional Studies* 36: 245–62.

———. 2004. 'Learning in projects, remembering in networks? Communality, sociality, and connectivity in project ecologies', *European Urban and Regional Studies* 11: 103–23.

Hall, H., and B. Donald. 2009. *Innovation and Creativity on the Periphery: Challenges and Opportunities in Northern Ontario*. Toronto: University of Toronto, Martin Prosperity Institute (Ontario in the Creative Age Working Paper Series).

Hall, P.A., and D. Soskice, eds. 2001. *Varieties of Capitalism: The Institutional Foundations of Comparative Advantage*. Oxford: Oxford University Press.

Harrington, J.W., and P.W. Daniels, eds. 2006. *Knowledge-Based Services, Internationalization and Regional Development*. Aldershot: Ashgate.

Harvey, D. 1990. *The Condition of Postmodernity*. Cambridge, Mass.: Blackwell.

Jenson, J. 1990. 'Representations in crisis: The roots of Canada's permeable Fordism', *Canadian Journal of Political Science* 23: 653–83.

Lash, S., and J. Urry. 1994. *Economies of Signs and Spaces*. London: Sage.

Leslie, D., and S. Brail. 2008. 'The role of quality of place in attracting and retaining fashion design talent', presented at the Joint ONRIS/MRI Workshop. Toronto, 6–7 Nov.

Ley, D. 1996. *The New Middle Class and the Remaking of the Central City*. Toronto: Oxford University Press.

Lloyd, R. 2006. *Neo-Bohemia: Art and Commerce in the Post-industrial City*. New York: Routledge.

Lowe, N., and M.S. Gertler. 2005. 'Diversity and the evolution of a life-science innovation system: The Toronto region in comparative perspective', in D.A. Wolfe and M. Lucas, eds, *Global Networks and Local Linkages: The Paradox of Cluster Development in an Open Economy*. Montreal and Kingston: McGill-Queen's University Press.

——— and ———. 2009. 'Building on diversity: Institutional foundations of hybrid strategies in Toronto's life sciences complex', *Regional Studies* 43: 589–603.

Lucas, M., A. Sands, and D.A. Wolfe. 2009. 'Regional clusters in a global industry: ICT clusters in Canada', *European Planning Studies* 17: 189–209.

Markusen, A. 2006. 'Urban development and the politics of a creative class: Evidence from the study of artists', *Environment and Planning A* 38:1921–40.

Marshall, J.N., and P. Wood. 1992. 'The role of services in urban and regional development: Recent debates and new directions', *Environment and Planning A* 24: 1255–70.

Maskell, P., and A. Malmberg. 1999. 'Localised learning and industrial competitiveness', *Cambridge Journal of Economics* 23: 167–85.

Molotch, H. 2002. 'Place in product', *International Journal of Urban and Regional Research* 26: 665–88.

Morgan, K. 2004. 'The exaggerated death of geography: Learning, proximity and territorial innovation systems', *Journal of Economic Geography* 4: 3–22.

Nelles, J., and D. Wolfe. 2009. 'The Waterloo myth? Governance in the clouds', presented at the annual meeting of the Innovation Systems Research Network, Halifax, 1 May.

Peck, J. 2005. 'Struggling with the creative class', *International Journal of Urban and Regional Research* 29: 740–70.

——— and N. Theodore. 2007. 'Variegated capitalism', *Progress in Human Geography* 31: 731–72.

Piore, M.J., and C. Sabel. 1984. *The Second Industrial Divide: Possibilities for Prosperity*. New York: Basic Books.

Porter, M.E. 1998. 'Clusters and the new economics of competition', *Harvard Business Review* (Nov.–Dec.): 77–90.

Power, D., and A.J. Scott, eds. 2004. *The Cultural Industries and the Production of Culture*. London: Routledge.

Pratt, A.C. 1997. 'Employment in the cultural industries sector: A case study of Britain, 1984–91', *Environment and Planning A* 29: 1953–76.

Rantisi, N.M. 2004. 'The designer in the city and the city in the designer', in Power and Scott (2004).

Rutherford, T., and J. Holmes. 2008a. 'Engineering networks: University–industry networks in southern Ontario automotive industry clusters', *Cambridge Journal of Regions, Economy and Society* 1: 247–64.

———— and ————. 2008b. 'The flea on the tail of the dog: Power in global production networks and the restructuring of Canadian automotive clusters', *Journal of Economic Geography* 8: 519–44.

Scott, A.J. 2001. *The Cultural Economy of Cities: Essays on the Geography of Image-producing Industries*. Oxford: Sage.

————. 2004. 'Cultural-products industries and urban economic development: Prospects for growth and market contestation in global context', *Urban Affairs Review* 39: 461–90.

————. 2007. 'Capitalism and urbanization in a new key? The cognitive-cultural dimension', *Social Forces* 85: 1466–82.

Storper, M., and A.J. Scott, eds. 1992. *Pathways to Industrialization and Regional Development*. London: Routledge.

———— and ————. 2009. 'Rethinking human capital, creativity and urban growth', *Journal of Economic Geography* 9: 147–67.

———— and A.J. Venables. 2004. 'Buzz: Face-to-face contact and the urban economy', *Journal of Economic Geography* 4: 351–70.

Tickell, A., and J. Peck. 1992. 'Accumulation, regulation and the geographies of post-Fordism: Missing links in regulationist research', *Progress in Human Geography* 16: 190–218.

Vinodrai, T. 2006. 'Reproducing Toronto's design ecology: Career paths, intermediaries, and local labor markets', *Economic Geography* 82: 237–63.

————. 2009. *The Place of Design: Exploring Ontario's Design Economy*. Toronto: University of Toronto, Martin Prosperity Institute (Ontario in the Creative Age Working Paper Series).

———— and M.S. Gertler. 2007. *Measuring the Creative Economy: The Structure and Economic Performance of Ontario's Creative, Cultural and New Media Clusters*. Toronto: Government of Ontario, Ministry of Culture.

Vosko, L., ed. 2006. *Precarious Employment: Understanding Labour Market Insecurity in Canada*. Montreal and Kingston: McGill-Queen's University Press.

Warrian, P. 2009. 'Biotech in lunch buckets: The curious knowledge networks in Steeltown', paper presented at the 10th annual meeting of the Innovation Systems Research Network, Halifax, 30 Apr.

———— and C. Mulhern. 2009. 'From metal bashing to materials science and services: Advanced manufacturing and mining clusters in transition', *European Planning Studies* 17: 281–301.

Wolfe, D.A., and A. Bramwell. 2008. 'Innovation, creativity and governance: Social dynamics of economic performance in city-regions', *Innovation: Management, Policy and Practice* 10: 170–82.

———— and M.S. Gertler. 2004. 'Clusters from the inside and out: Local dynamics and global linkages', *Urban Studies* 41: 1071–93.

Zukin, S. 1982. *Loft Living: Culture and Capital in Urban Change*. Baltimore: Johns Hopkins University Press.

CHAPTER 7

Economic Change in Canadian Cities: Locational Dynamics of Employment

Tom Hutton

Introduction: Employment and Urban Transition in Canada

Over the past quarter-century, Canada has developed not simply as an *urban* society, but more emphatically, as a *metropolitan* society. More than two-thirds of the national population and jobs are situated in the nation's 33 metropolitan areas. Over time, ever-increasing shares of the national population, employment, services, and advanced industrial capacity concentrate in the higher echelons of the Canadian urban system. Greater Toronto, Montreal, Vancouver, Ottawa–Gatineau, Calgary, and Edmonton—with metropolitan populations ranging from just over 1 million to 5.5 million—are increasingly influential in configuring Canada's national economic systems, social agenda, and polity.

Canada's metropolitan cities also function as the nation's global gateways (see Hoernig and Zhuang, Chapter 9), constitute the principal centres of higher education and research and development, and comprise the influential sites of productive diversity that underpin creativity and innovation. These same metropolitan regions attract a disproportionate share of international immigrants, who will in turn shape to a large extent the future growth of the labour force, as well as Canada's urban and regional economies, community morphologies, and multicultural identities (see

Bunting and Filion, Chapter 2; Filion and Bunting, Chapter 3; Hall, Chapter 4; Vinodrai, Chapter 6). At the same time, medium-size Canadian cities such as Waterloo, Saskatoon, and Halifax have found new vocations, notably in the knowledge economy. Overall, city-regions are the critical geospatial elements of the emerging Canadian economy.

Employment represents one of the key measures of analysis for understanding processes of growth and change in cities. The quantity and quality of jobs generated, and labour force participation rates, are leading indicators of the robustness of national, regional, and metropolitan economies. Employment is directly linked to incomes, and to socio-economic welfare and well-being. Employment provides both a metric of economic change among Canadian city-regions and an entrée into larger experiences of metropolitan transformation.

Processes of employment growth and change in Canadian cities are perhaps best understood as *multiscalar flows*. At the Canadian urban system level, employment change reflects structural tendencies, such as the shift from manufacturing to service employment, as well as factors contingent to particular regions, such as urban scale, path dependencies, industrial structure, and policy factors. Change at the national level also reflects the forces of globalization, notably in the concentrations of specialized services (head offices, intermediate finance, and business services) within cities at the peak of the national urban hierarchy.

At the regional scale, employment is an important metric in assessing changes in the metropolitan *space-economy*: the geography of work within urban areas. The metropolis typically comprises strategic-scale agglomerations of linked industries and labour, including major *regional clusters and nodes*, such as the corporate complex of the central business district (CBD), as well as industrial estates, secondary commercial centres, airports and ports, and universities and science parks.

In this chapter I present a concise digest of labour force data for selected city-regions, drawing out aspects of commonality as well as important contrasts. Next, I describe motive forces shaping new patterns of enterprise and employment in Canadian city-regions, and outline the new geography of employment, including the new economic spaces in the suburbs and discussion of the role of employment in the formation of the 'new inner city'. Synergies of social, market, and policy factors, we shall see, reproduce space and landscape. The conclusion is based on conjecture about future conditions of employment change in Canadian cities.

Industrial Structure and Employment in Canadian Cities

Certain structural elements are common to most Canadian cities and indeed to urban areas within the developed world generally, including, notably, the decline of basic manufacturing, and the concomitant rise of service industries, labour, and occupations. The service sector is made up of an exceptionally diverse array of industries, including those catering for 'final demand' (education, retail, personal, and most public-sector services), as well as for 'intermediate demand' (i.e., providers of inputs to other businesses and industries), for example commercial banking, and business services (Bryson and Daniels, 2007). Service

industries also contain a large and growing cultural industry component with another large portion attributed to an omnibus 'service' category, and smaller proportions engaged in 'working' (mostly manufacturing) and primary-sector industries. About 80 per cent of the labour force of the larger Canadian cities is comprised of services, with the goods-producing industries (principally manufacturing and construction) accounting for the bulk of the remainder. Most Canadian cities now encompass only very small primary sectors—the 'extractive' industries such as forestry, agriculture, and mining—essentially residuals of earlier periods of development. But there are important distinctions to acknowledge in the industrial structure of labour and employment among the Canadian cities derived from scale and specialization, path dependency (legacies of historical development, embedded competitive advantages), and confluences of domestic and global forces. The Ottawa–Gatineau city-region represents a striking example of path dependency specialization, with over one-fifth of the regional labour force (21.2 per cent) in the public administration category: roughly twice the proportion for Halifax (11.1 per cent), a provincial capital with high dependency on public-sector employment, and about six times the proportion of public administration workers in Toronto (3.4 per cent), a provincial capital city with significant cadres of federal and municipal workers (Spencer and Vinodrai, 2009).

Montreal was the first major Canadian city to experience industrialization, followed by Toronto, and these two cities (together with smaller industrial satellite communities in southern Ontario and along the St Lawrence) jointly comprised the industrial 'core' of Harold Innis's (1933) classic core–periphery model of the Canadian economy. The GTA encompasses by far the largest complex of manufacturing capacity in Canada, with a manufacturing labour force of 371,275 according to the 2006 census (Table 7.1), while Montreal's labour force includes 260,530 workers in manufacturing (Table 7.2). For both city-regions, manufacturing comprised 13.5 per cent of total labour force in 2006. But manufacturing continues a long process of decline, Toronto having lost 6.2 per cent of its manufacturing employment in the census period 2001–6, and Montreal 14.6 per cent. In contrast, Vancouver's manufacturing labour force, while only 8.5 per cent of the city-region total, experienced only a modest decline of 1.3 per cent from 2001 to 2006, holding relatively steady at around 100,000 workers over the last decade (Table 7.3). Vancouver never developed as a fully industrial city on the model of its larger (and older) central Canadian counterparts, but its mix of mostly small production industries (including food products, garment production, and niche-level advanced-technology industries) has proven relatively resilient.

Manufacturing remains important for some Canadian cities, but represents a small proportion of the metropolitan labour force for many and so the real story concerns performance in the services sector. As in others of the leading OECD economies, the fortunes of Canadian cities are shaped by the dimensions, degree of specialization, and growth dynamics of advanced services. While the GTA encompasses the largest industrial production sector in Canada, its primacy within the national economy,

Table 7.1 Labour Force by Industry Group, Toronto (GTA)

	# Labour Force	% Labour Force	% Change 2001–2006
All industries	2,758,695	100.0	9.4
11 Agr., forestry, fish. & hunt.	9,720	0.4	3.1
21 Mining & oil/gas extraction	4,660	0.2	74.9
22 Utilities	16,030	0.6	1.7
23 Construction	148,895	5.4	19.7
31–33 Manufacturing	371,275	13.5	–6.2
41 Wholesale trade	166,325	6.0	9.5
44–45 Retail trade	293,465	10.6	7.6
48–49 Transport & warehousing	140,205	5.1	13.9
51 Information & cultural ind.	101,850	3.7	1.1
52 Finance and insurance	193,760	7.0	9.3
53 Real estate/rental & leasing	66,115	2.4	16.2
54 Prof., sci. & tech. services	267,620	9.7	8.5
55 Mgmt. of companies	5,565	0.2	15.1
56 Admin./support, & wst. mgt.	143,265	5.2	17.9
61 Educational services	172,990	6.3	20.1
62 Health care & soc. assist.	222,140	8.1	17.3
71 Arts, entertainment & rec.	55,300	2.0	15.5
72 Accom. & food services	157,680	5.7	11.4
81 Other services	127,635	4.6	15.2
91 Public administration	94,195	3.4	11.3

Source: Spencer and Vinodrai (2009).

Table 7.2 Labour Force by Industry Group, Montreal

	# Labour Force	% Labour Force	% Change 2001–2006
All industries	1,923,970	100.0	8.2
11 Agr., forestry, fish. & hunt.	9,215	0.5	21.3
21 Mining & oil/gas extraction	2,040	0.1	38.3
22 Utilities	16,120	0.8	28.9
23 Construction	88,790	4.6	23.8
31–33 Manufacturing	260,530	13.5	–14.6
41 Wholesale trade	108,185	5.6	4.3
44–45 Retail trade	232,435	12.1	15.2
48–49 Transport & warehousing	94,995	4.9	3.6
51 Information & cultural ind.	70,170	3.6	1.1
52 Finance and insurance	89,190	4.6	12.9
53 Real estate/rental & leasing	35,020	1.8	16.9
54 Prof., sci. & tech. services	159,495	8.3	13.2
55 Mgmt. of companies	2,750	0.1	32.9
56 Admin./support, & wst. mgt.	81,545	4.2	14.2
61 Educational services	136,600	7.1	15.5
62 Health care & soc. assist.	206,860	10.8	18.6
71 Arts, entertainment & rec.	41,590	2.2	24.1
72 Accom. & food services	113,655	5.9	13.2
81 Other services	89,385	4.6	8.1
91 Public administration	85,400	4.4	5.4

Source: Spencer and Vinodrai (2009).

and its claims to **global city** status, rest on its platform of specialized service industries, firms, and labour. To illustrate, Toronto's professional, scientific, and technical services labour force—a key marker of socio-economic transformation as defined in Daniel Bell's seminal work on *The Coming of Post-Industrial Society* (1973)—exceeded a quarter of a million workers in 2006, comprising 9.7 per cent of the GTA's total, an increase of 8.5 per cent over the 2001–6 period, while the key financial, insurance, and real estate (FIRE) sector labour force (numbering just under 200,000 in 2006 or 7 per cent of the regional total) saw an increase of 9.3 per cent over the same five-year period. In Montreal the professional, scientific, and technical labour force stood at about 160,000 in 2006, enjoying a robust growth of 13.2 per cent since 2001.

Calgary, the growth leader among Canada's major cities, presents a storyline of spectacular expansion in its key industrial specialization, notably in the mining and oil and gas sector, where the labour force expanded by 52 per cent in the period 2001–6. Buttressing this propulsive sector of the Calgary economy, the professional, scientific, and technical labour force has expanded by 23.2 per cent since 2001, while financial and insurance

Table 7.3 Labour Force by Industry Group, Vancouver

	# Labour Force	% Labour Force	% Change 2001–2006
All industries	1,150,490	100.0	9.6
11 Agr., forestry, fish. & hunt.	13,890	1.2	4.7
21 Mining & oil/gas extraction	4,380	0.4	95.1
22 Utilities	5,700	0.5	–8.1
23 Construction	73,385	6.4	36.4
31–33 Manufacturing	97,800	8.5	–1.3
41 Wholesale trade	61,655	5.4	10.1
44–45 Retail trade	124,965	10.9	7.2
48–49 Transport & warehousing	65,600	5.7	–0.2
51 Information & cultural ind.	42,145	3.7	–5.0
52 Finance and insurance	55,640	4.8	2.8
53 Real estate/rental & leasing	29,575	2.6	19.6
54 Prof., sci. & tech. services	107,490	9.3	17.2
55 Mgmt. of companies	2,160	0.3	123.8
56 Admin./support, & wst. mgt.	53,725	4.7	18.0
61 Educational services	83,200	7.2	11.7
62 Health care & soc. assist.	107,065	9.3	7.8
71 Arts, entertainment & rec.	27,350	2.4	13.7
72 Accom. & food services	91,585	8.0	12.3
81 Other services	59,055	5.1	14.4
91 Public administration	44,120	3.8	–1.1

Source: Spencer and Vinodrai (2009).

labour experienced a 7.8 per cent growth over the same period.

As crucial as these specialized intermediate services are to advanced economies, in terms of productivity gains, wealth generation, and salaries, our necessarily succinct profile includes an acknowledgement of the very large labour force representations in the public sector. In the GTA, for example, the labour force in the educational services, and health care and associated social services, totalled about 400,000 in 2006, experiencing brisk growth rates of 20.1 per cent, and 17.3 per cent, respectively. In the Ottawa case, these two exceeded 100,000 workers, with growth rates of 11.1 per cent and 15.8 per cent, respectively (Table 7.4). Public services tend notably to be more resistant to recession than employment in private-sector services, acting to buffer local economies during downturns. This is especially true for smaller and mid-size urban communities, which in many cases do not encompass the large, specialized business and financial sectors of the major metropolitan cities.

Labour force in the construction sector represents a useful bellwether of urban growth. Until the abrupt check of the current downturn, the economic and physical growth of large Canadian

Table 7.4 Labour Force by Industry Group, Ottawa–Gatineau

	# Labour Force	% Labour Force	% Change 2001–2006
All industries	627,101	100.0	6.6
11 Agr., forestry, fish. & hunt.	3.690	0.6	–9.6
21 Mining & oil/gas extraction	645	0.1	118.6
22 Utilities	2,405	0.4	24.9
23 Construction	30,890	4.9	13.4
31–33 Manufacturing	28,760	4.6	–35.9
41 Wholesale trade	15,470	2.5	8.5
44–45 Retail trade	65,025	10.4	10.5
48–49 Transport & warehousing	20,965	3.3	1.0
51 Information & cultural ind.	18,870	3.0	–13.1
52 Finance and insurance	19,110	3.0	0.2
53 Real estate/rental & leasing	10,630	1.7	10.1
54 Prof., sci. & tech. services	59,650	9.5	–2.9
55 Mgmt. of companies	310	0.0	5.1
56 Admin./support, & wst. mgt.	29,705	4.7	14.4
61 Educational services	43,970	7.0	11.1
62 Health care & soc. assist.	61,905	9.9	15.8
71 Arts, entertainment & rec.	13,640	2.2	16.8
72 Accom. & food services	38,360	6.1	10.9
81 Other services	29,795	4.8	13.4
91 Public administration	133,200	21.2	18.8

Source: Spencer and Vinodrai (2009).

cities was facilitated by the expansion of the construction labour force. At the economic peak in 2006, the GTA's construction labour force numbered almost 150,000, comprising about 5 per cent of the regional labour force, and enjoying an expansion of almost one-fifth (19.7 per cent) during the period 2001–6. Montreal's construction labour force expanded by 23.8 per cent over this five-year period, reaching almost 89,000 by 2006. But the leading cities of Canada's 'New West' experienced the most dramatic growth in construction. In Calgary, the construction sector expanded by almost one-third (32.7 per cent) over the period 2001–6 (Table 7.5), while metropolitan Vancouver, in the middle of a housing boom and the massive pre-Olympic capital expenditure program, experienced 36.4 per cent growth in the construction labour force during the same period, attaining a level of 73,385 by 2006.

While labour force and employment have experienced impressive growth within the Canadian urban system, successive phases of innovation and restructuring over the last two decades have caused some destabilization of employment and labour force. Moreover, the current economic downturn most definitely will produce new structural

Table 7.5 Labour Force by Industry Group, Calgary

	# Labour Force	% Labour Force	% Change 2001–2006
All industries	653,505	100.0	15.9
11 Agr., forestry, fish. & hunt.	3,755	0.6	−8.5
21 Mining & oil/gas extraction	42,390	6.5	52.0
22 Utilities	6,625	1.0	35.9
23 Construction	53,670	8.2	32.7
31–33 Manufacturing	48,665	7.4	−1.7
41 Wholesale trade	31,445	4.8	11.2
44–45 Retail trade	68,570	10.5	12.6
48–49 Transport & warehousing	37,235	5.7	6.7
51 Information & cultural ind.	17,360	2.7	−8.3
52 Finance and insurance	25,035	3.8	7.8
53 Real estate/rental & leasing	14,660	2.2	16.3
54 Prof., sci. & tech. services	75,815	11.6	23.2
55 Mgmt. of companies	1,375	0.2	27.9
56 Admin./support, & wst. mgt.	27,135	4.2	12.0
61 Educational services	37,960	5.8	14.0
62 Health care & soc. assist.	55,860	8.5	22.0
71 Arts, entertainment & rec.	14,455	2.2	18.7
72 Accom. & food services	42,685	6.5	9.6
81 Other services	29,480	4.5	16.8
91 Public administration	19,320	3.0	19.2

Source: Spencer and Vinodrai (2009).

contractions as well as more transient effects. Fluctuations in the metropolitan labour force are accompanied in many cases by shifts in the location of enterprise and associated patterns of land use and city structure, a theme we shall turn to below.

Changing Dynamics of Employment Location in the Canadian City

As in other advanced societies, the location of employment in Canadian cities exhibits both continuities and disjuncture. Within the urban production sector (services as well as goods) firms and labour still congregate within agglomerations and clusters, to reduce input costs, and to benefit from knowledge spillovers and other externalities. Firms in the consumption sectors and industries, such as retail, personal, and many public services (such as K–12 education), still tend to 'follow the population' and are growing most rapidly in suburban areas with expanding residential communities. But a new mix of factors is reshaping the location of labour and, consequently, the geography of work, in Canadian cities.

Technological Innovation and the 'New Economy' of the City

Among factors of change we can acknowledge advances in *telecommunications technologies* and the digital revolution. Information and communications technologies haven't meant the 'death of distance'. Rather, information technologies have in some ways augmented the traditional advantages of locational proximity and face-to-face contact (Wheeler et al., 2000). But advanced telecom systems also open up new possibilities of conducting economic activity over extended space—for example, e-businesses, telecommuting, and use of the Internet for input sourcing, staff recruitment, and marketing. So, the effects of technological innovation in communications include both the *concentration* of activities in major centres, such as Toronto, which enjoy advantages of advanced telecommunications and other knowledge networks, as well as the *diffusion/dispersion* of work and employment, including teleworking and telecommuting (for an elaboration of these new forms and spatial patterns of work in the metropolis, see www.emergence.nu/news/growth.html).

Globalization and the New International Division of Labour

A second major bundle of change underpinning new employment patterns concerns *the international division of labour*. The 'new international division of labour' (Fröbel et al., 1980) was associated with the shift of industrial production from the developed West to the emergent economies of East Asia and Southeast Asia. More recently, a yet newer international division of labour entails larger, more diverse, and more complex movements of labour for short- or long-term periods across global space, shaped by bilateral trade agreements, sectoral treaties, corporate takeovers, strategic partnerships, and new approaches to supply chain management.

In Canadian cities, we see clear evidence of these trends, notably in (1) consolidation of the North American auto industry forcing labour contractions and plant closures in a number of southern Ontario cities; (2) reorganization of the pulp and paper sector, principally in Quebec and British Columbia, in response to global oversupply and the market decline (e.g., in the newspaper business); and (3) takeover of Hamilton's steel mills by foreign multinationals, owing to a range of factors including the intention of multinational steel corporations to capture critical expertise embedded in the Hamilton steel sector (Warrian, 2009). The rise of film and video production in Toronto, Montreal, and Vancouver also can be attributed to a range of factors, including Hollywood's desire to achieve operational economies (seen in the growth of 'runaway productions') and concentrations of local talent in each city.

International Immigration and the Reshaping of Urban Labour Markets

International immigration is well-established as a leading agent of transformation in urban-regional employment growth and change. For the largest cities, especially Toronto and Vancouver, immigration comprises a major growth component (UBC Metropolis website: riim.metropolis.net). The regional labour force has benefited greatly from inflows of workers, including large proportions of younger, working-age immigrants; significant numbers of entrepreneurial immigrants, with new energy and expertise; and investors who have expanded the pool of capital for business start-ups and development. New immigrants also contribute to the *global connectivity* of Canada's cities through communications and knowledge transfer operating between immigrants and co-ethnicists abroad: a clear competitive advantage in the knowledge economy.

Market Interdependencies and the Location of Employment

Another set of influences is subsumed within the general category of markets, specifically *housing*, *property*, and *consumption markets*. First, housing characteristics (price, supply, type) exert significant influences on employment location within urban-regional labour markets. In some of the larger cities, notably Toronto, Vancouver, and Calgary, inflationary housing markets have acted to reduce housing options for many, and to increase the spatial separation between place of work and place of residence. Second, the property market acts as a filter on the location of economic activity and employment. To illustrate: the insistent revalorization of inner-city property markets, a quarter-century following the collapse of the urban core's manufacturing sector, has displaced low-income populations and low-margin firms especially in Toronto, Vancouver, and Montreal. As a second example, relatively low land costs in suburban/exurban areas have tended to encourage the proliferation of low-density land use and employment formation, including business and industrial parks and retail strips. Clearly, a 'business case' (including considerations of cost and convenience) exists for these low-density employment sites. But they are difficult to service by public transit; are, thus, highly auto-dependent, so contribute to increased carbon emissions. Finally, trends in consumption influence certain kinds of activity and labour, e.g., the importance of consumption amenity for attracting cultural workers and the 'creative class'. Whether or not we're inclined to accept the ebullient prognosis of 'amenity as destiny' (Florida, 2002; for a counter-argument, see Storper and Scott, 2009), it seems clear that the distinctive amenity package of the urban core is important in attracting talent.

Policy Factors and the Location of Employment

Planning and policies directly, or indirectly, shape the metropolitan space-economy and labour force. At the regional level, these include not only plans for urban structure and land use, but also public investments in transportation (public transit as well as highway/road construction). At the local municipal level, public agencies exercise significant control over the location of enterprises and employment by means of zoning and land use policies and, less directly, through building and development guidelines, infrastructure provision, and fiscal policies. The past two decades have seen the emergence of more assertive economic development programs (including strategic economic development strategies, sector programs, and talent attraction/retention policies) in Canada (Spencer and Vinodrai, 2008), witnessing what David Harvey (1989) has referred to as a shift from 'managerial' to 'entrepreneurial' **governance**.

A spectrum of more localized programs influences metropolitan employment including (1) local regeneration programs, which promote investment and start-ups in specific locations, notably in low-income communities with inadequate access to jobs; (2) **business improvement area** (BIA) initiatives, designed to improve the local environment (taxation, amenities, information networks, and services), normally initiated by the local business community; (3) 'brownfield' redevelopment or retrofitting, involving older inner suburban districts in need of new investments and employment opportunities; (4) 'compact and complete' community programs designated for special zones of employment and residential concentration within urban-regional strategies; and (5) 'New Urbanism' projects on the metropolitan periphery, which may include new employment formation in business, retail, and personal services, alongside residential tracts.

The Changing Space-Economy and Patterns of Employment

Owing in large part to the mix of factors described above, the last quarter-century has seen significant changes in the industrial structure (mix of industries and employment), space economy and spatial divisions of labour in the Canadian metropolis. Canadian cities for the most part still present a profile that includes a relatively robust central area, including specialized industries, firms, and labour, and reinvestments in the core driven by new housing and residential districts, consumption, and specialized production. Indeed, a variant of the 'back to the city' movement has included the introduction of new social classes and cohorts in such cities as Vancouver, Toronto, Montreal, and Halifax, among others, facilitated in some cases by city planning and policies, including rezoning, housing policies, and public realm improvements. That said, the urban core has been essentially stripped of its manufacturing capacity, and the high growth trajectory of the downtown office sector of the 1970s and 1980s has slowed appreciably, so there has been both a relative and an absolute shift of employment growth to the suburbs.

The Rise of Suburban Employment in Canadian City-Regions

While manufacturing has been in secular decline in most Canadian city-regions since the 1970s, what remains is highly concentrated in the suburbs and exurbs. Suburban areas also encompass major concentrations of warehousing and distribution activity, taking advantage of the larger land parcels available on the periphery, as well as access to major regional transportation installations (ports, airports, rail systems, highways).

A second principal category of suburban employment within Canadian city-regions comprises the major final demand services linked to residential populations. These include local government and other public institutions; public education; large concentrations of retail services, including major shopping malls, retail strips, and more scattered retail outlets; and personal services, including professional services such as medical, dental, legal, and accounting services.

Over the past two decades or so suburban areas have attracted new industries, institutions, and associated labour, marking the continuing development and maturation of these zones of the metropolis. These include, for example, science parks and research and development (R&D)

facilities; universities and other tertiary educational institutions; secondary business, financial, and commercial centres, with specialized labour cohorts; film production and other cultural industries; and recreational and leisure development. International airports have expanded significantly, generating new suburban employment, notably at Pearson International Airport (in Mississauga) and Vancouver International Airport (in Richmond, in Vancouver's inner suburbs), each of which constitutes a major regional growth pole.

Contrasts and Commonalities in Urban Structure and Employment Patterns

While *suburbanization* as described represents a spatial feature of urban growth and change, the experience of labour formation in Canadian cities presents contrasts in the basic geography of enterprise, economic development, and employment. In the Greater Toronto Area (GTA), Canada's largest metropolis, the CBD's corporate office complex of specialized financial, head office, and commercial activities underpins that city's global status and functions. Over the past two decades, an internationally significant cultural economy has developed within Toronto's metropolitan core, including high design performance and exhibition space, a thriving cultural production sector (including film production as well as new media, graphic design, and music), and complementary amenities—each of which has generated substantial creative industry employment. That said, the fastest growing areas are in the GTA's suburban communities, including Mississauga (about 700,000 population, and almost half a million jobs), Scarborough (older industrial suburb), and Markham. While we tend to associate 'edge cities' (after Garreau, 1991) with the American urban experience, the GTA also encompasses representations of this particular phenomenon, with large commercial centres, business parks, and industrial estates distributed on the outer margins of the city-region, to an extent not seen anywhere else in Canada (Bourne et al., 2009).

In contrast, Montreal's economic structure exhibits a characteristically 'strong centre' profile, with employment (especially in terms of the most specialized industries and labour) still concentrated on the Island and in Laval. As is well known, Montreal lost its 'western gateway' functions (including transportation: Coffey, 1994), and, more decisively, its national primacy in banking and corporate control, to Toronto (Polèse and Shearmur, 2004) during the 1970s. But the metropolitan core has reconstituted itself over the last two decades as a zone of business services, tourism, and cultural industries and institutions, catering to regional and, selectively, to international markets. Indeed, as two specialists on Montreal's development have observed, the slow-growth trajectory of the 1970s and 1980s meant that 'the fabric of Montreal's downtown and central neighbourhoods has remained relatively intact'; preserving a built environment conducive to the arts, design, innovation, and convivial urbanism—all features of Montreal's renaissance since the 1990s (Shearmur and Rantisi, 2009). That said, Montreal's suburbs encompass significant manufacturing, including an important high-tech sector specializing in aerospace and transportation, following the decline of long-established inner-city districts such as Lachine in southwest Montreal, as well as port, warehousing, and distributional industries and employment.

The Ottawa–Gatineau metropolis presents still another city-region model of economic development and employment formation, shaped both by its national capital roles and by more recent developments. The metropolitan core of Ottawa is dominated by federal government institutions, agencies, and employment, with significant clusters of business services, retail, and consumption that cater to the large tourism sector as well to local demand; while the suburbs encompass important high-tech clusters. The steep decline of Nortel, one of Canada's truly propulsive advanced-technology corporations in the 1990s, has attenuated the growth trajectory of the suburbs, but

Kanata, in particular, boasts a nationally important cluster of biotech and telecommunications industries. Federal government and socio-cultural influences both are crucial to patterns of growth. In the words of leading scholars 'the city exemplifies the strong intersections between politics, ethno-cultural diversity, gender and industry location dynamics in a highly suburbanized metropolitan development. . . . Ottawa–Gatineau's weak manufacturing sector and strong service-based employment makes it a compelling example of the complex geographies of work in post-industrial urban Canada' (Andrew et al., 2008).

Much discussed in the media and elsewhere is a shift in national growth (population, employment, and investment) towards western Canada; aggregate data tend to support this thesis. But this growth is locationally selective, favouring Vancouver and Calgary, the leading urban centres of the 'New West'.

Vancouver offers a distinctive case study in sustained growth in the post-war period. Here the 1980s was a decade of transformative change, marked by the decline of Vancouver's role as centre of an expanding provincial resource economy, and the integration of Vancouver into the markets, capital flows, and migration patterns of the Asia-Pacific (Olds, 2001). The outcomes of these transformative processes of change are complex, but include the comprehensive redevelopment of the urban core, and the rapid growth of the suburbs. With regard to the first, global forces served to strip the city of much of its head-office and senior management functions. In 1991, the city's seminal *Central Area Plan* enabled a major reallocation of land resources in the core, consolidating the office complex within a smaller CBD, and privileging housing in most of the inner city though the central area still encompasses about 200,000 jobs (Vancouver, 2008). And the city still has about one-third of the approximately 900,000 jobs in the regional labour force. But the Vancouver suburbs are leading growth as about three-quarters of the regional population of 2.2 million are located beyond the City of Vancouver. The older inner suburbs, such as Burnaby, Richmond, and North Vancouver, possess major employment centres, such as designated regional town centres, universities and colleges (e.g., Simon Fraser University and British Columbia Institute of Technology in Burnaby), advanced-technology industries, and major gateway facilities (Vancouver International Airport, in Richmond) and enjoy a favourable jobs-to-residents balance. But the newer, outer suburbs—notably Surrey, but also Delta, Coquitlam, and Langley, are experiencing the highest growth rates: the population of Surrey approaches a half-million, and the municipality is experiencing significant job growth. On the whole, though, the outer suburbs are deficient in jobs, relative to the resident population, and much of the job growth is occurring in scattered, low-density developments, posing problems for regional and local planners.

Calgary represents the other leading urban growth pole in western Canada. At just over 1 million, it has a little less than half of Vancouver's population. Aside from this scalar difference are other, perhaps more consequential, contrasts between these two largest western metropolitan cities. Vancouver, stripped of many of its leading corporations, and embedded firmly in a 'post-staples' development modality, exhibits a classic entrepreneurial SME (small- to medium-size enterprise) economy and labour force, with industrial diversification including high education, gateway functions, film production and new media. Calgary presents the classic image of an essentially 'mono-cultural economy', driven by Alberta's oil and gas industry. Over the last decade or so, Calgary has supplanted Vancouver as the leading head-office centre west of Toronto. A leading physical consequence of this trajectory is the development of a major corporate head-office complex in Calgary's downtown, anchoring the regional economy, and reinforcing the core as the centre of economic gravity in the region. Moreover, the expansion of corporate head offices has generated ancillary expansion in the intermediate service sector, notably business and financial services, as well as the consumption amenities that constitute derived demand features of the office economy. Beyond the urban core Calgary presents a largely dispersed and diffused regional economy and labour force, albeit with Calgary International Airport and some substantial office and business parks constituting significant clusters.

The New Economy and the Remaking of the Central City

The 1970s and 1980s saw the dramatic rise of the CBD's corporate office complex, and an associated workforce of segmented office labour, as well as the collapse of Fordist manufacturing and blue-collar labour within the traditional industrial districts of the inner city (Bourne and Ley, 1993; Ley, 1996). More recently, successive phases of innovation and restructuring have transformed the economy and labour force of the metropolis, including a powerful cultural inflection (Scott, 1997), and have shaped change in the urban core's land use, social morphology, and housing markets (Hutton, 2008, 2009).

In many cities among advanced societies, including those in Canada, the initial cohorts to recolonize the inner city's post-industrial terrains of disinvestment and decline included artists and designers. In Toronto, Montreal, and Vancouver, new uses have included museums, galleries, studios, and exhibition spaces, presaging the full-blown cultural makeover of the core now seen as a marker of globalizing cities. Then, by the mid-1990s, the technology boom gave rise to a 'New Economy' construct in the inner city, comprising new media, digital art, and the ubiquitous dot.coms alongside more conventional cultural industries (Pratt, 2000; Hutton, 2004). The New Economy phenomenon displaced both low-income residents and marginal creative workers, a trajectory only attenuated by the collapse of the technology sector of 2000–1, in the wake of oversupply and the inflation of technology stocks.

The late twentieth-century New Economy has been replaced by a 'new cultural economy' of production and consumption in the city, a sector that has absorbed many of the innovations in production and communications technologies of the digital age, but taking in a far larger set of industries, institutions, and labour. The locational preferences of the new cultural economy are shaped in large part by *agglomeration*: the oft-replicated co-location of firms and labour that produces input and transactional cost savings, knowledge spillovers, and opportunities for co-operation and collaboration. So, important locational continuities span periods of development. Further, the cultural economy operates within the conventional economic parameters of markets, firms, production networks, and competition. But the attraction of the urban core—and more particularly the heritage landscapes of the post-industrial inner city—for cultural industries and creative labour involves a more diverse and complex array of factors, which these agencies and workers can draw on for creative production, as depicted schematically in Figure 7.1.

First, we can acknowledge the foundational elements of 'space and spatiality': the distinctive localized **micro-spaces of the core**, which provide ideal territories for intensive interaction, facilitating the exchange and transfer of tacit knowledge and information so critical to the functioning of the knowledge economy; and the unique built environment of the inner city, in the form of heritage industrial structures, older housing, and the new structures associated with the most recent phase of redevelopment in the city. The resonant landscapes of the inner city combine with the rich amenity package of the core to produce the well-known **milieu effects** conducive to creativity and innovation in the knowledge economy.

The urban core also possesses an important 'talent advantage' for creativity (Figure 7.1), in the form of the mix of human, social, and cultural capital concentrated within the inner city of the metropolis. These include the congeries of artists, designers, and 'neo-artisanal' workers (Norcliffe and Eberts, 1999) who value the commingling of social and economic worlds characteristic of the core, the distinctive 'social density' of core area neighbourhoods and communities, and the 'neo-bohemian' cohorts of creative producers and consumers widely acknowledged as drivers of the cultural economy. At the larger metropolitan level, the emergence of a 'new cultural economy' concentrated within the urban core reflects the emerging regional divisions of industrial activity and labour, a process Allen Scott has described as the continuing 'specialization of the internal spaces of the metropolis' (Scott, 1988).

The development of the cultural economy of the city is supported by heritage preservation, supports for artists and designers, and land use and zoning programs which preserve land resources for cultural production amid the high-development terrains of the urban core. But over the last decade cultural planning and programming for the creative sector have entered the policy mainstream, abetted by the energetic pitch of Richard Florida and his acolytes (Florida, 2008), who insist that the prosperity of cities is inextricably linked to the creative impulse. While others caution a more measured stance on the potential for culture and creativity as forces of urban regeneration (see Evans, 2001, 2009), or, in some cases, a rejection of the more excessive claims of the cultural economy as just the latest in a sequence of neo-liberal siren songs (Peck, 2005), the overall message of culture as panacea for urban stagnation has proven seductive to more and more cities.

The Cultural Economy in Canadian Cities: Reference Cases

Inner-city districts of Canadian cities have been recast as zones of creative innovation, production, and consumption. In medium-size cities such as

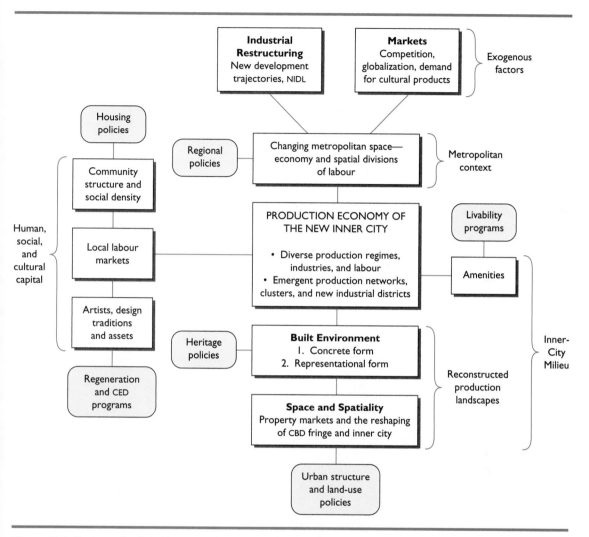

Figure 7.1 Factors shaping the production economy of the new inner city

Source: Hutton (2008). © Routledge/Taylor & Francis.

Halifax, Quebec City, and Victoria, for example, the old inner city has been redeployed as a site of creative production (artists, graphic designers, architects, and the like), consumption, and spectacle, drawing on the unique heritage landscapes and legacies of those cities, complementing the important capital, administrative, and public-agency functions long-established in central urban precincts.

In the major Canadian metropolitan cities, the scale of the 'new cultural economy' is commensurately larger, and inclusive of a greater range of specialized industries and labour. As noted earlier, the preservation of Montreal's core area urban fabric has provided an ideal environment for the nurturing of cultural industries in districts such as the Plateau and Mile End, complemented by the City's famously tolerant lifestyle orientation and vibrant consumption sector. Among Canadian cities generally, Montreal boasts perhaps the most generous state/public support for the creative sector, reflecting a well-established state commitment to fostering the city's unique cultural values, assets, and practices, as well as sector-specific programs for film, fashions, and architecture (Klein and Tremblay, 2008).

The situation of the new cultural economy within the dense and diverse landscapes and communities of the inner city especially generates a mixed palette of outcomes, including **dislocation** as well as **regeneration**, vividly illustrated by experiences in Toronto and Vancouver. In the case of the former, J.P. Catungal et al. have documented the development trajectory of Liberty Village, a 45-acre **brownfield site** on the western margins

of the downtown, and proximate to the old industrial community of Parkdale. The district included well-known manufacturing operations such as Inglis (electrical appliances) and Massey-Ferguson (agricultural equipment), an assortment of firms redolent of Toronto's heyday as a major Canadian industrial city. The formative years of Liberty Village's evolution as cultural precinct included the classic experience of recolonization of obsolescent industrial space by artists, as early as the 1970s, with the 'new cultural economy' phase beginning 'in earnest' in the mid-1990s, 'during the rise of the so-called dot-com industry, a technology-driven period of economic boom' (Catungal et al., 2009: 1099). The district's development was facilitated by the government's abolition of monopolies over local telecommunications services, which led to developers offering advanced Internet services

Figure 7.2 Yaletown

to attract new media firms; and also through the deregulation of municipal zoning laws, which served to facilitate the recycling of industrial buildings for new uses. On the face of it, Liberty Village stands as a major success story in cultural programming as the instrument of urban regeneration. But a more critical perspective suggests an at least partial failure of the project, in terms of addressing 'attendant urban problems such as gentrification, inequality, working poverty, and racialised exclusion' (Catungal et al., 2009: 1111).

At a finer grain of resolution we can identify more specific employment profiles and regeneration/dislocation outcomes within the inner city. In Vancouver the last decade and a half or so has seen the emergence of discrete, specialized production zones in the metropolitan core, differentiated by product sector, labour, and proximate impacts (Figure 7.3), producing a new geography of employment extending well beyond the CBD (Figure 7.4). At the top of the cultural production pyramid is Yaletown, a high-integrity heritage district between the Downtown South and Concord Pacific new residential communities. Yaletown enjoys the optimal location, richest amenity package, and most exclusive cachet of all the inner-city districts, reflected in its land prices and rents. Yaletown attracts many of the high-end, most successful, new-economy firms, including video game production, software designers, and Internet design and imaging concerns. The very high prices ensure a ruthless filtering process and continuous turnover of firms, but the scope of social dislocation is limited by the mostly upscale housing and consumption spaces in the area. On the other hand, Victory Square presents a grittier landscape 'look and feel', and attracts a more start-up and/or struggling firm profile relative to Yaletown, but nonetheless a significant upgrading experience, given its location proximate to the Downtown Eastside (Barnes and Hutton, 2009). The intimate juxtaposition of creative industry workers within the marginal communities of the Downtown Eastside

presents a vivid example of the increasingly finer-grained geography of employment in the city.[1]

Conclusion: Emergent Morphologies of Urban Employment

Important continuities are to be seen in the location of employment in Canadian cities, notably (at the urban system level) in the primacy of the Greater Toronto Area (GTA), and the growing dominance of the largest city-regions; and (at the intra-metropolitan scale) in the persistence of agglomeration in the location of specialized industries and labour. That said, in many other respects *change* is the defining trend in the structure and spatiality of employment in Canadian city-regions, driven by globalization, accelerated industrial restructuring, insistent technological change, intensified market competition, and a series of economic shocks and stresses over the past two decades.

At the national urban system level, Canada has six city-regions with populations exceeding one million—the GTA, Montreal, Vancouver, Ottawa–Gatineau, Calgary, and Edmonton. Common to each is an ascendant service sector, concentrations of specialized, advanced service labour, and a powerful 'professionalization' trajectory. But the development formula for each of these major city-regions presents important contrasts: Toronto and Montreal have large, but declining, manufacturing employment levels, while Vancouver has a small but relatively stable manufacturing sector. Within the dominant service economy, Toronto and Calgary have major head-office sectors, relative to the diminished corporate control presence in Montreal, and the almost 'post-corporate' character of Vancouver's economy.

Spatially, Toronto and Vancouver feature central-city, new industry formations in the cultural sector, combined with a residential trajectory

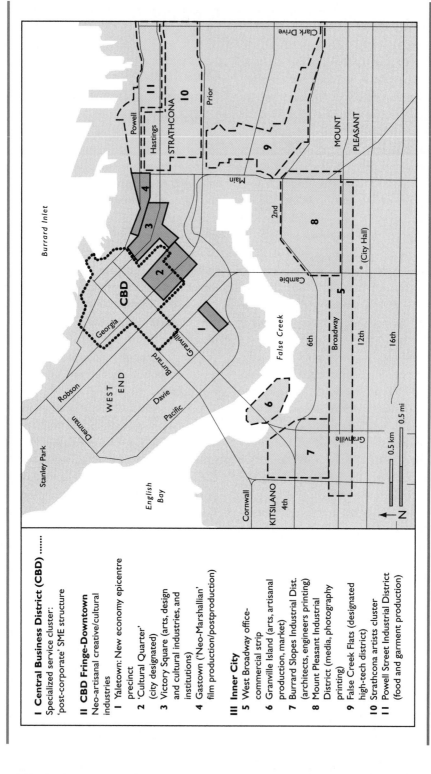

I Central Business District (CBD) ......
Specialized service cluster:
'post-corporate' SME structure

II CBD Fringe-Downtown
Neo-artisanal creative/cultural
industries

1 Yaletown: New economy epicentre
precinct
2 'Cultural Quarter'
(city designated)
3 Victory Square (arts, design
and cultural industries, and
institutions)
4 Gastown ('Neo-Marshallian'
film production/postproduction)

III Inner City

5 West Broadway office-
commercial strip
6 Granville Island (arts, artisanal
production, market)
7 Burrard Slopes Industrial Dist.
(architects, engineers printing)
8 Mount Pleasant Industrial
District (media, photography
printing)
9 False Creek Flats (designated
high-tech district)
10 Strathcona artists cluster
11 Powell Street Industrial District
(food and garment production)

Figure 7.3 Vancouver: central business district, downtown, and inner city

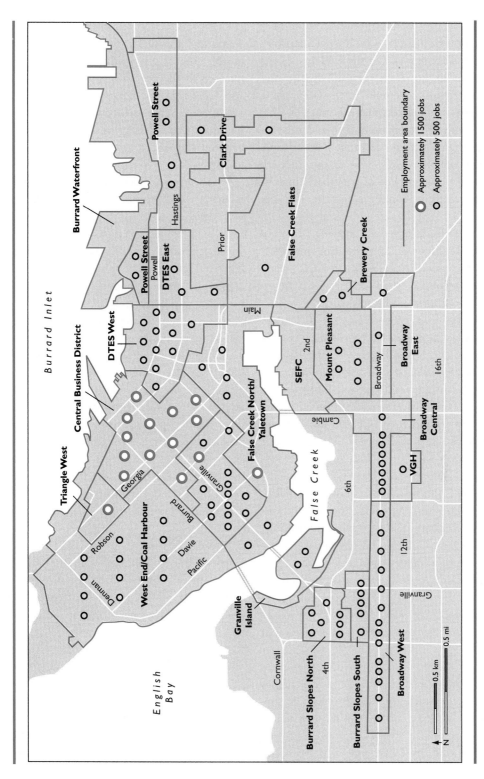

Figure 7.4 Vancouver: job density in CBD, downtown, and inner city

driven both by new condominium development and the adaptive re-use of obsolescent industrial and commercial buildings. On the evidence of the last decade or so, however, volatility in the mix of industries, firms, and employment within these new economy sites suggests that the inner city has become a zone of experimentation and innovation, rather than a domain of durable and deeply embedded production firms and labour. In each city-region, too, the growth of suburban employment is an important commonality, but there are major contrasts of scale and industrial mix, with the GTA exhibiting a suburban and edge/edgeless city formation almost on an American metropolitan scale, because it includes Mississauga, which would rank among Canada's 10 largest cities if considered as an autonomous settlement.

Below the level of the 'big six' Canadian city-regions we can identity large but relatively slower-growing cities characterized by a balance of services, manufacturing, and distribution industries (Winnipeg, Regina, Hamilton, Saint John), as well as smaller cities that have capitalized on an industrial niche and employment specializations. These include, notably, cities with important higher education and R&D vocations (Waterloo, Saskatoon, Halifax), high-amenity cities with considerable in-migration 'pull' (Kelowna, Niagara-on-the-Lake), and the 'satellite' manufacturing cities on and just beyond the periphery of the GTA, including Oshawa, Windsor, and Oakville. The latter have functioned as sites of advanced production and skilled, highly paid manufacturing labour consequent to the Canada–US auto pact. But the current shake-out in the North American auto industry certainly imperils the viability of these crucial employment centres.

The last three decades have witnessed a sequence of innovation and restructuring phases in Canadian cities, including the structural decline of basic manufacturing common to most advanced societies that produced a major downturn in central Canada in the late 1980s and early 1990s, the crash of the technology boom of 2000 and afterwards, and the deep contractions associated with (at the time of this writing) the economic downturn. As in the previous episodes of restructuring and recession, the downturn is generating both transitory contractions as well as permanent losses of capacity and labour in key industries, such as banking and finance, the auto sector, and primary industries.

As we project the conditions for employment formation and location likely to develop over the next decade and beyond, it is difficult to imagine how a more stable labour force could emerge within Canada's urban landscapes. To the destabilizing effects of technological substitution for labour and market competition, we can add migration, demographic trends, and property market pressures, although most concede that these factors can produce positive effects (innovation, creativity, efficiency) as well as negative impacts (structural unemployment, dislocation and displacement, **polarization**, and marginalization). Growing awareness and acceptance of climate change will force further changes to business practices, again producing in some cases significant job losses, but also stimulating the formation of new environmental industries, firms, and labour: part and parcel of the cycles of transition and the ever-shifting geography of employment in Canadian cities.

Acknowledgements

I would like to acknowledge the cartographic contributions of Mr Eric Leinberger, of the Department of Geography, University of British Columbia, in preparing and adapting the maps included in this chapter. I want to acknowledge Routledge (Taylor & Francis) for granting permission to include Figure 7.1. I also owe a debt of gratitude to a number of colleagues for statistical data and acute insights on employment structure and location in Canadian cities, notably Richard Shearmur (INRS-Montreal), Jim Simmons (University of Toronto and Ryerson), Greg

Spencer (University of Toronto), and Tara Vinodrai (University of Waterloo). Larry Bourne read an earlier draft of the paper and contributed useful ideas for revision.

Review Questions

1. What does change in the intra-urban location of employment imply about overall change in the structure of metropolitan areas across Canada?
2. Referring to a particular employment type, explain how its changing location can be seen to be brought about by multi-scalar flows.

Note

1. For data on inter- and intra-urban employment distribution, see also William J. Coffey and Richard G. Shearmur, 'Employment in Canadian Cities', Chapter 14 in Bunting and Filion (2006) at: <www.oup canada.com/bunting4e>.

References

Andrew, C., B. Ray, and G. Chiasson. 2009. 'Capital formation', paper prepared for 'Trajectories of Change in Canadian Urban Regions' project. Ottawa: University of Ottawa.

Barnes, T.J., and T.A. Hutton. 2009. 'Situating the new economy: Contingencies of regeneration and dislocation in Vancouver's inner city', *Urban Studies* 46, 5 and 6: 1247–69.

Bell, D. 1973. *The Coming of Postindustrial Society: A Venture in Social Forecasting*. New York: Basic Books.

Bourne, L.S., J.N.H. Britton, and D. Leslie. 2009. 'The Greater Toronto Region: The challenges of economic restructuring, social diversity and globalization', paper prepared for 'Trajectories of Change in Canadian Urban Regions' project. Toronto: Department of Geography, University of Toronto.

——— and D.F. Ley. 1993. *The Changing Social Geography of Canadian Cities*. Montreal and Kingston: McGill-Queen's University Press.

Bryson, J.B., and P.W. Daniels, eds. 2007. *The Handbook of Service Industries*. Cheltenham, UK: Edward Elgar.

Catungal, J.-P., D. Leslie, and Y. Hii. 2009. 'Geographies of displacement in the creative city: The case of Liberty Village, Toronto', *Urban Studies* 46, 5 and 6: 1095–114.

Coffey, W.J. 1994. *The Evolution of Canada's Metropolitan Economies*. Montreal: Institute for Research on Public Policy.

Evans, G. 2001. *Cultural Planning: An Urban Renaissance?* London and New York: Routledge.

———. 2009. 'Creative cities, creative spaces and urban policy', *Urban Studies* 46: 1003–40.

Florida, R. 2002. *The Creative Class: And How It's Transforming Work, Leisure, Community and Everyday Life*. New York: Basic Books.

———. 2008. *Who's Your City? How the Creative Class Is Making Where to Live the Most Important Decision of Your Life*. New York: Basic Books.

Fröbel, F., J. Heinrichs, and O. Kreye. 1980. *The New International Division of Labour*. Cambridge: Cambridge University Press.

Garreau, J. 1991. *Edge City: Life on the New Frontier*. New York: Doubleday.

Harvey, D. 1989. 'From managerialism to entrepreneurialism: Transformation in urban governance in late capitalism', *Geografiska Annaler Series B—Human Geography* 88B: 145–58.

Hutton, T.A. 2004. 'The new economy of the inner city', *Cities* 21, 2: 89–108.

———. 2008. *The New Economy of the Inner City: Restructuring, Regeneration and Dislocation in the Twenty-First-Century Metropolis*. London and New York: Routledge.

———. 2009. 'The inner city as site of cultural production *sui generis*: A review essay', *Geography Compass* 3, 2: 600–29.

Innis, H. 1933. *Problems of Staple Production in Canada*. Toronto: Ryerson.

Klein, J.-L., and D.-G. Tremblay. 2008. 'The actors of civil society and their role in metropolitan governance: Toward a more inclusive governance?', paper presented at the annual meeting of the Innovation Systems Research Network, Montreal, 30 Apr.–2 May.

Ley, D.F. 1996. *The New Middle Class and the Remaking of the Central City*. Oxford: Oxford University Press.

Norcliffe, G., and D. Eberts. 1999. 'The new artisan and metropolitan space: The computer animation industry in Toronto', in J.-M. Fontan, J.-L. Klein, and D.-G. Tremblay, eds, *Entre la métropolisation et la village global: Les scenes territoriales de la reconversion*. Quebec: Presses de l'Université du Québec.

Olds, K. 2001. *Globalization and Urban Change: Capital, Culture, and Pacific Rim Megaprojects*. Oxford: Oxford University Press.

Peck, J. 2005. 'Struggling with the creative class', *International Journal of Urban and Regional Research* 29: 740–70.

Polèse, M., and R. Shearmur. 2004. 'Culture, language and the location of high-order services functions: The case of Montréal and Toronto', *Economic Geography* 80: 329–50.

Pratt, A. 2000. 'New media, the new economy, and new spaces', *Geoforum* 31: 425–36.

Scott, A.J. 1988. *Metropolis: From Division of Labor to Urban Form*. Berkeley and Los Angeles: University of California Press.

———. 1997. 'The cultural economy of cities', *International Journal of Urban and Regional Research* 21: 323–39.

Shearmur, R., and N. Rantisi. 'Montreal: Case study of a strong-centre city region', paper for 'Trajectories of Change in Canadian Urban Regions' project. Montreal: INRS and Concordia University.

Spencer, G., and T. Vinodrai. 2008. 'Where have all the cowboys gone? Assessing talent flows between Canadian cities', paper presented to annual meeting of the Innovation Systems Research Network, Montreal, 30 Apr.–2 May.

——— and ———. 2009. 'Statistical digest of Canadian cities', prepared for Innovation Systems Research Network, Munk Centre, University of Toronto.

Storper, P., and A.J. Scott. 2009. 'Rethinking human capital, creativity and urban growth', *Journal of Economic Geography* 9, 2: 147–67.

Vancouver, City of. 2008. *Metropolitan Core Jobs and Economy Land Use Plan*. Planning Department, Central Area Division.

Warrian, P. 2009. 'Biotech in lunch buckets: The curious knowledge networks in Steel Town', paper presented to annual meeting of the Innovation Systems Research Network, Halifax, 29 Apr.–1 May.

Wheeler, J.O., Y. Aoyama, and B. Warf, eds. *Cities in the Telecommunications Age: The Fracturing of Geographies*. New York and London: Routledge.

CHAPTER 8

Life Course and Lifestyle Changes: Urban Change through the Lens of Demography

Ivan Townshend and Ryan Walker

Demographic change is an important driver of social, economic, political, and cultural change. It affects the evolution or emergence of lifestyles, consumption patterns, housing markets, intergenerational relations, types of social inclusion and exclusion, and diversity. In the urban environment we see manifestations of demographic change through time, and we see and experience the spatial outcomes of demography in our everyday lives.

Even a brief introduction suggests that urbanists are working in the dark if they do not have some appreciation for demographic, **life course**, and lifestyle changes affecting Canadian cities. Accordingly, in this chapter we outline a selection of demographic trends that are important structuring parameters of twenty-first-century urbanization in Canada. The first section examines some of the key features of Canadian demographic change. The second follows with a discussion of the life course transitions people typically experience and how these relate to lifestyle changes. The third section brings demographic, life course, and lifestyle changes together through a look at a selected series of effects on the built environment.

Key Forces of Social and Demographic Change in Canadian Cities

According to the 2006 census of population, Canada is home to 31.6 million people. By international standards we are a medium-sized country, ranked as only the thirty-ninth largest in population. By comparison, the population of China is 40 times larger and the population of the United States is more than nine times that of Canada. Nevertheless, Canada has experienced tremendous population growth in recent decades. The population has more than doubled since 1951. It is important to understand the sources of this growth. Demographers typically allocate population growth or change to three distinctive components: fertility (births), mortality (deaths), and migration (immigration).

Patterns of fertility in Canada have changed dramatically over the past few decades (Figure 8.1). If you are in your early twenties now, think back to your great-grandparents' or grandparents' generations—those born in the 1920s or 1930s. It was common for these people to be born into relatively large families, and, though family size dropped progressively throughout the twentieth century, by the time the women born after World War I were in their peak child-bearing years in the late 1940s to early 1960s, as a group they were producing a greater number of children than at any other time in Canada's history. In 1961, fertility rates peaked at almost four offspring per woman of child-bearing age (Statistics Canada, 2006). Children of the post-World War II period (your parents' generation if you are now in your twenties or thirties) are part of the biggest cohort of children ever born

Figure 8.1 Total fertility rate, Canada, 1926–2006

Source: Statistics Canada (2008b: 8).

in this country, the so-called 'baby boomers'. The baby boomers are a notable feature of Canada's age structure and have impacted society at every stage of their development in everything from their need for schools, jobs, demand for housing, and consumer preferences (Foot with Stoffman, 1996). Population pyramids, which depict the age-sex structure of a population, clearly show this large 'bulge' of baby boomers, and as they have aged, the bulge has gradually moved through the age structure through time (Figure 8.2). One analogy used to describe this process is the 'pig in the python', a symbolic statement of how, through time, the baby boomers are detectable within the age structure of the population (Dytchwald, 1990).

Figure 8.1 shows that following the introduction of new birth control methods such as the pill in 1961 (legalized for birth control in 1968) and the increased choices of women with respect to the termination of pregnancies, fertility rates began to plummet in the early 1960s and continued a systematic decline until the late 1980s. Despite a very modest increase in fertility during the late 1980s (the baby boom echo), fertility rates have continued to decline until the present, reaching the lowest levels around 2001 at just over 1.5 children per woman (Statistics Canada, 2008a). This pattern of change in fertility through time is common to most of the developed world, and is a significant feature of what has been called the second demographic transition (Bourne and Rose, 2001; Lesthaeghe, 1995). There are a number of implications of such fertility trends. Quite simply, fewer babies being born to each couple contributes to population aging, increasing the average (median) age, which has risen from 23.9 years in 1921 to 39.5 years in 2006. A second implication of low levels of fertility surrounds the issue of population replacement. Generally, given the chances of accidental death or infant death, the replacement rate is

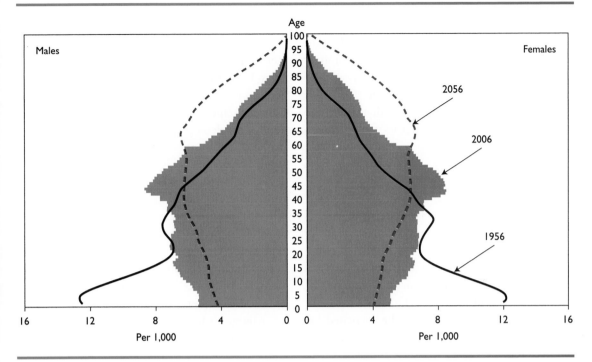

Figure 8.2 Population pyramids, 1956, 2006, and projection for 2056

Source: Statistics Canada (2008b: 25).

2.1 children born by every woman. In other words, to maintain a stable population size, every male and female couple needs to be replaced with 2.1 children. Given that fertility rates have been below replacement levels since the late 1960s and are significantly below replacement levels at present, in the absence of immigration the Canadian population would decline as baby boomers pass away over this and subsequent decades. Immigration is a critical feature of stabilizing population levels as well as population growth in Canada.

The low aggregate figures for fertility rates across the Canadian population mask a demographic issue of great transformative potential in some cities, particularly in western Canada. The fertility rate among Aboriginal women in 2006 was 2.6 children, considerably higher than among non-Aboriginal women (1.5 children) and above

the replacement rate (2.1 children). In cities like Saskatoon, Regina, and Prince Albert, Saskatchewan, children and youth (i.e., aged 24 years and younger) made up over half of the Aboriginal urban population in 2006 (Statistics Canada, 2008e). The Aboriginal population accounts for a significant part of the total urban population in each of these cities (approximately 9 per cent in both Saskatoon and Regina; 34 per cent in Prince Albert). According to population projections by Statistics Canada, by 2017 Aboriginal people aged 20–29 could make up 30 per cent of people in their twenties in Saskatchewan. The corresponding figure in Manitoba is 24 per cent and in the federal territories it ranges from 40 to over 80 per cent (Statistics Canada, 2008e).

The second component of population growth is death. Like most industrialized Western societies,

Canada has seen profound increases in life expectancy at birth during the nineteenth and twentieth centuries. The earliest estimates in Canada (for 1831) placed male life expectancy at only 38.3 years and female life expectancy at 39.8 years (Beaujot, 1991). A century later (1931), life expectancies at birth had increased dramatically to 60 years for males and 62.1 years for females—an increase of 57 and 56 per cent, respectively. For males, life expectancy at birth has risen from 60 in 1931 to 78 years in 2005, and for females from 62.1 in 1931 to 82.7 years in 2005 (Nagnur, 1986; Statistics Canada, 2008a). Patterns of survival have changed dramatically, primarily as a result of advances in health care as well as improved nutrition and lifestyles. Two key features of survival have contributed to changes in the demographic structure of Canadians: the decline and near elimination of infant mortality and the increased life expectancy at almost all ages, and especially in old age. Together, these two features have led to an increasing 'rectangularization' of the **survival curve** (Fries, 1980; Kraus, 1988; Simmons-Tropea and Osborn, 1993).

In the early part of the twentieth century (1931), infant mortality was a significant portion of death in Canada, with approximately 1 in 10 children not surviving to their first birthday (Statistics Canada, 2008a). By 2001 infant mortality had been virtually eradicated with fewer than 5.1 children per thousand dying before age one. The second aspect of survival, the increase in longevity, can be seen at almost every age. In Canada, a male aged 65 in 2005 can expect to live for another 17.9 years while a female aged 65 can expect to live for another 21.1 years (Statistics Canada, 2008a). It is important to note that the aggregate figures for the Canadian population mask a disparity between Aboriginal and non-Aboriginal peoples in life expectancy. In 2001 the life expectancy of Aboriginal men and women was about five years less than for non-Aboriginal Canadians (Statistics Canada, 2008e).

One notable issue is the disparity in life expectancies and survival patterns between men and women, with women surviving longer than men, although the gap has been closing in recent years. This situation is sometimes referred to as the 'feminization of survival', meaning that old age in Canada is increasingly characterized by widowhood. This demographic has implications for the size and composition of elderly households and living arrangements. Second, when coupled with declining fertility rates, greater life expectancy contributes to population aging, reflecting the fact that death is being compressed into a relatively short period at the end of the scale of life expectancy. Changes in survival patterns have had important impacts on life course transitions and lifestyles, aspects of society that will be discussed later in the chapter.

Migratory increase is the third component of population growth. Canada has long been recognized as a country of immigrants, and distinctive waves of immigration to this country have been documented (Boyd and Vickers, 2000; Hiller, 2006). Figure 8.3 shows the trends in the share of population growth attributable to natural increase and migration. Between the early 1980s and 2006 the share of migratory increase has risen dramatically, and in 2006 migration was responsible for more than half of population growth (Statistics Canada, 2008b). Demographic projections show that immigration is essential to the sustainability and growth of the Canadian population because, unless fertility rates begin to rise again, deaths in Canada will outnumber births (to be expected in an aging population) by about 2030, and the Canadian population could become entirely dependent on immigration for its growth. Immigration has helped reduce the rate of population aging that otherwise would have occurred. More than ever before, immigration has contributed to the rich social diversity of this country. In 2006, 6.2 million people, or one in five persons (19.8 per cent) in Canada, were immigrants, and 3.6 per cent (1.1

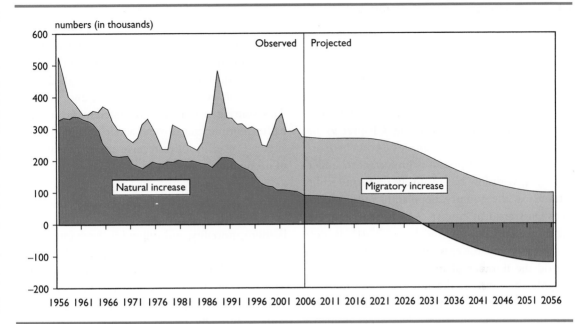

Figure 8.3 Migratory and natural increase in the Canadian population, 1956–2056
Source: Statistics Canada (2008b: 5).

million) of the population were recent immigrants, having arrived in the previous five years.

Earlier immigrant streams to Canada had been mainly European in origin so social distance within different waves of immigrants or between immigrants and the host Canadian society were minimized. However, since the 1970s other originating regions, such as Africa, the Caribbean, Latin America, and especially Asia, have been favoured. Nearly 80 per cent of Canada's new immigrants were from these regions by the 1990s (Bourne and Rose, 2001; Boyd and Vickers, 2000; Halli and Driedger, 1999). Immigrant streams have increasingly been visible minorities with higher levels of social distance, so that between 1981 (the first time visible minorities were enumerated) and 2006 the share of visible minorities in the population increased from 3.2 million (11.2 per cent) to almost 5 million (16.2 per cent) people. In short, social diversity on the basis of visible minority

status and ethnicity is increasing the social mix in Canada with important implications for urban social life in this country (Statistics Canada, 2008c). In this regard, too, it should be pointed out that Aboriginal Canadians are not considered 'visible minorities' in the accounting of Statistics Canada.

While Canada's population is aging overall, the migratory decisions made by younger and older Canadians regarding their ultimate choice of destination creates interesting city-specific differences (Murdie and Teixeira, 2006). Examples include the movement to retirement centres like Kelowna and Victoria or to relatively younger communities with greater employment opportunities and lower cost of living, such as Calgary and Saskatoon (see Rosenberg and Wilson, Chapter 21). Domestic migration trends, presented at a provincial scale in Table 8.1, are an important source of population redistribution and a structuring parameter of Canadian cities as well. Studies have shown that jobs in the mining, oil, and gas sectors and in public administration are key drivers of interprovincial migration (Statistics Canada, 2008a). Some provinces consistently gain from interprovincial migration, however, while others consistently lose (Table 8.1). Other chapters cover this in greater detail in relation to inter-city migration by immigrants (see Hoernig and Zhuang, Chapter 9) and by people of different age groups (Rosenberg and Wilson, Chapter 21), but it is important to understand that with low but relatively uniform fertility rates across the country, population redistribution between Canadian cities is a significant source of urban change (Bourne and Rose, 2001). Bourne and Rose found, for example, that over a five-year period the number of interprovincial moves was roughly the same as the number of immigrants to Canada. They also found that Canada's 25 largest metropolitan areas have had a net domestic migrant loss over the past couple of decades (the result of 'domestics' either moving to an alternative

metropolitan area—e.g., from an eastern metropolitan area to Calgary or Vancouver—or to outer parts of their own extended metropolitan area—e.g., from the Toronto CMA to outer parts of the Greater Toronto Area or GTA, to high-growth places such as Guelph, Kitchener, and Oshawa. This shifting is dramatically different from international immigrant streams where a handful of the largest Canadian cities receive the majority share of newcomers.

Life Course and Lifestyles

Fertility, mortality, and migration have introduced important changes in the ways the Canadian population has grown and will continue to grow. These factors point to a society that has age or generational imbalances in cohort sizes, is aging rapidly, and—by necessity, if not by design—is becoming more ethnoculturally diverse. Structural changes in our economy and society have also laid

Table 8.1 Net Interprovincial Migration Rates

	1966–71	1971–76	1976–81	1981–86	1986–91	1991–96	1996–01	2001–06
NF	–3.3	–1.4	–3.7	–3.1	–2.6	–4.3	–6.1	–1.3
PE	–1.0	2.3	0.0	1.4	–0.7	1.2	0.1	0.5
NS	–1.1	0.7	–1.1	0.8	–0.6	–0.8	–0.2	–0.9
NB	–1.3	1.6	–1.3	–0.2	–0.9	–0.3	–1.2	–1.5
PQ	–1.3	–1.1	–2.4	–1.1	–0.4	–0.6	–0.9	–0.2
ON	0.8	–0.7	–1.0	1.2	0.5	–0.5	0.5	–0.2
MB	–3.4	–2.9	–4.6	–0.2	–3.5	–1.9	–1.8	–2.0
SK	–7.6	–3.5	–0.7	–0.3	–6.4	–2.2	–2.7	–2.8
AB	1.7	4.0	11.3	–1.3	–1.1	0.1	4.7	3.1
BC	6.9	4.7	4.8	0.4	4.6	4.8	–0.7	0.6
YK	12.1	2.2	–2.6	–11.4	3.4	2.4	–9.4	–1.2
NWT	6.2	2.0	–5.0	–1.6	–3.4	–0.7	–8.6	–1.8
NU	NA	NA	NA	NA	NA	NA	–1.4	–1.3

Source: Compiled from Statistics Canada (2008a: 82). Rates expressed as percentage of base population for each period.

a foundation, however, for remarkable changes in how individuals negotiate life course changes, how they form households and families, and the rich diversity of lifestyles comprising our contemporary society.

A useful starting point for this discussion is the idea of the life course. Individuals typically go through a number of life course transitions, and researchers have devised numerous schema to depict some of the significant transition events and life course stages (Clark, 2007; Hareven and Adams, 1982; Murphy, 1987). Life course perspectives emphasize the importance and uniqueness of individual biography in relation to social and historical time, but at the same time they recognize that at any given time there will be a certain degree of synchronicity of individual and collective biographies. For example, the transition from child or adolescent to adult is typically marked by a series of transition events, such as leaving school, leaving the nest, entering full-time employment, and entering a conjugal union. Over the past century or so, the normative timing and characteristics

of such events have changed, as young people today generally spend a longer period of time in childhood and education prior to initiating the independent adulthood stage of life. Likewise, the institution of retirement, coupled with life expectancies in later life, has radically transformed the lives of elderly people and afforded new opportunities for lifestyles of leisure outside of years in the labour force (Atchley, 1992; Markides and Cooper, 1987). A useful schema is to consider a broad life course conceptualization, originally devised as part of the Theory of the Third Age (Laslett, 1987, 1991). This schema recognizes four major stages or 'ages' in life. Figure 8.4 is a simplification of some of the key features of these four ages.

The First Age is a time of dependence, socialization, immaturity, and education. Clearly, the First Age begins at birth, and at first it is characterized by babyhood, childhood, and initial instruction. It is a period of socialization and, for the most part, dependence on others. In general, this phase of the life course has been expanded through time as young people delay and elongate the timing

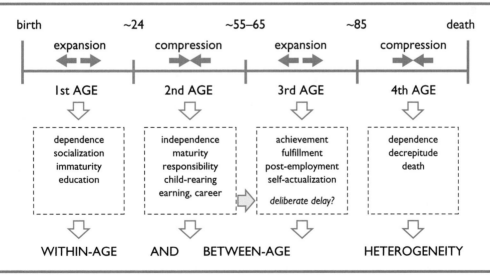

Figure 8.4 Summary conceptualization of the Third Age divisions of the life course

Source: Townshend (1997: 21).

of major transitions between the First Age and Second Age. They are staying in school longer, pursuing post-graduate education to a greater extent, delaying entry into the labour force, living in the parental home for a longer period, and postponing conjugal unions and child-bearing. In the early 1970s about three-quarters of Canadian young adults had left school by age 22, whereas by 2006 less than half had left by this age (Clark, 2007). While women often make these transitions younger than men, their delays in transition often are linked to post-secondary education. There have been significant increases in the educational attainment of Canadians, especially women, over the past few decades, and since the early 1990s women have outnumbered men in Canadian universities (Statistics Canada, 2008d).

The Second Age, beginning around 24 years for many (though the transition is individualistic and as noted is increasingly delayed), is a period of independence, maturity, responsibility, and labour force activity. This typically is a period when most people start a career, start a family, attain and maintain some degree of power and authority over others, and enjoy all of the social and legal privileges of adulthood. For the population as a whole, this life course phase has become truncated (in relative terms) over the past century. Increases in survival, coupled with the individual ability to cease work at earlier ages—primarily as a result of the institutionalization of personal, corporate, and public retirement savings schemes, and the social sanction of early retirement—have accentuated the tendency towards earlier retirement (McDonald and Wanner, 1984, 1990; Roadburg, 1985). In very recent years, however, we may be seeing a reversal of the trend to early retirement or, at least, episodic retirement, partial retirement, or various forms of bridge employment as people exit and re-enter the labour force (Hebert and Luong, 2008; Marshall and Ferraco, 2007; Singh and Verna, 2001; Stone, 2006). In general, however, the Second Age has become relatively compressed in time. Typically,

it occupies a smaller fraction of an individual's life course.

The same processes that have resulted in the compression of the Second Age of the life course have resulted in the relative elongation of the Third Age. This phase generally describes active and healthy seniors in society, but is marked by a movement out of the labour force, or at least partial retirement. For many, and particularly early retirees, nearly as much of their life will be spent in retirement as was spent in the Second Age. The Third Age is, demographically speaking, an unprecedented opportunity for a large share of society (and increasingly large in an aging society) to engage in many years of fulfillment outside of the labour force (Laslett, 1991). Considering advances in life expectancy at all ages, it is quite possible that someone who enters the labour force at 25 and retires at 55 will spend as many years in retirement as in the labour force. This concept has been marketed fervently by financial institutions (e.g., London Life's 'Freedom 55') and developers, and was embraced and planned for by many. The outlook on seniors' lifestyles can be seen in many cities in North America, most notably in new forms of exclusive retirement villages and age-restricted developments. Although the vast majority of Canadians do not work past the age of 65, and the majority of retirees are early retirees, the 1980s and 1990s may have been overly optimistic about the potential for universal early retirement. While there has been a systematic decline in the average age of retirement since the 1970s, retirement ages seem to have levelled off in the late 1990s, and recent evidence suggests that retirement is becoming increasingly complex, with multiple labour force exits and entries. In 2006 a record proportion of 60–64-year-olds were in the labour force (45 per cent) while the average retirement age was 61.5 years.

The Fourth Age, marked for many by the late seventies or early eighties, comprises the later years in life, or more generally when people have been called the 'old-old' (Neugarten, 1974). As of 1996, the disability-free life expectancy (for both sexes) of someone aged 65 in Canada was 11.7 years; it is not until the elderly are in their mid-eighties that we see a majority of seniors experiencing severe disabilities, a period akin to Shakespeare's seventh age of man, with increasing loss of mobility and independence, health concerns, senility, and decrepitude (Laslett, 1991; Norland, 1994). Either way, for most, this last phase of the life course has been pushed well past the normal (65-year-old) age of retirement, compressed, from an historical point of view, largely as a function of advances in health care and nutrition.

Considerable heterogeneity prevails in the timing of transitions within each of the four ages, as in the social composition of people within each of these phases of the life course, when people may be differentiated by age, socio-economic status, family structures, ethnicity, 'race', gender, and lifestyles. These differences no doubt create highly variable pathways through the life course, different opportunities and constraints, various inequalities, and different life experiences and biographies. As these generalized life course phases intersect with micro and macro differentiation in our society, they define the potential for myriad life paths and expressions of lifestyles.

One of the most important trends in Canadian society over the past quarter-century has been the revolutionary changes in the structure and composition of households and families. We have already seen how basic demographic changes, such as those of fertility and survival, have been influential in reshaping the age structure of the population, have led to fewer younger people, and have resulted in an aging population. These trends do not occur in isolation from broader forces of societal change, however, that is, changes in normative behaviours and life choices throughout the life course. The shifting choices of human agents with respect to household and lifestyle fundamentally affect the urban experience.

Smaller Households

In 1901, the average household in Canada had five people. Today the average is 2.5 people. The size of Canadian households has been declining for more than a century (Rose and Villeneuve, 2006). This transformation in the size of households is largely attributable to the growth (both absolute and relative) in the number of small households and the decline of large households of four or more people. Figure 8.5 shows how dramatic these changes have been, with the number of single-person households rising from 1.68 million in 1981 to 3.33 million in 2006, an increase of 98 per cent over this time period. The number of two-person households has also risen dramatically from 2.40 million in 1981 to 4.18 million in 2006, an increase of 74 per cent over the past quarter-century. The number of three- or four-person households has also increased, but only marginally; while the number of households with five or more people has declined. By 2006 the majority (60.3 per cent) of Canadian households were small households of one or two people (Figure 8.6).

Within the smaller household group, the fastest-growing segment has been single-person households. Since the household is generally considered the basic unit of consumption as well as a means of achieving economies of scale in living expenses, there are some important implications for the rise in single-person households, especially in terms of housing affordability (as a result of increased consumption of/demand for smaller units) and income distributions (Miron, 1993). The rise of the single-person household is linked to what some have described as a major shift in the propensity to live alone, and is one of the most significant demographic trends observed in Canadian society, with rates rising from 2.6 per cent in 1951 to 12.3 per cent in 2001 (Clark, 2007; Rose and Villeneuve, 2006). Living alone used to be a principal feature of rural, non-farming areas, but is now linked to urban lifestyles and the changing socio-cultural values surrounding the meaning of

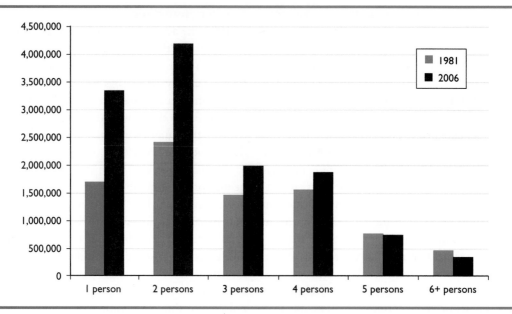

Figure 8.5 Number of households by household size, 1981 and 2006

Source: Compiled from data in Statistics Canada, 1981 and 2006 Census of Population.

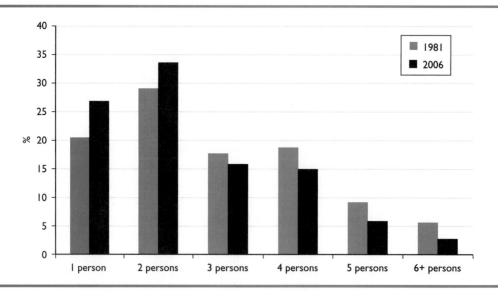

Figure 8.6 Percentage of households by size, 1981 and 2006

Source: Compiled from data in Statistics Canada, 1981 and 2006 Census of Population.

living alone. While living alone may have been a significant phase for people in the early stages of the Second Age (i.e., having just left the 'nest'), or of some in the later stages of the Third Age, it has become increasingly prevalent for people to live alone in mid-life as well as in later life (Laslett, 1991). Also, as noted earlier in the chapter, a feminization of survival in Canada means that old age increasingly is characterized by widows living alone in single-person households.

A further reason for the rise in single-person households has been changes in the values and meanings surrounding marriage and family formation, and delays in conjugal unions that are linked to the later transition from the First to the Second Age. The age of first marriage for both brides and grooms has been rising steadily since the 1960s, reaching approximately 30 years for men and 28 years for women by 2003 (Clark, 2007; Statistics Canada, 2008a). Considerably fewer young adults are entering marriage or other conjugal relationships at an early age. In 1971, for example, 65 per cent of men and 80 per cent of women were in or had been in a conjugal relationship by the age of 25. By 2001 these rates were almost halved, dropping to 34 and 49 per cent, respectively. Young adults, or those in the early stages of the Second Age, more often are choosing to live alone.

New Family Configurations and Fluid Conjugal Relationships

Just as there have been significant changes in the size and composition of households, so have there been major transformations in the meaning and composition of families in Canada. The most prevalent family form is a married couple with children living at home (38.7 per cent of families), followed by married couples without children at home (29.9 per cent). Especially noteworthy about changing family configurations are the changes in the relative shares of different family forms (Figure 8.7). Married couples with children living at home are the only type of family to have experienced negative growth over the last quarter-century. This

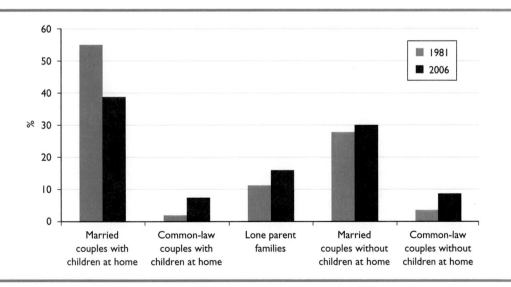

Figure 8.7 Percentage of census families by structure, 1981 and 2006

Source: Compiled from data in Statistics Canada, 1981 and 2006 Census of Population.

decline has been offset by the rise of alternate family forms, especially common-law families—both with or without children—and the dramatic rise of single-parent families. It is estimated that by 2020 as many people will be living in common-law unions as in marriages (Milan, 2000). Together, the share of families without children at home has risen from 32 per cent in 1981 to 38 per cent in 2006, thereby contributing to the declining household sizes.

The fluidity with which people move in and out of various types of families also has increased. In part due to separation and divorce, but also because common-law unions are less stable than marriages, people more frequently move in and out of relationships throughout their life course, most notably during the Second and Third Age (Milan, 2000). Less than 10 per cent of marriages end in divorce within the first five years of marriage, whereas about half of common-law unions dissolve within that time frame (Milan, 2000; Statistics Canada, 2008a). Despite the fact that about four in 10 marriages end in divorce, which also contributes to the formation of single-person households, approximately 75 per cent will remarry or recouple, creating an episodic pattern of marriage and remarriage. This pattern gives rise to a variety of configurations of blended and step-families, either married or common-law.

New forms of family and household formation can also be seen among same-sex couples. Recent years have seen a growing social acceptance and institutionalization of the rights of same-sex couples, including the legalization of same-sex marriages in 2005. Canadian same-sex couples were first enumerated with voluntary identification in 2001, and while undoubtedly under-reported, it was estimated that there were just over 34,000 same-sex couples in Canada, about 0.5 per cent of all couple families. By 2006 this count had risen to about 45,000, or about 0.6 per cent of all couples. These couples are not necessarily all small households or couples without children at home,

although the vast majority are. Nine per cent of same-sex couples reported having children under the age of 24 living at home in 2006.

The Complexity of Lifestyles

Geographers and planners have long been interested in the ways that people use their time, how time is allocated to different functions in space, how space is segmented with different types of people or consumers, and how lifestyles are manifest in time–space interactions (Chapin, 1965). While a relative paucity of spatially oriented lifestyle studies existed in the 1980s and 1990s, in recent years a resurgence of interest in lifestyle has produced a number of studies, in particular about the ways in which human agents create lifestyle spaces or spatial lifestyles (Schnell and Benjamini, 2001, 2002; Townshend and Davies, 1999). Schnell and Benjamini (2002) remind us that lifestyles are an intersection of time, activity, and space patterns, and so lifestyles can differ on such things as the radius of daily activity, the importance of telecommunications, types and intimacy of social relations, extent of orientation to a home base, and differences in the value placed on family, social life, work, and leisure.

Lifestyles are often associated with or labelled in regard to population subgroups or in terms of market-based geo-demographic segments, so it makes little sense to try to define First Age, Second Age, and Third Age lifestyles. Within each of these phases of the life course multiple lifestyles will manifest themselves. It is not uncommon, for example, for market researchers to develop complex segmentation models and typologies to define groups of young people based on their consumption preferences (Cahill, 2006). Many of these typologies attempt to link key lifestyle traits with behaviours, outlook on life, types of social relationships, and especially consumption preferences of particular generations. Foot (with Stoffman, 1996), for example, has shown how some of these lifestyle traits have changed through Canadian generations, impacting everything from people's preferences for different types of social relationships to demands for housing.

Others have focused specifically on selected generations, such as the 'Millennials' (Alsop, 2008; Howe and Strauss, 2007). While there is no agreed-on range of birth dates for the millennial generation, they are loosely considered to be those born between the early 1980s and 2000. They are the children of the late baby boomers and early baby-bust generation, and sometimes are called the 'most wanted' generation in history. This generation of people is said to exhibit very different lifestyle orientations from previous generations (Howe and Strauss, 2007). They have few siblings, have always been considered special or important to their parents, have lived sheltered lives, have a sense of entitlement, are electronically savvy, value tight social networks, appreciate and value social and ethnic diversity, are group- or team-oriented rather than individualistic, and eschew preferential treatment. They also are high achievers and career-oriented. Some have claimed that these are 'trophy kids', both as trophies to their parents (almost all are planned pregnancies) and because they have grown up in school and sports activities where everyone received a prize or trophy for participating. From a demographic and especially geographical perspective, these individuals, as a subset of those in their First Age, have a unique set of preferences, behaviours, form of social networking, consumption potential, and propensity to be involved in civic affairs. As they move through the various transitions and stages of the life course they will leave their own generation and lifestyle impact.

Lifestyles are highly differentiated in the Second Age. While researchers have studied the values and lifestyles of the hippie generation and the yuppies (young urban professionals), and have explored the activities and social relations of swingles (swinging singles) in large metropolitan areas, a more recent example of the popularization of lifestyles as a demographic by-product can be seen in the book and film, *Bridget Jones's Diary* (*Economist*, 2001; Fielding, 1996; Partridge, 1973; Smith, 1987; Starr and Carns, 1972). While the film is a light-hearted romantic comedy, a very serious side to the urban lifestyle is portrayed (Maguire, 2001). Bridget Jones is a 30-something, single, highly educated professional working in the creative economy in a major metropolitan area (London). She lives alone in a small flat—a single-person household—and apart from juggling time at the gym and work, has plenty of time, money, and a passion for spending on things fashionable, frivolous, and fun. Her lifestyle is consumption-oriented—dining out, upscale wine bars, weekend holiday getaways, etc. She strives to enjoy her career, to find romance, and perhaps at some point to marry. Relatives and coupled friends constantly remind her that 'the clock is ticking' and her delay in marriage and child-bearing may become problematic. Bridget longs for acceptance outside her tightly knit group of friends, all of whom live similar lifestyles and are similarly fun- and consumption-oriented. True as the Bridget character is to life, it has been argued that many singles and other single-person households adopt their friendship network (which for Bridget is a heterogeneous group of professionals and artists, of diverse ethnic backgrounds and sexual orientation) as a surrogate family, in what some have suggested is a new form of 'urban tribe' or 'fictive kin' (*Economist*, 2001; Martin-Mathews, 2001; Rose and Villeneuve, 2006).

Meanwhile, economists have noted that the 'Bridget Jones Economy' is an urban reality. This demographic and the associated lifestyle have both a spatial and economic footprint. They are clustered predominantly in central-city apartments, often in close proximity to urban nightlife and in areas with high densities of singles. They are key drivers of emerging goods and services in these areas, including functions such as dry cleaners, health food stores, specialty delicatessens, supermarkets specializing in small portions and prepared foods, Internet kiosks, dating services, travel agencies, wine bars, specialty clothing boutiques, and health professionals. In other words, the Bridget Jones factor may be a lifestyle in the sense that it incorporates a particular constellation of identity, taste, consumption, style, and leisure. Most importantly, it is very much a spatial lifestyle. Such lifestyles and living arrangements are also triggers of change in the social complexity and social ecology of cities (Davies and Murdie, 1991; see Walks, Chapter 10; Bain, Chapter 15; Walker and Carter, Chapter 20).

The Third Age also is marked by social and lifestyle diversity, and the concept of 'seniors' as a homogeneous group with similar traits fails to capture this diversity. As the large baby boomer cohort progresses through the stages of transition to retirement or Third Age, the diversity and consumption potential of this group will be even more noticeable (Foot with Stoffman, 1996). In fact, the diversity of this population is fuel for market researchers and housing developers who try to isolate distinctive lifestyle segments and cater to variations in demands for goods, services, and housing. While websites are dedicated to the Third Age, successful aging, and advice on matters on everything from seniors' lifestyles and finance to relationships and sex (e.g., www.ThirdAge.com), market research professionals have also tried to isolate distinctive segments of third-agers. Acronyms such as SUPPIES (Senior Urban Professionals), WOOPIES (Well-Off Older People), OPALS (Older People with Active Lifestyles), RAPPIES (Retired Affluent Professionals), or DUMPIES (Destitute Unprepared Mature People) have been used to capture some of the market and lifestyle differences (Dytchwald, 1990; Karpel, 1995). Housing developers have recognized that the elderly are not a homogeneous group, and have employed labels such as the Go-Go segment (active healthy retirees), Slow-Go segment (not very active, failing health), and the No-Go segment (extremely frail, or Fourth Age) to identify subgroups with respect to housing needs (Seiler, 1986). While these may be gross generalizations, they are based on recognizable differences in lifestyle, behaviour, and consumption.

Demography, Lifestyle, and the Built Environment

There is little doubt that age, life cycle stage, and lifestyle have become increasingly commodified, not just in geo-demographic marketing but in the built environment. Developers have responded to a variety of age, preference, and lifestyle niche markets, the result being increasing diversity in the residential landscape. Two of the biggest trends in this area are the growth in condominium ownership and the rise of residential niche communities.

Given the rise in small households without children at home, more housing options today offer a smaller, more convenient, and often maintenance-free lifestyle. Condominium ownership, a type of common interest development that became prevalent in the 1970s, has seen a dramatic increase over the past few decades, with the share of households owning condominiums rising from 3.5 per cent in 1981 to 10.9 per cent in 2006 (Harris, 2006; Lo, 1996; McKenzie, 1994, 2003; Rea et al., 2008; see Walker and Carter, Chapter 20). Condominium living is considerably more prevalent in metropolitan areas, with rates as high as 31 per cent in Vancouver, 18.6 per cent in Toronto, and 13.3 per cent in Montreal. In some cities, such as Toronto, condominium developments now account for the majority of new housing starts.

A second trend in response to the rise in differentiated lifestyles and preferences among those in their Second or Third Age has been innovation in the way residential communities are designed and developed. Many of these changes are linked to new expressions of the privatization of space and communities. Whereas the typical planned community of the 1950s or 1960s may have followed neighbourhood unit design principles, entire subdivisions today are being developed in response to changes in demography and lifestyle preferences, taking on a variety of expressions of product differentiation, gating, and exclusivity (Hodge and Gordon, 2007; Perry, 1929). The rise of gated communities (a type of common interest development) in various countries, including Canada, has been studied in some detail, with many authors expressing concern about the consequences of such developments, especially the loss of the public realm to private space (see Grant and Filion, Chapter 18; Blakely and Snyder, 1999; Blandy et al., 2003; McKenzie, 1994; Punter, 1990; Townshend, 2006; Webster et al., 2002). Blakely and Snyder's (1999) pioneering study of gated communities in the United States identified three main types: lifestyle communities (including retirement communities, golf course communities, and new towns), prestige communities, and security zone communities. Others have attempted to define typologies of gated communities in Canada and internationally (Coy and Pohler, 2002; Glasze, 2002; Grant, 2005; Grant and Mittelsteadt, 2004; Landman, 2002; Raposo, 2006). Despite difficulties in defining such typologies, it seems clear that the function of these designs is to create landscapes of exclusivity and exclusion, and as one author noted, these proprietary neighbourhoods are consumed as club goods within a kind of 'shareholder democracy' (Glasze, 2002). The supply of and demand for private and gated residential communities doubtless will continue. Likely, also, is increased product differentiation as developers refine these commodities to appeal to more tightly defined lifestyles and market segments (Lang and Danielsen, 1997). This means that the geographical manifestation of private neighbourhoods in the city will become more complex and more differentiated.

Other forms of residential community privatization are evident in the built environment. Townshend (2006), for example, has discussed the trend towards the thematic development of subdivisions since the 1960s. Initial expressions of these, such as the development of golf course communities, were perhaps a direct response to lifestyle demands and the pursuit of leisure. Along similar lines, a number of developers experimented with 'lake communities' in which a private lake, with controlled access, forms the recreational nexus of the community. By the 1980s other types of niche communities of a smaller scale began to appear in Canadian cities in the form of retirement villages specifically designed and marketed to the 55-and-over population (i.e., appealing to the Third Age). Many of these developments, such as 'Horizon Villages' with age-restrictive covenants, actively advertised the concept of Freedom 55 and offered a commodified form of community in which residents could have condominium ownership, maintenance-free living, and a socially homogeneous set of neighbours. The popularity of these types of retirement communities, with particular appeal to elderly single persons, has grown rapidly. In Calgary, for example, almost all residential subdivisions contain at least one of these developments and almost all new subdivisions incorporate plans for at least one retirement community (Townshend, 2006).

A range of different thematic foci for new residential areas has become common. While developers have continued to build lake and golf course communities, at least four other types occur. One is the development of New Urbanist or neo-traditional (NTD) communities. In response to problems of conventional suburban development patterns, these communities attempt to foster a more traditional urban lifestyle, despite being constructed in new suburban regions (see Grant and Filion, Chapter 18). Following the ideas of leading proponents of New Urbanism, such as Duany and Plater-Zyberk, features of these developments include the return to a smaller or more human scale of mixed housing development and mixed land uses, the return to a narrow grid and more pedestrian-friendly street network, and a reduction in vehicular traffic (Katz, 1994). NTD communities usually include provision of a town centre with commercial and local administrative facilities, and aim to promote increased social interaction through design features such as front porches on the houses and the placement of garages at the rear of lots. McKenzie Towne in Calgary and Cornell in Markham, just outside the Toronto CMA, are excellent examples of early efforts to adopt New Urbanist principles in Canadian cities.

A second type of new community has explicitly focused on environment in its design. In these the developer retains some natural environmental features within the design scheme (e.g., wetlands, natural prairie grassland, forest, bird sanctuary). The share of open recreational space in these communities is considerably higher than in other neighbourhood districts. It also is typically more than what is legally required for subdivision approval, with some communities boasting as much as 40–50 per cent of the land area of the district retained as natural environment. Developers seem to have capitalized on the recent trend towards environmental stewardship, so that residents of these areas pay compulsory homeowner association fees to ensure appropriate maintenance and use of the natural features, and to ensure they remain intact in perpetuity. In some cases, a private residents' clubhouse with meeting or sports facilities has been constructed as part of the overall plan.

A third type of thematic community development may be called wired or e-communities (Townshend, 2006). High technology is the primary selling feature of these areas, which are designed to deliver a well-connected community with well-connected 'smart homes'. All homes in the development contain in-home network wiring for data, audio, and video distribution, and for connection to smart appliances. These communities are wired for broadband Internet connection; but a key feature of these places is the presence of a local community intranet or web portal that functions as an electronic town hall or community centre for residents of the area to post and share information about local events or to buy and sell from each other. These developments are perhaps a response to a technology-oriented lifestyle and locality

orientation, geared towards residents or consumers that demand 'fast spaces' and connectivity within the city (McGuirk, 2003). Again, it is not uncommon for such communities to require membership and payment to a homeowner association for the maintenance of this infrastructure.

The fourth type may be called multiple foci communities, consisting of different combinations of recreational amenities, environmental focus, telecommunication infrastructure, and so on. Thus, some of the planned subdivisions include a lake or golf course in addition to being constructed as e-communities. Still others include some environmental focus, but also provide other private amenities such as a wading pool, splash pool, or skating park for children. Others have some elements of New Urbanism in addition to private recreational ponds or resident clubhouses. New communities are evolving in such a way that any combination of specialized recreational, environmental, historical, and architectural design features (New Urbanism), high technology (e-community), or other feature can form the basis for the common interest development.

Conclusion

This chapter has presented some of the key demographic features of change in Canadian cities and linked them to the malleable boundaries of significant life course transitions. The conceptualization of life course as four 'ages' is a useful rubric for tying demographic and lifestyle trends together and for proposing how these might relate to the differentiation and marketing of the built environment. The futures of Canadian cities are tied to a greatly diversifying population. Much of that diversity is now quite visible in the form of culture, ethnicity, and Aboriginality, with the importance of each varying regionally. In the Prairie provinces, for example, those cities that do not embrace the rich contemporary cultures of their growing young First Nations and Métis populations will be left behind culturally, socially, and economically in the twenty-first century (Walker, 2008).

Other forms of diversity relate to household sizes, conjugal relationships, and the greater propensity to live alone than has been the case in past decades. How is the Bridget Jones Economy likely to continue affecting our inner cities and suburban areas? Finally, an aging society means that no form of lifestyle diversity, whether based on culture, ethnicity, Aboriginality, or household preferences, can be understood without inserting the centrality of the age dimension and the rising opportunities being taken by those in their Third Age of life. How will our built environment continue to change to serve the pig in the python as it moves through its elongated Third Age? Will those in their Third Age continue to advance suburban development that they have become accustomed to throughout their lives, even if modified to suit new lifestyle preferences (see Grant and Filion, Chapter 18)? Will the millennial generation opt for newer urban forms that turn away from conventional development patterns and instead promote better environmental stewardship (see Rees, Chapter 5; Connelly and Roseland, Chapter 14; Blay-Palmer, Chapter 24) and qualitatively 'urban' places (see Bain, Chapter 15; Lynch and Ley, Chapter 19; Gilliland, Chapter 23)? Most of the other chapters in this book contemplate the impacts of diversity on Canadian cities in more specific ways, such as immigration (Hoernig and Zhuang, Chapter 9), social polarization (Walks, Chapter 10), youth and aging (Rosenberg and Wilson, Chapter 21), and Aboriginality (Peters, Chapter 22). Consider this chapter a primer on the demographic and lifestyle changes that run through all phenomena that constitute urban Canada.

Review Questions

1. Describe at least three of the key forces of social and demographic change affecting contemporary Canadian cities.

2. What are some of the impacts changing lifestyles may have on how we plan and develop our built environment?

References

Alsop, R. 2008. *The Trophy Kids Grow Up: How the Millennial Generation is Shaking Up the Workplace*. Toronto: Jossey-Bass.

Atchley, R.C. 1982. 'Retirement as a social institution', *Annual Review of Sociology* 8: 263–87.

Beaujot, R. 1991. *Population Change in Canada: The Challenges of Policy Adaptation*. Toronto: McClelland & Stewart.

Blakely, E., and M. Snyder. 1997. *Fortress America: Gated Communities in the United States*. Washington: Brookings Institution Press.

Blandy, S., D. Lister, R. Atkinson, and J. Flint. 2003. *Gated Communities: A Systematic Review of the Research Evidence*. Bristol, UK: Economic and Social Research Council, Centre for Neighbourhood Research, CNR Paper 12.

Bourne, L.S., and D. Rose. 2001. 'The changing face of Canada: The uneven geographies of population and social change', *Canadian Geographer* 45: 105–19.

Boyd, M., and M. Vickers. 2000. 'One hundred years of immigration to Canada', *Canadian Social Trends* (Fall); reprinted in *100 Years of Canadian Society*, Teachers Kit.

Bunting, T., and P. Filion, eds. 2006. *Canadian Cities in Transition: Local Through Global Perspectives*, 3rd edn. Toronto: Oxford University Press.

Cahill, D. 2006. *Lifestyle Market Segmentation*. Toronto: Routledge.

Chapin, F.S. 1965. *Urban Land Use Planning*, 2nd edn. Urbana: University of Illinois Press.

Clark, W. 2000. 'One hundred years of education', *Canadian Social Trends* (Winter); reprinted in *100 Years of Canadian Society*, Teachers Kit.

———. 2007. 'Delayed transitions of young adults', *Canadian Social Trends* (Winter).

Coy, M., and M. Pohler. 2002. 'Gated communities in Latin American megacities: Case studies in Brazil and Argentina', *Environment and Planning B: Planning and Design* 29: 355–70.

Davies, W.K.D., and R.A. Murdie. 1991. 'Measuring the social ecology of Canadian cities', in D. Ley and L. Bourne, eds, *The Changing Social Geography of Canadian Cities*. Montreal and Kingston: McGill-Queen's University Press.

Dytchwald, K. 1990. *Age Wave*. Toronto: Bantam Books.

Economist, The. 2001. 'The Bridget Jones economy', 361, 8253: 68–70.

Fielding, H. 1996. *Bridget Jones's Diary*. London: Picador.

Foot, D., with D. Stoffman. 1996. *Boom, Bust and Echo: How to Profit from the Coming Demographic Shift*. Toronto: Macfarlane Walter and Ross.

Fries, F. 1980. 'Aging, natural death, and the compression of morbidity', *New England Journal of Medicine* 303: 130–5.

Glasze, G. 2002. 'Gated housing estates in the Arab world: Case studies in Lebanon and Riyadh, Saudi Arabia', *Environment and Planning B: Planning and Design* 29: 321–36.

Grant, J. 2005. 'Planning responses to gated communities in Canada', *Housing Studies* 20: 277–89.

——— and L. Mittelsteadt. 2004. 'Types of gated communities', *Environment and Planning B: Planning and Design* 31: 913–30.

Halli, S., and L. Driedger, eds. 1999. *Immigrant Canada: Demographic, Economic, and Social Challenges*. Toronto: University of Toronto Press.

Hareven, T.K., and K.J. Adams, eds. 1982. *Aging and Life Course Transitions: An Interdisciplinary Perspective*. New York: Guilford Press.

Harris, R. 2006. 'Housing: Dreams, responsibilities, and consequences', in Bunting and Filion (2006).

Hebert, B.P., and M. Luong. 2008. 'Bridge employment', *Perspectives on Labour and Income* 9, 11. Statistics Canada Catalogue no. 75–001–X.

Hiller, H. 2006. *Canadian Society: A Macro Analysis*, 5th edn. Toronto: Pearson Prentice-Hall.

Hodge, G., and D. Gordon. 2007. *Planning Canadian Communities*, 5th edn. Scarborough, Ont.: Nelson Education.

Howe, N., and W. Strauss. 2007. *Millennials Go to College*, 2nd edn. Ithaca, NY: Paramount Books.

Karpel, C. 1995. *The Retirement Myth*. New York: HarperCollins.

Katz, P. 1994. *The New Urbanism*. New York: McGraw Hill.

Kraus, A.S. 1988. 'Is a compression of morbidity in late life occurring? Examination of death certificate evidence', *Canadian Journal on Aging* 7: 58–70.

Landman, K. 2006. 'Privatising public space in post-apartheid South African cities through neighbourhood enclosures', *GeoJournal* 66: 133–46.

Lang, R., and K. Danielsen. 1997. 'Gated communities in America: Walling out the world?', *Housing Policy Debate* 8: 867–99.

Laslett, P. 1987. 'The emergence of the third age', *Ageing and Society* 7: 133–60.

———. 1991. *A Fresh Map of Life: The Emergence of the Third Age*. London: Weidenfeld and Nicolson.

Lesthaeghe, R. 1995. 'The second demographic transition in Western societies: An interpretation', in K.O. Mason and A.M. Jensen, eds, *Gender and Family Change in Industrialized Countries*. Toronto: Oxford University Press.

Lo, O. 1996. 'Condominium living', *Canadian Social Trends* (Summer): 41.

McDonald, P.L., and R.A. Wanner. 1984. 'Socioeconomic determinants of early retirement in Canada', *Canadian Journal on Aging* 3: 105–16.

——— and ———. 1990. *Retirement in Canada*. Toronto: Butterworths.

McGuirk, P. 2003. 'The future of the city: A geography of connection and disconnection', *Geodate* 16: 5–9.

McKenzie, E. 1994. *Privatopia: Homeowner Associations and the Rise of Residential Private Government*. New Haven: Yale University Press.

Maguire, S., dir. 2001. *Bridget Jones's Diary*. Miramax Films.

Markides, K.S., and C.L. Cooper, eds. 1987. *Retirement in Industrialized Societies*. Toronto: John Wiley and Sons.

Marshall, K., and V. Ferraco. 2007. 'Participation of older workers', *Perspectives on Labour and Income*, Statistics Canada Catalogue no. 75–001–XIE.

Milan, A. 2000. 'One hundred years of families', *Canadian Social Trends* (Spring 2000); reprinted in *100 Years of Canadian Society*, Teachers Kit.

Miron, J.R. 1993. 'Demography, living arrangement, and residential geography', in L.S. Bourne and D. Ley, eds, *The Changing Social Geography of Canadian Cities*. Montreal and Kingston: McGill-Queen's University Press.

Murdie, R., and C. Teixeira. 2006. 'Urban social space', in Bunting and Filion (2006).

Murphy, M. 1987. 'Measuring the family life cycle: Concepts, data and methods', in A. Bryman, B. Bytheway, P. Allatt, and T. Keil, eds, *Rethinking the Life Cycle*. London: Macmillan.

Nagnur, D. 1986. 'Longevity and historical life tables: 1921–81 (abridged)', *Canada and the Provinces*. Ottawa: Statistics Canada.

Neugarten, B. 1974. 'Age groups in American society and the rise of the young-old', *Annals of the American Academy*: 187–98.

Norland, J.A. 1994. *Profile of Canada's Seniors*. Ottawa: Statistics Canada, Focus on Canada Series, Catalogue no. 96–312E.

Partridge, W. 1973. *The Hippie Ghetto: The Natural History of a Subculture*. New York: Holt, Rinehart and Winston.

Perry, C. 1929. 'The neighborhood unit', in *Regional Survey of New York and Its Environs, Regional Survey*, vol 7. New York: Neighborhood and Community Planning.

Punter, J.V. 1990. 'The privatisation of the public realm', *Planning Practice and Research* 5: 9–18.

Raposo, R. 2006. 'Gated communities, commodification and aestheticization: The case of the Lisbon metropolitan area', *GeoJournal* 66: 43–56.

Rea, W., D. McKay, and S. LeVaseur. 2008. *Changing Patterns in Canadian Homeownership and Shelter Costs*. Ottawa: Statistics Canada Catalogue no. 97–554–X.

Roadburg, A. 1985. *Aging: Retirement, Leisure and Work in Canada*. Toronto: Methuen.

Rose, D., and P. Villeneuve. 2006. 'Life stages, living arrangements, and lifestyles', in Bunting and Filion (2006).

Schnell, I., and Y. Benjamini. 2001. 'The socio-spatial isolation of agents in everyday life spaces as an aspect of segregation', *Annals, Association of American Geographers* 91: 622–36.

——— and ———. 2002. 'Measuring spatial lifestyles in Israel: A Tel-Aviv–Jaffa case study', in W.K.D. Davies and I. Townshend, eds, *Monitoring Cities: International Perspectives*. Calgary and Berlin: International Geographical Union, Urban Commission.

Seiler, S.R. 1986. 'How to develop retirement communities for profit', *Real Estate Review* 16: 70–5.

Simmons-Tropea, D., and R. Osborn. 1993. 'Disease, survival and death: The health status of Canada's elderly', in V.W. Marshall, ed., *Aging in Canada: Social Perspectives*. Markham, Ont.: Fitzhenry & Whiteside.

Singh, G., and A. Verma. 2001. 'Is there life after career employment? Labour-market experience of early "retirees"', in V.W. Marshall, W.R. Heinz, H. Kruger, and A. Verma, eds, *Restructuring Work and the Life Course*. Toronto: University of Toronto Press.

Smith, N. 1987. 'Of yuppies and housing: Gentrification, social restructuring, and the urban dream', *Environment and Planning D: Society and Space* 5: 151–72.

Starr, J., and D. Carns. 1972. 'Singles in the city', *Society* 9: 43–8.

Statistics Canada. 2007. *Immigration in Canada: A Portrait of the Foreign-born Population*. Catalogue no. 97–557–XIE.

———. 2008a. *Report on the Demographic Situation in Canada 2005 and 2006*. Catalogue no. 91–209–X.

———. 2008b. *Canadian Demographics at a Glance*. Catalogue no. 91–003–XIE.

———. 2008c. *Canada's Ethnocultural Mosaic*. Catalogue no. 97–562–X.

———. 2008d. *Educational Portrait of Canada*. Catalogue no. 97–560–X.

———. 2008e. *Aboriginal Peoples in Canada in 2006: Inuit, Métis and First Nations*. Catalogue no. 97–558–XIE.

Stone, L. 2006. *New Frontiers of Research on Retirement*. Ottawa: Statistics Canada Catalogue no. 75–511–XIE.

Townshend, I.J. 1997. 'An Urban Geography of the Third Age', Ph.D. dissertation, University of Calgary.

———. 2002. 'Age segregated and retirement communities in the third age: The differential contribution of place-community to self-actualization', *Environment and Planning B: Planning and Design* 29: 371–96.

———. 2006. 'From public neighbourhoods to multi-tier private neighbourhoods: The evolving ecology of neighbourhood privatization in Calgary', *GeoJournal* 66: 103–20.

——— and W.K.D. Davies. 1999. 'The derivation of shopper typologies in business revitalization zones', *Papers and Proceedings of the Applied Geography Conferences* 22: 132–45.

Trovato, F. 1987. 'A longitudinal analysis of divorce and suicide in Canada', *Journal of Marriage and Family* 49: 193–203.

Walker, R.C. 2008. 'Improving the interface between urban municipalities and Aboriginal communities', *Canadian Journal of Urban Research* 17 (suppl.): 20–36.

Webster, C., G. Glasze, and K. Frantz. 2002. 'Guest editorial: The global spread of gated communities', *Environment and Planning B: Planning and Design* 29: 315–20.

CHAPTER 9

New Diversity: Social Change as Immigration

Heidi Hoernig and ZhiXi (Cecilia) Zhuang

This chapter considers how immigration influences the contemporary evolution of Canadian cities. Over the past 15 years, immigration rates have averaged 0.8 per cent of the Canadian population, which translates into 200,000 to 250,000 new arrivals annually (CIC, 2008a). Of these, 97 per cent of immigrants arriving between 2001 and 2006 and 95 per cent of Canada's total foreign-born population live in urban areas (Chui, Tran, and Maheux, 2007). Given these numbers, it is not surprising that immigration is a major source of social transformation in cities across the country. But immigrants do not randomly distribute themselves across Canada's urban system. Complex and dynamic constellations of variables determine how immigrants settle into particular places and what will be the consequences of this settlement for cities and residents, both immigrant and native-born.

Immigration is a long-term, complex process. Factors at all geographic scales—global, national, regional, and local—influence the origins, flows, and destinations of immigrants. The socio-spatial patterns that emerge from different points along migrants' journeys often extend across generations. We have organized this chapter along the three main stages of the immigrant's journey: arrival, settlement, and integration. For each stage, we explore the experiences of immigrants and their contributions to both continuity and change in Canadian cities. We conclude with a spatial overview of the Canadian geography of immigration presented as the culmination of these factors and interacting dynamics.

What's New about Immigration and Canadian Cities?

Looking back through Canadian history, it is clear that migration and inter-ethnic relations have always contributed significantly to transformation in Canadian settlements. This has been true for Canada's Aboriginal peoples, which before European contact included more than 150 distinct ethnolinguistic groups distributed across the country in a vast system of settlements and seasonal camps (Dickason with McNab, 2009; Morrison and Wilson, 2004). The arrival of Europeans radically and, too often, tragically transformed this settlement landscape. Many Aboriginal settlements were obliterated or displaced, first, through the devastating effects of disease and then through the colonial and federal policies and treaties that opened the way for continuing waves of immigrants. Through the colonial and the post-Confederation periods, existing Aboriginal socio-spatial patterns were transformed into the thousands of villages, towns, and cities that today constitute Canada's urban system (see Peters, Chapter 22). The past 400 years of change in Canada's urban geography have been dominated by the story of immigration.

We begin with this brief historical reference to underscore the central themes of this chapter. First, several dimensions of the current trends in immigration are not new. Regardless of the historical period, immigration introduces new languages, religions, economic activities, resources, technologies, and social institutions to Canada. Furthermore, the settlement and integration experiences and impacts of new arrivals are contingent not only on what immigrants have brought with them—their skills and resources—but also the context into which they arrive. Just as the flourishing or demise of early explorers and traders corresponded to their capacities, prevailing conditions, and supports from local peoples, the contemporary experiences of newcomers reflect a similar set of interacting variables. Accordingly, one central theme in the following is how the broad spectrum of experiences and outcomes for immigrants and cities corresponds with relationships between what immigrants have to offer and the social and economic conditions into which they arrive.

A second and related node of comparison includes the themes of continuity, change, and fusion. When we consider the major historical waves of immigration, the integration of new arrivals into the Canadian urban landscape certainly has brought about tremendous changes. But in most cases, these changes did not take place immediately. Rather, newcomers adapted and integrated into existing economic, social, and political systems, and, over time, they introduced new ideas and ways of doing things. The current portrait of urban geographies of immigration displays a similar pattern. The effect of immigration on cities is often a story of the augmentation or expansion of existing geographic dynamics and patterns, in addition to those elements of both change and fusion.

Leaving Home and Entering Canada

The decision to uproot oneself and one's family from familiar circumstances in order to travel to and settle into an unknown and unpredictable society, economy, and environment is not a decision most people take lightly. Nor are global migration patterns arbitrary. Immigration theorists like Massey (1999: 50–2) argue that international migration responds to four interlocking elements: (1) the restructuring circumstances that promote emigration from countries of origin, conditions, known as **push factors**; (2) those factors attracting people into a host country, including formal policies, programs, and recruitment campaigns, referred to as **pull factors**; (3) the social and psychological characteristics and decision-making of migrants themselves; and (4) the social, political, and economic relationships between areas of out-migration and in-migration.

Other spatial and temporal dimensions of the immigration process are important determinants in urban geographies of newcomers. Immigration entails much more than the simple relocation of people from one geographic area to another. Financial assets, material resources, and social and economic relationships precede, accompany, and follow these movements. The journey is not necessarily a single, one-way trip. Some people experience complicated trajectories, spending time in more than one country before settling at a destination. Many refugees, for example, spend months, years, even decades, in refugee camps in third countries before being sponsored by a host country like Canada. Some migrants do not remain permanently in the municipality in which they initially settle. For example, for the 2000–2 period 4.8 per cent of immigrants made an interprovincial move and 4.9 per cent moved between CMAs (Newbold, 2007). Some international migrants elect to return home or to move on to other countries: about 4.3 per cent for the decade 1990–2000 (Dryburgh and Hamel, 2004). Many establish **transnational** and sometimes mobile lifestyles, maintaining strong economic, social, or political networks across two or more national borders (Ghosh and Wang, 2003; Hiebert and Ley, 2003; Hyndman and

Walton-Roberts, 2000; Mensah, 2008; Satzewich and Wong, 2006; Veronis, 2007; Walton-Roberts, 2001, 2003; Wong and Ng, 2002).

The contemporary Canadian urban system is deeply rooted in these characteristics of immigration. Since European contact, immigration was encouraged and facilitated first by colonial administrators and then federal governments to increase the national population, to settle specific areas of the country, to develop staples economies like agriculture and forestry, and to meet labour demands of projects like canals and railroads, industrialization, and urban development. Onerous living conditions have been a catalyst for emigration across numerous time periods and places as well as the present day. For example, during the late nineteenth century, difficult economic circumstances in southern China and in the Prussian, Austro-Hungarian, and Russian empires spurred the emigration of several millions from these four regions, bringing 17,000 Chinese, 110,000 Poles, 12,000 Hungarians, 150,000 Ukrainians, and 175,000 Germans into Canada. Other triggers were the American Revolution (1775–83), the War of 1812–14, the Irish famine (1845–9), World Wars I and II, as well as more recent conflicts in such areas as Eastern Europe (1950s and 1960s), Chile, Asia, and Haiti (1970s), central America (1980s), and the Balkans and Africa (1990s).

Many of these movements are easily identifiable in the contemporary Canadian urban landscape, while others have been erased, muted, or were always relatively invisible. Contemporary national patterns of settlements and neighbourhoods of English- and French-speaking residents across the country testify to the historical numerical, economic, and social dominance of the British and French. Barely salient today, English, Scottish, and Irish neighbourhoods at one time were very significant in the residential patterns in nineteenth-century Montreal (Germain and Rose, 2000). Historical immigrant settlements and neighbourhoods include black loyalist, refugee, and Caribbean immigrant settlements in the former Africville in present-day Preston, Nova Scotia; Hungarians in Winnipeg, Regina, and Hamilton; Ukrainians in Winnipeg and across the Prairies; Greeks, Italians, and Portuguese in Toronto and Montreal; Germans in Kitchener and throughout the country; Sikhs in Victoria and Vancouver; Japanese in Vancouver and in small BC towns like Steveson and Chemainus; and Chinese in cities like Victoria, Vancouver, Calgary, Winnipeg, Toronto, and Montreal (McQuillon, 1993).

Until relatively recently, Canadian immigration policy almost exclusively favoured European, predominantly British, migrants through racist policies limiting, taxing, or barring migration from non-European countries. Beginning in the 1960s, continuing labour needs, the slowing supply of Western European migrants, and advances in international and national human rights legislation led to the removal of geographic restrictions. In 1967, the points system established admission criteria such as education and language (Green and Green, 2004). Over the next 40 years, immigration policy was strongly influenced by factors that remain with us today—declining fertility, the decline of industrial-manufacturing economies, and rise of service, knowledge, assets, and technology sectors (Simmons, 1999).

Pursuant to the 2002 Immigration and Refugee Protection Act, current immigration policy works towards three main objectives: 'to support Canada's economy and competitiveness, support family reunification and uphold Canada's humanitarian commitments' (CIC, 2008a: 9) that correspond to the three major categories: economic, family class, and refugee. Each class of immigrants is associated with particular policy rationales, migration origins, flows, destinations, and settlement and integration experiences. Since 1996, economic immigrants have been more than double (58 per cent) the number of family immigrants (28 per cent) compared to a 14 per cent intake of refugees (CIC, 2005, 2008b). Uneven rates of urban growth in Canada over the past half-century have funnelled most immigrants (63 per cent in 2006) into the largest metropolitan regions: Montreal, Toronto, and Vancouver (MTV) (Chui, Tran, and Maheux, 2007). Second- and third-tier cities have generally had greater challenges attracting and retaining immigrants, although this pattern is slowly changing (Pruegger and Cook, 2009; Radford, 2007).

A brief examination of each of the immigrant categories illustrates the uneven and dynamic urban geographies of immigration. The economic immigrant class is a diverse group, including skilled workers (overwhelmingly the largest class of immigrants), business-class immigrants, and provincial/territorial nominees. Highly educated—more than a third have university degrees as compared to 14 per cent of adult Canadians (CIC, 2005)—most economic immigrants settle in MTV (between 75–80 per cent between 1986 and 2000). The latest data on permanent residents in the economic class show that since 1999, Toronto has received more than 50,000 per year and Vancouver flows have been consistent, averaging 22,000 per year; whereas Montreal has experienced a steady increase of economic immigrants from almost 13,000 in 1999 to more than 23,000 per year since 2004 (CIC, 2008b). The business-class program targets those whose business experience, risk capital, and entrepreneurship can augment economic development (Ley, 2000). Through the 1990s, most business-class migrants were East Asian (e.g., from Taiwan and Hong Kong) and settled primarily in the Toronto and Vancouver CMAs.

The Canada–Quebec Accord and the Provincial/Territorial Nominee Program (PNP) responded to slow growth rates in second- and third-tier cities, the uneven regional distribution of immigration, and the demand for regional economic development (CIC, 2001). Through the PNP, provinces participate in the recruitment and selection of immigrants to meet specific regional requirements, such as labour market needs or French-speaking immigrants destined for minority francophone communities (Belkhodja, 2008). The impact of the PNP has been increasing rapidly; for example, in Manitoba, the proportion of PNP migrants has jumped from 30 per cent in 2002 to 70 per cent in 2007 (Allan, 2009).

Temporary migrants include live-in caregivers, foreign workers, and students, each group experiencing very different geographies and settlement trajectories. Live-in caregivers, more than three-quarters of whom are Filipina women, are hired primarily by middle-class Canadian families across the urban system as child-care and domestic workers (Bonifacio, 2009). The **temporary foreign workers** (TFW) program was designed to provide short-term solutions to labour shortages in specific sectors of the economy, but it has been steadily increasing over the past decade. In 2007 more than 200,000 TFWs entered the country, only 15 per cent fewer than those in the permanent categories (CIC, 2007). Unlike permanent immigrants, high proportions (40 per cent) of TFWs go to smaller urban centres: for example, to work in the oil sands near Fort McMurray, Alberta, or at meat-packing plants in Brandon, Manitoba. Some question whether the TFW program is truly temporary, given that agricultural TFWs have been supporting Canada's horticulture sector for 40 years (Ferguson, 2007). International students are significant populations for second- and third-tier cities, yet, as Walton-Roberts (2008) notes, universities and municipal governments could do more to incorporate international students into regional labour markets. The New Canadian Experience class, created in late 2008, has been designed to facilitate this type of transition for students and TFWs.

Family-class immigrants are sponsored by immigrants with independent status and are associated with the geographic patterns of those they join. The refugee program includes government and privately sponsored refugees and landed refugees. With less choice concerning their destination, sponsored refugees tend to be more evenly distributed across the country, with significant proportions settled outside of MTV: between 52 per cent and 32 per cent between 1986 and 2000. Particular groups of refugees often are associated with sponsorship-related settlement patterns, for example, Kosovar settlement in northern Alberta (Derwing and Mulder, 2003).

Finally, the frequently overlooked refugee category concerns people with precarious immigration status (Goldring, Berinstein, and Bernhard, 2007), those who for a variety of reasons, and often through the 'legal production of illegality', are caught in the precarious zone between legal and illegal status. This complex policy area involves several troubling issues including the geographic isolation of many of these people, the gendered and racialized processes contributing to their vulnerability, the inaccessibility of services, and a lack of data describing these people and their experiences.

Woven together with immigrant class, many other factors such as country of origin, language, and religion also influence the flows of immigrants into and across the country. The most significant shift since the 1960s has been the decrease in European immigration and the tremendous increase in the diversity of immigrants coming from regions throughout the world, particularly Asia. During the past two censuses, the People's Republic of China, India, the Philippines, and Pakistan have been the top four countries sending immigrants to Canada (Statistics Canada, 2008) (see Table 9.1). However, the translation of these trends into the specific profiles of Canadian cities is complex. For example, for 2007 (CIC, 2008b), the lead source country of immigrants varied significantly across Canada's first-, second-, and third-tier cities:

India: Calgary, Edmonton, Kitchener, and Toronto
China: Halifax, Hamilton, Ottawa–Gatineau
    (Ontario), and Vancouver
Philippines: Winnipeg and Saskatoon
France: Montreal and Quebec
Colombia: London and Ottawa–Gatineau
    (Quebec)
United States: Oshawa, St Catharines–Niagara,
    Victoria, and Windsor

Table 9.1 Top 10 Countries of Birth of Recent Immigrants, 1981–2006

Order	2006 Census	2001 Census	1996 Census	1991 Census	1981 Census
1	People's Republic of China (PRC)	PRC	Hong Kong	Hong Kong	United Kingdom
2	India	India	PRC	Poland	Viet Nam
3	Philippines	Philippines	India	PRC	USA
4	Pakistan	Pakistan	Philippines	India	India
5	USA	Hong Kong	Sri Lanka	Philippines	Philippines
6	South Korea	Iran	Poland	United Kingdom	Jamaica
7	Romania	Taiwan	Taiwan	Viet Nam	Hong Kong
8	Iran	USA	Viet Nam	USA	Portugal
9	United Kingdom	South Korea	USA	Lebanon	Taiwan
10	Colombia	Sri Lanka	United Kingdom	Portugal	PRC

Source: Statistics Canada (2008: 10).

Trends in top source countries tend to shift from year to year and from city to city, highlighting the local specificity of immigrant settlement. While national trends of inflows are more easily summarized, they are less directly transferable to municipal settings.

Despite this variability, some broad trends are identifiable (CIC, 2005). During the 1995–2001 period, Toronto's profile tended to mirror the total national picture, with greater proportions of South and Central Asian and Caribbean immigrants. Vancouver's immigrants were predominantly Asian-born, one-half of whom were from East Asia, namely Hong Kong and China. Montreal drew larger proportions of its immigrants from francophone countries. Seen as a group, the highest proportion of immigrants arriving in second-tier cities came from Southeast Asia and the Pacific (18.3 per cent) and Eastern Europe (17.5 per cent), whereas, for those migrating to third-tier cities, the largest proportions were from East Asia (17.7 per cent) and Western Europe (14.5 per cent). Interestingly, the largest proportion of immigrants from the United States (40.7 per cent) and the United Kingdom (34.6 per cent) went to communities outside Canada's first-, second- and third-tier cities. There are numerous implications of this 'diversity of diversity' for cities, particularly the importance of monitoring annual fluctuations in immigrant settlement patterns at regional, municipal, ward, and neighbourhood levels.

Settlement

Settlement refers to the initial stages following the arrival of an immigrant or refugee. A spectrum of urban issues and outcomes relates to this phase. The average university student has experienced the many steps required to settle into a new community: sorting out housing, employment, and transportation, orienting to the new local geography, locating basic goods and services, and establishing a new support network. The immigrant and refugee experience of these aspects of settlement is more arduous in numerous respects and, for many, will take months or years rather than days or weeks to accomplish. Many more tasks are required than in the Canadian-born student's settlement experience, combined with the disorienting experience of adapting to a foreign culture, a new language,[1] a different climate, and an unknown urban environment (Mwarigha, 2002; Omidvar and Richmond, 2003). For refugees fleeing dangerous circumstances these challenges are further compounded by recovery from physical and psychological trauma.

Suitable employment and housing are essential dimensions of settlement and greatly determine immigrants' capacities to deal with the other elements of the settlement process. While some immigrants successfully fulfill both needs, these are major issues of concern for many others, particularly for visible minorities. Beginning with employment, the lack of recognition of foreign credentials and racial discrimination are significant problems (Bauder, 2003). Foreign credential recognition is a complex policy area, with multiple barriers that include the capacity of institutions and professional associations to assess credentials, skeptical or prejudiced attitudes of employers despite recognized equivalence, large proportions of the labour market belonging to non-regulated occupations (Hawthorne, 2007), and protectionism within professions (Girard and Bauder, 2007). The recent Foreign Credential Recognition Program has been developed by Human Resources and Development Canada to address these issues, but there is considerable debate concerning the most appropriate solutions. Compounding the issue, recent studies have pointed to the declining incomes and higher poverty rates for visible minorities. Data from 2002 show a 23.2 per cent gap in household income between visible minorities and whites, and in 2001 racial minorities experienced poverty levels almost twice as high as the rest of the population (Reitz and Banerjee, 2007: 492). Gender further aggravates these differences as there has been a persistent, albeit poorly under-stood, gap between immigrant women's economic performances and those of their male counterparts (Tastsoglou and Preston, 2005). The co-ordination of a coherent, multi-stakeholder strategy to address these employment equity issues is a tremendous challenge (Hawthorne, 2007) but a key element in the successful settlement. Some cities, through organizations like the Hamilton Immigrant Work-force Integration Network (Orme, 2007), are play-ing a more active role in the co-ordination of this much needed reform.

Housing issues for immigrants display con-siderable geographic variation because of sig-nificant differences in housing markets and demographics from city to city (Hiebert et al., 2006; see Walker and Carter, Chapter 20). Key issues for **core housing needs** of vulnerable groups, particularly low-income, visible minority immigrants, remain relatively constant across the urban system and include accessibility to affordable housing, discrimination, and housing stock suit-ability (e.g., units for large families).

Settlement services help adapt immigrants' skills and knowledge to Canadian contexts, respond to specialized needs, and strengthen relationships between newcomer and host communities (Can-adian Council for Refugees, 1998). Two main pro-grams, Language Instruction for Newcomers to Canada (LINC) and the Immigrant Settlement and Adaptation Program (ISAP), are funded both fed-erally and provincially and are delivered by local governments, the not-for-profit sector, and the private sector. In addition to these formal services, religious organizations, immigrant agencies, social networks, universities, and co-ethnic support are instrumental in immigrant settlement.

Geographic issues, particularly the challenge of matching the location of services and infra-structure with communities of need, are common to many immigrant settlement service provid-ers (Apparicio and Séguin, 2006; Truelove, 2000), particularly in smaller urban centres or rural areas

(Reimer, 2007). Recent settlement of immigrants in suburban rather than traditional central-city neighbourhoods requires the simultaneous sub-urbanization of settlement services (Hiebert, 2000; Ley and Murphy, 2001). Other settlement service gaps exist for addressing successful labour market participation, including entrepreneurship (Teixeira, Lo, and Truelove, 2007); for poor living and work conditions for TFWs and live-in caregivers (Alberta Federation of Labour, 2009; Bucklaschuk, Moss, and Annis, 2009; Ferguson, 2005; Pratt, 1997); and for support during transitional periods from temporary to permanent residency and/or during family unification (Pratt, 2006). Further exacer-bating the situation, the **non-governmental organizations** are increasingly under pressure due to growing demands, diminished and short-term funding sources, increased competition for fund-ing, greater accountability reporting requirements, fewer volunteers, and insufficient staff/administra-tive support (Richmond and Shields, 2004). These pressures are especially onerous for smaller organ-izations and for those addressing structural issues related to racism and poverty (Viswanathan et al., 2003; Viswanathan, forthcoming).

Integration

Integration refers to the long-term process through which immigrants come to participate fully in their new society. Integration issues invariably are tied to broader discourses concerning those social differences commonly associated with recent immigrants: culture, language, ethnicity, 'race', and, increasingly, religion.[2] Questions about how soci-ety should address social difference are at the base of debates concerning multiculturalism, immigra-tion, and diversity, and while these debates are often national in scale, their greatest consequences are felt at the local level. During earlier periods, immi-grants (and Aboriginal peoples) were expected to assimilate and acculturate, that is, to shed their languages, cultures, and ethnic affiliations and adopt the cultural variants of mainstream society (Fleras and Elliot, 2003; Kobayashi, 1993). These approaches devalued cultural differences, ignored the daily experiences and structural foundations of inequity, and generally lacked imagination for creating alternative modes of society-building.

Since the early 1970s, **multiculturalism** has been an influential, albeit controversial, model for addressing social difference (Kymlicka, 1998; Ley, 2007). The multicultural approach argues that minority groups can participate fully in Canadian society while also maintaining distinctively different social values, practices, and institutions provided they protect human rights and freedoms. Interculturalism, Quebec's corollary concept, proposes greater emphasis on interchange between minority communities and Quebec society through use of the French language in the public sphere, while integrating minority needs and interests through Quebec institutions (McAndrew, 2007; Salée, 2007). While many have lauded the success of Canada's approach to multiculturalism, some have feared that it will lead to a splintering of Canadian society (Karim, 2002), while others have questioned its capacity to create an equitable society and to challenge deep structural issues related to racism and discrimination (Abu-Laban, 2007; Goonewardena and Kipfer, 2005; Henry and Tator, 1999; Wood and Gilbert, 2005).

These criticisms are related to two broad concepts at the local level: urban citizenship and urban governance. Taken together, they address the means by which immigrants integrate over the long term and participate in the development of their cities. Briefly, **citizenship** describes a view of the city from 'below' and refers to the rights and responsibilities associated with membership in a political community (Isin and Wood, 1999). Citizenship practices include formal, state-recognized practices like voting in a municipal election as well as informal practices like organizing a community event. Examining the relationships among citizenship, cities, and place, urban scholars observe that various dimensions of identity (gender, ethnicity, etc.) and particular places (cities, neighbourhoods) have important consequences for the ways in which people are able to negotiate their sense of belonging in the city (Desforges, Jones, and Woods, 2005; Holston, 1999; Lister, 2007).

Immigrant municipal electoral participation, for example, is one key area of urban citizenship research. Scholars have explored the factors that influence rates of participation for different immigrant groups and across the range of immigrant-receiving municipalities, demonstrating a range of issues and experiences (Andrew et al., 2008; Simard, 2001). Human capital, age, language, and religion are significant factors in electoral participation. Geographic factors such as length of residence and neighbourhood composition have been shown to correlate with higher voter turnout (Tolley, 2003: 14).

Several ongoing issues are related to representation and legitimacy in political discourse and decision-making. One universal concern is what Andrew et al. (2008: 18) refer to as the archetype of the Canadian elected official: male, white, middle-class, middle-aged, Christian, Canadian-born, and majority-language-speaking. While some cities are shifting slowly towards more diverse representation, overall a chronic under-representation of minority groups has prevailed. To address these problems, scholars argue for greater public education on the topics of elections, campaigning, and civic literacy alongside improved accessibility (physical and linguistic) to electoral systems at all levels.

Veronis's recent work (2006, 2007) on Toronto's Latin American community provides a second illustration of the relationships between urban citizenship and long-term integration. She examined the different approaches this heterogeneous immigrant community has used to 'negotiate their internal differences, build a local collective identity' (Veronis, 2007: 468), and establish a material and political presence in the city. She studied community practices (2006), such as Toronto's

Canadian Hispanic Day parade, as well as imagined geographies of identity (2007), like planning efforts towards creating a Latin American community centre (*casa*) and the collective dream of establishing a Latin American neighbourhood (*barrio latino*). Identifying how community members link these spatialized strategies with the need for a more concrete sense of collective belonging in the city, Veronis has demonstrated that the creation of urban ethnic places is a meaningful strategy towards a more inclusive urban citizenship.

Recent immigrant experiences of developing minority places of worship provide a third example of urban citizenship. Like ethnic community centres, places of worship are important community resources for settlement and integration, the renegotiation of religious and ethnic identities, and resistance to the marginalization that results from discrimination along multiple axes (Tettey, 2007). The construction of a place of worship is an important way minority groups stake their claim to belonging in cities (Isin and Siemiatycki, 1999). Despite the fact that freedom of religion is guaranteed by the Canadian Charter of Rights and Freedoms, minority groups have faced numerous challenges in Toronto and Montreal due to land-use planning issues and local social and political dynamics (Germain, 2006; Germain and Gagnon, 2003; Qadeer and Chaudhry, 2000). Hoernig's research (2006, 2009) in Brampton, Markham, and Mississauga shows that minority religious groups (Buddhist, ethnic-minority Christian, Hindu, Muslim, and Sikh) address these challenges using multiple strategies throughout the development process for establishing a place of worship. These include adapting religious practices and spatial needs to various constraints, accommodating the concerns of neighbours in site planning and day-to-day practices, avoiding potential conflicts through location decisions (e.g., to locate in industrial zones), and employing different tactical strategies to engage neighbours, local politicians, and municipal staff to maximize support and minimize conflict.

In contrast to citizenship, governance takes a view of the city from above. **Governance** is the exercise of power through authority and control and the allocation of resources. Recent research has emphasized that, in addition to government, many public, private, and civil society organizations are involved in governance, through a tremendous variety of structures, relationships, roles, and processes (Hajer, 2003; Pierre, 2005). Questions related to urban governance and immigration ask to what extent urban decision-making reflects the needs and interests of immigrant and minority groups.

A growing body of work examines the responses of municipal governments to diversifying populations, exploring various factors across the spectrum of active to inactive municipalities. Cities face several challenges in addressing immigrant issues (Frisken and Wallace, 2003: 157–8) because they have no constitutional obligation to address immigration and are largely constrained by the dictates of their provincial masters. Meanwhile, national and provincial governments provide little guidance on handling municipal issues related to immigration.

A series of studies at the turn of the century (Edgington and Hutton, 2002; Germain et al., 2003; Moore Milroy and Wallace, 2002) found among the Montreal, Toronto, and Vancouver CMAs that central cities were much more responsive to diversity issues than their suburban counterparts. Furthermore, despite numerous policies, many municipal efforts were ad hoc rather than proactive in addressing the broad range of cultural and equity issues in municipal programming and service delivery. Five years later, great variation in municipal responses continues, including in language services (translation and interpretation), multiculturalism and anti-racism initiatives, funding, public consultation, and human resources policies and initiatives like employment equity, monitoring, and outreach (Good, 2007; Tossutti, 2009b). Toronto, Vancouver, and more recently Edmonton have been the most active in these areas. In Tossutti's (2009a, 2009b) analysis, municipal responses to diversity are explained by a number of factors, including economic drivers, local political values and cultures, immigrant community mobilization, municipal corporate philosophies, and resistance within bureaucracies and public-service unions. Edmonton, for example, in response to a high economic need to attract and retain immigrants for its local labour force, has developed a more proactive stance towards immigration, establishing a municipally based immigration and settlement policy and developing its diversity and inclusion office. However, issues like the low representation of minorities within municipal workforces continue to be significant challenges for all Canadian municipalities, signalling the need for improvement in many areas.

Another way to approach governance is to examine municipal responses to particular types of urban developments associated with immigrants. Zhuang's (2008, 2009) research on ethnic retail demonstrates the complexity of urban governance when dealing with ethnocultural diversity. Her findings reveal that, among the various players involved, municipal government has been the least motivated and plays only a reactive role in addressing ethnic commercial activity and urban development. Other players, such as ethnic entrepreneurs, community agencies, and developers, have demonstrated much greater determination and leadership in shaping ethnic retail areas.

The **revitalization** of East Chinatown in Toronto illustrates this dynamic. East Chinatown retail business has been declining gradually because of demographic change, suburbanization, and the poor physical condition of the business area itself. In 1998, the local voluntary business association, the Chinese Chamber of Commerce, East Toronto (CCCET), initiated the China Gate project in hopes of promoting local business, boosting tourism development, and revitalizing the business environment. Using political resources, the CCCET eventually gained support for their project, securing City Council's approval and sponsorship through donated land, financial contributions, and city maintenance

of the China Gate. Community participation has also been high for related projects such as the East Chinatown capital improvement plan and a beautification mural (see Figure 9.1). But it has taken more than 10 years to accomplish the China Gate project, and meanwhile, many other fundamental business issues, such as cleanliness, parking, and safety, remain unresolved. While merchants struggle to survive and have worked hard for many years to revitalize this

Figure 9.1 [Top] The Chinese Archway was officially opened in 2009. [Bottom] Murals in East Chinatown beautification efforts. (Cecilia Zhuang)

declining retail area, the city has responded only reactively to the community's concerns.

Continuity, Change, and Fusion: Understanding Urban Geographies of Immigration

This chapter concludes with a spatial overview of city–immigration relationships, highlighting the ways in which urban growth patterns and immigration trends are tightly interwoven and mutually constitutive: cities draw immigrants to particular places; immigrants in turn contribute to the continuation of some development patterns while bringing significant changes to others.

Immigration settlement trends at the national, regional, and local scales tend to reflect general urban growth trends. Data for 2001–6 on each city's proportion of recent immigrants relative to the city's share of the total Canadian population reveal several important patterns (Chui, Tran, and Maheux, 2007: 20–1). Toronto maintains its primary position as an immigrant-receiving metropolitan region with a 2.5 times larger share of recent immigrants than its share of the total Canadian population. Vancouver has twice its share and Calgary and Montreal have 1.5 and 1.3 times their shares, respectively. Importantly, Kitchener, Windsor, Winnipeg, and Abbotsford also show ratios greater than one, demonstrating the significant influx of recent immigrants to these communities during the time period. A third group of cities, Ottawa–Gatineau, Hamilton, Guelph, London, and Edmonton, at 0.8 and 0.9, are receiving close to their proportion of the national population in recent immigrant flows, but a large group of small cities across the country, especially those with low or negative growth rates, are receiving very low proportions of recent immigrants—the lowest ranking (at 0.1 and 0.2) includes St John's, Moncton, Saint John, Saguenay, Trois-Rivières, Peterborough, Greater Sudbury, and Thunder Bay.

At the regional scale, too, trends mirror general metropolitan growth patterns, reflecting the broad 'donut' pattern of suburban development. Traditionally, inner-city neighbourhoods played an important role in immigrant reception. Centrally located ethnic concentrations of residences and businesses capitalized on mutual support from co-ethnic people to address numerous challenges, including discrimination experienced outside of these enclaves (Anderson, 1991). Today, the inner city is no longer the only trajectory for immigrants. Following the trends of many Canadian-born residents, some recent immigrants bypass the inner city and settle directly in suburban areas (Chui, Tran, and Maheux, 2007), while others relocate there later on (Teixeira, 2007). In MTV, immigrant populations have settled in suburban municipalities like Brampton, Mississauga, and Markham (Toronto CMA), and Burnaby, Richmond, and Surrey (Vancouver CMA). In the Montreal CMA, immigrant settlement continues to be primarily concentrated on the island of Montreal, but more so in inner suburban municipalities like Côte-Saint-Luc while slowly beginning to disperse to the outer municipalities of Laval, Longueil, and Brossard (Figure 9.2). Several Canadian suburban municipalities such as Markham, Richmond, and Burnaby now have populations with greater than 50 per cent foreign-born and visible minority populations, turning upside down the very concept of visible *minority*.

Within these broad regional patterns, the distribution of immigrant groups across census tracts and neighbourhoods is complex, characterized by patterns of both dispersion and concentration. As Ray (2009: 84) emphasizes, 'immigrants, whether new arrivals or long-established, live in a vast array of neighbourhood types—from inner city ethnic "villages" to quintessential post-war suburbs.' Although immigrant groups and visible minority groups experience lower incomes than the rest of the Canadian population, in those areas where they are concentrated they tend to live in socially

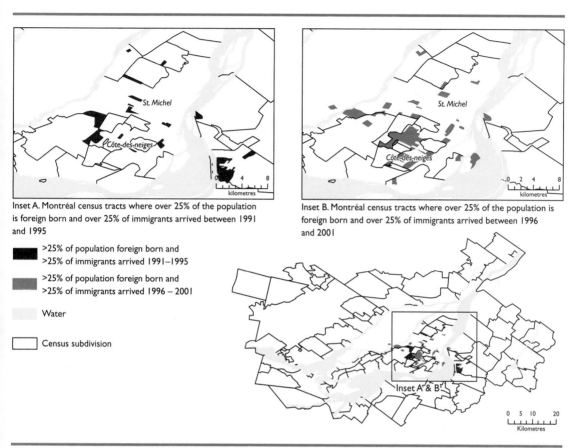

Figure 9.2 Areas of high foreign-born and new immigrant population concentration, Montreal CMA, 2001

Source: Statistics Canada, 2001 Census, Citizenship and Immigration Canada custom tabulations, 2006, Brian Ray; also in Ray (2009: 86).

heterogeneous neighbourhoods of both low- and medium-income households (Hiebert et al., 2007). Settlement patterns within neighbourhoods can also be dynamic where multiple waves of immigrants contribute to shifting local social, economic, and political relations over time. Montreal's Parc Extension, for example, reflects a complex mix of social and institutional practices, having been transformed from an area of high Greek concentration in the 1960s to a more evenly mixed multi-ethnic neighbourhood in the early to mid-1990s, shifting again at the turn of the century as South Asian immigrants gained prominence (Poirier, 2006).

Traditional spatial analyses of intra-urban immigrant or visible minority patterns assume that the spaces in and around people's homes and local neighbourhoods are the primary places of social interaction and that ethnicity, 'race', and measures of economic status are key variables in these settlement patterns and social interaction (Apparicio, Leloup, and Rivet, 2006; Mendez, 2008; Walks and Bourne, 2006). More recent areas of geographical inquiry demonstrate that, while residential areas are indeed important urban spaces (Germain et al., 1995), the lived experiences of cities are complex. Multiple dimensions of people's identities (gender, ethnicity, etc.) play out in different interacting ways across time and space, what Pratt (1998) refers to as the multiple 'grids of difference' including the various home, work, social, and public spaces people traverse throughout their day as well as across the shifting stages of their life cycles (Valentine, 2007).

This broadening understanding of how the city is experienced affirms recent research streams. One has been the exploration of spaces of inter-ethnic interaction beyond the residential neighbourhood, such as Ray and Bergeron's (2007) work examining recent immigrant workplace geographies; this group also includes a number of recent studies looking at ethnic and multi-ethnic commercial areas (Radice, 2008; Preston, 2008; Zhuang, 2008, 2009), and others exploring public space (Germain et al., 2008; Germain, Liégeois, and Hoernig, 2008). A second stream explores the ways in which

religion contributes to immigrant and ethnic minority place-making (Agrawal, 2008, 2009; Germain, 2009; Hoernig, 2009) (Figures 9.3 and 9.4). A third area includes those studies emphasizing the diversity and complexity of broad ethnic, linguistic, and 'racial' categories (Lo, 2006; Wang and Lo, 2004). For example, Ghosh's research comparing Indian and Bangladeshi Bengali-speaking immigrants demonstrates how cultural identity, religion, and cultural capital diversify settlement patterns and housing trajectories within a single linguistic group. Finally, a growing body of work builds on network theories of sociability and emphasizes the variety of ways in which immigrants organize and coalesce, in Webber's (1964) words, in 'communities without propinquity' (Laux and Thieme, 2006; Mitropolitska, 2008; Zelinsky and Lee, 1998). Viswanathan's (forthcoming) research, for instance,

Figure 9.3 Punjabi retail development, 2005, beside the Sikh temple, Sri Guru Singh Sabha Gurdwara, Airport Road, Malton, Ontario. The sign reads 'Great Punjab Business Centre—Where Business and Culture Come Together'. (Heidi Hoernig)

Figure 9.4 Residential development incorporating the Slovak Cathedral of Transfiguration (Slovaks of the Byzantine Catholic rite) on Woodbine Avenue, Markham, Ontario. (Heidi Hoernig)

highlights Toronto's Alternative Planning Group's (AGP) relational understanding of the city as a complex web of social networks with numerous nodes of interaction, including homes, workplaces, schools, religious institutions, family networks, and other social and cultural organization.

Another important spatial reading of city–immigration relationships is the urban consequences of particular immigrant policies. For example, the business immigrant program resulted in more than $4 billion of investment and the creation of tens of thousands of jobs between the mid-eighties and 1998 (Ley, 2000). Suburban development in both Toronto and Vancouver CMAs, particularly growth in places like Markham and Richmond, has been directly, although certainly not exclusively, related to these arrivals and

accompanying investments (Lo, 2006). Growth and change in the housing market (Hiebert et al., 2006; Ley and Tutchener, 2001), banking and financial services (Edgington, Goldberg, and Hutton, 2006), and retailing (Wang, 1999; Wang and Lo, 1994) are some of the many dimensions of this development. But there have been negative consequences, too. For example, Ley's (2000) study describes the significant impediments, disappointments, and losses experienced by participants in the business-class program. Several high-profile incidents of social tension and land-use conflict related to xenophobia, racism, and cultural conflict have arisen, also (Ley, 1995; Ley 2000; Li, 1994; Ray, Halseth, and Johnson, 1997; Rose, 1999), demonstrating the strong resistance of some residents to immigrant-related change.

Shifts in conventional approaches to municipal planning and regulation of land use have accompanied recent immigration trends. While immigrant developers must adapt to the Canadian land-use planning system, land-use planners are also increasingly challenged to reconsider their definition of 'public interest', and to look for hidden cultural values embedded in planning practice. The degree and nature of community participation within planning practices have come under review (Qadeer, 1997; Sandercock, 2003). Ethnic retail development illustrates these mutual adjustments. For example, Asian shopping centres are a rapidly growing type of suburban retail centre (Preston and Lo, 2000; Wang, 1999). Several features set them apart from mainstream retail forms: condominium ownership (versus leasehold), the absence of conventional anchors such as the Bay or Wal-Mart, a catalytic development strategy, variable store hours, and site layout (e.g., smaller store size). When these centres first emerged, a series of conflicts erupted regarding land-use planning and public concern. Conventional land-use planning policies used traditional hierarchical criteria such as floor area, numbers of stores, market size, and sales to classify shopping centres. Because this classification system was irrelevant in the assessment of Asian malls, municipalities had to rework their policies in order to regulate this new form of development.

In conclusion, continuity, change, and fusion are outcomes of immigration to Canadian cities. We return again to the example of ethnic retail to summarize how these effects emerge across evolving urban landscapes (Zhuang, 2008). Ethnic retail falls within mainstream urban development patterns of planned suburban centres and inner-city retail strips. Ethnic retail has contributed in numerous ways to the diversification of streetscapes through urban design (e.g., street furniture, festival space), decorations (e.g., banners, facades) and architectural elements such as materials (e.g., ceramic tiling, wrought iron), cultural symbols (e.g., colours, iconography), and creative uses of space (e.g., patios, plazas, awnings, gardens).

These features sometimes are enhanced further to create official municipal tourist or heritage districts. But the boundaries between what was historically defined as 'ethnic' and 'mainstream' retail are becoming blurred. For instance, the use of patio space for an outdoor café was treated as an 'ethnic' practice in the 1960s when the City of Toronto issued its first outdoor café permit. Over time, some ethnic businesses have shifted from catering exclusively to co-ethnic customers to expanding their retail strategies to reach broader markets, for instance, by carrying more general merchandise and diluting visible ethnic expression. Meanwhile, mainstream chain stores, such as Wal-Mart and No Frills, have begun to target ethnic markets. The ethnic retail sector is no longer segregated socially, economically, or spatially from the general economy; instead, a two-way penetration between both 'ethnic' and 'mainstream' markets has occurred.

Tremendous exchange and fusion of ideas, resources, and networks can result from immigration. Canadian cities can benefit enormously from this interaction. But this potential will be needlessly squandered if everyday practices of leaders and citizens do not recognize the pivotal role that host communities play in the successful integration of newcomers. As many have cautioned, the celebration of the benefits of immigration and diversity must not camouflage or become sterile euphemisms for long-standing issues related to xenophobia, racism, and linguistic, religious, and ethnic discrimination. Key issues like credential recognition, employment equity, housing accessibility, electoral representation, and equitable urban governance need to be substantially addressed. Efforts across all levels of government and throughout public, private, and civil society organizations will need to support positive day-to-day interactions of Canadian-born and immigrant residents in neighbourhoods, workplaces, and public spaces. Canadian cities will truly flourish when all residents are able to actively engage and participate in the evolution of their communities.

Review Questions

1. How and why has immigration changed in recent decades?
2. How/why does immigration raise issues associated with citizenship and governance?
3. Using your own knowledge of the community in which you grew up, give one or more examples as to how cultural diversity is expressed on the urban landscape.

Notes

1. Even when migrants speak the same language as their host community, major adjustments are necessary due to different accents, vocabularies, expressions, slang, and other place-specific socio-linguistic norms.
2. For example, Quebec's recent debate on reasonable accommodation (Bouchard and Taylor, 2007).

References

Abu-Laban, Y. 2007. 'History, power and contradictions in a liberal state', in Banting et al. (2007).

Agrawal, S.K. 2008. 'Faith-based ethnic residential communities and neighbourliness in Canada', *Planning Practice and Research* 23: 41–56.

———. 2009. 'New ethnic places of worship and planning challenges', special issue, 'Welcoming communities: Planning for diverse populations', *Plan Canada*: 55–9.

Alberta Federation of Labour. 2009. *Entrenching Exploitation*. Edmonton: Alberta Federation of Labour.

Allan, N. 2009. 'Foreign worker recruitment and protection: The role of Manitoba's Worker Recruitment and Protection Act', *Our Diverse Cities: The Prairies—Frontiers of Migration* 6: 31–6.

Anderson, K. 1991. *Vancouver's Chinatown: Racial Discourse in Canada 1875–1980*. Montreal and Kingston: McGill-Queen's University Press.

Andrew, C., J. Biles, M. Siemiatycki, and E. Tolley, eds. 2008. *Electing a Diverse Canada: The Representation of Immigrants, Minorities, and Women*. Vancouver: University of British Columbia Press.

———, ———, ———, and ———. 2008. 'Introduction', in Andrew et al. (2008).

Apparicio, P., X. Leloup, and P. Rivet. 2006. *La Répartition Spatiale des Immigrants à Montréal: Apport des Indices de Ségrégation Résidentielle*. Publication IM no. 28.

Montreal: Centre Métropolis du Québec. Immigration et métropoles.

——— and A.-M. Séguin. 2006. 'Measuring the accessibility of services and facilities for residents of public housing in Montréal', *Urban Studies* 43: 187–211.

Banting, K., T.J. Courchene, and F.L. Seidle, eds. 2007. *The Art of the State III: Belonging? Diversity and Recognition and Shared Citizenship in Canada*. Montreal: Institute for Research on Public Policy.

Bauder, H. 2003. '"Brain abuse" or the devaluation of immigrant labour in Canada', *Antipode* 35: 699–717.

Beiser, M. 2004. 'Refugees in Canada', *Canadian Issues/Thèmes canadiens* (Spring): 54–6.

Belkhodja, C. 2008. 'Immigration and diversity in francophone minority communities: Introduction', *Canadian Issues: Thèmes canadiens* (Spring): 3–5.

Bonifacio, G. 2009. 'From temporary workers to permanent residents: Transitional services for Filipino Live-in Caregivers in southern Alberta', *Our Diverse Cities: Prairies Region* (Spring): 136–41.

Bouchard, G., and C. Taylor, 2007. *Building the Future: A Time for Reconciliation*. Quebec: Commission de Consultation sur les Pratiques d'Accommodement Reliées aux Différences Culturelles.

Bucklaschuk, J., A. Moss, and R. Annis. 2009. 'Temporary may not always be temporary', *Our Diverse Cities: The Prairies—Frontiers of Migration* 6: 64–9.

Canadian Council for Refugees. 1998. *Best Settlement Practices*. Montreal: Canadian Council for Refugees.

Chui, T., K. Tran, and H. Maheux. 2007. *Immigration in Canada: A Portrait of the Foreign-born Population, 2006 Census*. Ottawa: Statistics Canada.

Citizenship and Immigration Canada. 2001. *Towards a More Balanced Geographic Distribution of Immigrants*. Ottawa: Citizenship and Immigration Canada.

———. 2005. *Recent Immigrants in Metropolitan Areas, Canada: A Comparative Profile Based on the 2001 Census*. Ottawa: Citizenship and Immigration Canada.

———. 2008a. *Annual Report to Parliament on Immigration*. Ottawa: Citizenship and Immigration Canada.

———. 2008b. *Facts and Figures 2007. Immigration Overview: Permanent and Temporary Residents*. Ottawa: Citizenship and Immigration Canada.

Dempsey, C., and S. Yu. 2004. 'Refugees to Canada: Who are they and how are they faring?', *Canadian Issues/Thèmes canadiens* (Spring): 5–10.

Derwing, T., and M. Mulder. 2003. 'The Kosovar sponsoring experience in northern Alberta', *Journal of International Migration and Integration* 4, 2: 217–36.

Desforges, L., R. Jones, and M. Woods. 2005. 'New geographies of citizenship', *Citizenship Studies* 9, 5: 439–51.

Dickason, O., with D. McNab. 2009. *Canada's First Nations: A History of Founding Peoples from Earliest Times*, 4th edn. Toronto: Oxford University Press.

Dryburgh, H., and J. Hamel. 2004. 'Immigrant in demand: Staying or leaving', *Social Trends* (Statistics Canada Catalogue no. 11–008) (Autumn): 12–17.

Edgington, D., M. Goldberg, and T. Hutton. 2006. 'Hong Kong business, money and migration in Vancouver, Canada', in Li (2006).

——— and T. Hutton. 2002. *Multiculturalism and Local Government in Greater Vancouver*. Vancouver: Research on Immigration and Integration in the Metropolis.

Ferguson, N. 2007. 'The seasonal agricultural workers program: Considerations for the future of farming and the implications of managed migration', *Our Diverse Cities: Rural Communities* 3: 189–93.

Fleras, A., and J.L. Elliot. 2003. *Unequal Relations: An Introduction to Race and Ethnic Dynamics in Canada*, 4th edn. Toronto: Prentice-Hall.

Frisken, F., and M. Wallace. 2003. 'Governing the multicultural city-region', *Canadian Public Administration* 46: 153–77.

Germain, A. 2006. 'Le municipal à l'épreuve de la multiethnicité: aménagement des lieux de culte dits "ethniques" et crise du zonage à Montréal', in A. Bourdin, M.-P. Lefeuvre, and P. Melé, *Les Règles du Jeu Urbain*. Paris: Descartes et Cie.

———, J. Archamault, B. Blanc, J. Charbonneau, F. Dansereau, and D. Rose. 1995. *Cohabitation interethnique et vie de quartier*. Québec: gouvernement du Québec, ministère des Affaires internationales, de l'Immigration et des Communautés culturelles.

———, F. Dansereau, F. Bernèche, C. Poirier, M. Alain, and J.-E. Gagnon. 2003. *Les Pratiques Municipales de Gestion de la Diversité à Montréal*. Montréal: Institut national de la recherche scientifique.

——— and J.E. Gagnon. 2003. 'Minority places of worship and zoning dilemmas in Montreal', *Planning Theory and Practice* 4: 295–318.

———, L. Liégeois, and H. Hoernig. 2007. 'L'espace public à l'épreuve des religions: des paysage pluriels à négocier?', in A. Da Cunha and L. Matthey, eds., *La Ville et l'Urbain: Des Saviors Émergents*. Lausanne: Presses Polytechniques and Universitaires Romandes.

———, ———, and ———. 2008. 'Les espaces public en contexte multiethnique: Religion, visibilité et pasteurization', in Leloup and Radice (2008).

——— and D. Rose. 2000. *Montréal: The Quest for a Metropolis*. Chichester: John Wiley and Sons.

Ghosh, S. 2007. 'Transnational ties and intra-immigrant group settlement experiences: A case study of Indian Bengalis and Bangladeshis in Toronto', *Geojournal* 68: 223–42.

———. 2007. *'We Are Not All the Same': The Differential Migration, Settlement Patterns and Housing Trajectories of Indian Bengalis and Bangladeshis in Toronto*. Ottawa: Canada Mortgage and Housing Corporation.

——— and L. Wang. 2003. 'Transnationalism and identity: A tale of two faces and multiple lives', *Canadian Geographer* 47: 269–82.

Giriard, E., and H. Bauder. 2007. 'The making of an "arcane" infrastructure: Immigrant practitioners and the origins of professional engineering regulation in Ontario', *Canadian Geographer* 51: 233–46.

Goldring, L., C. Berinstein, and J. Bernhard. 2007. *Institutionalizing Precarious Immigration Status in Canada*. Working Paper No. 61. Toronto: CERIS—Ontario Metropolis Centre.

Good, K. 2007. 'Multiculturalism policy and the importance of place: An uneven policy and jurisdictional landscape', Immigration, Minorities and Multiculturalism in Democracies Conference, Montreal, 5–7 Oct.

Goonewardena, K., and S. Kipfer. 2005. 'Spaces of difference: Reflections from Toronto on multiculturalism, bourgeois urbanism and the possibility of radical urban politics', *International Journal of Urban and Regional Research* 29: 670–8.

Graham, K., and S. Phillips. 2007. 'Another fine balance: Managing diversity in Canadian cities', in Banting et al. (2007).

Green, A., and D. Green. 2004. 'The goals of Canada's immigration policy: A historical perspective', *Canadian Journal of Urban Research* 13: 102–39.

Hajer, M. 2003. *Deliberative Policy Analysis: Understanding Governance in the Network Society*. Cambridge: Cambridge University Press.

Hawthorne, L. 2007. 'Foreign credential recognition and assessment: An introduction', *Canadian Issues / Thèmes canadiens* (Spring): 3–13.

Henry, F., and C. Tator. 1999. 'State policy and practices as racialized discourses: Multiculturalism, the Charter and employment equity', in P.S. Li, ed., *Race and Ethnic Relations in Canada*. Toronto: Oxford University Press.

Hiebert, D. 2000. *The Social Geography of Immigration and Urbanization in Canada: A Review and Interpretation*. Vancouver: Research on Immigration and Integration in the Metropolis.

——— and D. Ley. 2003. *Characteristics of Immigrant Transnationalism in Vancouver*. Working Paper Series, No. 03–15. Vancouver: Vancouver Centre of Excellence for Research on Immigration and Integration in the Metropolis.

————, A. Germain, R. Murdie, V. Preston, J. Renaud, D. Rose, E. Wyly, V. Terreira, P. Mendez, and A.-M. Murnaghan. 2006. *The Housing Situation and Needs of Recent Immigrants in the Montréal, Toronto and Vancouver CMAs: An Overview*. Ottawa: Canada Mortgage and Housing Corporation.

————, N. Schuurman, and H. Smith. 2007. *Multiculturalism 'on the ground': The Social Geography of Immigrant and Visible Minority Populations in Montréal, Toronto, and Vancouver, Projected to 2017*. Working Paper no. 07–12. Vancouver: Vancouver Centre of Excellence. Research on Immigration and Integration in the Metropolis.

Hoernig, H. 2006. 'Worship in the Suburbs: The Development Experience of Recent Immigrant Religious Communities', Ph.D. thesis, University of Waterloo.

————. 2009. 'Planning amidst cultural diversity: Lessons from religious development', special issue, 'Welcoming communities: Planning for diverse populations', *Plan Canada* 55–9.

Holston, J. 1999. 'Spaces of insurgent citizenship', in J. Holston, ed., *Cities and Citizenship*. Durham, NC: Duke University Press.

Hyndman, J., and M. Walton-Roberts. 2000. 'Interrogating borders: A transnational approach to refugee research in Vancouver', *Canadian Geographer* 47: 244–58.

Isin, E.F., and M. Siemiatycki. 1999. *Fate and Faith: Claiming Urban Citizenship in Immigrant Toronto*. Working Paper Series, Report No. 8. Toronto: CERIS—Ontario Metropolis Centre.

———— and P. Wood. 1999. *Citizenship and Identity*. London: Sage.

Karim, K. 2002. 'The multiculturalism debate in Canadian newspapers: The harbinger of a political storm', *Journal of International Migration and Integration* 3: 439–55.

Kelly, P. 2003. 'Canadian-Asian transnationalism', *Canadian Geographer* 47: 209–18.

Kobayashi, A. 1993. 'Multiculturalism: Representing a Canadian institution' in J. Duncan and D. Ley, eds., *Place/Culture/Representation*. London: Routledge.

Kymlicka, W. 1998. *Finding Our Way: Rethinking Ethnocultural Relations in Canada*. Toronto: Oxford University Press.

Laux, H.D., and G. Thieme. 2006. 'Koreans in greater Los Angeles: Socioeconomic polarization, ethnic attachment and residential patterns', in Li (2006).

Leloup, X., and M. Radice, eds. 2008. *Les Nouveaux Territoires de l'Ethnicité*. Quebec: Les Presses de l'Université Laval.

———— and ————. 2008. 'Introduction. Revisiter les liens entre espace et ethnicité: La nécessaire reterritorialisation de la question ethnique', in Leloup and Radice (2008).

Ley, D. 1995. 'Between Europe and Asia: The case of the missing sequoias', *Ecumene* 2: 185–210.

————. 2000. *Seeking 'Homo Economicus': The Strange Story of Canada's Business Immigration Program*. Working Paper No. 00–02. Vancouver: Vancouver Centre of Excellence, Research on Immigration and Integration in the Metropolis.

————. 2000. 'Multicultural planning: Whose city? Whose identity?', in K. Dosen and I. Molina, eds, *Best Practices for the Social Inclusion of Ethnic Minorities in Local Communities*. Norrkoping: Partnerknap for Multietnisk Intergration (PfMI).

————. 2007. *Multiculturalism: A Canadian Defence*. Working Paper No. 07–04. Vancouver: Vancouver Centre of Excellence, Research on Immigration and Integration in the Metropolis.

———— and P. Murphy. 2001. 'Immigration in gateway cities: Sydney and Vancouver in comparative perspective', *Progress in Planning* 55: 119–94.

———— and J. Tutchener. 2001. 'Immigration, globalization and house prices in Canada's gateway cities', *Housing Studies* 16: 199–223.

Li, P.S. 1994. 'Unneighbourly houses or unwelcome Chinese: The social construction of race in the battle over "Monster Homes" in Vancouver, Canada', *International Journal of Comparative Race and Ethnic Studies* 1: 14–33.

Li, W., ed. 2006. *From Urban Enclave to Ethnic Suburb: New Asian Communities in Pacific Rim Countries*. Honolulu: University of Hawaii Press.

Lister, R. 2007. 'Inclusive citizenship: Realizing the potential', *Citizenship Studies* 11, 1: 49–61.

Lo, L. 2006. 'Suburban housing and indoor shopping: The production of the contemporary Chinese landscape in Toronto', in Li (2006).

McAndrew, M. 2007. 'Quebec's interculturalism policy: An alternative vision', in Banting et al. (2007).

McQuillan, A. 1993. 'Historical geography and ethnic communities in North America', *Progress in Human Geography* 17: 355–66.

Massey, D. 1999. 'Why does immigration occur? A theoretical synthesis', in C. Hirschman, P. Kasinitz, and J. DeWind, eds., *The Handbook of International Migration: The American Experience*. New York: Russell Sage Foundation.

Mendez, P. 2008. *Immigrant Residential Geographies and the 'Spatial Assimilation' Debate in Canada*. Working paper No. 08–07. Vancouver: Vancouver Centre of Excellence, Research on Immigration and Integration in the Metropolis.

Mensah, J. 2008. 'Religious transnationalism among Ghanaian immigrants in Toronto: A binary logistic regress analysis', *Canadian Geographer* 52: 309–30.

Mitropolitska, N. 2008. 'Les réseaux immigrants "virtuels": De l'aspatial au territorial', in Leloup and Radice (2008).

Moore Milroy, B., and M. Wallace. 2002. *Ethnoracial Diversity and Planning Practices in the Greater Toronto Area*. Toronto: CERIS—Ontario Metropolis Centre.

Morrison, R.B., and C.R. Wilson. 2004. *Natives Peoples: The Canadian Experience*, 3rd ed. Toronto: Oxford University Press.

Mwarigha, M.S. 2002. *Towards a Framework for Local Responsibility: Taking Action to End the Current Limbo in Immigrant Settlement*. Toronto: Maytree Foundation.

Newbold, B. 2007. 'Secondary migration of immigrants to Canada: An analysis of LSIC wave 1 data', *Canadian Geographer* 51: 58–71.

Omidvar, R., and T. Richmond. 2003. *Immigrant Settlement and Social Inclusion in Canada*. Toronto: Laidlaw Foundation.

Orme, L. 2007. 'Immigrant workforce integration: The Hamilton experience', *Canadian Issues/Thèmes canadiens* (Spring): 74–8.

Ornstein, M. 2002. *Ethno-racial Inequality in the City of Toronto: An Analysis of the 1996 Census*. Toronto: City of Toronto and CERIS—Ontario Metropolis Centre.

Pierre, J. 2005. *Governing Complex Societies: Trajectories and Scenarios*. Basingstoke: Palgrave.

Poirier, C. 2006. 'Parc Extension: Le renouveau d'un quartier d'intégration à Montréal', *Urban Diversité* 6: 51–68.

Pratt, G. 1997. 'Stereotypes and ambivalence: The construction of domestic workers in Vancouver, British Columbia', *Gender, Place and Culture* 4: 159–77.

——. 1998. 'Grids of difference: Place and identity formation', in R. Fincher and J. Jacobs, eds, *Cities of Difference*. New York: Guilford Press.

——. 2006. 'Separation and reunification among Filipino families in Vancouver', *Canadian Issues/Thèmes Canadiens* (Spring): 46–9.

Preston, V. 2008. 'Le cadre bâti comme métaphore de l'inclusion des immigrants', in Leloup and Radice (2008).

—— and L. Lo. 2000. '"Asian theme" malls in suburban Toronto: Land use conflict in Richmond Hill', *Canadian Geographer* 44: 182–90.

Pruegger, V., and D. Cook. 2009. 'An analysis of immigrant attraction and retention patterns among western Canadian CMAs', *Our Diverse Cities: The Prairies—Frontiers of Migration* 6: 44–9.

Qadeer, M. 1997. 'Pluralistic planning for multicultural cities: The Canadian practice', *Journal of the American Planning Association* 63: 481–94.

—— and M. Chaudhry. 2000. 'The planning system and the development of mosques in the Greater Toronto area', *Plan Canada* 40: 17–21.

Radford, P. 2007. 'A call for greater research on immigration outside of Canada's three largest cities', *Our Diverse Cities: Rural Communities* 3: 47–51.

Radice, M. 2008. 'Les rue commerçantes en contexte pluriethnique: Entre la confort et la différence', in Leloup and Radice (2008).

Ray, B. 2009. 'So where is ethnocultural diversity in Canada', *Plan Canada* special issue, 'Welcoming communities: Planning for diverse populations': 83–8.

—— and J. Bergeron. 2007. 'Geographies of ethnocultural diversity in a second-tier city: Moving beyond where people sleep', *Our Diverse Cities: Ontario* 4 (Fall): 44–7.

——, G. Halseth, and B. Johnson. 1997. 'The changing "face" of the suburbs: Issues of ethnicity and residential change in suburban Vancouver', *International Journal of Urban and Regional Research* 21, 1: 75–99.

Reimer, B. 2007. 'Immigration in the new rural economy', *Our Diverse Cities* 3: 3–8.

Reitz, J., and R. Banerjee. 2007. 'Ethnocultural communities: Participation and social cohesion', in Banting et al. (2007).

Richmond, T., and J. Shields. 2005. 'NGO–Government relations and immigrant services: Contradictions and challenges', *Journal of International Migration and Integration* 6: 513–24.

Rose, J. 1999. *Immigration, Neighbourhood Change and Racism: Immigrant Reception in Richmond, B.C.* Working Paper No. 99–15. Vancouver: Research on Immigration and Integration in the Metropolis.

Rossiter, M. 2005. 'Slavic brides in rural Alberta', *Journal of International Migration and Integration* 6: 493–512.

Salée, D. 2007. 'The Quebec State and the management of ethnocultural diversity: Perspectives on an ambiguous record', in Banting et al. (2007).

Sandercock, L. 2003. *Cosmopolis II: Mongrel Cities in the 21st Century*. London: Continuum.

Satzewich, V., and L. Wong, eds. 2006. *Transnational Identities and Practices in Canada*. Vancouver: University of British Columbia Press.

Séguin, A.-M., and G. Divay. 2002. *Urban Poverty: Fostering Sustainable and Supportive Communities*. Ottawa: Canadian Policy Research Networks.

—— and P. Apparicio. 2002. 'La division de l'espace résidentiel en fonction de la langue maternelle: Apport des indices de ségrégation résidentielle', *Canadian Journal of Urban Research* 11: 265–78.

Siemiatycki, M. 2008. 'Reputation and representation: Reaching for political inclusion in Toronto', in Andrew et al. (2008).

Simmons, A. 1999. 'Immigration policy: Imagined futures', in S. Halli and L. Driedger, eds, *Immigrant Canada:*

Demographic, Economic, and Social Challenges. Toronto: University of Toronto Press.

Simard, C. 2001. *La Représentation des Groupes Ethniques et des Minorités Visibles au Niveau Municipal: Candidats et Élus*. Rapport de recherche, Remis au Conseil des relations interculturelles. Montréal: Université du Québec à Montréal.

Statistics Canada. 2008. *Canada's Ethnocultural Mosaic, 2006 Census*. Catalogue no. 97–562–X. Ottawa: Minister of Industry.

Tastsoglou, E., and V. Preston. 2004. 'Gender, immigration and labour market integration: Where are we and what we do still need to know?', in J. Seager and L. Nelson, eds, *A Companion to Feminist Geography*. Oxford: Blackwell.

Teixeira, C. 2007. 'Residential experiences and the culture of suburbanization—A case study of Portuguese homebuyers in Mississauga', *Housing Studies* 22: 495–521.

———, L. Lo, and M. Truelove. 2007. 'Immigrant entrepreneurship, institutional discrimination, and implications for public policy: A case study in Toronto', *Environment and Planning C: Government and Policy* 25: 176–93.

Tettey, W. 2007. 'Transnationalism, religion and the African diaspora in Canada: An examination of Ghanaians and Ghanaian Churches', in J. K. Olupona and R. Gemignani, *African Immigrant Religions in America*. New York: New York University Press.

Tolley, E. 2003. 'Expressing citizenship through electoral participation: Values and responsibilities', *Diversity/Diversité canadienne* 2:17–19.

Tossutti, L. 2009a. 'The language of municipal government in global communities: Corporate communications policies in six cities', special issue, 'Welcoming communities: Planning for diverse populations', *Plan Canada*: 60–3.

———. 2009b. 'Canadian cities and international migration: Municipal corporate responses to ethnocultural diversity', *Metropolis Brown Bag* (Ottawa: Projet Metropolis Project), 23 Apr.

Truelove, M. 2000. 'Services for immigrant women: An evaluation of locations', *Canadian Geographer* 44, 2: 135–51.

Valentine, G. 2007. 'Theorizing and researching intersectionality: A challenge for the feminist geographer', *Professional Geographer* 59: 10–21.

Veronis, L. 2006. 'The Canadian Hispanic Day parade, or how Latin American immigrants practice (sub)urban citizenship in Toronto', *Environment and Planning A* 38: 1653–71.

———. 2007. 'Strategic spatial essentialism: Latin Americans' real and imagined geographies of belonging in Toronto', *Social and Cultural Geography* 8: 455–73.

Viswanathan, L. Forthcoming. 'Contesting racialization in a neoliberal city: Cross-cultural collective formation as a strategy among alternative social planning organizations in Toronto', *Geojournal*.

———, U. Shakir, C. Tang, and D. Ramos. 2003. *Social Inclusion and the City: Considerations for Social Planning*. Toronto: Alternative Planning Group.

Walks, R.A., and L.S. Bourne. 2006. 'Ghettos in Canada's cities? Racial segregation, ethnic enclaves and poverty concentration in Canadian urban areas', *Canadian Geographer* 50: 273–97.

Walton-Roberts, M. 2001. *Returning, Remitting, Reshaping: Nonresident Indians and the Transformation of Society and Space*. Working Paper Series No. 01–15. Vancouver: Vancouver Centre of Excellence for Research on Immigration and Integration in the Metropolis.

———. 2003. 'Transnational geographies: Indian immigration to Canada', *Canadian Geographer* 47: 235–50.

———. 2008. *Immigration, the University and the Tolerant Second Tier City*. CERIS Working Paper 69. Toronto: CERIS—Ontario Metropolis Centre.

Wang, S. 1999. 'Chinese commercial activity in the Toronto CMA: New development patterns and impacts', *Canadian Geographer* 43: 19–35.

——— and L. Lo. 2004. *Chinese Immigrants in Canada: Their Changing Composition and Economic Performance*. CERIS Working Paper No. 30. Toronto: CERIS—Ontario Metropolis Centre.

Webber, M. 1964. 'The urban place and the non-place urban realm', in M. Webber, ed., *Explorations into Urban Structure*. Philadelphia: University of Pennsylvania Press.

Wong, L., and M. Ng. 2002. 'The emergence of small transnational enterprise in Vancouver: The case of Chinese entrepreneur immigrants', *International Journal of Urban and Regional Research* 26: 508–30.

Wood, P.K., and L. Gilbert. 2005. 'Multiculturalism in Canada: Accidental discourse, alternative vision, urban practice', *International Journal of Urban and Regional Research* 29: 679–91.

Zelinsky, W., and B. Lee 1998. 'Heterolocalism: An alternative model of the sociospatial behaviour of immigrant ethnic communities', *International Journal of Population Geography* 4: 281–98.

Zhuang, Z.C. 2008. 'Ethnic Retailing and the Role of Municipal Planning: Four Case Studies in the Greater Toronto Area', Ph.D. thesis, University of Waterloo.

———. 2009. 'Ethnic retailing and implications for planning multicultural communities', special issue, 'Welcoming communities: Planning for diverse populations', *Plan Canada* 79–82.

CHAPTER 10

New Divisions: Social Polarization and Neighbourhood Inequality in the Canadian City

R. Alan Walks

One mark of a healthy society is how well it ensures equality and equity among its citizens. It is important in this regard to understand that Canadian metropolitan areas have been growing less equal and increasingly segregated spatially over the last decade. While not yet displaying the level of problems witnessed in its neighbour to the south, the Organisation for Economic Co-operation and Development (OECD) notes that over the preceding decade income inequality has grown faster in Canada than anywhere else in the developed world except (by a slight margin) Finland (OECD, 2008). Inequality and **polarization** are growing at multiple spatial scales, and are felt most acutely in the urban realm where the factors driving social polarization are most salient.

This chapter examines the trajectory of inequality and polarization in Canadian urban areas. It begins by detailing the extent of shifts in income inequality and poverty between and within metropolitan areas. It interrogates the factors producing growing inequality, both those articulated structurally between households and individuals (including homelessness, which is not dealt with at any length here; see Bourne and Walks, Chapter 25) and spatially between neighbourhoods. The potential importance of place and neighbourhood in contributing to, or ameliorating, poverty and inequality is then discussed, and the effects of segregation are examined. The relationship between

ethnic and racial segregation and concentrated neighbourhood poverty, relevant for determining the potential for ghettoization, is then analyzed. The chapter ends with a discussion of the implications and possible policy solutions for reversing creeping social polarization.

Trajectories of Inequality and Polarization

Canada has grown to be less egalitarian and more polarized.[1] According to Heisz's (2007) careful calculations, 1989 marked the time when Canada's overall income distribution was at its most equal and least polarized since reliable income data became available. From that time until 2004, however, indices of both income inequality and the polarization of income increased substantially.[2] These shifts occurred despite declining inequality and polarization among seniors—all the growth in inequality was felt among those younger than retirement age.

Canada's growth in income inequality has its origins in social divisions that are geographically articulated at multiple spatial scales. Canada has always maintained a strong regionalism not only in its cultural and political identities, but in its industrial, trade, and agricultural policies, and in the resulting spatial articulation of concentrations of poverty and affluence among metropolitan areas.

Strikingly different resource bases coupled with highly uneven development has produced increasingly diverging average incomes among Canada's census metropolitan areas (CMAs). At the same time, income inequality and polarization have grown within each major metropolitan area in Canada at the household level since the 1980s. One feature of polarization is a declining middle, while the population becomes increasingly grouped around two separate poles (Esteban and Ray, 1994). This is a distinct characteristic of Canadian cities. Middle-income households have declined as a proportion of the total across virtually all metropolitan areas (Figure 10.1). Since 2000, the only metropolitan areas to see slight gains in their proportions of middle-income households are located in the province of Quebec. In 2005, the Toronto region, Canada's most 'global' city, had the smallest proportion of its households in this middle-income range, and thus arguably the most polarized income structure among CMAs.

While inequality has grown due to both increasing poverty and higher incomes for the wealthy, trends in inequality and polarization from 1989 to 2006 have been particularly influenced by the latter. The rate of poverty, at least as indicated

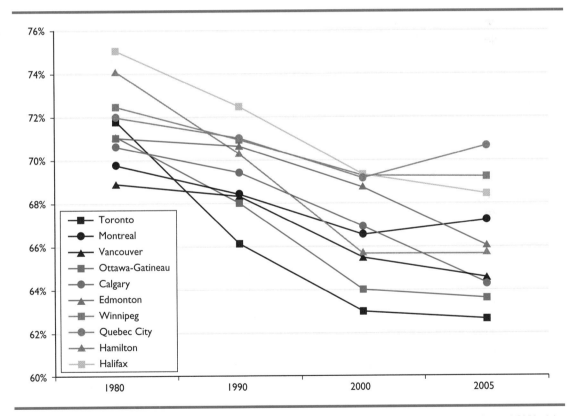

Figure 10.1 The declining middle: proportion of households with middle-range incomes*, selected CMAs**, 1980–2005

*$20,000 to $99,999 in constant 2000 dollars.

**Nine largest CMAs plus Halifax.

Source: Calculated by author from Statistics Canada, Census 1971–2006, custom tabulations, Tables E1171, E982.

by the rate of low income, has waxed and waned according to the economic fortunes of each metropolitan area, although the general trend for most CMAs is slow growth upward. While not strictly a poverty line, Statistics Canada's low-income cut-offs (LICOs) are often used as a proxy for such a line. Most CMAs had slightly higher rates of poverty in the 2000s than they did in the early 1980s when Canada went through a deep recession that ended only shortly before 1985. And some CMAs, including Toronto, have seen increases in poverty during the booming 2000s (from 16.7 per cent to 18.4 per cent).

Homelessness, perhaps the most severe articulation of extreme poverty, has grown in many Canadian cities, particularly those with tight rental markets such as Calgary (Figure 10.2). This sets the scene for far greater problems in the future as economic cycles move in the opposite direction, since economic downturns always impact low-income earners more severely than others.

Meanwhile, the proportion of households with high incomes increased. Between 1970 and 2005, the proportion of all households with incomes above $100,000 (in 2000 constant dollars) grew from an average across all CMAs of only 3 per cent to 16 per cent. Furthermore, the rich have become richer. After a long period of stability dating back to 1945, the income share of the top 5 per cent of earners grew dramatically from 24 per cent to 29 per cent of Canada's total income after 1990 (Saez and Veall, 2005). Even more dramatic has been the almost doubling in the income shares of both the wealthiest 0.1 per cent and 0.01 per cent of the population between 1990 and 2000, a trend that mirrors the situation in the United

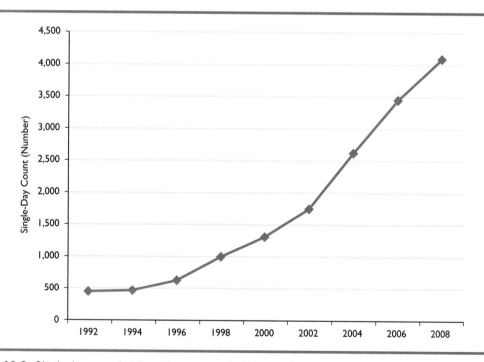

Figure 10.2 Single-day snapshot homeless counts in Calgary, 1992–2008

Note: Data for Calgary reflect counts of both shelter users and those sleeping rough on a single night in May or November of each study year.

Source: Biennial Count of Homeless Persons in Calgary: 14 May 2008. City of Calgary, Community and Neighbourhood Services, Social Research Unit. Copyright © 2008 The City of Calgary. All rights reserved. Used with permission.

States (ibid., 835–6). This enriching of the wealthy has occurred alongside the integration of Canada's economy with that of the United States, a unification accelerated by the North American Free Trade Agreement and the growing ability of high-profile CEOs on both sides of the border to attain exceedingly high levels of compensation.

Factors Driving Growing Inequality and Social Polarization: Globalization, Neo-Liberalism, and Shifting Demographics

Multiple factors have been driving the growth of income inequality and social polarization in Canada's cities. One important factor is the effect of globalization on regional economic specialization and on labour market segmentation. As Peter Hall demonstrates in Chapter 4, globalization and continentalization have had profound effects on Canada's cities. While some large Canadian metropolitan regions, including Toronto and Vancouver, and to a lesser extent Montreal, represent second-tier 'global' or 'world' cities specializing in **producer services** with strong connections to networks of global capital, information, and migration flows (see Taylor, 2002), many smaller cities that specialized in what have become declining industrial sectors or as transportation depots for agricultural products have seen their economies grow weaker under the onslaught of globalization. Alternatively, some urban regions have become drawn into functional production complexes that span the international border with the United States. While driving new flows of investment and providing local employment booms, continental integration also introduces a level of vulnerability for local workers who, like those in Windsor, face the threat of plant relocation and restructuring of the production complexes to reduce border delays, vertically integrate production, or take advantage

of protectionist policies or sentiments for made-at-home products in the US.

The expectation of the work of Sassen (2001) and Castells (2000) is that globalization should be accompanied by increasing labour market bifurcation and polarization. Those tied to favoured industries including international finance and producer services see their earnings rise, while those working in low-level personal services and manufacturing see their wages slip under competition from low-skilled local migrants and developing nations, respectively. Likewise, the efforts of those in what Florida (2002) terms the 'creative class' to attain increasing rewards for their skills in a globalizing economy also produce labour-market polarization. As Tara Vinodrai demonstrates in Chapter 6, those employed in such 'creative class' occupations, particularly entrepreneurs, managers, and design/marketing professionals located in the global cities, have been able to demand higher levels of compensation and a greater portion of the overall pie, even when they have not been responsible for increasing productivity or profitability. A culture of compensation paid in yearly bonuses within the financial services industry and some other sectors has produced a sense of entitlement among a new managerial elite, which has become increasingly detached from the realities of most workers or the public good and which increasingly resembles the parasitic leisure class examined originally by Veblen (2007 [1899]). The practice of paying even mid-level management with shares, bonuses, and stock options turns them into quasi-entrepreneurs whose economic interest is furthered by exploiting the labour of those working under them (Sennett, 1998).

The census data confirm this hypothesis of an urban polarization of employment incomes by occupation. While the relative incomes of professionals of various stripes (engineers, teachers, barristers and solicitors, medical professionals, etc.) residing in the largest CMAs have remained fairly consistent, managers and administrators have seen their earnings balloon relative to everyone else, while those employed in manufacturing, sales, and services all have suffered declining relative earnings. Managers in the global cities, some capital cities, and in metropolitan Alberta have seen their incomes rise most drastically. Between 1990 and 2005, employment incomes for managers increased by the most in Calgary (59 per cent), Toronto (58 per cent), Hamilton (45 per cent), Edmonton (39 per cent), Kitchener–Waterloo (39 per cent), Quebec City (31 per cent), and Ottawa (30 per cent). In other CMAs managerial incomes began the period at lower levels, and grew by on average 21 per cent. Simultaneously, in cities like Ottawa and Toronto workers in service occupations saw their relative earnings drop by almost 20 per cent. These results appear to confirm Sassen's hypothesis regarding occupational polarization. However, income losses among manufacturing workers were most stark (by between 18 and 22 per cent) in more peripheral cities, including Halifax, St Catharines, and Winnipeg, whereas they fell between 2 and 13 per cent in the larger industrial metros like Montreal (–2.1 per cent) and Toronto (–12.8 per cent).[3]

Immigration

Another factor related to globalization that is central to understanding social polarization dynamics in Canadian cities is immigration. As Hoernig and Zhuang discuss in Chapter 9, many Canadian cities rank very high internationally in terms of rates of immigration and levels of ethnic diversity. However, there is a distinct geography to immigration flows, which overwhelmingly favours the largest 'global' cities, particularly Toronto and Vancouver, while other cities, such as Quebec City and many smaller cities, receive very few immigrants. Overall, immigrants to Canada have seen their incomes decline relative to native-born Canadians over time. Relative to the incomes of native-born Canadians, recent immigrants (arriving in the previous 10 years) in Canada's major metropolitan areas have seen their relative incomes drop by approximately 15 per cent between 1981

when they earned on average 90 per cent of what native-born Canadians earned, and 2001 when they earned only 65 per cent. According to Wang and Lo (2005) it now takes more than 20 years for Chinese immigrants to close the earnings gap with the rest of the population, up from 15 years in previous decades. This produces a pronounced divergence in income between members of visible minorities and the white population in Canada, since ethnic and racial diversity in Canadian cities is tightly tied to flows of immigrants, whose source countries are now mainly in East and South Asia, the Middle East, Africa, and Latin America (Figure 10.3). As Galabuzi (2006) has demonstrated, the racialization of poverty and the escalating relationship between race and economic success are producing new forms of social exclusion, most visible in Canada's cities. This alienating force is most acute in those metropoli with the highest rates of immigration and levels of ethnic diversity, Toronto and Vancouver. By 2001, these two 'global' cities exhibited the lowest incomes for recent immigrants of Canada's CMAs.

Neo-Liberalism

An important factor explaining increasing social inequality in Canadian cities involves the restructuring of the **welfare state** and the embracing of neo-liberal policies on behalf of upper levels of government, and their effects both on the incomes of urban residents and on the spatial structure of Canada's cities. In Chapter 12, Allahwala, Boudreau, and Keil explain the basic tenets of neo-liberalism, describe how neo-liberal policies have been adopted by Canadian governments at different scales, and discuss the implications of neo-liberalism for future urban **governance**, ecological sustainability, labour market integration, and increasingly uneven development. In Canada, neo-liberalism has been part and parcel of the drive to integrate Canada's economy with that of the United States, and involves reducing barriers to trade and flows of labour and capital between

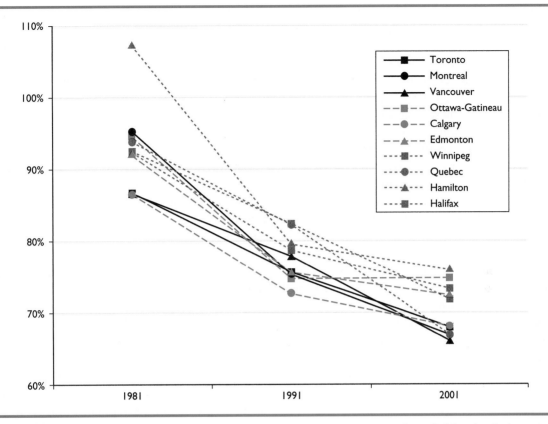

Figure 10.3 Average per capita incomes of visible minorities as a proportion of non-visible minority incomes, largest census metropolitan areas

Notes: Visible minorities are self-identified in the census. Three largest CMAs are noted by solid lines, medium-sized CMAs by segmented lines, smaller CMAs by dotted lines.

Source: Calculated by the author with the help of Richard Maaranen from Census of Canada Public Use Microsample Files, 1981, 1991, 2001.

the two countries. After the North American Free Trade Agreement (NAFTA) was implemented, a number of authors argued that Canada would then be compelled to bring its social policies in line with those of the United States in order to maintain the competitiveness of its industries in relation to its much larger neighbour (see Hurtig, 2002).

Indeed, from the mid-1990s onward, both the federal government and many provincial governments have sought to limit or reduce the range of social protections inherent in Canada's welfare state, to cut taxes, and to restructure the system of

benefits. The federal government first began to 'roll back' the federal state in the late 1980s, but more aggressively moved to restructure the welfare state system in the 1990s with the ending of conditional transfers to the provinces for health and education, reductions in employment insurance benefits, and income tax reductions. The election of right-wing governments committed to neo-liberal reforms of the welfare state, first in Alberta (1993), Ontario (1995), and then British Columbia (2001), led to drastic reductions in (and the restructuring of) welfare benefits, **social housing** and housing subsidies, public transit subsidies, and in some cities public-sector wages, public education funding, and transfers to municipalities.

As a result, Canada's welfare state and tax structure became less redistributive and equitable than it had been during the previous 15 years. Heisz (2007: 46) shows that while the redistributive effect of government taxes and transfers grew over the period from 1976 to 1994, largely offsetting the growth in market-based income inequality, there has been a persistent reduction in the redistributive effects of the welfare state and taxation since 1995. This is precisely the point in time that the richest 20 per cent of Canada's population saw their relative income and their proportionate share of income grow, while the relative incomes of the other 80 per cent of Canadians remained stagnant or declined (ibid., 33, 43). As a result, after-tax income inequality continued to increase.[4] Neo-liberal government policies are thus implicated in exacerbating the level of social inequality.

Household Structure and Shifting Demographics

Layered onto the effects of globalization, economic restructuring, immigration, and shifts in the welfare state are the changes in the composition of households, living arrangements, aging, and demographics that Townshend and Walker examine in Chapter 8 of this volume. As they highlight, such shifts have implications for where households prefer to be located, for predominant development forms in the inner cities and suburbs, and for the privatization of space. The proliferation of single-person households and the growth of female lone-parent families (from 12 to 16.7 per cent of all families between 1981 and 2006, on average), coupled with the decline of families with children (from 56 to 38.7 per cent of all families between 1981 and 2006), have introduced greater vulnerability and inequality into the urban household sector. Such changes in family composition have not been distributed evenly across metropolitan areas. While St John's and Toronto have witnessed a rapid decline of couple families with children at home (30 and 26 per cent, respectively), Kelowna and Victoria have seen the share of this sort of family grow since 1981 (albeit only slightly). While the majority of families in Toronto were childless in 2006 (53 per cent), in the two aforementioned CMAs in British Columbia, childless families made up just over one-third of families.

From the perspective of inequality, a key development has been the polarization of households based on employment status (Table 10.1). The pairing of highly educated and well-paid men and women of educated and wealthy parents is part and parcel of the class polarization that has been enveloping Canadian cities. Those less-educated and less well-off find it more difficult to meet prospective partners from different socio-economic groups, and increasingly partner with those of similar class background and educational opportunities. As family composition becomes more polarized, those who can draw on two or more incomes are better able to compete in housing markets, hence driving up housing costs and forcing those with access to fewer earners and resources into the rental housing sector. There has been a massive growth in families with two or more earners, which by 2006 represented 70 per cent of all couple-family households, while no-earner households have increased during tough times (doubling between 1971 and 1996, but falling slightly with the growth of employment at least until 2006). The single-earner household, evoking the stereotypical image of the male breadwinner supporting a nuclear family with a stay-at-home housewife, perhaps the dominant image of the Canadian family even today, has declined substantially from 43 per cent to a small minority of only 18 per cent of couple-family households.

One result of these new divisions in urban society (polarization in household structure, employment earnings, and between the native-born and recent immigrants, and neo-liberal policies favouring those with high income) is a

Table 10.1 Changing Composition of Couple-Family Households by Number of Earners (%), Canada 1971–2006

	1971	1981	1991	1996	2001	2006	Change 1971–2006
No earners	8.2	11.8	15.0	16.8	15.1	12.6	+ 4.4
One earner	43.1	30.7	23.0	23.1	19.6	17.8	−25.3
Two or more earners	48.6	58.5	62.1	60.5	67.3	69.7	+21.1

Sources: Bourne (1993); Census of Canada, 1991, 1996, 2001, 2006.

polarization in incomes between increasingly poorer tenants, who must find housing in a residualized rental sector, and those who own their own homes. The income *gap* between the median household incomes of homeowners and tenants grew from $20,770 in 1977 to $36,099 in 2005 (in inflation-adjusted 2005 constant dollars). Polarization in wealth is even starker. While homeowners saw their median net worth increase (in real terms) by 24 per cent between 1984 and 1999, renters saw it decline by 48 per cent (Hulchanski, 2004: 83). Walker and Carter, in Chapter 20, discuss how homeownership has been one way those with above average income have been able to accumulate wealth. Of course, social polarization based on housing tenure, occupation, race, and immigration status has *spatial* implications, because rental apartments, visible minorities, recent immigrants, and different kinds of workers are concentrated in particular places and neighbourhoods within Canada's metropolitan areas.

The Spatial Polarization of the City

Growing social polarization in Canadian society, which has been mirrored in the spatial polarization of Canadian cities, is most evident at the scale of the neighbourhood. Of course, it is difficult to define exactly what a neighbourhood is, as the

process of delineating neighbourhoods is a subjective one that differs at the individual level based on each resident's locally defined needs, social networks, and lived experience. Often, the census tracts defined by Statistics Canada[5] are used as proxies for neighbourhoods.[6]

The spatial distribution of household income among neighbourhoods (census tracts) in every Canadian CMA has become more unequal, or segregated, since 1970. Regardless of the index used, the pattern is found to be the same.[7] In each decade (or half-decade as the case may be) over the study period, the spatial distribution of household income among neighbourhoods (census tracts) became less equal, and more polarized, in virtually every CMA (Table 10.2).[8] Toronto, Calgary, Montreal, Hamilton, Vancouver, and (to a lesser extent) Winnipeg are the metropolitan areas that consistently score highest in regard to spatial inequality in census tract incomes, while smaller cities in central and eastern Canada exhibit the lowest scores. When the real estate market determines access to space, managers and those with high and rising incomes are able to outbid others for housing and location, thereby displacing low-income households from desirable neighbourhoods, driving up housing values, and contributing to greater neighbourhood segregation.

Increasing income segregation at the neighbourhood level does not imply, however, that

Table 10.2 WT Index of Neighbourhood Income Polarization, Canadian Cities, 1970–2005

CMA	1970	1980	1990	2000	2005
Calgary	0.431	0.458	0.492	0.508	0.608
Edmonton	0.418	0.454	0.471	0.494	0.528
Halifax	0.384	0.417	0.438	0.500	0.527
Hamilton	0.390	0.447	0.492	0.554	0.573
Kitchener–Waterloo	0.334	0.385	0.444	0.481	0.507
London	0.393	0.488	0.505	0.528	0.529
Montreal	0.440	0.484	0.513	0.545	0.562
Oshawa	0.374	0.387	0.421	0.453	0.479
Ottawa–Gatineau*	0.447	0.477	0.490	0.538	0.543
Quebec	0.437	0.442	0.484	0.528	0.537
Regina	0.413	0.477	0.508	0.550	0.557
Saint John	0.393	0.440	0.493	0.565	0.563
Saskatoon	0.394	0.419	0.446	0.510	0.539
St John's	0.436	0.414	0.460	0.475	0.506
Sudbury	0.316	0.399	0.487	0.533	0.571
Thunder Bay	0.336	0.423	0.443	0.473	0.492
Toronto	0.416	0.459	0.491	0.547	0.586
Trois-Rivières	0.403	0.449	0.473	0.522	0.542
Vancouver	0.418	0.476	0.502	0.529	0.568
Victoria	0.434	0.473	0.497	0.496	0.532
Windsor	0.384	0.461	0.495	0.550	0.569
Winnipeg	0.442	0.472	0.508	0.544	0.552
Average (non-weighted)	0.402	0.446	0.479	0.519	0.544

*With the January 2002 amalgamation of five municipalities in the Outaouais region of Quebec—Gatineau, Hull, Aylmer, Buckingham, and Masson-Angers—Gatineau was designated as the name of this larger municipality.

Note: Values of zero indicate complete spatial equality of income, while a value of 1.00 indicates complete segregation (all the income is found in only one census tract). Higher values thus indicate higher levels of spatial polarization.

Source: Calculated by the author from the Census of Canada, custom tabulations, Tables E1171, E982.

areas within the inner cities are always becoming poorer, or that the suburbs are always richer. Indeed, in comparison with their counterparts in the United States (though not necessarily those found in Europe), the Canadian inner city is relatively wealthier and has experienced significant **gentrification** since the 1960s (Ley, 1986, 1996; Walks and Maaranen, 2008b; Bain, Chapter 15). Furthermore, while many Canadian central cities have witnessed a slow decline, in recent decades inner cities in

Table 10.3 Ratio of Central City to CMA Average Household Income, 1970–2005

	1970	1980	1990	2000	2005
Calgary	108.2	100.0	99.1	98.1	97.3
Edmonton	99.9	96.3	94.7	92.8	92.2
Halifax*	103.1	100.0	98.4	93.1	88.9
Hamilton*	92.4	87.6	81.8	76.5	76.6
Hull*	84.5	80.1	75.5	71.5	74.2
London	100.9	98.6	97.8	98.1	97.9
Montreal*	86.1	80.6	78.7	78.3	77.9
Oshawa	99.1	97.1	91.6	87.4	86.3
Ottawa*	103.9	95.3	93.4	89.5	89.5
Quebec*	87.5	81.3	80.4	78.8	77.5
Regina	100.1	99.8	99.5	98.7	98.4
Saint John	98.8	94.1	89.4	85.7	84.8
Saskatoon	107.2	100.0	99.3	98.0	97.3
St Catharines	99.9	100.7	100.8	97.7	97.6
St John's	107.2	101.9	99.1	96.9	96.7
Sudbury	101.6	97.8	96.9	98.1	85.6
Thunder Bay	100.4	99.5	98.6	98.0	97.8
Toronto*	90.8	89.0	91.3	100.9	100.5
Vancouver	93.8	89.8	89.1	91.9	93.2
Victoria	84.5	76.0	74.1	76.2	74.5
Windsor	99.4	96.3	91.2	87.7	86.9
Winnipeg	83.1	99.6	99.0	97.6	97.7
Average	97.8	95.1	93.2	92.0	93.7

*These central-city municipalities no longer exist, having been amalgamated with other municipalities in the late 1990s or early 2000s. However, in these cases the census tract files were used to reconstruct the old central-city municipal boundary and to estimate the average household income using this boundary.

Source: Calculated by the author from the Census of Canada, 1971, 1981, 1991, 2001, 2006.

Toronto and Vancouver, and Ottawa–Gatineau have seen a turnaround in their relative incomes (Table 10.3). Economic restructuring, the emergence of neo-liberalism as the guiding policy framework for local regimes, and even amalgamation have accelerated gentrification processes in these cities, as predicted by Keil (2000) and others (Lees, Slater, and Wyly, 2007; Allahwala, Boudreau, and Keil, Chapter 12). As manufacturing work disappears from the inner cities and is replaced in part by high-end jobs in finance, real estate, producer-services, and other 'quaternary' sector industries, the established and gentrifying neighbourhoods in the inner cities become the focus of the residential careers of the new workers, as well as the site of new condominium developments (Ley, 1986, 1996; Walks, 2001; Walks and Maaranen, 2008b).

Inner-city restructuring presents a problem for the older suburbs, however. The inner cities traditionally have provided affordable rental housing for low-income households, immigrants, and migrant workers in the city. With gentrification of the housing stock, conversion of rental to owner occupation, and lack of significant new investments in social housing or rental apartments in the inner city, low-income households and in-migrants are forced to search for affordable housing in the suburbs, particularly the nearby suburban ring built up during the early post-war period. In many of the

larger Canadian cities, a sizable proportion of the social housing and of the higher-density market rental housing stock built during the 1960s and 1970s was located in the inner suburbs. These areas now house the greatest proportion of low-income households and are functioning as the new immigrant-reception neighbourhoods.

The gentrification of the inner cities, coupled with the decline of the older post-war suburbs, has led to a new spatial patterning of urban disadvantage, perhaps most evident within the amalgamated City of Toronto (Figure 10.4). While elite neighbourhoods and gentrifying inner-city neighbourhoods have seen average per capita incomes grow substantially, much of the inner suburban area, particularly in the northeast and northwest of the amalgamated city that are most isolated from the core and from public transit, has suffered income decline over the last 35

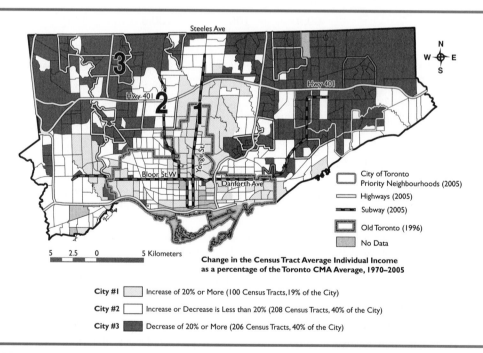

Figure 10.4 The three cities within the City of Toronto: collections of neighbourhoods with increasing, decreasing, or stable incomes over the period 1970 to 2005.

Source: Created by Richard Maaranen at the University of Toronto Cities Centre as an update to the map published in Hulchanski (2007).

years. Hulchanski calls this pattern the new 'city of disparities' (Hulchanski, 2007: 10), in which the ever-improving neighbourhoods of city #1 (Figure 10.4), where many managers and professionals are concentrating, stand in contrast to the declining areas of city #3, where ever-greater proportions of the poor, single-parent households, seniors, and recent immigrants are being housed. A similar, although more muted, situation is evident in many other Canadian metropolitan areas, particularly those that have witnessed significant inner-city gentrification.

Neighbourhood Segregation and Concentrated Poverty: Ghettos in Canada's Cities

The spatial polarization of the city raises the possibility that where one lives may be increasingly important for determining the life chances of urban residents. Neighbourhoods provide a basis for the formation of social relationships and networks, determine accessibility to jobs and information, provide amenities and services, and are a major context for living and understanding daily life. When poverty becomes concentrated in place, increased social needs may put pressure on local schools, potentially affecting the quality of public education for local children. If neighbourhoods in an urban municipality disproportionately concentrate residents who rely more on public services, this population density puts pressure on the municipality either to raise property taxes in order to expand the level of service or to find itself overwhelmed with social needs it cannot meet. Higher-poverty neighbourhoods cannot support as many retail establishments, impacting local residents' access to quality food and other resources (see Blay-Palmer, Chapter 24). When they are geographically isolated, high-poverty neighbourhoods may limit access to employment or social contact with other urban residents. The spatial context and

social composition of the neighbourhood then become independent factors determining future opportunities. In the United States, concentrated neighbourhood poverty is associated with heightened levels of crime and insecurity, which may provoke out-migration of those who are able to flee, deepening the level of concentration and the potential vulnerability of the local population (Jargowsky, 1997).

When neighbourhoods become independently important for determining life chances, even after taking into account the socio-demographics of their residents, cities witness the operation of what are termed 'neighbourhood effects'. The presence of neighbourhood effects on life chances becomes particularly troubling when correlated with ethnic and racial segregation. The worry is that in countries like Canada, where immigration is such a prominent feature and a driver of urbanization and economic growth, the integration of new immigrants and visible minorities into Canadian society may break down, leading to increasing racial divisions, distrust, discrimination, and social exclusion. The growing income gap between visible minorities and the white population raises the possibility of heightened segregation occurring in Canadian cities, potentially along the lines of the American experience where significant proportions of blacks and Latinos are segregated from white society and forced to live in undesirable, high-poverty neighbourhoods (see Jargowsky, 1997). Whether such a thing could occur in urban Canada is debated. While Adams (2007) paints the picture of an 'unlikely utopia' of racial and cultural pluralism where immigrants have integrated faster and more smoothly than in any other industrialized nation, Galabuzi (2006) suggests that the racialization of poverty and racial discrimination are leading to a new form of economic and cultural apartheid in Canada.

The work of Kazemipur and Halli (2000) was the first to raise the spectre of ghettoization in Canadian cities. However, they employed the word 'ghetto' outside its usual meaning. From its origins in Europe to its usual application in the United States, **ghettos** have been defined as residential areas that both concentrate and contain a particular ethnic or racial group as a consequence of discrimination by the host community, which forces that group to live within their boundaries (Johnston et al., 2003; Logan et al., 2002; Marcuse, 1997; Massey and Denton, 1993; Peach, 1996; Philpott, 1978; Walks and Bourne, 2006). Ghettos thus defined are ethnic or racial concentrations produced through racial discrimination in housing and labour markets, which may or may not be modified by the state. They are not the same as poor neighbourhoods or slums, which concentrate the poor, and which may or may not also contain concentrations of particular ethnic or racial groups. However, Kazemipur and Halli (2000) use the word 'ghetto' for a neighbourhood with a poverty rate above 40 per cent, parroting Wilson's (1987) discussion of the 'ghetto-poor' and the 'underclass' in discussing high-poverty neighbourhoods in the United States. While it may be mostly true that the highest-poverty neighbourhoods in the United States are indeed racial ghettos, such a case is by no means evident in Canada. Furthermore, the use of 'ghetto' to mean a high-poverty neighbourhood impedes clarity when what we are interested in knowing is whether there is any *relationship* between ethnic concentration and concentrated poverty.

Substantial debate surrounds the benefits and drawbacks of ethnic and racial segregation. While traditional models see any degree of segregation as reflecting social exclusion and a breakdown in the assimilation process, the more recent literature contrasts the effects of discrimination with ethnic enclaves in which residency appears voluntary and from which residents receive some benefit. Logan et al. (2002) distinguish between the traditional immigrant enclave, seen as a temporary neighbourhood of convenience until immigrants have attained the social contacts and resources necessary

to assimilate and move out, and an emerging new form they term the 'ethnic community', which is the desired and chosen residential destination of prosperous immigrants and cultural minorities who wish to live among others from the same group. The 'ethnoburbs' (Li, 1998), discussed by Hoernig and Zhuang (Chapter 9), are one variant on this form of enclave. Enclaves may provide access to ethnic resources, social contacts, and amenities and services in ethnic languages, and thus may promote cultural identity, employment opportunities, community well-being, and the fostering of social integration (Boal, 2005; Peach, 1996; Qadeer, 2005).

We wish to know not only the degree to which visible minorities and Aboriginal people have become increasingly concentrated in particular neighbourhoods within Canadian cities, but similarly whether there is a relationship between the concentration of minorities and the concentration of poverty. If there is a strong correlation between racial concentration and neighbourhood poverty, this information helps to confirm the fears of Kazemipur and Halli (2000) and Galabuzi (2006). The confluence of the racial ghetto and class polarization, according to Marcuse (1997), reflects the emergence of a new form of neighbourhood in the globalized United States, the 'outcast ghetto' in which the poorest segments of racialized groups who are marginal to current production processes become trapped in neighbourhoods formed through discrimination and the abdication of state responsibility to the poor.

The census tract data, however, do not support the contention that there might be ghettos in Canada. Table 10.4 updates the work of Walks and Bourne (2006), classifying census tracts in Canadian metropolitan areas based on the visible minority proportions of their populations using data from the 2006 census, via a classification scheme developed by Poulsen et al. (2001). Areas with less than 20 per cent visible minorities are termed 'isolated communities', while areas having 20–50 per cent visible minorities are 'non-isolated' and those

Table 10.4 Proportion of CMA Population Residing in Tracts of Each Neighbourhood Type, 2006

CMA	% of Population / (# of census tracts)					
	Isolated	Non-isolated	Pluralistic	Mixed	Polarized	Ghettos
Segregated/Segmented						
Toronto	23.3 (270)	36.9 (369)	19.6 (184)	17.9 (150)	2.3 (22)	0
Vancouver	20.4 (87)	38.1 (156)	25.9 (105)	10.9 (41)	5.0 (20)	0
Abbotsford	47.1 (17)	37.6 (14)	8.7 (2)	0	6.7 (1)	0
Winnipeg	52.1 (88)	34.6 (56)	11.6 (19)	1.7 (4)	0	0
Calgary	50.3 (110)	41.0 (79)	7.6 (11)	1.1 (2)	0	0
Montreal	67.8 (579)	27.2 (240)	4.2 (34)	0.8 (7)	0	0
Relatively Pluralistic						
Ottawa–Gatineau	65.9 (168)	31.7 (76)	2.4 (5)	0	0	0
Edmonton	44.3 (99)	55.3 (116)	0.4 (2)	0	0	0
Windsor	60.6 (41)	39.4 (29)	0	0	0	0
Relatively Integrated						
Kitchener–Waterloo	74.9 (72)	25.1 (19)	0	0	0	0
Hamilton	78.7 (141)	21.3 (35)	0	0	0	0
Regina	79.7 (39)	20.3 (13)	0	0	0	0
London	79.9 (84)	20.1 (19)	0	0	0	0
Saskatoon	80.4 (40)	19.6 (11)	0	0	0	0
Victoria	85.1 (59)	14.9 (10)	0	0	0	0
Oshawa	88.5 (66)	11.5 (7)	0	0	0	0
Thunder Bay	90.2 (30)	9.8 (3)	0	0	0	0
Halifax	96.1 (83)	3.9 (3)	0	0	0	0
St Catharines	97.6 (91)	2.4 (2)	0	0	0	0
Homogeneous						
Chicoutimi–Jonquière	100 (38)	0	0	0	0	0
Quebec	100 (166)	0	0	0	0	0
Kelowna	100 (37)	0	0	0	0	0
Saint John	100 (46)	0	0	0	0	0
Sherbrooke	100 (42)	0	0	0	0	0
St John's	100 (47)	0	0	0	0	0
Sudbury	100 (42)	0	0	0	0	0
Trois-Rivières	100 (38)	0	0	0	0	0

Note: Isolated = less than 20% visible minorities and/or Aboriginals; non-isolated = between 20% and 50% visible minorities and/or Aboriginals; pluralistic = between 50% and 70% visible minorities and/or Aboriginals; mixed = over 70% visible minorities and/or Aboriginals but no group dominates; polarized = over 70% visible minorities and/or Aboriginals and one group makes up over 66%, but less than 30% of this group lives in such neighbourhoods; ghettos = over 70% visible minorities and/or Aboriginals *and* one group makes up over 60% *and* at least 30% of this group lives in such neighbourhoods.

Source: Calculated by the author from Census of Canada, 2006.

with 50–70 per cent are 'pluralistic/assimilation enclaves'. Neighbourhoods (tracts) with greater than 70 per cent of their populations from visible minority groups are classified into 'mixed-minority enclaves' (no single minority group dominates), 'polarized enclaves' (at least two-thirds derive from a single minority group that dominates), and 'ghettos' (at least 60 per cent are from a single group and at least 30 per cent of that group must reside in that neighbourhood). This complex definition of the 'ghetto' derives from criteria developed over many years (Johnston et al., 2002; Marcuse, 1997; Massey and Denton, 1993; Peach, 1996; Philpott, 1978).

Not only are there no ghettos in Canada's cities, but polarized neighbourhoods dominated by a single visible minority group (the next closest thing to a ghetto) are rare. In most Canadian metropolitan areas, the majority of visible minorities live in neighbourhoods that are over 80 per cent white. In only three metropolitan areas—Toronto, Vancouver, and Abbotsford, precisely the three CMAs with the highest proportions of visible minorities in Canada—are there any racially polarized tracts. However, such neighbourhoods are dwindling, not spreading. The numbers of ethnically polarized neighbourhoods actually fell between 2001 and 2006, from 57 tracts to 43 (compare with Walks and Bourne, 2006: 284), while the proportion of the population in both Toronto and Vancouver living in such neighbourhoods also declined. Unlike the situation in the United States, none of these ethnically polarized neighbourhoods concentrates either blacks or Latinos. Instead, the vast majority of such neighbourhoods (18 of Toronto's 22, and 13 of Vancouver's 20) contain East Asians (mainly Chinese), with the remaining 'polarized' tracts, including the one in Abbotsford, dominated by South Asians. Becoming increasingly prevalent, meanwhile, are the non-isolated and pluralistic communities, in particular, the mixed-minority enclaves in which visible minorities and Aboriginal people are highly concentrated but no one single group dominates. The mixed-minority enclaves in 2006 make up over 15 per cent of the Toronto region's tracts (and contain almost 18 per cent of its population), and just over 10 per cent of the tracts and population in Vancouver. They are found in Winnipeg, Montreal, and Calgary, also.

Even in the absence of actual ghettos, a correlation between segregation of visible minorities (or Aboriginal people) and concentrated neighbourhood poverty could suggest movement along a trajectory towards a more exclusionary neighbourhood structure. Table 10.5 provides, for each type of enclave, the aggregate incidence of low income (per cent below the LICO, often used as a measure of poverty, as noted above) and the number of census tracts with rates of poverty of 40 per cent or greater (as per Walks and Bourne, 2006). The 40 per cent threshold usually is adopted as the main indicator of a high-poverty neighbourhood (Jargowsky, 1997; Kazemipur and Halli, 2000; Wilson, 1987). While higher proportions of visible minorities typically are associated with higher rates of low income, the polarized neighbourhoods are not the poorest; rather, the mixed-minority enclaves generally are the poorest, often by a significant degree. Indeed, more high-poverty neighbourhoods are found in the non-isolated neighbourhoods than in the ethnically polarized category, and average household incomes in many polarized tracts are above the national average.

What appears to be occurring in many CMAs is that middle-class members of visible minorities are increasingly able to choose where they live, and many select middle-class 'ethnic communities' containing high proportions of middle-class residents of their own group. This evidence of 'locational attainment' among particular ethnic groups (Myles and Hou, 2004) supports the hypothesis that, at least for the middle classes, the general settlement pattern largely matches the hypothesis of cultural pluralism (Adams, 2007; Hiebert and Ley, 2003).

On the other hand, as social polarization becomes a broad feature of Canadian society, the

Table 10.5 Incidence of Low Income by Neighbourhood Type (Plus Number of Concentrated Low-Income Tracts), 2006

CMA	Incidence of Low Income as Per Cent and Number of Tracts with Low Income Rate Equal To or Greater Than 40 Per Cent						
	Total %/No.	Isolated %/No.	Non-Isolated %/No.	Pluralistic %/No.	Mixed %/No.	Polarized %/No.	Ghetto %/No.
Segregated/Segmented							
Toronto	8.3 36	9.7 0	17.4 1	22.6 10	26.2 24	24.4 1	0
Vancouver	20.8 17	11.4 0	20.3 4	24.8 8	27.4 41	28.0 2	0
Abbotsford	14.0 0	12.1 0	17.0 0	11.8 0	0 0	13.4 0	0
Winnipeg	18.8 19	11.9 0	22.5 7	33.0 8	64.1 4	0	0
Calgary	13.4 1	11.2 0	14.8 1	15.0 0	15.7 0	0	0
Montreal	21.1 119	16.1 34	28.6 56	45.8 22	54.8 7	0	0
Relatively Pluralistic							
Ottawa–Gatineau	14.7 16	10.9 3	20.3 9	47.2 4	0	0	0
Edmonton	14.1 2	9.4 0	18.0 2	9.1 0	0	0	0
Windsor	14.0 4	8.8 0	22.2 4	0	0	0	0
Relatively Integrated							
Kitchener–Waterloo	10.5 1	9.4 0	13.7 1	0	0	0	0
Hamilton	15.6 10	12.4 1	27.5 9	0	0	0	0
Regina	13.5 2	9.8 0	27.8 2	0	0	0	0
London	13.7 0	12.0 0	20.3 0	0	0	0	0
Saskatoon	16.3 4	12.9 0	30.1 4	0	0	0	0
Victoria	13.2 0	12.3 0	18.0 0	0	0	0	0
Oshawa	9.3 0	9.5 0	7.2 0	0	0	0	0
Thunder Bay	12.8 1	11.8 1	23.1 0	0	0	0	0
Halifax	14.3 2	13.7 1	25.2 1	0	0	0	0
St Catharines	12.5 0	12.3 0	21.7 0	0	0	0	0

Note: Definitions as in Table 10.4.
Source: Calculated by the author from Census of Canada, 2006.

poorest members of each minority group are forced, due to their lower purchasing power within the real estate market, to find housing in the least desirable neighbourhoods, many of which contain social housing and/or high-density market-rent apartments built during the 1960s and 1970s. These neighbourhoods can mainly be found in the inner suburbs, as well as in some inner-city areas. The concentration of poor minorities from across different ethnic and racial groups in the mixed-minority enclaves thus occurs through their reduced purchasing power and limited choice in the housing market. These findings suggest that for low-income minorities, the processes producing their residential settlement patterns may be more exclusionary (Galabuzi, 2006). Furthermore, neighbourhood location within the inner suburbs, spatial accessibility, and isolation play important additive roles in shaping the experience of concentrated neighbourhood poverty among low-income minorities and immigrants (Smith and Ley, 2008).

Conclusion

Canadian cities are at a crossroads. The combined effect of globalization, economic restructuring, continentalization, the racialization of poverty, demographic shifts, and neo-liberal policies has been to increase the level of social inequality at multiple scales throughout urban Canada. Although the unemployment rate has moved up and down according to local fortunes, poverty and homelessness have become more entrenched, while the number of rich, as well as their level of relative wealth, has grown. As the middle declines, the mutual feeling that everyone is in the same boat wanes while fears of those in other classes grow. Social polarization spurs a series of self-reinforcing feedback loops, in which the responses of individuals feed into the system to further aggravate the problems caused by polarization. As they become wealthier, rich households attain more power to influence the political process to push for policy

reforms in their favour, such as tax and public-spending cuts, which further drive the inequality facilitating the transfer of power to the wealthy. As the rich enrol their children in private schools or move to a few wealthy suburbs, the quality of urban (and inner-suburban) education declines, spurring other wealthy families to do the same and making poor districts all the more undesirable. The racialization of poverty fuels the stereotyping and scapegoating of visible minorities and immigrants, reducing their access to resources and increasing their likelihood of falling into poverty, which in turn leads to a greater concentration of poverty in areas with poorer segments from all minority groups. The residualization of the older rental housing stock makes renting less desirable, limiting rent levels and the incentives to build new rental properties, at the same time fuelling the filtering of housing and concentrating the poor in increasingly stigmatized rental districts. These developments feed on each other, driving a complex dialectical process that interweaves trajectories of social and neighbourhood polarization together in urban space.

To reverse the growth of social polarization and inequality, and ameliorate the problems it has caused, a multi-pronged, multi-scaled response is needed from both the public and private sectors. Nations and cities in continental Western Europe provide a number of potential models for such a reversal. The private sector needs to re-equalize salaries and wages, and reverse decades of excessive compensation among executive and managerial staff, which has not been in the long-term interest of public or private firms. At the same time, investment by the private sector needs to flow in productive directions that provide sustainable jobs and an economic base upon which new enterprise can flourish—heavy and light manufacturing, utilities, transportation, infrastructure, and in particular energy, referred to as the 'old' economy—and away from non-productive enterprises that rely on income produced elsewhere, regressively redistribute income upward, and encourage rent-taking. Information technology, finance, advertising, real estate, and other consumer service sectors are among the 'non-productive' parts of the economy. Governments need to increase the redistributive function of the tax system and public services. Instead of tax cuts for the middle and upper classes, under the hope that some will 'trickle down' to the poor, the tax system needs to be reformed so that it is fully progressive, with top marginal tax rates increased substantially and minimum wages and welfare benefits considerably increased (to a level facilitating the meeting of, at the very least, basic daily needs). Governments should work with the private sector to establish apprenticeship programs for young workers, and should require some minimal levels of civic participation in order to enrich the pool of social capital. Items of collective consumption and public infrastructure, on which the majority of urban residents depend, should be maintained so that they are universally accessible, and privatizations of strategic public infrastructure should be reversed. Revenue generation, particularly for social services and other 'soft' services, should be shifted from property taxation and other regressive forms of taxation to the progressive income tax. Special efforts will be required to counter the declining incomes of new immigrants and Aboriginal people, and the increase of racialization of poverty in larger cities, as well as to enhance gender and racial equity. Property rights must be democratized, and the right to the city, and to full participation in society, must be implemented.

The future for Canadian metropolitan areas remains unclear and uncertain. The prospect of de-globalization and/or declines in international trade has serious implications for a country whose population is highly urbanized and so heavily dependent on trade in commodities and on imports of food. As Bill Rees points out in Chapter 5, use of the globe's natural resources has surpassed its current carrying capacity. Urban **ecological footprints**

larger than the entire land area of the nations within which they reside may portend drastic reductions in urban populations at some point in the near future. Such an energy-dependent nation as Canada will have to find ways of meeting the needs of its urban populations. The solutions will either be progressive (equalizing) or regressive (polarizing). Continued economic restructuring risks heavy impacts on the poor, on new immigrants, and on the young. Serious attention to the problems of both social and environmental sustainability is required now in order to prevent Canada's cities from becoming more unequal and polarized.

Review Questions

1. What are the major contributors to social polarization?
2. Do patterns of polarization vary from city to city?
3. Discuss the policy implications of increasing social polarization.

Notes

1. Social polarization and inequality are highly related but distinct concepts. Perhaps the easiest way to distinguish the two is to visualize the shape of their respective distributions, with inequality matching a pyramid shape, while polarization resembles an hourglass. Polarization is seen as producing more social conflict as a result of a divided society in which those within each group identify only with fellow group members. Inequality, on the other hand, is said to result in less within-group identification, but the larger size of the impoverished segment gives it potentially greater revolutionary power (see Duclos et al., 2004; Esteban and Ray, 1994). In practice, however, indices of inequality and polarization typically paint the same picture and are closely related. Under the Pigou-Dalton axiom inequality is always reduced if income is transferred from a richer person to a poorer person. Polarization is, on the other hand, driven by the bipolarity axiom, which states that increasing

concentration of a population towards each pole must increase polarization, even if it simultaneously reduces inequality (see Esteban and Ray, 1994).

2. To be precise, the Gini coefficient of family income inequality grew persistently in virtually every year, from a level under 0.28 to almost 0.32; at the same time, the Wolfson polarization index increased from a low of 0.235 to 0.265 (Heisz, 2007).

3. Calculated by the author from the Census of Canada, Public Use Microsample Files, Statistics Canada, various years.

4. However, it is not completely clear what the precise effects of tax changes have been for different groups. According to Murphy, Roberts, and Wolfson (2007), the share of income taxes paid by the wealthiest income earners has grown between 1992 and 2004. Despite the drop in the effective marginal tax rate for high-income earners from 29 to 27 per cent, which gave the top 5 per cent of income earners a significant break in the taxes they otherwise would have paid, due to the fact that they saw their incomes rise so significantly, their overall share of income taxes paid also rose from 30 per cent to 36 per cent.

5. Census tracts contain between 2,000 and 8,000 people, and have been constructed in order to contain a similar social demographic with boundaries based on recognizable physical features and transportation routes, although it should be noted that census tracts are not likely to map to the social and physical activity spaces of local residents. Importantly, however, the fact that census tract boundaries have largely been maintained over time (if a tract is split, the outer boundary is typically preserved) makes them the best available unit for examination of neighbourhood changes.

6. For a discussion of models used to depict the distribution of social variables across metropolitan regions, see Murdie and Teixeira, Chapter 9 in Bunting and Filion (2006), at: <www.oupcanada.com/bunting4e>.

7. Originally, three different indices were calculated. The exponent (EXP) index of inequality is more sensitive to the lower end of the income distribution, and so is more likely to pick up increasing segregation of low-income households, while the squared coefficient of variation (CV^2) is more sensitive to the distribution of high-income households. The WT index is an index of polarization, and so takes into account segregation at both the upper and lower ends of the distribution, although it is more sensitive to the upper end. The WT index is the only one of the polarization indices that can be easily calculated from data aggregated at the census tract level (for more explanation, see Walks and Maaranen, 2008a).

8. The severe recession that gripped Canada in the early 1990s was responsible in particular for raising levels of poverty and producing heightened residential income segregation in many Canadian metropolitan areas (see Ross et al., 2004). This was layered onto the patterning of income polarization dating from the 1980s (Townshend and Walker, 2002; Walks, 2001).

References

Adams, M. 2007. *Unlikely Utopia: The Surprising Triumph of Canadian Pluralism*. Toronto: Viking.

Bacher, J. 1993. *Keeping to the Marketplace: The Evolution of Canadian Housing Policy*. Montreal and Kingston: McGill-Queen's University Press.

Boal, F.W. 2005. 'Urban ethnic segregation and the scenarios spectrum', in Varady (2005: 62–78).

Bourne, L.S. 1993. 'Close together and worlds apart: An analysis of changes in the ecology of income in Canadian cities', *Urban Studies* 30, 8: 1293–1317.

Calgary, City of. 2002. *The 2002 Count of Homeless Persons—2002 May 15*. Calgary: City of Calgary Community Vitality and Protection, Community Strategies, Policy and Planning Division.

———. 2008. *Biennial Count of Homeless Persons in Calgary 2008 14 May*. Calgary: Community and Neighbourhood Services Social Research Unit.

Castells, M. 2000. *The Rise of the Network Society*, 2nd edn. Oxford: Blackwell.

Duclos, J., J.M. Esteban, and D. Ray. 2004. 'Polarization: Concepts, measurement, estimation;, *Econometrica* 72, 6: 1737–72.

Esteban, J.M., and D. Ray. 1994. 'On the measurement of polarization', *Econometrica* 62, 4: 819–51.

Federation of Canadian Municipalities (FCM). 2005. *Incomes, Shelter, and Necessities: Quality of Life in Canadian Communities, Theme Report #1*. Ottawa: Federation of Canadian Municipalities.

Florida, R.L. 2002. *The Rise of the Creative Class: How It's Transforming Work, Leisure, Community and Everyday Life*. New York: Basic Books.

Galabuzi, G.-E. 2006. *Canada's Economic Apartheid: The Social Exclusion of Racialized Groups in the New Century*. Toronto: Canadian Scholars' Press.

Heisz, A. 2007. *Income Inequality and Redistribution in Canada: 1976 to 2004*. Ottawa: Statistics Canada Catalogue no. 11F0019MIE, Research Paper No. 298.

Hiebert, D., and D. Ley. 2003. 'Assimilaton, cultural pluralism, and social exclusion among ethnocultural groups in Vancouver', *Urban Geography* 24, 1: 16–44.

Hulchanski, J.D. 2004. 'A tale of two Canadas: Home owners getting richer, renters getting poorer', in Hulchanski and Shapcott, eds, *Making Room: Policy Options for a Canadian Rental Housing Strategy*. Toronto: CUCS Press, 81–8.

———. 2007. *The Three Cities in Toronto*. Toronto: University of Toronto Centre for Urban and Community Studies Research Bulletin 41.

Hurtig, M. 2002. *The Vanishing Country*. Toronto: McClelland & Stewart.

Jargowsky, P. 1997. *Poverty and Place: Ghettos, Barrios, and the American City*. New York: Russell Sage Foundation.

Johnston, R., J. Forrest, and M. Poulsen. 2002. 'Are there ethnic enclaves/ghettos in English cities?', *Urban Studies* 39, 4: 591–618.

Kazemipur, A., and S.S. Halli. 2000. *The New Poverty in Canada: Ethnic Groups and Ghetto Neighbourhoods*. Toronto: Thompson Educational Publishing.

Keil, R. 2000. 'Governance restructuring in Los Angeles and Toronto: Amalgamation or secession?', *International Journal of Urban and Regional Research* 24, 4: 758–81.

Lees, L., T. Slater, and E. Wyly. 2007. *Gentrification*. London: Routledge.

Ley, D. 1986. 'Alternative explanations for inner-city gentrification: A Canadian assessment', *Annals, Association of American Geographers* 76, 4: 521–35.

———. 1996. *The New Middle Class and the Re-Making of the Central City*. New York: Oxford University Press.

Li, W. 1998. 'Anatomy of a new ethnic settlement: The Chinese ethnoburb of Los Angeles', *Urban Studies* 35, 3: 479–501.

Logan, J., R. Alba, and W. Zhang, W. 2002. 'Immigrant enclaves and ethnic communities in New York and Los Angeles', *American Sociological Review* 67, 2: 299–322.

Marcuse, P. 1997. 'The enclave, the citadel, and the ghetto: What has changed in the post-Fordist U.S. city?', *Urban Affairs Review* 33, 2: 228–64.

Massey, D., and N. Denton. 1993. *American Apartheid: Segregation and the Making of the Underclass*. Cambridge, Mass.: Harvard University Press.

Murphy, B., P. Roberts, and M. Wolfson. 2007. *A Profile of High-Income Canadians*. Ottawa: Statistics Canada, Income Research Paper Series Paper No. 006, Catalogue no. 75F0002MIE–006.

Myles, J., and F. Hou. 2004. 'Changing colours: Spatial assimilation and new racial minority immigrants', *Canadian Journal of Sociology* 29, 1: 29–58.

Organisation for Economic Co-operation and Development (OECD). 2008. *Growing Unequal? Income Distribution and Poverty in OECD Countries*. Zurich: OECD, Report 8108051E.

Peach, C. 1996. 'Does Britain have ghettos?', *Transactions of the Institute of British Geographers* 21, 2: 216–35.

Philpott, T.L. 1978. *The Slum and the Ghetto: Neighbourhood Deterioration and Middle Class Reform, Chicago 1880–1930*. New York: Oxford University Press.

Poulsen, M., R. Johnston, and J. Forrest. 2001. 'Intraurban ethnic enclaves: Introducing a knowledge-based classification method', *Environment and Planning A* 33, 10: 2071–82.

Qadeer, M.A. 2005. 'Ethnic segregation in a multicultural city', in Varady (2005: 49–61).

Ross, N., C. Houle, J. Dunn, and M. Aye. 2004. 'Dimensions and dynamics of residential segregation by income in urban Canada, 1991–1996', *Canadian Geographer* 44, 4: 433–45.

Saez, E., and M.R. Veall. 2005. 'The evolution of high incomes in Northern America: Lessons from Canadian evidence', *American Economic Review* 95, 3: 831–49.

Sassen, S. 2001. *Global City: New York, London, Tokyo*, 2nd edn. Princeton, NJ: Princeton University Press.

Sennett, R. 1998. *The Corrosion of Character: The Personal Consequences of Work in the New Capitalism*. New York: Norton.

Smith, H., and D. Ley. 2008. 'Even in Canada? The multiscalar construction and experience of concentrated immigrant poverty in gateway cities', *Annals, Association of American Geographers* 98, 3: 686–713.

Taylor, P. 2002. 'Measurement of the world city network', *Urban Studies* 39, 13: 2367–76.

Townshend, I., and R. Walker. 2002. 'The structure of income residential segregation in Canadian metropolitan areas', *Canadian Journal of Regional Science* 25, 1: 25–52.

Varady, D.P., ed. 2005. *Desegregating the City: Ghettos, Enclaves, and Inequality*. Albany: State University of New York Press.

Veblen, T. 2007 [1899]. *Theory of the Leisure Class*. Oxford: Oxford University Press.

Walks, R.A. 2001. 'The social ecology of the post-Fordist/global city? Economic restructuring and socio-spatial polarization in the Toronto urban region', *Urban Studies* 38, 3: 407–47.

——— and L.S. Bourne. 2006. 'Ghettos in Canada's cities? Racial segregation, ethnic enclaves and poverty concentration in Canadian urban areas', *Canadian Geographer* 50, 3: 273–97.

——— and R. Maaranen. 2008a. 'Gentrification, social mix, and social polarization: Testing the linkages in large Canadian cities', *Urban Geography* 29, 4: 293–326.

——— and ———. 2008b. *The Timing, Patterning, and Forms of Gentrification and Neighbourhood Upgrading in Montreal, Toronto, and Vancouver, 1961–2001*. Toronto: University of Toronto Centre for Urban and Community Studies/Cities Centre, Research Paper 211.

Wang, S.G. and L. Lo. 2005. 'Chinese immigrants in Canada: Their changing composition and economic performance', *International Migration* 43, 3: 35–71.

Wilson, W.J. 1987. *The Truly Disadvantaged: The Inner City, the Underclass, and Public Policy*. Chicago: University of Chicago Press.

CHAPTER 11

The Canadian City at a Crossroads between 'Passage' and 'Place'

Anthony Perl and Jeffrey Kenworthy

The Canadian Approach to Urban Transportation Planning

Canadian cities grapple with the same urban dilemmas that confront large human settlements across the globe. Among these is trying to satisfy humans' considerable appetite for mobility while simultaneously accommodating their desire for healthy and attractive places to live. This urban transportation conundrum has been a factor in Canadian city planning since the 1950s.

The simultaneous desire for easy and cheap mobility and **livable cities** poses a conflict that rarely troubles Canadians, except when a new road, rail, port, or airport infrastructure is proposed in or near their neighbourhood. It should come as no surprise that citizens expect local governments to satisfy their competing preferences for mobility and community. What becomes noteworthy is the extent to which Canadian cities have been able to address both demands without compromising either their transport system's performance or the viability of urban communities.

This is a considerable achievement considering the predominant transportation planning paradigm, which informs what is sometimes referred to as the 'predict and provide' approach to assessing mobility needs. Such techniques, generically called four-step gravity models, would form the core content of many chapters in urban transportation planning and engineering textbooks (Meyer and Miller, 2001; Ortúzar and Willumsen, 2001). Students using these texts would learn how to forecast mobility needs based on data collected from household travel surveys. The trip patterns revealed by such data are seen to arise from the particular urban land uses and population densities, which the models then examine according to alternative land-use scenarios. The resulting predictions, predominantly for travel by car, are often used to justify new high-capacity road infrastructure. Cities where this methodology has been embraced tend to experience a spiral of increasing car use, congestion, more road-building, more car use, and so on.[1]

The alternative approach, which Canada's largest cities have experienced, and which will be examined in this chapter, is sometimes referred to as a 'debate and decide' approach to planning. Rather than allowing the city's future to be mechanistically determined by statistical models, this technique adopts a community-based engagement where citizens establish visions of the future city and see what would be required of transportation and land-use policy to achieve such visions. **Transportation demand management** (TDM) usually contributes to making more efficient use of established road infrastructure, once it is recognized that cities cannot continue to cater to open-ended travel demand by simply increasing the supply of automotive infrastructure.

This chapter will explore how Canadian planners and policy-makers have expanded urban mobility while also retaining the attributes that make cities desirable. The results demonstrate how 'balanced transportation' planning can add value to urban communities. In Canada, such a balance has meant not just providing a choice between the means of motion (e.g., auto, bus, train, boat, bike, and foot), but also weighing the value of enhancing passage through a city against the resulting diminution of place-based attributes that could ensue. While Canadians have yet to solve the urban conundrum of simultaneously expanding transport capacity while enhancing the quality of life, their attempts at trying to do so offer lessons in dealing with the challenges that city planners and public officials face.

We begin with data on recent mobility trends in five major Canadian cities: Montreal, Toronto, Calgary, Ottawa, and Vancouver. International comparisons will situate the recent performance of these cities' transportation systems. In many aspects, these Canadian cities are situated between the auto-dominated urbanism of Australia and the United States and the more transit-oriented urban experience found in Europe and affluent Asian cities. For some, Canada's middle ground represents the best of both worlds, while others see it as combining the deficiencies of both auto- and transit-oriented mobility. But as these data will show, no other urban environment juxtaposes the automotive- and transit-oriented urban transportation systems to the degree found in Canada.

We go on to consider the institutional context in which Canada's urban transportation policies and plans have developed. We feature the urban outcomes of Canada's middle ground in transportation in Toronto and Vancouver, cities where balanced transportation planning set the stage for Canada's success stories of higher quality of life in communities that made different (i.e., more modest) accommodation for total mobility during the 1970s and 1980s. Considering the effect of neoliberal policies during the 1990s when spending on transit was reduced while freeway-building picked up, with corresponding increases in auto dependence, we finally examine the Canadian federal government's return to urban policy in the early twenty-first century, evidenced through the first major spending on urban infrastructure in a generation, raising the question of how Canada's balance between passage and place could evolve in coming years (see Allahwala, Boudreau, and Keil, Chapter 12). We conclude by examining energy and climate factors that could drive future change.

Transportation in Canadian Cities, 1996–2006: An International Comparison

We begin with an overview of how five major Canadian metropolitan areas compare in transportation and land-use factors in the year 2006 with a collection of other cities from the US, Australia, and a handful of European cities. The overview also examines how the Canadian cities have changed in key factors describing the land-use–transport system over the 1996–2006 decade, relative to how other cities have changed. Results represent research in progress and not all cities are completely documented at the time of this writing. Major aspects of the land-use–transport system are measured using a total of 14 separate variables on each city. The cities used for these comparisons are presented in Table 11.1.

Private Transport Use Levels

Private transport use is perhaps best measured by the distance people travel in cars (passenger kilometres per capita). Table 11.2 provides this critical urban transportation factor.

The Canadian cities are significantly lower in car use than both the US and Australian samples. European cities are represented only by Frankfurt and Zurich and the Canadian cities collectively are some 7 per cent higher than these cities in car

Table 11.1 Cities in the Study

US	Canada	Australia	Europe
Atlanta	Calgary	Brisbane	Berlin
Chicago	Montreal	Melbourne	Frankfurt
Denver	Ottawa	Perth	Hamburg
Houston	Toronto	Sydney	Munich
Los Angeles	Vancouver		Zurich
New York			
Phoenix			
San Diego			
San Francisco			
Washington			

Table 11.2 Private Transportation Mobility in an International Sample of Cities, 1995–2005

Cities	Car Pass. km per Person 1995–6	Car Pass. km per Person 2005–6	% Change over 10 Years
Canada			
Calgary	11,203	11,038	–1.5
Montreal	7,597	6,453	–15.1
Ottawa	8,298	8,708	+4.9
Toronto	6,818	6,290	–7.7
Vancouver	9,310	9,987	+7.3
Average	**8,645**	**8,495**	**–1.7**
US	18,155	18,703	+3.0
Australia	12,114	12,447	+2.7
Europe	7,401	7,913	+6.9

Note: The percentage change column for each variable is calculated on the same group of cities for 1995 and 2005, depending on which 2005 data are finalized. All 1995 data are available whereas for 2005 some data for some cities are missing. Averages where there are insufficient completed cities for 2005 need to be used cautiously as these will change as more data are finalized.

Sources: Data for this and subsequent tables are derived from many sources, including the official census of the country under scrutiny, e.g., Statistics Canada. In most cases data have been acquired from a variety of government sources, including planning departments, transportation departments, municipal agencies, and centralized sources such as the Canadian Urban Transit Association (CUTA), the US National Transit Development Program, and so on.

use per person. The Canadian cities here reduced this factor by 1.7 per cent in the time frame, while the other groups of cities all increased over the 1995–2005 period.

Within the Canadian sample, Calgary is by far the leader with 40 per cent higher car use than the average for Montreal, Toronto, Ottawa, and Vancouver. Montreal and Toronto are below even Frankfurt and Zurich in their car use per capita, despite these latter cities' better-performing transit systems (which are discussed later). Calgary, Montreal, and Toronto all declined in car use per capita, while Vancouver crept up by 7 per cent despite many positive features of urban development over the decade. Ottawa increased by 5 per cent.

Public Transport Service and Use Levels

Having looked at car use, it is important to also consider transit use and service levels. In Table 11.3, per capita use of transit is examined in terms of trips made (boardings) and the distance people travel on transit (passenger kilometres). Transit service levels are contained in Table 11.3 as passenger kilometres of service supplied per capita.[2]

Annual 'transit boardings per person' offers a measure of how much the transit system is used. US cities rose by 11 per cent in transit boardings per capita, despite the declines in Atlanta, Houston, and San Francisco. Australian cities rose by 6 per cent, the Canadian cities by nearly 8 per cent, and the European counterparts by almost 18 per cent (Berlin, Hamburg, Munich, and Zurich).

This transit-use measure reflects Canada's more balanced development of urban transportation. Canadian cities register 151 annual transit boardings per capita, right between a low of 67 trips per person annually in US cities and a high of 447 transit trips per person each year in European cities. Australian cities are significantly better than American cities in transit use (96 trips per capita), but fall well below the Canadian cities, most likely due to lower-density, less transit-oriented urban form. While these data reveal much work ahead

in building transit's role in North America, Canada's ability to balance mobility with community in urban development is evident. Clearly, transit is playing a more important role in Canadian urban systems with, on average, 126 per cent higher trip-making than in US cities. Montreal and Toronto are Canada's most transit-intensive cities, consistent with their also being the lowest users of cars. But recent trends suggest that earlier momentum in developing public transit during the 1970s and 1980s may have abated. In Montreal, transit use has stagnated with identical levels in 2006 as compared to 1996, while Toronto saw an actual decline of transit ridership of 3 per cent. The data suggest that transit has faced hard times of late in Canada's two largest cities, whereas in Calgary, Ottawa, and Vancouver it appears to be maintaining an upward trajectory, but from a lower base.

Vancouver inaugurated Canada's latest urban rail transit expansion. The new 'Canada Line' connecting the CBD, Vancouver International Airport, and the City of Richmond opened on 17 August 2009. Estimated patronage just from this new line is projected to add around 13 trips per capita to the city's transit use, without considering new transit boardings on feeder services. If such trends continue, Vancouver could possibly overtake Toronto in transit trips per capita.

Annual transit passenger kilometres per person measures the actual distance people travel each year on transit. At 571 annual transit passenger kilometres per capita, US cities are still way below the average distances travelled by transit in other cities (Australian 1,075, Canadian 1,031, European 2,391 passenger km). Increasing passenger kilometres can result from ridership going up, but also from passengers travelling longer distances on each trip. It seems the latter is mostly in evidence because the data in Table 11.3 reveal that passenger kilometres have bigger percentage increases than boardings—even in Montreal where transit trips were stagnant, and in Toronto where they declined, passenger kilometres per capita rose.[3]

Table 11.3 Public Transportation Mobility in an International Sample of Cities, 1995–2005

Cities	Transit Boardings per Person 1995–6	Transit Boardings per Person 2005–6	% Change over 10 Years	Transit Pass. km per Person 1995–6	Transit Pass. km per Person 2005–6	% Change over 10 Years
Canada						
Calgary	113.5	131.1	+15.5%	925	1,130	+22.2%
Montreal	206.3	206.3	0.0%	993	1,122	+13.0%
Ottawa	104.8	129.0	+23.1%	851	849	–0.2%
Toronto	158.2	153.7	–2.8%	1,050	1,125	+7.1%
Vancouver	118.2	133.6	+13.0%	767	928	+21.0%
Average	**140.2**	**150.7**	**+7.5%**	**917**	**1,031**	**+12.4%**
US	60.1	66.7	+11.0%	492	571	+16.1%
Australia	90.4	95.6	+5.8%	966	1,075	+11.3%
Europe	380.5	447.3	+17.6%	2,077	2,391	+15.1%

Sources: See Table 11.2.

Interestingly, transit usage levels do not strongly follow service supply in terms of seat kilometres. Certainly, the Canadian cities are higher than the US cities in seat kilometres per capita, but only by 26 per cent, whereas per capita passenger kilometres are 81 per cent higher in the Canadian cities. Toronto has over 1,000 more seat kilometres per capita, or nearly 50 per cent more than in Montreal, and yet transit travel distances per capita are almost identical. This means that Montreal's urban transit is among the most crowded in North America, something that regular users will attest to. Explaining both private and public transportation patterns clearly requires some examination of other factors.

Urban Form

Population data show that the majority of cities in Australia, Canada, and the US exhibit a somewhat upward trend in density (Table 11.4), an important break from the trajectory following World War II, when densities declined continuously under the influence of 'predict and provide' transport planning and sprawling development. Increasing density is needed if cities are to control their car use (see Bunting and Filion, Chapter 2). On the other hand, the European cities here still are showing minor declines in density or close to stable densities. This is mainly due to population growth in many European cities being very modest or in some cases negative, alongside declining household sizes, meaning that new dwelling construction provides for an increasing number of smaller households.

The US cities increased on average from 14.9 persons per hectare to 15.4 persons per hectare (ha), over a 3 per cent increase. The Australian cities went from 13.3 per ha to 14.0 per ha, about a 5 per cent increase in density. Australian cities have had significant re-urbanization or 'urban consolidation' policies for many years and these have strengthened recently. Most US cities seem to be going through a similar process. The four Canadian cities for which final data are available show a decrease in population density, dropping on average from 26.4 per ha to 25.5 per ha, a 3.3 per cent decline. Montreal declined significantly, by 19.2 per cent, while Calgary and Ottawa declined modestly, by 1.4 per cent and 1.6 per cent, respectively. Vancouver went in the opposite direction, increasing its population density by 17 per cent, a significant turnaround in metropolitan-scale density considering that density increases are focused in selected areas around transit and in inner areas, while much suburban development continues to be at lower densities.

Total land use or 'activity intensity' represents population plus job densities. When jobs are considered we find that, overall, there is hardly a city in the sample that has not had a net increase in total land-use intensity. Even those that declined in population density mostly made up for it in job density. We can conclude that the majority trend, at least in this sample of cities, is for overall intensification of land uses. US cities increased 5.4 per cent in activity density from 22.4 to 23.6 people and jobs per ha. Australian cities rose by 8 per cent, Canadian cities by 1.5 per cent, and even European cities rose by 1 per cent. Within Canada, the intensification increase is strongest in Vancouver, where a 22.4 per cent growth in activity intensity overshadowed more modest increases of 1.9 per cent in Calgary and 3 per cent in Ottawa and offset the decline of 15.3 per cent in Montreal.

Activity density is very important as it means less land being taken for destructive urban sprawl and implies gradual strengthening in the potential to become less automobile-dependent. Denser land uses mean shorter travel distances for walking and cycling (and shorter car trips), and greater potential for public transport.

Centralization of jobs is measured by the proportion of metropolitan jobs located in the central business district (CBD). Although jobs are increasing in cities there has been a trend towards decline in the pre-eminence of the central area and for jobs to disperse across metropolitan areas. US cities declined on average from 9.7 per cent of metropolitan jobs in the CBD to 8.8 per cent of jobs. In Australian cities CBD jobs have declined from 13.3 per cent to 12.7 per cent. In the Canadian cities, where decentralization of jobs is less than in both the US and Australia, the average decline was from 15.7 per cent to 15.0 per cent of jobs in the CBD (Chapter 7, Hutton). Montreal actually rose, Toronto was almost stable, and Calgary and Vancouver declined. There is, however, a huge range in this factor, with Calgary having the most centralized employment market (21 per cent of jobs) and Toronto (perhaps surprisingly) languishing at 6 per cent, or significantly below even the US average. Even Atlanta and Denver have 7 per cent of jobs in their CBDs. Only two European cities have final data, but both declined (Frankfurt from 20.5 per cent to 17.3 per cent and Zurich from 12.2 per cent to 10.3 per cent). It appears that a variety of forces are at work and the trends are not consistent, though the overall pattern is for a decline in the relative importance of CBDs as job locations. In a few cities **revitalization** strategies are bringing more jobs to the main centre, some of which will be linked to the global information economy. In other cities, the CBD jobs are still growing but declining relative to the growth in jobs elsewhere. In some cities, the number of jobs in the CBD is actually experiencing a net decline as other places around the metropolitan area take precedence as employment locations.

In summary, the decline in importance of the CBD will tend to work against sustainable transportation, unless the new jobs are locating into strong centres serviced by good transit (decentralized concentration). If they are going to auto-oriented places, then this will increase car commute journeys. Transit systems still are focused mostly on central areas and can play a much greater role in commuting if more jobs are located there or along the transit lines, in particular, in rail station precincts.

Table 11.4 Urban Form in an International Sample of Cities, 1995–2005

Cities	Urban Density (persons per ha) 1995–6	Urban Density (persons per ha) 2005–6	% Change over 10 Years	Activity Density (persons + jobs per ha) 1995–6	Activity Density (persons + jobs per ha) 2005–6	% Change over 10 Years	% of Jobs in the CBD 1995–6	% of Jobs in the CBD 2005–6	% Change over 10 Years
Canada									
Calgary	20.8	20.5	–1.4%	31.9	32.5	+1.9%	23.0%	20.9%	–9.1%
Montreal	31.7	25.6	–19.2%	45.0	38.1	–15.3%	16.3%	17.8%	+9.2%
Ottawa	31.3	30.8	–1.6%	47.0	48.4	+3.0%	20.1%	19.1%	–5.0%
Toronto	25.5	nf	nf	38.3	nf	nf	6.5%	6.4%	–1.5%
Vancouver	21.6	25.2	+16.7%	32.1	39.3	+22.4%	12.6%	11.0%	–12.7%
Average	**26.4**	**25.5**	**–3.3%**	**39.0**	**39.6**	**+1.5%**	**15.7%**	**15.0%**	**–4.5%**
US	14.9	15.4	+3.4%	22.4	23.6	+5.4%	9.7%	8.8%	–9.3%
Australia	13.3	14.0	+5.3%	18.8	20.3	+8.0%	13.3%	12.7%	–4.5%
Europe	48.4	47.2	–2.5%	78.1	78.6	+0.6%	16.3%	13.8%	–15.3%

Note: nf means data are not finalized yet. It does not mean data are not available. Strictly, it is not possible to calculate a percentage change from two percentages. For the purposes here, this statistic is illustrative only and should be interpreted cautiously.

Sources: See Table 11.2.

Private Transport Infrastructure and Performance

As shown in Table 11.5, car ownership is one measure of the importance of cars in a city, although not as important as actual car use. One would expect this factor to have universally risen as incomes have grown and motorization has continued, but some cities have seen a decline in car ownership. The 10 US cities have an average increase in car ownership of 9 per cent from a very high 587 per 1,000 up to 640 per 1,000, though still a fraction less than the Australian cities today (647/1,000 people), which rose 9.5 per cent over the decade. In the five Canadian cities, two cities declined in car ownership (Calgary and Vancouver, from 703 to 632 cars per 1,000 people and 520 to 506 per 1,000 people, respectively), with an overall average decline for the five cities of 1.5 per cent, Montreal and Toronto having risen by 4 per cent

and 4.5 per cent, respectively. The five European cities in this study rose significantly in car ownership (431 cars per 1,000 people up to 481, or a 12 per cent increase).

Overall, 19 out of the 24 cities show increases in car ownership. However, two of the four cities showing declines were in Canada. The prevalence of cars in cities, even wealthy already highly motorized cities, is still on the ascendancy, especially in suburbanized zones. This obviously works against sustainable transportation, but is also probably linked to the changing demographics of urban populations, with a trend towards older population profiles and more and smaller households, many of which will need a car, rather than larger families sharing fewer vehicles.

Freeway supply per capita reveals changes in the supply of infrastructure over the last 10 years. It is a potentially important factor as freeways represent a highly symbolic as well as concrete (and asphalt) commitment to the automobile, and can only be realized at great financial expense and significant environmental and social costs. Changes in this factor are split right down the middle. In 12 cities freeway provision per capita has increased relative to the populations served, one city (Frankfurt) has remained identical, while 12 cities have declined.

In Canadian cities, unfortunately, it appears that freeway provision is on the rise with a 30 per cent average increase in the five Canadian cities and, quite extraordinarily, the average freeway provision per capita is now a tiny fraction higher than the average for the US cities, whereas in 1996 it was 22 per cent less. Considering that Canadian cities have never had a central government-funded freeway program like the Highway Trust Fund in America, it is remarkable that cities like Calgary and Ottawa are now substantially above the average US urban level of freeway provision and Montreal has the same level as an average US city. One explanation may lie in the Canadian 'balance' of building urban freeway infrastructure, but then not changing the land use to fully embrace sprawl and scatter.

In contrast, Australian cities have on average increased by a much lower 17 per cent (largely due to an increase in Perth of nearly threefold from 0.043 to 0.125 metres of freeway per person). In the five European cities, there is on average no change in per capita freeway provision and, likewise, the US cities have remained stable in this factor. Overall, one might say that the 'freeway paradigm' in the US is loosening its stranglehold, though within this sample some cities still increased in freeway provision while others declined.

The number of parking spaces per 1,000 CBD jobs is another somewhat symbolic factor as it measures the commitment central areas have to accommodating the automobile by how much parking they provide relative to the number of jobs. CBDs are the easiest places in cities to prioritize transit as they represent the single biggest concentration of jobs in any metropolitan area. CBDs should be where policies reduce the pre-eminence of the car, and the easiest way to do this is to reduce the parking ratio relative to jobs.

Unfortunately, data relating to downtown parking are difficult to gather. However, we have information for 17 out of the 24 cities in Table 11.1. Of these 17 cities, downtown parking has declined in 11 and risen in six. In the US cities there has been a 16 per cent average decline, from 645 to 539 parking spaces per 1,000 CBD jobs. In Australia, there has been an overall decline of 19 per cent, from 367 to 298 spaces per 1,000 jobs, with all cities having declined by over 20 per cent (except for Sydney, which rose by 5 per cent). The four completed Canadian cities have uniformly declined (13 per cent on average), with a big reduction in Toronto of 27 per cent. In Europe, Frankfurt has decreased and Zurich has increased CBD parking supply, but both post very low levels compared to most of the other cities. It would appear that a majority of these wealthier cities are providing less CBD parking than in previous decades. Certainly, the four Canadian cities all are heading in this direction.

Public Transportation Infrastructure and Performance

As presented in Table 11.6, several variables can be used to assess the extent to which high-quality transit routes are provided in the city, namely: (1) dedicated rights-of-way for transit in the form of rail lines and bus-only lanes; (2) how the extent of these compare to high-quality infrastructure for cars in the form of freeways; and (3) how the transit system performs in terms of the time it takes for a trip of a certain distance, especially in relation to the car.

Length of reserved route per person covers all transit routes that are fully protected from the incursion of road traffic. This is generally a 'good

Table 11.5 Private Transport Infrastructure in an International Sample of Cities, 1995–2005

Cities	Cars per 1,000 Persons 1995–6	Cars per 1,000 Persons 2005–6	% Change over 10 Years	Freeway Length per Person 1995–6	Freeway Length per Person 2005–6	% Change over 10 Years	Parking Spaces/ 1,000 CBD Jobs 1995–6	Parking Spaces/ 1,000 CBD Jobs 2005–6	% Change over 10 Years
Canada									
Calgary	703	632	–10.1%	0.163	0.282	+73.0%	465	401	–13.8%
Montreal	429	446	+4.0%	0.145	0.156	+7.6%	455	nf	nf
Ottawa	532	542	+1.9%	0.144	0.187	+29.9%	348	333	–4.3%
Toronto	464	485	+4.5%	0.080	0.089	+11.2%	239	174	–27.2%
Vancouver	520	506	–2.7%	0.077	0.069	–10.4%	444	389	–12.4%
Average	**530**	**522**	**–1.5%**	**0.122**	**0.157**	**+29.7%**	**374**	**324**	**–13.4%**
US	587	640	+9.0%	0.156	0.156	0.0%	645	539	–16.4%
Australia	591	647	+9.5%	0.086	0.101	+17.4%	367	298	–18.8%
Europe	431	481	+11.6%	0.076	0.076	0.0%	200	205	+2.5%

Sources: See Table 11.2.

Table 11.6 Public Transport Infrastructure in an International Sample of Cities, 1995–2005

Cities	Length of Reserved Transit Route per 1,000 Persons (m) 1995–6	Length of Reserved Transit Route per 1,000 Persons (m) 2005–6	% Change over 10 Years	Ratio of Reserved Transit Route to Freeway 1995–6	Ratio of Reserved Transit Route to Freeway 2005–6	% Change over 10 Years	Ratio of Average Transit System Speed to Road Speed 1995–6	Ratio of Average Transit System Speed to Road Speed 2005–6	% Change over 10 Years
Canada									
Calgary	40.4	45.5	+12.6%	0.25	0.16	−36.0%	0.55	0.53	−3.6%
Montreal	68.5	119.2	+74.0%	0.47	0.76	+61.7%	0.58	0.67	+15.5%
Ottawa	26.7	33.6	+25.8%	0.19	0.18	−5.3%	0.52	0.59	+13.5%
Toronto	92.0	80.1	−12.9%	1.15	0.90	−21.7%	0.47	nf	nf
Vancouver	53.7	55.5	+3.4%	0.70	0.81	+15.7%	0.74	0.56	−24.3%
Average	**56.3**	**66.8**	**+18.6%**	**0.55**	**0.56**	**+1.8%**	**0.60**	**0.59**	**−1.7%**
US	48.7	71.7	+47.2%	0.41	0.56	+36.7%	0.52	0.52	0.0%
Australia	169.7	158.0	−6.9%	2.18	1.67	−23.4%	0.75	0.78	+4.0%
Europe	181.2	236.7	+30.7%	5.45	6.14	+12.7%	0.95	1.23	+29.5%

Note: nf means data are not finalized yet. It does not mean data are not available.

Sources: See Table 11.2.

news' story for sustainable transportation. In the 22 cities where data on this factor have been finalized, only four showed declines in the per capita availability of high-quality transit lines. In one city, Phoenix, it remained as zero. In all groups of cities, except those in Australia, this factor has risen by significant levels. The Canadian cities have gone up by 18.6 per cent, from 56.3 to 66.8 metres per 1,000 persons, though the variability is immense, with Montreal almost doubling its network, while Toronto actually declined by over 13 per cent in per capita transit reserved route, due to only a small addition to its network in the form of the 6.4 km Sheppard subway line in 2002. The European cities increased by 31 per cent from a much greater figure of 181.2 to 236.7 metres per 1,000 persons, or 3.5 times the Canadian level. Notwithstanding the decline in the Australian cities, their transit reserved length per person is still 158 metres, or 2.4 times higher than in the Canadian cities. However, within the Australian cities, only Perth has increased, from 82.0 to 89.4 metres per 1,000 persons (a 9 per cent increase). The US cities have increased by a very healthy 47 per cent in reserved transit route per capita and are now fractionally higher than the Canadian cities, at 71.7 metres per 1,000 persons compared to 66.8, whereas in 1995 they had a little less reserved route per person than their northern neighbours (48.7 compared to 56.3 meters per 1,000 persons). Even though US cities now have more reserved transit route length per person than Canadian cities, their use of transit is much lower. Overall, there is a strong trend towards providing more high-quality transit lines in many cities around the world and this is generally true of the Canadian cities. This result is good news for public transportation systems as it means that along particular corridors where lines have been built or extended, there will be more effective competition between the car and public transport, though service levels may not be frequent, as for example along much of Montreal's new suburban network (Kenworthy and Townsend, 2009).

The ratio of reserved transit route to freeways is a useful indicator of the priority that cities are giving to transit by measuring how much high-quality transit there is relative to how much freeway. In the five Canadian cities, the overall average for this ratio actually improved slightly (2 per cent), though only Montreal and Vancouver contributed to this increase. In Toronto, Ottawa, and Calgary transit/freeway ratio fell due to quite significant freeway construction, which overshadowed the addition of transit reserved routes.

Overall, these five Canadian cities provide only a little more than half (0.56 per cent) as much reserved transit route as they do freeway. This figure is now identical to the US cities, but the US cities improved much more over the decade, increasing by 37 per cent. Australian cities, despite declining by 23 per cent in this factor, still had some 1.67 times as much reserved transit route as freeways, a much better result than for the Canadian cities, while European cities top the scale with 6.14 times as much after an increase of 13 per cent over the 10 years. The fact that Canadian cities still perform a lot better than Australian cities in transit use, despite having much less transit reserved right-of-way relative to freeway, is something of a testament to their significantly higher density and therefore more effective use of the transit service and infrastructure provided. Also, the fact that Canadian cities are not much different from the US cities in their relative supply of transit and freeway infrastructure, and yet have some 2.3 times higher transit trips per capita, suggests that Canadian cities have transit-supportive spatial factors at work (and in all likelihood, better co-ordinated and interconnected transit services).

The final factor here, the ratio of transit system speed to general road traffic speed, is a measure of the competitiveness of transit systems with the car. US cities remained steady in this factor over the 10 years, at the lowest value of all cities (0.52). In the Australian cities there was a slight improvement in this factor of 4 per cent, up to 0.78. Canadian cities dropped slightly by 1.7 per cent, down to 0.59, while European cities increased by 29 per cent, up to 1.23 (but this was only for Berlin and Munich). In general, the speed competitiveness of Canadian transit systems is relatively poor, with Montreal, the highest transit user, performing best. Vancouver and Calgary still struggle to develop transit systems that compete overall with the car's speed and it would appear that the only way this will improve is for those cities to develop larger rail systems and to increase the number of streets where buses have dedicated lanes and, of course, not to build more high-capacity roads to further raise traffic speed.

Transport Fatalities

The number of transport fatalities per 100,000 people measures the mortality rate in cities from transport-related causes (Table 11.7). Most transport fatalities are in private—auto and truck—transport, as well as involving cycling and walking, as opposed to public transit. Fatalities are an important aspect of the sustainability of transport systems, since loss of life has huge economic costs to society, not to mention the immense emotional and non-quantifiable losses. This factor shows the clearest and strongest trend of all variables towards improvement. Only in Calgary, Canada's most auto-oriented urban area, did transport deaths per capita rise between 1996 and 2006. This outcome goes against conventional traffic engineering wisdom, which assumes that freeways are the safest types of road. The answer to this conundrum may be found in traffic speed.

Average traffic speed in Calgary increased from 47.7 km/h to 50.1 km/h over this period, and higher traffic speeds usually correlate with increased transport fatalities. All other cities showed generally substantive declines. US cities went from a very high figure of 12.7 to 9.5 deaths per 100,000 people, a 25 per cent decrease over the 10 years. Australian cities have declined from 9.1 to 6.2 deaths per 100,000 people, a 32 per cent

fall. The five Canadian cities declined slightly from 6.5 to 6.3 per 100,000 people (3 per cent decline), despite the substantive rise in Calgary. Three European cities (Zurich, Berlin, and Hamburg) have continued to fall from an already very low average level of 5.3 to 2.6, a 51 per cent decrease.

Overall across all cities for which data are available, trends in traffic fatalities have been in decline—a positive **urban sustainability** indicator.[4] The only negative is that US cities still continue to experience significantly higher death rates in transport than other comparably well-off cities, though they are much lower in the more transit-oriented cities such as New York than in the extreme auto-oriented cities such as Atlanta.

Explaining Canada's Urban Transport Anomalies

From the data presented above, Canada's urban transportation performance appears somewhat anomalous. Canadians have been using public transit more than Americans and Australians for an extended period, but they have more recently built urban freeways while the growth of transit has levelled off in North America's two most transit-oriented cities, Montreal and Toronto. Yet, Canadians have not abandoned transit, which tends to operate with a higher occupancy than is found in either the United States or Australia. How is it that Canadians have been able to advance both transit-city enhancements and auto-city sprawl in tandem?

The simplest and most persuasive explanation for this outcome is that the people shaping Canadian cities found ways to pursue both the auto- and transit-oriented urbanism because they could. Urban planners and municipal politicians in Canada possessed greater freedom to invent transportation solutions compared with their American counterparts because Canadian federalism, specifically the rivalry between the national and provincial governments that intensified during the 1970s, barred the door to a national urban policy that

Table 11.7 Metro GDP, Public Transport Service Supply, and Transport Deaths in an International Sample of Cities, 1995–2005

Cities	Metro GDP per Person (US$ 1995) 1995–6	Metro GDP per Person (US$ 1995) 2005–6	% Change over 10 Years	Transit Seat km per Person 1995–6	Transit Seat km per Person 2005–6	% Change over 10 Years	Transport Deaths per 100,000 People 1995–6	Transport Deaths per 100,000 People 2005–6	% Change over 10 Years
Canada									
Calgary	23,983	39,086	+63.0%	2,249	2,369	+5.3%	3.8	10.2	+168.4%
Montreal	16,066	26,685	+66.1%	2,371	2,048	−13.6%	7.7	6.1	−20.8%
Ottawa	18,827	24,936	+32.4%	2,089	2,193	+5.0%	8.6	4.4	−48.8%
Toronto	19,456	24,083	+23.8%	2,803	3,056	+9.0%	6.1	5.3	−13.1%
Vancouver	25,793	29,582	+14.7%	1,936	2,173	+12.2%	6.5	5.3	−18.5%
Average	**20,825**	**28,874**	**+38.7%**	**2,290**	**2,368**	**+3.4%**	**6.5**	**6.3**	**−3.1%**
US	31,386	42,547	+35.6%	1,560	1,874	+20.1%	12.7	9.5	−25.2%
Australia	20,226	28,401	+40.4%	3,997	4,077	+2.0%	9.1	6.2	−31.9%
Europe	44,043	36,945	−16.1%	5,620	6,665	+18.6%	5.3	2.6	−50.9%

Sources: See Table 11.2.

could have dictated a standardized approach (Cullingworth, 1987).

American cities were reshaped by the '**urban renewal**' policy mechanisms established in the Federal Housing Act of 1954 and even more so by the designs and objectives for a national freeway network that were prescribed in the 1956 National Interstate and Defense Highways Act, one impetus for which was the desire for quick evacuation of US cities in the event of catastrophe, specifically, nuclear war. These national programs came with significant funding incentives—two-thirds of the cost of urban renewal projects and 90 per cent of Interstate Highway expenditures were paid from the federal treasury. These fiscal inducements made it virtually impossible for local officials to reject projects and thus deprive their city of considerable resources. Along with the projects came national mandates, such as the design guidelines for urban freeways codified in the *Highway Capacity Manual*. These regulations took no account of local context and often imposed unnecessarily disruptive freeways on American cities to meet a 'one size fits all' set of engineering standards. Instead of the federal mandates and 'free money' projects that were advancing the auto city south of the border, Canadian cities were left to their own devices in meeting transportation needs during the second half of the twentieth century. Neither financial assistance nor national programs were in place that would constrain local initiative, with the exception of federal funding for the Trans-Canada Highway for urban areas in its path. In Toronto, a momentous decision was made to cash in the transit system's operating *surplus* to build Canada's first subway line, which opened in 1954. In contrast to neighbouring American cities, Toronto found its first (and only) 'easy money' to meet post-war demands for mobility in the transit surplus that had accumulated during wartime gasoline rationing when transit ridership had soared.

Toronto's initiative, and especially the plans for Expo in 1967, spurred Montreal to launch a long-planned rapid transit system, the first line of which opened in 1966. The Montreal Metro differentiated itself from Toronto's subway in both style and substance. The Metro adapted the French approach of integrating mobility into urban architecture through an 'underground city' whereas the sole purpose of the Toronto subway's pedestrian passages was to take travellers to the street level; likewise, Toronto's first steel-wheeled subway cars were imported from England while Montreal opted for rubber-tired trains that had been designed to serve Paris. By the 1960s, Canada had two rapid transit systems that reflected the particularities of its largest cities. Toronto had a frugal and functional subway while Montreal had an avant-garde metro that provided the skeleton for its underground city.

Canada's lack of a national transportation program meant that local and provincial politicians had little cover to deflect public criticism generated by protests against urban freeway infrastructure. Canada had never authorized a national network of freeways that was intended to defend against a Communist attack. American urban planners like Robert Moses could thus forge ahead with immensely destructive urban freeways such as New York's Cross Bronx Expressway,[5] ignoring local protests in the name of protecting cities that would soon be torn apart. Local protests succeeded in stopping the threat to inner-city communities posed by Toronto's planned Spadina Expressway and Vancouver's proposed Strathcona freeway (Lee, 2007; Zielinski and Laird, 1995). These urban highway revolts helped trigger municipal council realignments in which reform councillors gained power to assert local land-use priorities during the 1970s and 1980s (Sewell, 1972). Weighing the value of mobility against the benefits of community worked out much differently when pursued from the bottom up by local councils as compared to top-down master plans created by national and provincial governments. The results yielded urban success stories that Toronto and Vancouver are now celebrated for. Prominent international livability surveys consistently place Vancouver and Toronto at the top of their scales. *The Economist* (2008), for example, rates the two cities first and fifth respectively, while Mercer (2008) rates the two cities fourth and fifteenth. No US city rates close on either scale. This admiration is mirrored in popular writing (Montgomery, 2006) and academic literature (Punter, 2003), while many cities around the globe seek to emulate Canada's urban virtuosity.

Canada's successful balancing of passage and place has been a source of inspiration for New Urbanism (Calthorpe and Van der Ryn, 1986; Duany et al., 2003) and sustainable urbanism (Satterthwaite, 1999, Timmer and Seymoar, 2005). The latter posits a renewal of classical urban ideals of limited scale, community comprehensibility, dense housing, and mixed-use neighbourhoods (see also Grant and Filion, Chapter 18). During the 1980s within the pre-World War II urban envelope, Toronto and Vancouver both demonstrated that investing less in motorized transport could yield considerable benefits to quality of life and inner-city **regeneration**. However, it also needs to be recognized that these inner-city lessons were less appreciated by Canada's other major urban transportation actors, the provincial transport departments with responsibility for movement on a larger scale.

Cities in Canada are constitutionally subordinate to provincial governments, which in recent years have abolished and reconstituted municipal jurisdictions in both Ontario and Quebec. Provincial governments worked with municipalities in shaping urban transit initiatives during the 1960s and 1970s (particularly once freeway protests generated a powerful political backlash). But when it came to moving people and goods outside cities, the provincial transport departments were fully committed to freeway expansion that differed from American practices only in the speed and scale of its deployment. With a few minor exceptions, provinces have primary jurisdiction over Canada's inter-city freeway networks and little involvement in railway infrastructure, which falls under federal jurisdiction. When the Canada–US Free Trade Agreement came into effect in 1989 and truck traffic between Canada and the United States grew considerably (Woudsma, 1999), it was hardly surprising that provinces sought to accelerate their freeway building projects to keep up with rising goods movement. These trends accelerated with the North American Free Trade Agreement (NAFTA), which came into effect in 1994. Accommodating more freight movement along Canada's freeways meant expanding road networks in and around cities. Although provincial freeway expansion rarely meant adding new road infrastructure in the urban core, road expansion in less developed urban fringes was not challenged by smaller municipalities eager for the resulting land development. Thus, new provincial freeways stimulated greenfield development that brought Canadian cities further into auto-oriented development at the same time that transit-oriented intensification was being pursued with great success in the urban core. Suburban housing subdivisions and office parks expanded Canada's auto-city profile, not by hollowing out the urban cores as had occurred during 1970s 'white flight' in the United States, but by juxtaposing a heterogeneous mix of high- and low-density urban form. This morphology was described by Toronto transportation planner Juri Pill as 'Vienna surrounded by Phoenix' (1990: A27). However, it was during this period of provincial freeway building that public transit operating subsidies became 'burdensome' so that subsidy cuts, fare increases, and ridership declines began occurring by the early 1990s (Perl and Pucher, 1995).

In Canada, then, the lack of a national urban transportation framework encouraged policy innovations within cities by municipal planners and elected officials, and that same policy vacuum at the federal level allowed provincial transportation planners to establish how much highway capacity would be added. The US Interstate Highways program had prescribed urban freeway capacity, but this standard also provided an upper limit. In Canada, only the financial capacity of government limited the scope of freeway expansion. What kept provincial road-building within reason was the budget limit of most provinces. But when money became available through either fiscal windfalls, like Alberta's oil and gas boom, or external transfers, like the federal government's recent spending under the 'New Deal for Cities and Communities', freeways were built faster. A good example of this recent temptation to grow mobility at the expense of community can be found in British Columbia's Gateway Program, which aims to link the Port of Vancouver to the Trans-Canada Highway and other provincial freeways through a multi-billion dollar road and bridge expansion around the urban region funded by a mix of federal, provincial, and private investments (Hall, Chapter 4). Ottawa is committed to expanding freight movement through important trade gateways like Vancouver, while British Columbia has added on local transport agendas that include expanding commuter capacity in the auto-dependent municipalities of Surrey and Langley (Ginnell et al., 2008). The incompatibility of these national and provincial transport goals will eventually be recognized, but probably not before the new freeway infrastructure has added to the Vancouver region's auto dependence. The long-term impacts of such an emphasis of passage over place remain to be seen, but Vancouver's much vaunted livability is likely to suffer.

Conclusion: What of 'Transport Revolutions' and Future Urban Mobility?

As we have seen, in the absence of a national policy framework on what urban mobility should look like and how it should be paid for, Canada's development of urban transportation has been eclectic and idiosyncratic. The balance between mobility and community that Canadian cities are lauded for, and which is reflected in the quantitative indicators that we have presented, results from an oscillation between projects and programs that advance the 'debate and decide' paradigm of sustainable urban communities and those that, intentionally or otherwise, bring Canadian urban spaces closer to the auto dependence advanced by 'predict and provide' freeway development. Given the economic and ecological uncertainty now challenging Canada's cities, we conclude this chapter with a brief reflection on how the Canadian approach to urban transportation might deal with changes ahead.

An accumulation of evidence suggests approaches to mobility are going to change more during the coming 10 years than they have in the past three decades (Newman, 2007; see also Filion and Bunting, Chapter 1; Bunting and Filion, Chapter 2; Filion and Bunting, Chapter 3). These 'transport revolutions' will be driven by an imperative to reduce the use of oil as a transportation fuel, to attain either or both of energy security or climate protection goals. Transport revolutions are defined as a change of 50 per cent, either up or down, in the movement of goods or people by a particular technology and organizational arrangement (Gilbert and Perl, 2008). Unlike the incremental mobility change that is more familiar to people these days, transport revolutions occur in time frames of less than 25 years. Some examples include the arrival of the steam railway in England, the rationing of gasoline and rubber tires during World War II, the introduction of jet aircraft on transatlantic passenger routes, the reinvention of high-speed passenger trains in Japan and France, and the inauguration of guaranteed overnight delivery by Federal Express. It would be prudent to anticipate a change of revolutionary proportions in urban mobility before long.

Suggestions that we are now on the brink of a new wave of innovation, which some have ventured to suggest might be termed 'the age of sustainability' (Hargroves and Smith, 2005), can be heard. This new wave may be associated with new mobility technologies such as electric vehicles providing a reduced level of private motorized mobility due to infrastructure and technical constraints, but also a revival in the green modes of transport and their accompanying walking-city and transit-city forms, which will help to transform auto-oriented regions into more polycentric, sustainable urban structures.

Oil now fuels 95 per cent of the world's mobility, and 60 per cent of the world's oil consumption goes towards moving people and goods. Given that global oil production will soon be approaching its peak, it behooves society to find alternative ways to power mobility, sooner rather than later. As one of the world's most oil-intensive societies, Canada would be at the leading edge of such change.

Fortunately, Canada's urban passenger travel is particularly well suited for a rapid shift off oil since the grid-connected vehicles (GCVs) that currently provide the majority of transit trips in Canada's five largest urban areas represent a mature technology. Electricity that is mostly generated from renewable sources already powers the GCVs—trolley buses, streetcars, and subways in service across Canada. Unlike some of the more complex and costly alternatives to oil-fuelled internal combustion vehicles that are yet to be commercialized (e.g., hydrogen fuel cells), GCVs are literally ready to roll. Electric traction is also quieter and less polluting of the local environment than oil-fuelled vehicles. But are Canada's urban planning and development policies up to the challenge of implementing rapid and large-scale change along these lines?

Canada's history reveals a facility in changing course with urban transportation development, which bodes well for the kinds of radical shifts that could lie ahead. Stopping highway projects in midstream, as occurred with the Spadina Expressway, and starting up new highway programs that were not part of regional plans and priorities, as is now happening with British Columbia's Gateway

Program, illustrate the ways in which planners and public officials can move quickly in response to perceived opportunities and threats. While the benefits of anticipating change and getting ahead of the need to reduce urban mobility's carbon footprint are considerable, Canada might also be able to adapt rapidly if an oil shock were to hit before anticipatory measures are in progress. Decades of balancing between transit and auto-oriented urban development have left Canadian cities with less sprawl and less auto dependence to recover from than their American counterparts. This could provide an important advantage in the transition to sustainable cities that lies ahead.

Review Questions

1. In what ways are cities, and parts of cities, crossroads between 'passage' and 'place', and what conflicts can arise from this?

2. How does Canada compare to other countries in the attempt to encourage alternative modes of transportation?

3. Why do the authors believe that Canadian cities have been able to plan for both private and public transportation?

4. In your opinion, what kinds of challenges face transportation planning in Canada?

Notes

1. For a critical discussion of the data and methodologies used in urban transportation planning, see Eric J. Miller, 'Transportation and Communication', Chapter 6 in Bunting and Filion (2006), at: <www.oup canada.com/bunting4e>.

2. Seat kilometres are better indicants of service than vehicle kilometres as they take account of vehicle size. Where rail modes play a significant transit role, seat kilometres are higher because of the high capacity of passenger coaches—up to 144 seats on GO Transit trains serving southern Ontario and the West Coast Express in Vancouver, along with the number of coaches per train.

3. Only Ottawa bucked the trend, indicating shorter trip lengths by transit users. While boardings per capita in Canadian cities varied quite a lot, per capita transit passenger kilometres are remarkably consistent, certainly in Calgary, Montreal, and Toronto and even the Australian cities were hardly different from the Canadian cities on this factor. Australians use transit less but they travel longer distances due to lower-density cities. US cities, again, were the big laggards in this perspective with European cities the leaders.

4. They are 53 per cent higher than Australian cities, 83 per cent higher than Canadian cities, and 313 per cent higher than two cities in Germany. It is interesting to note that the highest transit-use city in the US, New York, has only 6.4 deaths per 100,000 people, or 33 per cent below the average for the 10 US cities. The other more transit-oriented metro regions, such as San Francisco (6.5 deaths per 100,000) and Chicago (7.5), also have much lower deaths rates in transport compared to heavily auto-oriented cities such as Phoenix (14.6 deaths per 100,000 or 54 per cent above the US average) and Atlanta (12.5 or 32 per cent above the average).

5. For an evocative discussion of how the Cross Bronx Expressway was built, see Caro (1975: chs 36–8).

References

Bode, P.M., S. Hamberger, and W. Zängl. 1986. *Alptraum Auto: Eine hundertjährige Erfindung und ihre Folgen*. München: Raben Verlag.

British Columbia. 2009. *Gateway Program: Improving Roads and Bridges for People, Goods and Transit throughout Greater Vancouver*. Victoria: Province of British Columbia. At: <www.gatewayprogram.bc.ca/>. (7 July 2009)

Calthorpe, P., and S. Van der Ryn. 1986. *Sustainable Communities: A New Design Synthesis for Cities, Suburbs and Towns*. San Francisco: Sierra Club Books.

Caro, R. 1975. *The Power Broker: Robert Moses and the Fall of New York*. New York: Vantage Books.

Cullingworth, J.B. 1987. *Urban and Regional Planning in Canada*. New Brunswick, NJ: Transaction.

Duany, A., E. Plater-Zyberk, and R. Alminana. 2003. *The New Civic Art: Elements of Town Planning*. New York: Rizzoli International.

Economist, The. 2008. Liveability ranking, 'Urban idylls', 28 Apr. 2008. At: <www.economist.com/markets/rankings/displaystory.cfm?story_id=11116839>. (7 July 2009)

Gehl, J., and L. Gemzøe. 1996. *Public Spaces, Public Life*. Copenhagen: City of Copenhagen.

Gilbert, R., and A. Perl. 2008. *Transport Revolutions: Moving People and Freight without Oil*. London: Earthscan.

Ginnell, K., P. Smith, and H.P. Oberlander. 2008. 'Making biggest bigger: Port Metro Vancouver's 21st century re-structuring—Global meets local at the Asia Pacific Gateway', *Canadian Political Science Review* 2, 4: 76–92.

Hargroves, K., and M.H. Smith. 2004. *The Natural Advantage of Nations: Business Opportunities, Innovation and Governance in the 21st Century*. London: Earthscan.

Kenworthy, J.R., and G. Hu. 2002. 'Transport and urban form in Chinese cities: An international and comparative policy perspective with implications for sustainable urban transport in China', *DISP* (Zurich) 151: 4–14.

———, F.B. Laube, et al. 1999. *An International Sourcebook of Automobile Dependence in Cities, 1960–1990*. Niwot, Colo.: University Press of Colorado.

——— and C. Townsend. 2002. 'An international comparative perspective on motorisation in urban China: Problems and prospects', *IATSS Research* 26, 2: 99–109.

Kostof, S. 1991. *The City Shaped: Urban Patterns and Meaning through History*. New York: Little, Brown.

Lee, J.A. 2007. 'Gender, ethnicity, and hybrid forms of community-based urban activism in Vancouver, 1957–1978: The Strathcona story revisited', *Gender, Place & Culture* 14, 4: 381–407.

Mercer. 2008. 'Mercer's Quality of Living Survey Highlights, 2008'. At: <www.mercer.com/qualityofliving#Top_5 _cities_in_each_region>. (7 July 2009)

Meyer, M., and E. Miller. 2001. *Urban Transportation Planning: A Decision-Oriented Approach*, 2nd edn. New York: McGraw-Hill.

Montgomery, C. 2006. 'Sustainable cities: Futureville', *Canadian Geographic* 126, 3: 44–61.

Newman, P. 2007. 'Beyond peak oil—Will our cities collapse?', *Journal of Urban Technology* 4, 2: 15–30.

——— and J.R. Kenworthy. 1999. *Sustainability and Cities: Overcoming Automobile Dependence*. Washington: Island Press.

———, ———, and P. Vintila. 1992. *Housing, Transport and Urban Form*. Background Paper 15 + Appendices for the National Housing Strategy, Commonwealth of Australia, Canberra.

Ortúzar, J.D., and L.G. Willumsen. 2001. *Modelling Transport*, 3rd edn. Chichester, UK: John Wiley & Sons.

Perl, A., and J. Pucher. 1995. 'Transit in trouble: The policy challenge posed by Canada's changing urban mobility', *Canadian Public Policy* 22, 3: 261–83.

Pill, J. 1990. 'Metro's future: Vienna surrounded by Phoenix?', *Toronto Star*, 15 Feb., A27.

Punter, J.V. 2003. *The Vancouver Achievement: Urban Planning and Design*. Vancouver: University of British Columbia Press.

Satterthwaite, D., ed. 1999. *The Earthscan Reader in Sustainable Cities*. London: Earthscan.

Sewell, J. 1972. *Up Against City Hall*. Toronto: James, Lewis and Samuel.

Timmer, V., and N.-K. Seymoar. 2005. *The Livable City*. Vancouver: Vancouver Working Group, International Centre for Sustainable Cities.

Toronto Transit Commission. 2009. '2008 Quick System Facts Toronto'. At: <www3.ttc.ca/About_the_TTC/Operating_Statistics/2008.jsp>. (7 July 2009)

Vancouver, City of. 2009. *2008 Pedestrian Volume and Opinion Survey—Commercial Streets*. March 2009. At: <http://vancouver.ca/engsvcs/transport/guiding Documents/documents/2008PedMain%20Report.pdf>. (7 July 2009)

Williams, H. 1991. *Autogeddon*. London: Jonathon Cape.

Woudsma, C. 1999. 'NAFTA and Canada—US cross-border freight transportation', *Transport Geography* 7: 105–19.

Zielinski, S., and G. Laird, eds. 1995. *Beyond the Car: Essays on the Auto Culture*. Toronto: Steel Rail Publishing.

CHAPTER 12

Neo-Liberal Governance: Entrepreneurial Municipal Regimes in Canada

Ahmed Allahwala, Julie-Anne Boudreau, and Roger Keil

On 1 January 1998 a new era dawned in Canadian municipal politics. The amalgamated 'Megacity' of Toronto emerged from a contested process in which six municipalities and a regional government (the Municipality of Metropolitan Toronto) were consolidated to form one singular city of almost 2.5 million people. The birth of the new city was a watershed in terms of a number of important events. For example, it marked the moment in which Canadian public policy embraced in an unprecedented way the notion that urban regions had become key to the global competitive economy and that they had to be 'dressed up' for success to compete in this context. It also demonstrated that the Canadian tradition of merging local democracy, cultural diversity, and social solidarity, which had made the country's metropolitan centres the envy of observers in the United States and other parts of the world, was going to be put to a serious test if not called into question altogether; and that the shift to a larger municipality in Toronto, which supposedly was to create efficiencies, economies of scale, and concomitant savings in local government involved downloading responsibilities from the province to lower-level government while taking political powers effectively from cities and regions. In response to these pressures, and with the disappearance of classical state-centred managerial urban political options that were called wasteful,

inefficient, and slow, local government started to increasingly entwine its public governance processes with actors and influences from the private sector and civil society.

The Toronto development was played out in similar fashion, albeit in different form, in other municipalities across the country, as communities were forced to adapt to the changing conditions imposed on them by three linked processes. *Globalization* opened markets, civil societies, and governments to the influence of allegedly placeless and hard-to-control flows of capital, culture, and people. *Neo-liberalization* entailed the—ideologically driven—'liberation' of markets from 'constraints' of governmental regulation and workforce 'entitlements'. *Post-Fordism* meant the replacement, to a large extent, of the post–World War II system of 'Fordist' (mass) production and (mass) consumption, in which Taylorist work processes were combined with social compromises between the working class and the employers, resulting in a social wage that was linked to the rate of economic growth and that implied the safeguarding of individual and collective welfare through the Keynesian interventionist state. Instead, with increasingly flexible production processes and social disintegration, even heightened insecurities were considered important preconditions for growth in a post-Fordist regime of flexible accumulation. As a globalizing and neo-liberalizing post-Fordist reality took shape and imposed its discipline, municipalities also experienced increasing pressures from federal and provincial governments to put austerity measures into place. In general, there were three major shifts in city politics during the transition to post-Fordism. First, there is an increased engagement of local authority in economic development. Second, social consumption is being restructured and subordinated to the overall objective of supply-side economic competitiveness. Third, the local political sphere is being expanded to include new actors in less exclusively state-centred bargaining systems. The last point also relates to the proliferation of public-private partnerships as a key feature of entrepreneurial city politics (Kipfer and Keil, 2002; Mayer, 1994). At the same time, it is worth remembering that urban and regional politics operate in a dynamic and complex social space. In fact, urban and regional politics are multiscalar, potentially universalist, and, most importantly, potentially transformational (Kipfer, 1998: 177–8; 2009).

Related to these developments, is the shift from govern*ment* to govern*ance*, which has been viewed as a central feature of urban political realities in the past two decades. It entails the opening of governmental institutions and processes to the influence and the thinking of private actors such as corporations, but also of civil society. Triggered by the neo-liberal idea that states generally fail to safeguard the welfare of citizens and legitimized by the pluralist postulate of involving all relevant stakeholders, **governance** offers the opportunity to many unheard voices to come to the table and be heard. Others have noted the tendency of non-accountable, powerful actors to take advantage of the new possibilities to influence policy. Consequently, it has been pointed out that governance is a barely masked process by which public debate is depoliticized and subjected to the constraints of market efficiencies leading to an effective diminution of political debate (Keil, 1998; Swyngedouw, 2005; Walters, 2004).

The Changing Political-Economic Context of Urban Governance

Since the early 1970s, the political economy and geography of advanced capitalist nations have undergone significant transformation. These shifts have fundamentally altered not only the political and institutional context of urban governance but also its overarching objectives (Brenner, 2004). In this restructuring process, urban and regional governments have often been at the receiving end of decisions made at higher levels of government or larger scales of economic rationality but they have also been actors and drivers of globalization, neo-liberalization, and the shift towards post-Fordism. Cities and regions have been victims of, actors in, and arenas for these processes all at the same time as metropolitan regions have grown explosively and traditional city-suburban distinctions have waned (Hoffmann-Martinot and Sellers, 2005: 11). This extra-metropolitan expansion requires reform of the centre-suburban logic of traditional urban government and an expansion of governance to the diffused urban region in order to deal with complex material policy issues such as transportation, economic development, housing, welfare, and environmental protection. It is at the heart of debates on regional growth and sustainability (Filion, 2010). The political fragmentation that results from this situation is considered a problem of public policy and is to be addressed by the institutions of neo-liberal urban and regional governance that have been put in place in recent years (Kübler and Heinelt, 2005). For the most part, this is the situation we find in the exploding urban regions of Canada, where the boundary and scale-stretching effects of the globalizing space economy and accompanying demographic expansion and diversification have led to dramatic tensions between territory and government. Accordingly, local and regional states across Canada have been attempting to find variegated solutions to the governance of these incongruent political spaces.

These developments point to new territorial structures as well as politics and everyday practices that contest official urban and regional spatial fixes (Harding, 2007; Jonas and Ward, 2007).

We cannot offer a detailed account of these momentous shifts here but, in broad strokes, we can provide an outline. The general starting point is that the specific form of market economies present in Western countries after 1945 was **Fordism**. British social theorist Bob Jessop (2002: 55) defines Fordism as an 'accumulation regime based on a virtuous autocentric circle of mass production and mass consumption secured through a distinctive mode of regulation that was discursively, institutionally, and practically materialized in the Keynesian welfare national state.' A key objective of Keynesian regulation was to alleviate uneven economic growth at the subnational level. This 'spatial Keynesianism' had redistributive and stabilizing effects on regional and local economies (Brenner, 2004). Despite significant cross-national differences in the specific nature of intergovernmental relations and the constitutional standing of municipalities vis-à-vis higher levels of government, cities were an integral part of the governing matrix of the national Keynesian welfare state. Within this framework of economic, social, and territorial regulation, cities played a somewhat passive role and were largely responsive to the policy directives of higher levels of government (Castells, 1972; Peterson, 1981).

As a response to the deep economic crisis of the early 1970s, the political support of the business sector for the Keynesian social and territorial compromise and the institutional arrangements supporting it weakened significantly. In an effort to restore profitability, corporations, conservative think-tanks, public intellectuals, and increasingly politicians (Margaret Thatcher, Ronald Reagan, Brian Mulroney) embraced neo-liberal ideas of economic and social regulation and advocated for deregulation, privatization, and decentralization (Harvey, 2005). The new economic mantra coincided with an aggressive push to internationalize economies and to open national markets to globalized flows of increasingly financialized capital, which ostensibly lowered the control of national governments and other economic actors over capital flows and favoured so-called 'global cities' as the new centres of the world economy (Sassen, 1991). Small capitalists and large corporations alike combined the new availability of globalized (venture) capital and the deregulation of (labour) markets to introduce new production processes. The 'virtuous' cycles of the Fordist compromise were replaced with new regionally constituted economic accumulation regimes which were both globalized and localized, based as much on the infusion of international capital and unbounded markets as on local conditions in regional production networks (Storper and Scott, 1989).

In contrast to the spatially redistributive character of the Keynesian welfare state, current state-spatial management is 'spatially selective' (Jones, 1997). Differences and even competition among urban regions are now considered a necessary and virtuous precondition to successful economic and social development (Brenner, 2004). The overarching objective is no longer to alleviate uneven development subnationally but to strengthen the economically powerful regions of the country and to position them beneficially in the global economy ('metropolitanization') (see Bunting and Filion, Chapter 2; Hall, Chapter 4). Canada's economic territorial management, for example, has moved from a paradigm of equal opportunity for all to a paradigm that allows policy-makers to pick winners. In many ways, this regime favours Toronto as the country's 'economic engine' (Boudreau et al., 2007: 43) although other territorial interests compete with Toronto's predominance.

Neo-liberalization in particular was an important force in this transformation. It refers to the process of geographically specific institutional adaptation and political struggles associated with a new market-dominated governing paradigm

(Bakker and Gill, 2003: 25). **Neo-liberalism** as a hegemonic governance framework structures contemporary processes of urbanization across the world. These urban processes, in turn, reinforce the hegemonic character of neo-liberalism as a global governance framework. How the two relate has been of particular interest to urban scholars and social theorists since the onset of the recent global financial and economic crisis when the dogma of neo-liberalism was put to the test on several fronts (Brenner, Peck, and Theodore, 2009; Harvey, 2009; Keil, 2009; Wallerstein, 2008).

The neo-liberalization of state and society can be understood as a process of *creative destruction*. Peck and Tickell (2002: 37) have differentiated between two phases of neo-liberalization. The initial *roll-back phase* of the 1980s was mostly concerned with dismantling the multi-scalar institutional architecture of the Keynesian welfare state. At the municipal level, this translated into severe funding cuts for social programs and the privatization of urban infrastructures, or, more broadly speaking, the subordination of collective consumption—a key task of municipal government in the post-war Keynesian era—to the imperatives of market-driven growth. In the 1990s, the *roll-out phase* of neo-liberalism was an attempt to establish and consolidate new regulatory mechanism in an attempt to mediate and alleviate some of the contradictions and tensions triggered by the earlier phase of urban entrepreneurialism (Mayer, 2007: 91). We might now add a third concept to roll back and roll out: *roll-with-it neo-liberalization*. In this phase, neo-liberal subjects, even critical ones, 'co-construct, sustain and also contest a now normalized neo-liberal social reality' (Jessop, 2008; Keil, 2009). While in the first instance, the three concepts—roll-back, roll-out, roll-with-it—refer to distinct historical periods, they also point to coexisting moments and emerging contradictions in neo-liberalization. Neo-liberal thinking has become a general point of reference for urban political debate and governance. Simultaneously, the unfolding crises of

neo-liberalized regulation call for new directions in and beyond neo-liberal policies that involve 'reformed' neo-liberal elite practices often in direct reaction to widespread contestation and social unrest (Harvey, 2008; Keil, 2009). (On the social consequences in Canadian cities of the shift to neo-liberalism, see Walks, Chapter 10.)

Entrepreneurial Urban Governance in Neo-Liberal Policy Landscapes

Early debates about globalization suggested that advances in communication and information technology would render questions of geography—and of urbanization more specifically—increasingly irrelevant. Partly in response to these debates, new thinking emerged in economics and geography that stresses the continuing, if not heightened, importance of place—most importantly of city-regions—in an age of economic globalization (see, e.g., Scott, 2001). Despite significant theoretical and methodological differences, these literatures constitute a powerful response to the 'death of distance' argument (Cairncross, 1997) that has been considerably discredited in recent years. In the planning and governance field, a *new regionalism* has taken hold that advances the idea that metropolitan regions are the core of the globalizing economy and require new forms of regionally specific governance (Brenner, 2002; Deas and Ward, 2000; MacLeod, 2001; MacLeod and Goodwin, 1999; for Canada, see Sancton, 2001, 2005). In the economics literature, *new growth theory* has emphasized the importance of city-regions as strategic economic clusters and sites of diversity that stimulate innovation and growth. Interestingly, the work of Canadian urban theorist Jane Jacobs has been particularly influential in this context (1969, 1984; see also Nowlan, 1997). Similarly, the *institutionalist economic geography* associated with the Los Angeles School sees regions as sites of untraded interdependencies (Scott, 1988; Storper, 1997). Finally, John Friedmann and Saskia Sassen laid the groundwork for a burgeoning literature that emphasized the importance of *global cities* as strategic co-ordinating sites of the global economy (Friedmann, 1986; Sassen, 1991; for an overview, see Brenner and Keil, 2006).

This new thinking points to a globalism–regionalism nexus and the constructed relation between a globalized economy and city-regions as functional territorial units within it. Consequently, city-regions are revalorized as strategic sites for economic regulation and political governance (for a general overview, see MacLeod and Jones, 2007; for a neo-liberal version of this argument, see Ohmae, 1995; for the Canadian debate, see Courchene, 2007). If urban and regional spaces are the core of the new globalized economy, the role of city-regions has become more important and politics related to these spaces are considered central. There is now a concerted effort to find post-national solutions to regulatory problems and the urban and regional level is identified as an increasingly important site for economic governance and social regulation. The role of central government in regional and local planning has changed accordingly (Cox, 2005). The question that has arisen from these insights is whether urban and regional politics are nested—although not in an uncontested way—at a particular place in a scaled system of governance (Brenner, 2004) or whether regions and their politics are 'unbounded' territorially. Jones and MacLeod offer the distinction between economic flows as an unbounded space and the space of political action more likely bounded and territorial (Jones and MacLeod, 2004: 437; see also Painter, 2008: 347). Following this line of thinking, there is continuous tension here between notions of territorial boundedness and scales, on the one hand, and more relational, scalar ideas of regional space, on the other. It is, in fact, from this ongoing tension that the need for political regulation of regional space emerges in the first place.

The inward-looking posture of much municipal politics during the Keynesian period is replaced by an outward-looking strategy. The overarching objective of contemporary entrepreneurial city politics is to secure the competitiveness of the city vis-à-vis global flows of capital and people. The economic policy emphasis now falls on competition and innovation rather than on full employment and planning (Jessop, 2002: 459). Accordingly, we have seen the emergence of new *economic localisms*, i.e., the mobilization of city space for economic growth and competitiveness even at the accepted cost of greater socio-economic disparities. This gives birth to four strategies of the 'entrepreneurial city' to compete internationally (Harvey, 1989: 11): First, the exploitation of particular (endogenous) advantages for the production of goods and services; second, the introduction of a new spatial division of consumption (e.g., **gentrification**); third, a fierce struggle over the acquisition of key control and command functions in high finance, government, or information-gathering and -processing; and, finally, the seeking of a competitive edge with respect to the redistribution of surpluses through central governments. All of these aspects could be observed in Canadian urban governance in the past two decades but one added particularity of urban entrepreneurialism in this country—as a result of the constitutional subordination of municipalities—is its focus on improving intergovernmental relations and the inclusion of cities in policy decisions relevant in and for cities.

What, then, are the characteristics of urban development and city politics during the period of globalization, neo-liberalization, and post-Fordist restructuring? One of the key aspects in cities' strategies to attract new consumers, new highly skilled workers of the knowledge economy, and new businesses is a focus on the quality of life. As opposed to the welfarist vision of the 'standard of living' of the Fordist period, the idea of quality of life in the neo-liberalizing period privileges first and foremost the availability of consumer choices. City dwellers are offered a choice of features that, in aggregate forms, are considered to provide a distinctive quality of life. Public spaces are remodelled into open-air shopping malls, focusing on outside cafés and the creation of ambiances. This corresponds to the idea of a commodified, festive, and controlled urban environment (Zukin, 1995). Ethnic neighbourhoods are marketed for their distinctive flair. Cultural events and entertainment become central to urban marketing strategies. Mega-projects such as waterfront developments, high-cultural institutions (opera houses, museums, etc.), and sports stadiums are seen as priority investments by municipalities. These are mostly realized through public–private partnerships.

In this context, branding strategies, particularly around the motto of creativity, are mobilized. The work of Richard Florida has provided a powerful framework for linking cultural development policies and urban economic competitiveness. In his wildly successful book, *The Rise of the Creative Class* (2002), Florida argues that we have entered a new era of knowledge-based production driven by the 'creative class', referring to people with high educational backgrounds who work in 'creative' industries such as arts and design, fashion, and new media. These people are extremely mobile and prefer places that are rich in lifestyle amenities, characterized by ethnic diversity, green and cycling-friendly neighbourhoods filled with cafés, patios, and art galleries (see Vinodrai, Chapter 6). With this creative city discourse, the pessimism of the neo-liberal rollback period is replaced by an optimistic technological futurism that exemplifies the roll-with-it period (Peck, 2005). The local state is re-legitimated as an enabler of creativity (Grundy and Boudreau, 2008; Lehrer, 2005; Lehrer and Laidley, 2008; Peck, 2005).

Despite these common features, a neo-liberal model of urban governance offers no single path to entrepreneurial governance and no unidirectional convergence thereto. The institutional and political response of cities to global neo-liberalism is variegated and path-dependent and comparative analyses of neo-liberal urban governance need to take into account the contextually specific pathways of regulatory restructuring as well as the translocal commonalities of the neo-liberal project. Jessop (2002) proposes a useful typology to facilitate a comparative analysis of 'actually existing' neo-liberalization (Peck and Tickell, 2002). In addition to neo-liberal adjustment strategies, which encourage market-led economic and social restructuring, Jessop proposes three ideal-typical responses to global neo-liberalism: First, neo-statism 'involves a market conforming but state-sponsored approach to economic and social restructuring whereby the state seeks to guide market forces in support of a national economic strategy.' Second, neo-corporatist responses to global neo-liberalism 'involve a negotiated approach to restructuring by private, public, and third-sector actors that aims to balance competition and cooperation'. Finally, neo-communitarian strategies 'emphasize the contribution of the "third sector" and/or the "social economy" (both located between market and state) to economic development and social cohesion' (Jessop, 2002: 461–3). In the remainder of this chapter, we will use this typology to highlight the path-dependent and variegated nature of the political response to neo-liberalism in Canada's two largest city-regions, Toronto and Montreal. (See Table 12.1 for a synopsis of the argument of this chapter.)

Spaces of Neo-Liberalization in Canada

In Canada, as elsewhere, the very notion of what constitutes metropolitan governance has undergone changes during the period of globalization, neo-liberalization, and the shift to post-Fordism. While in most cases, as Frisken (2007) has found for the history of governance in Toronto, the provinces have remained the decisive players in regional policy (with no significant new regionalized

Table 12.1 Transitions Affecting Municipal Regimes

Regime of Accumulation Fordism	→	Post-Fordism
International Regime American hegemony, national models	→	New global order, a world of regions
Mode of Political Regulation Keynesianism	→	Neo-liberalism
Mode of Social Regulation/Societalization Collectivization/class compromise	→	Individualization/deregulation
Mode of Spatial Regulation Spatial Keynesianism	→	Spatial (strategic) selectivity
Mode of Urban Regulation Urban managerialism	→	Urban entrepreneurialism

institutions put in place), the boundaries of regions have been redefined geographically and politically, and internal conflict among actors has characterized the search for new modes of regional governance, which have been part of a more general process of state re-scaling in Canada (Boudreau et al., 2006, 2007).

The emergence of **entrepreneurial municipal regimes** and the increased 'policy activism' of Canadian municipalities and city-regional actors more generally can be interpreted as a response to neo-liberal state restructuring and the municipal fiscal crisis this has engendered. The apathy of both federal and provincial levels of government towards urban issues and the absence of a national urban policy in Canada prompted urban-based actors (mayors, business leaders, etc.) to launch a campaign to improve the standing of Canadian municipalities within the Canadian federation. Their aim was initially to solve the fiscal crisis of Canadian cities and metropolitan regions and to raise their power vis-à-vis higher levels of government. The initial push for a new deal for cities by big city mayors and the Federation of Canadian Municipalities under the presidency of Jack Layton, as well as by many civic organizations and

activists, which was taken up later by the federal government of Paul Martin supported by the NDP opposition, had many progressive elements and was explicitly understood as a counterpoint to upper-level governments' disinvestment in urban infrastructures, housing, and social welfare (such policy activism remained fairly weak in Montreal). Metropolitan regions such as Toronto and (much less so) Montreal became uneasy companions and often sites of vocal opposition to the neo-liberalization of Canadian society and state. Let us now have a closer look at these two cases.[1]

Toronto

In Ontario, the Progressive Conservative government under Mike Harris was at the forefront of neo-liberal state restructuring in Canada. The so-called Common Sense Revolution (CSR) was the policy cornerstone of the regime that Harris and his successor, Ernie Eaves, imposed on the province between 1995 and 2003. This platform combined market liberalism with authoritarian measures that created long-term societal change in Ontario. Some of the most notorious restructuring efforts of the time included: welfare cuts and the introduction of a punitive provincial workfare program; the euphemistically labelled Safe Streets Act directed against the presence of squeegee kids and panhandlers; the reduction, consolidation, and redesign of local government; cutbacks in the number of provincial staff; the legalization of the 60-hour work week; the deregulation of planning leading to intended (sub)urban sprawl; the elimination of funding for public housing; the deregulation of environmental protection; attacks on trade union rights; the systematic hollowing out of public education; etc. (Boudreau, Keil, and Young, 2009).

Since 2003, a new Liberal government has adopted a less aggressive form of neo-liberalization. It has rescinded some of the Harris reforms and introduced more subtle but perhaps even more far-reaching policies affecting urban life. Premier Dalton McGuinty's government has expanded the notion of 'the region' to include all of the Greater Golden Horseshoe. The framework for this policy shift is set by new planning legislation, which outlines and protects a Greenbelt on the one hand and identifies Places to Grow for intensified urban development on the other. Internally, the Liberals have responded to the socio-spatial and infrastructural bottlenecks of the region by funnelling conflicts around transportation planning into a regulated institutionalized process of governance through the founding of the Metrolinx regional transportation agency. This new regional planning body has now started to lay out a detailed list of projects that will change the transportation landscape in southern Ontario. The push by Metrolinx for more physical infrastructures of mobility is met with the TransitCity plans (for an extensive light-rail transit scheme) devised by the City of Toronto, although political and philosophical contradictions persist between politicians in the central city and their regional and suburban counterparts (Keil and Young, 2007).

After amalgamation, Toronto's civic activists and economic elites regrouped and adapted to the newly created socio-territorial realities. There was

no appetite for a sustained struggle to bring the old Toronto governance struggle back. Instead, social justice and environmental activists, political opinion-makers, urban thinkers, city hall insiders, and others contemplated other ways forward to establish new forms of governance that would address the chief issues of the new situation. Among the most influential ideas floated at the time was that of a city charter. Proponents of a charter demanded new rights and powers for Toronto to raise revenues, autonomy in how it conducts its business and delivers services, and the freedom to enter direct relationships with the federal government, as well as the private and non-profit sectors. A variety of business, civic, and state actors were part of this discussion. On the business side, the Toronto Board of Trade espoused policy notions that favoured urban autonomy in the face of growing challenges. On the civic side, the web-based newsletter *Local Self-Government* provided a forum for debate. But social movement and advocacy groups also participated in the debate. The Toronto Environmental Alliance (TEA), for example, published a pamphlet called *Towards a Greater Toronto Charter and the Environment* in the year 2000. In the same year, the City of Toronto's Chief Administrator's Office in a report requested that Toronto be granted a charter. The particular push towards a charter and more local and regional autonomy did not lead to a concrete result. Yet, these discussions in and outside the several levels of the local and provincial states ultimately led to the City of Toronto Act in 2005, which regulates the relationships between Toronto and Ontario in a new way. It gives the City a limited amount of freedom in creating fees and taxes to fund programs, and it changes the modalities under which the mayor governs the City (Boudreau, Keil, and Young, 2009: ch. 4).

Meanwhile, another process took hold that has been very much in line with the principles of neo-liberalized urban governance by which state-centred power is somewhat disseminated into areas of private and civic authority. In June 2002, the first

Toronto City Summit took place at the initiative of Mayor Mel Lastman. The result of this was the creation of the Toronto City Summit Alliance (TCSA), self-described as a coalition of 'civic leaders in the Toronto region' (www.torontoalliance.ca). While the TCSA speaks of itself as an inclusive organization, it is dominated by civic elites, despite the participation of two labour groups and the United Way. It has gained high visibility in Toronto and even on the Canadian national scene, perhaps because of the resources it can access through its elite membership. The TCSA policy priorities are related to fiscal arrangements to the benefit of cities: infrastructure development (especially regional transportation and the waterfront); tourism and city marketing; research and commercially viable knowledge-production; the integration of skilled immigrants in the economy; affordable housing; and arts and culture. The TCSA embraces a vision of Toronto that is decisively regional. It can be characterized as a new 'civic regionalism', dominated by the business and the philanthropic sector (e.g., the Maytree Foundation). This is not surprising given that Ontario has a long tradition of civic philanthropy and that Toronto is home to the most important financial firms and industrial corporations in Canada.

Following Jessop's typology, Toronto's new civic regionalism could qualify as a 'neo-corporatist' form of governance that 'involves a negotiated approach to restructuring by private, public, and third-sector actors and aims to balance competition and co-operation. It is based on commitment to social accords as well as the pursuit of private economic interests in securing the stability of a socially embedded, socially regulated economy' (Jessop, 2002: 462). Toronto has evolved since 2003 (the election of David Miller as Mayor; the defeat of the Tories at the provincial level) towards a governance regime of 'neo-reformism' characterized by the merging of centre-left progressivism inherited from the 1970s and neo-liberal initiatives aimed at enhancing economic competitiveness (Boudreau and Keil, 2006). The TCSA, which

is reminiscent of the traditional 'red Tory' political culture of conservatism in Ontario, fits well into this framework. One of the most important aspects of neo-corporatist governance is the multiplication of actors, in contrast to traditional corporatism under Fordism (trade unions, interventionist state, big business). These actors work as a network to frame policies and ensure that their implementation becomes voluntary and that they self-regulate their activities. This has the effect of depoliticizing urban governance. With its emphasis on stakeholders' language, collaboration, and civic consensus, the TCSA has effectively softened and made toothless the confrontation of political ideas on which the democratic electoral system is based.

Montreal

The shift to a neo-liberal policy environment occurred more subtly in Montreal, with the discrete involvement of the federal government in the restructuring of a metropolis harshly struck by the economic crises of the 1980s and early 1990s. Responding to the city's inability to provide basic services, the federal government created the Consultative Committee on the Development of the Area of Montreal (1986), which initiated a process of strategic planning (known as the Picard Report). This shifted political horizons in Montreal towards neo-liberalism (Boudreau et al., 2006). Despite traditionally tensed relations between the federal and the Quebec governments, particularly on municipal issues (an exclusive jurisdiction of the province), the Quebec Liberal government of the time was in favour of such a process because it expected to receive financial transfers for Montreal. Ottawa was thus freer to use its spending capacity in response to the demands of municipal and provincial elected officials who were confronted with an unprecedented increase of the unemployment rate. The Picard Report insisted on investing in winning sectors and neighbourhoods, a decisively neo-liberal shift from the territorial redistributive policies of the past. The main strategy was to internationalize Montreal's economy by attracting international organization headquarters and developing the sectors of information and communication technologies, biotechnologies, and aeronautics.

At the provincial level, the government did not hesitate to use public enterprises to shape the economy. Through tax devices or by forcing public enterprises such as the Société Générale de Financement and the Caisse de Dépôts et de Placements du Québec to invest in Montreal's economy, the Quebec government was actively shaping Montreal's metropolitan landscape.

At the municipal level, a regime dependent on public funds coalesced, supported by major trade unions in the construction sector, which was hit at the beginning of the 1990s by the international crisis in real estate. By the mid-1990s, both the federal and provincial governments offered tax subsidies to priority high-tech economic sectors. Progressive social movements at the time did not oppose this neo-liberal shift. Nevertheless, they thrived in terms of number and visibility and were able to elect the Montreal Citizens Movement to City Hall as of 1986, at the cost of their subversive capacity (Lustiger-Thaler and Shragge, 1998).

While the Common Sense Revolution was profoundly shaking Ontario, the Picard Report in Montreal was structuring more discretely but as forcefully a multi-scalar urban regime, an alliance of all three levels of government with a plurality of private actors. The strategy consisted of stimulating the service sector in the central city (mostly through office space construction supported, of course, by the construction workers unions), while subsidizing the development of the knowledge economy in the suburbs. This legitimization of post-Fordism was eased by the striking absence of socio-political contestation, as opposed to Toronto's more polarized environment. Social movements remained quite active at the neighbourhood scale, but emphasized social service provision rather than protest.

All cities on the Island of Montreal were amalgamated in 2002, while the Montreal Metropolitan Community (responsible for city-regional planning and co-ordination) and boroughs within the amalgamated city were simultaneously created. The Montreal Metropolitan Community did not raise much public debate before and after its implementation. It is very weak because of its restricted budget and conflicts with a plethora of other institutions that have overlapping jurisdictions. It has largely been a failure (Boudreau and Collin, 2009).

As in Toronto, mergers were forcefully opposed, most notably (but not exclusively) in anglophone municipalities. This struggle contributed to the defeat of the Parti Québécois in the provincial elections of 2003. The newly elected Liberal government immediately introduced legislation enabling municipalities to demerge. Many of them did in 2004. Bill 9 created a Council of Agglomeration composed of representatives of both the reconstituted cities and the central city in order to clarify the division of labour between the overlapping institutions created by this complex process. Through this Council of Agglomeration, the remaining 'megacity' still holds much power, being responsible for land evaluation, public safety, the elimination and valorization of residual waste, public transit, streets and roads management, **social housing**, etc., even in de-amalgamated municipalities. These responsibilities represent a budget twice as large as that of the former Communauté Urbaine de Montréal. Despite having recovered their municipal status, therefore, suburban municipalities were considerably weakened.

The institutional structure of Montreal remains highly decentralized. Boroughs were reinforced with Bill 33, the Act to Amend the Charter of Ville de Montréal (2003), replacing borough presidents with borough mayors, while increasing their borrowing capacities. These boroughs, more and more, are the direct interlocutors of social movements (Alain, 2007). Paradoxically, Bill 33 increases the power of the boroughs' elected officials and closes more doors to the channels of direct citizen involvement that were gained from the 1980s by the Montreal Citizens Movement (in power, 1986–94). Mayor Gérald Tremblay's party, the Union des citoyens de l'île de Montréal, was created from the old Montreal Citizens Movement and was elected on the promise of decentralizing the amalgamated city structure. He did so at the expense of the traditional channels of participation for civil society actors. This change was exacerbated by the creation (by the newly elected provincial government) of the Conférence Régionale des Élus (CRE), which contributes to the political fragmentation of the city-region. These CREs replace the Regional Councils of Development created in the mid-1990s. The previous structure was conceived as a deliberative model incorporating civil society. But in an attempt to transform this social democratic political culture, the Liberal government of Quebec decided to diminish the influence of trade unions and all civil society components.

Thus, while amalgamation in Toronto was followed by a regional convergence towards more power for the city (crystallized in the charter debate), the scene in Montreal was further fragmented. In June 2002, Montreal, too, organized a summit: the second Citizens' Summit of Montreal. The difference with Toronto's TCSA was that it was established and controlled by a network of community organizations centred around the Centre for Urban Ecology (working on participatory democracy and sustainable development), leftist academics (both French- and English-speaking), and Alternative (an international solidarity organization very involved locally). In June 2009, the fifth summit took place. Thus, it has become an important tradition of political mobilization in Montreal, which also has international connections. For instance, the AITEC (Association internationale de techniciens, experts et chercheurs), a Paris-based organization mobilizing on urban issues, has imitated the Montreal model to initiate debate on regional governance reform in the French capital.

Unlike the TCSA, the Citizens' Summit does not have a permanent staff and does not act as a well-defined collective actor on the political scene between the yearly summits. Governance in Montreal is characterized mostly by its lack of coherent regionalism. Other civic actors have been more successful in acting regionally, such as Montreal International (focused on business attraction) and Culture Montreal. While the former is far from being as important politically as the TCSA, the latter has a significant impact on Montreal politics, although it is mostly restricted to the scale of the city and not to the metropolitan area. Montreal's civil society activities are not turned towards philanthropy and business actors are not as well organized. Instead, a closely knit network of small community organizations, heavily dependent on provincial funding, characterizes Montreal's civil society.

Hence, the public debate in Montreal is not structured regionally as it is in Toronto under the impetus of the TCSA. Even the yearly Citizens' Summit is focused on micro-local issues. The themes revolve around democratic participation, social justice, the environment, the economy, and culture. While these could be regional in scope, they are almost always examined through a very local lens. This micro-localism can be explained by the fact that political debates in Montreal have long been organized by a network of community organizations that do not work very well together. In the 1960s and 1970s, like the reform movement in Toronto, they coalesced on the municipal electoral scene under the umbrella of the Montreal Citizen Movement. But unlike Toronto's evolution towards neo-reformism under the leadership of David Miller, amalgamation in Montreal has meant the end of that municipal party and its fragmentation into hundreds of community organizations. In the 1970s, these community organizations were created with strong support from the provincial government, which was, at the time, constructing a **welfare state** and seeking autonomy from the heavy hand of the Church, in the wake of the 1960s Quiet

Revolution. With the neo-liberal turn of the 1980s and 1990s, the provincial government downloaded many responsibilities to municipalities and community organizations became increasingly involved in service provision (compensating for what the province and municipalities used to do). This has meant an increasing dependence on the provincial and local governments in terms of funding and performance reviews. It has also meant a turning inward to the neighbourhood scale (proximity services) and the abandonment of a socially transformative agenda that could be more regional in scope (Boudreau et al., 2008). In addition, the multiple municipal institutional reforms between 2001 and 2005 (and still ongoing!) have paradoxically decentralized governance despite amalgamations.

Another crucial governance institution in Montreal is the neighbourhood forum (Table de concertation). In 2006, United Way, the Public Health Division, and the City of Montreal decided to invest annually $1.65 million to finance 30 forums in poorer neighbourhoods in order to work towards social local development (understood as pertaining to health, urban planning, environment, education, economy, housing, transportation, safety, employment, food security, culture, and recreation). While its budget is not very important, the program has had the effect of structuring the political debate on quality of life and all urban issues, including those related to globalization (employment, for instance) at a micro-local level. As there is virtually no regionally coherent discourse on neo-liberalism and economic competitiveness, community organizations in Montreal are not oriented towards a confrontation with such an elite-driven collective actor as the TCSA. They are locally oriented towards collective consumption issues and, just like in Toronto, the result is a depoliticization of urban governance. Indeed, working on local service provision, community organizations meet regularly in their neighbourhood forum, which functions on the principle of consensus and pragmatic problem-solving. Revealing that a forum is entangled in a

confrontation of ideas and conflicts can jeopardize its funding. Close relationships with the boroughs also favour the rise of a new clientelism.

In sum, neo-liberalism in Montreal is closer to what Jessop (2002) calls neo-communitarianism, characterized by an enhanced role for civil society (not only in its elitist form as in Toronto), an expansion of the social economy, and an emphasis on use-value, as well as its role as a buffer against free competition. However, given the long history of close links between community organizations and the provincial state, this neo-communitarianism is combined with elements of neo-statism. Most strikingly, the new relationship is characterized by increased auditing of performance and partnership under the guidance of the central city and the provincial state. Despite these centralizing features of the post-Keynesian regime in Montreal, the absence of a regional political space in the metropolis produces a highly fragmented, service-oriented, and even conformist (rather than socially transformative) regime.

Conclusion

Canadian municipalities and regions have been strongly affected by neo-liberal state restructuring and the austerity measures introduced both by provincial and by federal governments. This restructuring has contributed to the crisis of Canadian municipalities and the fiscal imbalance/fiscal squeeze. But cities were not just passive *sites* for processes initiated elsewhere (such as at higher levels of governments and scales of social organization) and *recipients* of downloaded responsibilities of social policy. They were the places where the 'dirty work' of globalization and neo-liberalization was done, pushing agendas of social welfare reform, administrative efficiencies, and so forth (Keil, 2000). Partly, this is a direct function of their weak position in the Canadian state architecture. Dependent entirely on provincial policy and largely also funding as they have no autonomous constitutional standing, Canadian municipalities have no source of income over which they have full control. They rely primarily on property taxes, user fees, and intergovernmental transfers and typically have no access to the types of taxes that grow with the economy. The strains associated with this situation have made cities innovators of governance tools that often anticipate or, at a minimum, dovetail with larger-frame neo-liberal policy—cost-cutting, efficiency-finding, performance-oriented, market-directed public policy has been the name of the game in municipal politics for at least two decades.

While now generalized and global, the process of neo-liberalization generates path-dependent outcomes and, as such, neo-liberalism is a place-, territory-, and scale-specific project (Brenner and Theodore, 2002). The forms that these global transformations take at the scale of city-regions are not identical, but rather reflect the historical geographies, political economies, and traditions of state–society relations in particular locales. Cities and city-regions are adapting to and co-producing neo-liberalism in distinct ways. As we illustrated with two case studies, Toronto and Montreal, local processes of responding to the demands of a global neo-liberal governance framework do not result in a uniform neo-liberal model of urban governance across the world, or even across Canada.

Acknowledgements

Funding for the research that underlies this chapter was provided by the Social Sciences and Humanities Research Council through a project on regional governance in Toronto, Montreal, Paris, and Frankfurt (Roger Keil, Principal Investigator, with Julie-Anne Boudreau, Pierre Hamel, and Stefan Kipfer).

Review Questions

1. What are the main features of neo-liberalism? How does neo-liberalism affect Canadian cities?

2. How are globalization, neo-liberalism, and post-Fordism interconnected? What is their joint impact on cities?

3. Compare responses to neo-liberalism in Toronto and Montreal. How do you explain the differences between the two urban areas?

Note

1. For a discussion of the municipal amalgamation phenomenon, see Andrew Sancton, 'City Politics: Municipalities and Multi-Level Governance', Chapter 17 in Bunting and Filion (2006), at <www.oupcanada.com/bunting4e>.

References

Alain, M. 2007. 'Gouvernance infra-municipale et gestion du social: Analyse multi-échelons de la construction politique des arrondissements montréalais', Ph.D. dissertation, Montreal, INRS–UCS.

Bakker, I., and S. Gill. 2003. 'Ontology, method, and hypotheses', in I. Bakker and S. Gill, eds, *Power, Production and Social Reproduction*. London: Palgrave Macmillan.

Boudreau, J.-A., and J.-P. Collin. 2009. 'Épilogue: L'espace métropolitain comme espace délibératif?', in G. Sénécal and L. Bherer, eds, *La métropolisation et ses territoires*. Québec: Presses de l'Université du Québec.

———, A. Germain, A. Rea, and M. Sacco. 2008. *De l'émancipation à la conformité culturelle? Changements de paradigme dans l'action sociale dans les quartiers défavorisés à Bruxelles (Belgique) et à Montréal (Québec)*. Montreal: QMC–IM (Working paper No.31).

———, P. Hamel, B. Jouve, and R. Keil. 2006. 'Comparing metropolitan governance: The cases of Montreal and Toronto', *Progress in Planning* 66, 1: 7–59.

———, ———, ———, and ———. 2007. 'New state spaces in Canada: Metropolitanization in Montreal and Toronto compared', *Urban Geography* 28: 30–53.

——— and R. Keil. 2006. 'La réconciliation de la démocratie locale et de la compétitivité internationale dans le discours réformiste à Toronto: Essai d'interprétation sur le néolibéralisme normalisé', *Politiques et Sociétés* 25, 1: 83–98.

———, ———, and D. Young. 2009. *Changing Toronto: Governing Urban Neo-liberalism*. Toronto: University of Toronto Press.

Brenner, N. 2002. 'Decoding the newest "metropolitan regionalism" in the USA: A critical overview', *Cities* 19: 3–21.

———. 2004. *New State Spaces: Urban Governance and the Rescaling of Statehood*. Oxford: Oxford University Press.

——— and R. Keil, eds. 2006. *The Global Cities Reader*. London: Routledge.

———, J. Peck, and N. Theodore. 2009. 'Variegated neo-liberalization: Geographies, modalities, pathways', *Global Networks* 9.

——— and N. Theodore, eds. 2002. *Spaces of Neo-liberalism: Urban Restructuring in North America and Western Europe*. Oxford: Blackwell.

Cairncross, F. 1997. *The Death of Distance*. Boston: HBS Press.

Castells, M. 1972. *La question urbaine*. Paris: Maspero.

Courchene, T. 2007. 'Global futures for Canada's global cities', *IRPP Policy Matters* 8, 2.

Cox, K. 2005. 'The politics of local and regional development', *Space and Polity* 9: 191–200.

Deas, I., and K. Ward. 2000. 'From the "new localism" to the "new regionalism"? Interpreting regional development agencies', *Political Geography* 19: 273–92.

Filion, P. 2010. 'Reorienting urban development? Structural obstruction to new urban forms', *International Journal of Urban and Regional Research* 34: 1–19.

Florida, R. 2002. *The Rise of the Creative Class*. New York: Basic Books.

Friedmann, J. 1986. 'The world city hypothesis', *Development and Change* 17: 69–83.

Frisken, F. 2007. *The Public Metropolis: The Political Dynamics of Urban Expansion in the Toronto Region, 1924–2003*. Toronto: Canadian Scholars' Press.

Grundy, J., and J.-A. Boudreau. 2008. '"Living with culture": Creative Citizenship Practices in Toronto', *Citizenship Studies* 12: 347–63.

Harding, A. 2007. 'Taking city regions seriously? Response to debate on "city-regions": New geographies of governance, democracy and social reproduction', *International Journal of Urban and Regional Research* 31: 443–58.

Harvey, D. 1989. 'From managerialism to entrepreneurialism: The transformation in urban governance in late capitalism', *Geografiska Annaler B* 71: 3–17.

———. 2005. *A Brief History of Neo-liberalism*. Oxford: Oxford University Press.

———. 2008. 'The right to the city', *New Left Review* 53 (Sept.–Oct.): 23–40.

———. 2009. 'Why the U.S. stimulus package is bound to fail', *The Bullet. Socialist Project E-Bulletin* No. 184, 12 Feb.

Jacobs, J. 1969. *The Economy of Cities*. New York: Random House.

———. 1984. *Cities and the Wealth of Nations*. New York: Random House.

Jessop, B. 2002. 'Liberalism, neo-liberalism, and urban governance: A state-theoretical perspective', *Antipode* 34: 452–72.

———. 2008. 'The crises of neo-liberalism, neo-neo-liberalism and post-neo-liberalism', presentation, University of Toronto, 19 Sept.

Jonas, A., and K. Ward. 2007. 'Introduction to a debate on city-regions: New geographies of governance, democracy and social reproduction', *International Journal of Urban and Regional Research* 31: 169–78.

Jones, M. 1997. 'Spatial selectivity of the state? The regulationist enigma and local struggles over economic governance', *Environment and Planning A* 29: 831–64.

——— and G. MacLeod. 2004. 'Regional spaces, spaces of regionalism: Territory, insurgent politics and the English question', *Transactions of the Institute of British Geographers* 29: 433–52.

Keil, R. 1998. 'Globalization makes states: Perspectives of local governance in the age of the world city', *Review of International Political Economy* 5: 616–46.

———. 2000. 'Third way urbanism: Opportunity or dead end?', *Alternatives* 25, 2 (Apr.–June).

———. 2009. 'The urban politics of roll-with-it neo-liberalization', *City*.

——— and D. Young. 2007. 'Transportation: The bottleneck of regional competitiveness in Toronto', *Environment and Planning C* 26: 728–51.

Kipfer, S. 1998. 'Urban politics in the 1990s: Notes on Toronto', in R. Wolff, A. Schneider, C. Schmid, P. Klaus, A. Hofer, and H. Hitz, eds, *Possible Urban Worlds: Urban Strategies at the End of the 20th Century*. Basel: Birkhaeuser.

———. 2009. 'Why the urban question still matters: Reflections on rescaling and the promise of the urban', in R. Keil and R. Mahon, eds, *Leviathan Undone? Towards a Political Economy of Scale*. Vancouver: University of British Columbia Press.

——— and R. Keil. 2002. 'Toronto Inc? Planning the competitive city in the New Toronto', *Antipode* 34: 227–64.

Kübler, D., and H. Heinelt. 2005. 'Metropolitan governance, democracy and the dynamics of place', in H. Heinhelt and D. Kübler, eds, *Metropolitan Governance: Capacity, Democracy and the Dynamics of Place*. Oxford: Routledge.

Lehrer, U. 2005. 'The Spectacularization of the building process: Berlin, Potsdamer Platz', *Genre: Forms of Discourse and Culture* 26: 383–404.

———— and J. Laidley. 2008. 'Old mega-projects newly packaged? Waterfront redevelopment in Toronto', *International Journal for Urban and Regional Research* 32: 786–803.

Lustiger-Thaler, H., and E. Shragge. 1998. 'The new urban left: Parties without actors', *International Journal of Urban and Regional Research* 22: 233–44.

Macleod, G. 2001. 'The new regionalism reconsidered: Globalization, regulation, and the recasting of political-economic space', *International Journal of Urban and Regional Research* 25: 804–29.

Macleod, G., and M. Goodwin. 1999. 'Space, scale and state strategy: Rethinking urban and regional governance', *Progress in Human Geography* 23: 503–27.

———— and M. Jones. 2007. 'Territorial, scalar, networked, connected: In what sense a "regional world"?', *Regional Studies* 41: 1177–91.

Mayer, M. 1994. 'Post-Fordist city politics', in A. Amin, ed., *Post Fordism: A Reader*. Oxford: Blackwell.

————. 2007. 'Contesting the neo-liberalization of urban governance', in H. Leitner, J. Peck, and E. Sheppard, eds, *Contesting Neo-liberalism: Urban Frontiers*. New York: Guilford Press.

Nowlan, D.M. 1997. 'Jane Jacobs among the economists', in M. Allen, ed., *Ideas That Matter*. Toronto: Ginger Press.

Ohmae, K. 1995. *The End of the Nation State: The Rise of Regional Economies*. New York: Free Press.

Painter, J. 2008. 'Cartographic anxiety and the search for regionality', *Environment and Planning A* 40: 342–61.

Peck, J. 2005. 'Struggling with the creative class', *International Journal of Urban and Regional Research* 29: 740–70.

———— and A. Tickell. 2002. 'Neo-liberalizing space', in N. Brenner and N. Theodore, eds, *Spaces of Neo-liberalism: Urban Restructuring in North America and Western Europe*. Oxford: Blackwell.

Peterson, P. 1981. *City Limits*. Chicago: University of Chicago Press.

Sancton, A. 2001. 'Canadian cities and the new regionalism', *Journal of Urban Affairs* 23: 543–55.

————. 2005. 'The governance of metropolitan areas in Canada', *Public Administration and Development* 25: 317–27.

Sassen, S. 1991. *The Global City: New York, London, Tokyo*. Princeton, NJ: Princeton University Press.

Scott, A.J. 1988. *New Industrial Spaces: Flexible Production Organization and Regional Development in North America and Western Europe*. London: Pion.

————, ed. 2001. *Global City-Regions: Trends, Theory, Policy*. Oxford: Oxford University Press.

Sellers, J., and V. Hoffmann-Martinot. 2005. 'Introduction: Metropolitanization and political change', in V. Hoffmann-Martinot and J. Sellers, eds, *Metropolitanization and Political Change*. Wiesbaden: Verlag fuer Sozialwissenschaften.

Storper, M. 1997. *The Regional World: Territorial Development in a Global Economy*. New York: Guilford Press.

———— and A.J. Scott. 1989. 'The geographical foundations and social regulation of flexible production complexes', in J. Wolch and M. Dear, eds, *The Power of Geography: How Territory Shapes Social Life*. Winchester, Mass.: Unwin Hyman.

Swyngedouw, E. 2005. 'Governance innovation and the citizen: The Janus face of governance-beyond-the-state', *Urban Studies* 42: 1991–2006.

Wallerstein, I. 2008. 'The demise of neo-liberal globalization', *Commentary* 226, 1 Feb.

Walters, W. 2004. 'Some critical notes on governance', *Studies in Political Economy* 73 (Spring–Summer): 27–46.

Zukin, S. 1995. *The Cultures of Cities*. Cambridge: Blackwell.

CHAPTER 13

The Economics of Urban Land

Andrejs Skaburskis and Markus Moos

You can fly over Canada for hours and see nothing but trees, lakes, and a few snaking roads; in some parts even the roads disappear. We have lots of land. So why pay a million dollars for a very small lot in Vancouver? What factors determine this price and how do land prices affect the development of the city? Who is involved in setting the price and what are their interests in shaping the city? What are the social consequences of these processes and what role does public policy play in shaping urban land markets? These questions are considered in this chapter.

Beginning with an overview of the factors that shape the value of land, we move on to discuss the characteristics of urban land markets in more detail. We distinguish between the rent and the price of land. We highlight the structure of ownership and discuss the role of planning and politics in shaping land markets. We introduce in more detail some of the models economists use to understand the factors that affect urban land markets, and we discuss demographic, labour market, political, and societal shifts that are changing both real estate costs and land-use patterns. As we shall see, changing land markets have efficiency and equity implications.

The Value of Location

When we think of expensive locations, London, Paris, New York, and Tokyo might come to mind. In Canada, Vancouver has become almost synonymous with expensive real estate and Calgary is not far behind. High-rise condominium apartment towers dominate the Vancouver skyline. In 2011, Vancouverites will see another luxury condominium development added to their skyline, the Ritz-Carlton. Gary Mason (2008) in the *Globe and Mail* observes that in the spring of 2008, the Ritz-Carlton was already 60 per cent sold out at an average cost of $2,300 per square foot: 'That's right, a 1,000-square-foot condo in downtown Vancouver that you can't get into until 2011 will cost you $2.3 million.' The penthouse suite is going for $29 million. Yet it is not just the luxury condos that are highly priced in Vancouver. In 2008, the *average* home in this city sold for over $700,000. Older and seemingly modest single-family dwellings are commonly sold for over $1 million, illustrative of the high cost of land (Figure 13.1). It is hard to believe these figures, especially considering that elsewhere in Canada, such as in a Quebec town on the beautiful Gaspé shoreline (Figure 13.2), a house can sell for under $50,000; one could own more than 20 homes for $1 million!

What explains the geographic differences in the price of houses and the land they are built on? The quick answer to most questions about the value of land is the realtor's mantra, 'location, location, location'. We need to know more, though, about the aspects of location that matter to people

Figure 13.1 A Vancouver million-dollar house in the Dunbar neighbourhood. (Andrejs Skaburskis)

and firms and the forces that determine the attractiveness of a location. The attributes of location that are of most interest to us in this chapter are formed by the relationship a place has with other places within a city and the factors that influence this relationship. These factors vary from the local to the global. The value of land and its use are determined by the conditions in the city but also by what is happening in other parts of the world. The recent economic crisis has led to devaluation, thus injecting risk into the valuation of property. Paying a high mortgage on a Vancouver home suddenly looks less attractive when housing values begin to decline without an immediate rebound in sight. However, the economic cycles do not necessarily alter the factors that shape *relative* valuation.

The social and economic relationships that tie cities together into complex urban systems shape the relative value of location. For instance, Montreal's declining status as Canada's dominant economic centre helps explain why its property values increased no faster than the national average, whereas Toronto, Canada's economic powerhouse, had the highest property value appreciation in the years before the recent economic recession—along with the emerging regional centres of Vancouver, Calgary, and Edmonton. Historically, Montreal had an advantage in terms of the location of its port that provided access to the West, but with the development of the railroads and the opening of the Panama Canal, western Canadian trade was increasingly captured by Vancouver. Toronto gained manufacturing industries over Montreal after World War II, partly due to the former's proximity to the US industrial belt. The out-migration of the English during the 1970s reduced housing demand and hurt Montreal's property values considerably. Economic strength gained by an advantageous location matters in the valuation of land.

Richard Hurd (1903), a noted land economist, tells us that the value characteristics of a location are determined by two sets of attributes: those related to *proximity* and those related to *accessibility*. People may be willing to pay more for locations close to parks, views, and quiet surroundings and less for locations close to municipal landfills or highway noise. Firms may also cluster together (proximity) to share infrastructure, local services, and communication opportunities among decision-makers. However, in many cases access rather than

proximity is valued. 'Access' refers to the ease of getting from one place to another. We might want to live by a lake but cannot on account of our need to access work, shops, and other facilities. Improvements in transport infrastructure raise the value of land in its vicinity as highways and public transit extensions enhance the accessibility of the locations connected by the system (see Perl and Kenworthy, Chapter 11). Between 1986 and 1996 vacant land prices in the vicinity of Vancouver's SkyTrain stations increased by 251 per cent as compared to 133 per cent for prices as a whole (Landcor, 2008). Store owners want to locate near transport hubs as well as each other to gain exposure to customers who are more likely to come to the shopping centres or main streets that let them reduce their overall travel costs. Polluting factories are nuisances and we do not want to live near them, but workers may have

Figure 13.2 $50,000 houses on the Gaspé Peninsula, Quebec. (Andrejs Skaburskis)

to live close by to access their workplace. The value of a location is a function of the advantages it offers in terms of proximity and in terms of its accessibility *relative* to other locations.

As was the case of the historic centre of cities whence urban development spread outward, proximity and accessibility to the downtown have remained most valued, and the highest per unit land prices are in the centre. The scatter-plots in Figure 13.3 show how the value of urban land in 2001 in Toronto, Montreal, and Vancouver decreases with distance from the centre. The centre continues to provide primary employment and service functions in Canadian cities to this day, and this explains why the price of land continues to be highest near the downtown and falls towards the periphery. Increases in business-related travel have made airport locations more important to many firms than downtown locations, and sub-centres have developed around the major airports that partially explain the secondary peaks in land values (Figure 13.3). Land values also increase around 'suburban downtowns' and emerging employment nodes in suburban locations where firms move for lower rents as well as to locate closer to an increasingly suburban labour force as cities continue to spread outward (Filion and Gad, 2006).

Factors external to the city also affect the value of land. The Vancouver land market heated up in the late 1980s as offshore investors and immigrants bought real estate largely due to worries about the return of Hong Kong to Chinese rule. The globalization of financial markets has made it easier for foreign investors to buy real estate in distant places (see Hall, Chapter 4). Immigration has dramatically increased in recent years and expands the demand for land and housing in the large Canadian cities (Ley and Tutchener, 2003). Labour markets are closely related to housing demand, the latter being a function of household formation, income, wealth, and preferences for housing space. Our demand for housing and land is related to how much income we earn currently, also called a monetary income,

and how much we expect to earn in the long run. Due to changes in immigration policies that evaluate migrants on a points system, recent immigrants to Canada tend to have higher permanent incomes that increase their housing consumption (see Hoernig and Zhuang, Chapter 9). If immigrants hold strong preferences for particular locations on account of their ethnic makeup, the growth in demand for these neighbourhoods increases their land values. In this context it would be fair to say that the federal government's immigration policy has become an important factor in shaping land markets in our largest cities.

Land markets also are influenced by planning efforts and municipal infrastructure projects that determine where public facilities are built, which waterfronts are beautified, what parts of the city are connected with roads and transit systems, and what parts of the city receive the landfills and other nuisances. Zoning, in particular, is a highly political process that sets out the pattern of land-use activities and builds expectations over the potential for land value appreciation. Decisions about zoning changes often are released only after being finalized to avoid creating speculative behaviour in land markets created by uncertainty. The role of government is highlighted by Charles Schultze's (1977: 30) bold assertion that the free market is 'made by government': landownership and its transactions are possible only within the protected environment formed by government.

The economic models that explain urban land values assume that the resulting land uses reflect the 'highest and best use'. In other words, land markets promote the efficient use of land by ensuring that the potential users of a parcel of land are those who value and can pay the most for its location attributes. Fairness issues are raised when the market brings changes that hurt the more vulnerable populations by forcing them to move or by reducing the supply of lower-priced housing. Nick Blomley (2004) argues that our system of property ownership can seem definitive and even natural

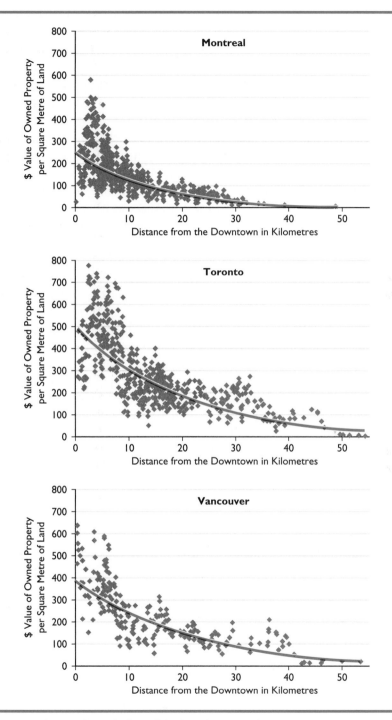

Figure 13.3 Property value gradients in Canada's three largest cities

Source: Authors' calculations from 2001 census data.

whereas, in fact, it is made possible by a regulatory system that favours property owners. Renters often have little to no claim to their space and are potentially harmed by changes in land prices that bring windfalls to the owners. The societal value of land is determined not just by the price someone is willing and able to pay but also by the occupants' emotional, cultural, and affective ties to place (see Lynch and Ley, Chapter 19).

The Characteristics of Urban Land Markets

Markets facilitate the selling and buying of goods and services, create incentives for firms to provide the goods and services consumers want most, and to do so efficiently. Changes in prices tell producers how demand is changing. The price of a good or service is determined in the market by the interplay of demand and supply. A rise in demand unmatched by an increase in supply raises prices. Land markets are unique in that the supply of land at any location is fixed. This means that any increase in demand for land at a location will increase its price. The supply of urban land can increase as more rural and agricultural lands are converted to urban uses but this expansion makes the inner-city locations even more desirable *relative* to the retreating peripheral locations and, due to the fixed supply, inner-city land prices increase with city growth. As land becomes more expensive, developers use less in creating real estate and city densities increase. The demand for land affects its price and, therefore, how it is used, and how the city looks and functions. Understanding land markets is a key to understanding urban geography. Knowledge of property values and understanding of the dynamics that drive them are prerequisite to informed land-use planning.

Rent and Prices

In examining land and property markets, economists distinguish between the value of *using* the property from the value of *owning* the property. In this sense there are two markets in land: one for its use and another for its ownership. When considering the value of the use of a property, we use the concept of a 'rent'. The rent is the amount a household or a firm pays or would be willing to pay for the use of a property for a period of time. Homeowners are regarded as paying 'rent' to themselves for the use of their property. When the property is sold a 'price' is paid. The price reflects the value of owning the property and usually is determined by the expected value of the use of the property in the future, or in other words, by the rent it is expected to generate in the future.

The distinction between these two markets is important because the factors determining price can be different from those determining current rents, but the two measures of value are linked. In a world in which the future is known and nothing changes, the highest price someone would offer would be the present value of the expected future rents—determined by the capitalization process. If a property would yield rents starting next year, for all years into the future, the price of the land would be the rent divided by the discount rate:

(1) Price = annual rent/discount rate

If we estimate the discount rate at 6 per cent and the rent on a property is $1,200 per month, then the price for that property as determined by equation (1) would be:

(2) Price = ($1,200 per month ×
12 months)/0.06 = $240,000

When we look at a growing city, we see much higher prices: one cannot buy a condominium unit that would rent for $1,200 a month in any of our major cities for as little as $240,000. The price is much higher because the rents are expected to increase in the future with continued city growth.

The formula needs to capitalize the future increase in rents by using the 'net of growth' discount rate and the simplified formula becomes:

(3) Price = annual rent/(discount rate − expected growth rate in rents)

Using the above numerical example, but adding an expected growth in rents of 2 per cent a year, changes the price calculation as follows:

(4) Price = ($1,200 × 12)/(0.06 − 0 .02) = $360,000

The growth rate is not known but is guessed at by the prospective buyers. The difference in the price of the property—the difference between $240,000 and $360,000—is due to the *expectation* of a 2 per cent annual growth in rent. Equation 3 can help us illustrate two important features of the land market that affect the way cities evolve. First, since the growth rate is not known but is guessed at, people's expectations of their city's future affect growth rates and subsequently **urban form**. If people expect the city to keep growing, then this expectation will drive up land prices and create the incentive for developers and builders to use less land in producing real estate. Anticipated future changes in land use affect the market's view of future rents and, therefore, affect its current price. Announcing a transportation improvement, for example, offers to increase the value of the location in the future and therefore the current price. New plans that will change the use of land affect its price at the time they become known. For example, when Ontario announced plans for a greenbelt around Toronto, the price of the farmland outside the greenbelt dropped because this growth boundary eliminated the possibility of gaining future urban rents.

The second feature revealed by Equation 3 relates to the stability of the market. The amount people will be willing to pay for land increases as their expectations of future growth rates increase. When they base their expectations on recent past trends, they may be induced to buy larger houses and increase the amount they can gain from future price increases. Prices rise and create expectations of an even higher future growth rate that is unrealistic. As the denominator in Equation 3 decreases in size, the price bubble forms until people recognize that there is no longer a connection between prices and realistic rents. The bubble bursts and prices tumble. People who had taken mortgages they can no longer afford and households who had refinanced their homes to make other purchases find themselves with mortgage debt that is greater than the value of their homes. As in the US housing market meltdown that triggered the economic crisis, the viability of the banks and other financial institutions holding the now almost worthless mortgages are threatened and the crisis broadens to affect the whole economy.

Of course, the recent economic crisis was not solely triggered by misguided expectations on the borrowers' part. It has been the worst downturn since the Great Depression of the 1930s and has drawn worldwide attention to the US sub-prime mortgage sector. Lending institutions actively pursued households who would not traditionally qualify for a mortgage in order to make money on the initial sale of the mortgage. Loans were made to working-class and racially marginalized communities with hidden costs in schemes commonly called 'predatory' (Wyly et al., 2009). Most of these loans were securitized immediately and sold to investors for money that was used to make more loans. While housing prices kept climbing, the system remained stable. A household about to default on their mortgage could sell their house at a higher price and cover all costs. However, once prices stabilized and houses were being taken back by the mortgage lenders it was realized that mortgages exceeded the value of their properties and the financial system began to collapse. The increase in foreclosures caused house prices to drop further, making the problem even worse by bringing bankruptcies in other sectors, loss of life savings, and widespread unemployment. Land, property markets, and the institutions that govern them can have very far-reaching consequences.

The Structure of Landownership

There are broad consequences to the way society determines who owns land and how to gain access to it. An understanding of the structure of ownership is important for understanding the consequences of how land markets operate. 'A man without land is nothing', Duddy Kravitz is told by his grandfather in Mordecai Richler's (1959) novel about the young Duddy growing up on Montreal's St Urbain Street. Duddy becomes obsessed with the idea of attaining property, doing anything regardless of its legality or morality to attain it. Since the mid-twentieth century, the majority of young households in North America have shared his goal. Our lending institutions have engineered financial instruments to permit and encourage households to achieve homeownership. The notion that property ownership comes with elevated rights, privileges, and wealth potential has pervaded scholarly and popular Western thought for many years (Ronald, 2008); indeed, the first concept of 'universal' suffrage in Western democratic societies limited voting to male property owners. There is some justification for this widespread sentiment; certainly we know today that households at risk of homelessness or coping with other forms of poverty are more likely to be renters than owners (Hulchanski and Shapcott, 2004). The advocates for a tax on rent, sometimes called a Ricardian land tax, argue that the value of a location, and thus rent, is not inherent in land but is socially constructed, thus raising questions as to whether the gains should accrue purely to owners. In the late nineteenth century, Henry George was among the most vocal advocates of a land tax. He expected his proposed single tax to finance *all*

government activity, sparking controversy in the US where government intervention generally was resented.

Clearly, property rights and the ability to gain profit from one's possession are essential characteristics of any efficiently operating market. In order to engage in the selling and buying of goods and services we must know who owns the good or service in question. But if the ability to attain property is unequally distributed and the status of owning property comes with elevated rights and privileges, we can begin to ask questions about the role of public policy in addressing these potential inequalities. Democratic rights are, of course, no longer defined by property ownership as they once were, but observers continue to argue that due to differential tax treatments, wealth gain potential, and the social status of owning, renters are at a disadvantage simply because of their tenure.

Conceptualizing the Spatial Structure of Urban Land Markets

The conceptual frameworks we use to understand land markets today have evolved from classic economic theory derived from the thinking of the eighteenth-century economist, Adam Smith. Examining the rents charged to tenant farmers, Smith (1970 [1776]) views rent as the residual value: the difference between the cost of growing produce and the price paid for the produce at the market. He observes that, regardless of what was grown, land rents near the market are higher than rents further from the market on account of the added cost of transportation. David Ricardo, recognized as the founder of rent theory, defines **land rent** as 'that portion of the produce of the earth, which is paid to the landlord for the use of the original and indestructible powers of the soil' (Ricardo, 1969 [1817]: 33). The price of produce is set by the cost of growing it on the least productive land. This land yields no residual value for the landowner. Since all produce fetches the same price in the market regardless of where it is grown, the owners of more fertile land can extract a rent from their tenant farmers equal to the difference between costs and price. 'Ricardian rent' levels throughout a city are set by the differences in the relative attractiveness of sites and neighbourhoods. 'Fertility' can also be transformed into the 'accessibility' attribute—the more accessible sites are similar to the more 'fertile' plots in Ricardo's model. The base rent for urban uses is set by the least attractive land that has to be used to house the population. From a commuting point of view these locations are at the periphery of the city. From here the urban land rent profile starts to rise from the agricultural rent level, increasing towards the centre to reflect the households' valuation of the reduced commute, as illustrated in Figure 13.4.

This very simple model still pertains to the way we can view the city today. If all other conditions were equal and if employment were concentrated in the centre of the city, as it was historically until mid-twentieth century in most Canadian and American cities, then some households would have shorter commutes than others and this difference would affect the amount they would pay for housing (see Hutton, Chapter 7). Land prices near the centre are driven up by people who want to reduce their commute costs. The neo-classical models of land use and urban density are built on the classical ideas. William Alonso (1964), for instance, develops the concept of a 'bid-rent' map with contours that trace the amount that a household would be willing to pay at each location while keeping its satisfaction constant (Figure 13.5). The

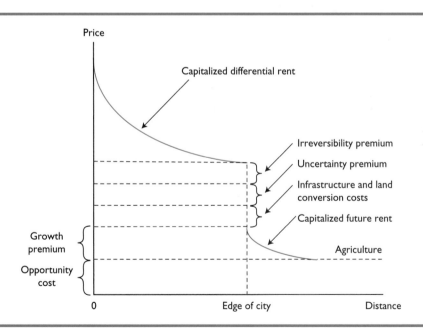

Figure 13.4 Hypothetical land price gradient

The price of an inner-city lot is set by the present value of the agricultural rents at the periphery plus the growth, uncertainty, and irreversibility premiums. It is set by the developer's cost of providing the on-site infrastructure plus the development cost charges for the public facilities and off-site infrastructure. Prices inside the city also are determined by the capitalized differential rents due to the relative increase in the attractiveness of locations closer to the centre.

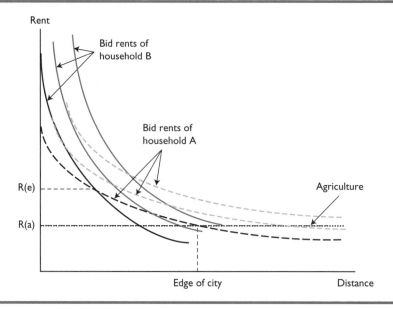

Figure 13.5 Alonso's bid-rent map

The figure assumes there are two households, A and B, each trying to bid for one of the two sites from agriculture for their own use. The solid lines are the bid rents of household B, and the broken lines are those of household A. Higher bid-rent curves, as indicated in blue, give lower welfare to a household because the household pays more rent for the site. Lines in black represent the highest household welfare, but are the minimum rent to landowners. Land rent from agriculture is denoted by the dotted line.

bid-rent curves depict the rent/distance trade-off that would make the householders indifferent as to their location. Households maximize their well-being by making a trade-off between better access to the city centre and lower-priced land. Since higher-income households tend to want larger lots, they are drawn towards the periphery by the lower land costs. Lower-income households occupy the more expensive inner-city land at higher densities to reduce their commute costs. Since they buy less housing and occupy less land, they would not benefit by the lower price of suburban land as much as higher-income households. Alonso's (1964) model explains the growth of the suburbs and the spread of the urban region as a function of increasing incomes and the preference for large houses. Richard Muth (1969) and Edwin Mills

(1969) expand the theory and explicitly introduce housing markets.

The 'New' Urban Economics

The 'new' urban economics introduces the time dimension to the classical and neo-classical economic models and helps answer such questions as: why do we find parking lots on very expensive downtown land when they clearly cannot generate the land value of a condominium tower? Why is suburban development discontiguous in places and leapfrogging into the countryside to the chagrin of most city planners? Why is the market price of agricultural land at the edge of the city so high compared to the capitalized agricultural rents? We start examining these questions through the capitalization process that links the expected future land rents with land prices and then discuss the costs and benefits of delaying development.

The market price of land at the edge of the city is set by its value in agricultural use and by the present value of the expected future growth in rents at that location. In a growing city, the boundary that sets the base rent for the rest of the city is continuously being pushed further from the centre, making the land within the built-up part of the city *relatively* more attractive. Since buyers recognize that the land at the city's current edge will command higher rents in the future, they are willing to pay a premium now to gain the future increases as described by Equation 3 above. Another complication is introduced by the possibility that development options can change over time. If a vacant lot became trapped within an expanding city, its eventual development would be different from the development that would take place when the land is at the very edge of the city. The intensity with which vacant land inside a growing city can be developed increases over time when zoning permits; and this intensity, in turn, increases the residual value of the land. It can, therefore, benefit an owner to hold the land vacant for an extended period of time before converting it to more intense use.

Hotelling's (1931) model builds the basis for understanding the process of land valuation over time. His model starts by recognizing that equilibrium in investment markets requires that assets, such as land, appreciate in value at a rate equal to the relevant interest rate. If the rate at which the asset's price increases is lower than the rate available in other equally risky investments, then its price drops as investors sell to buy more profitable assets. The reduction in the price of an asset increases the *rate* of return created by a given increase in price. Wicksell (1935) explains the profit-maximizing of timed decision-making with the example of a landowner waiting for trees to grow before cutting them down to sell their lumber. Waiting allows the

trees to grow larger and yield more lumber when they are cut, but waiting precludes the use of the funds that would be gained by the sale of the lumber. The most profitable time to harvest the trees is when the rate of growth in their value drops to equal the rate of return on the alternative investment, as predicted by Hotelling's model. Thus, in the context of city development the models predict that the owner of a downtown lot may delay its redevelopment until demand for space in an office tower has grown enough to justify the new building. While a smaller condominium on the site might be profitable now, waiting for the site to 'ripen' and allow the construction of the office tower might offer even larger profits.

Shoup (1970: 40) specifically applies the profit-maximizing timing model to urban land use and dispels the notion that 'development or redevelopment would or should occur as soon as the development value of a site, net of clearance costs, exceeds the value of the existing improved property, as is sometimes stated.' The most profitable development time is when the rate of change in the value of the development that can take place on a site is equal to the interest rate on equally risky alternative investments (Wicksell, 1935). Holding land vacant, or using it as a parking lot, is worthwhile when the rate of increase in the present value of the most profitable project that can be placed on the land exceeds the rate of return on alternative investments. The change in the profitability of development is due to the change in what would be built on the land and due to the fact that development freezes that use for a long time. If the development potential is not changing over time, then a site will be developed now or never. Arnott and Lewis (1979) show that waiting for development changes the density of development. As owners hold vacant land and the suburbs grow around them, land prices increase and less is used in developing real estate: higher land prices encourage higher-density development. Keeping some land vacant can help curb 'sprawl' in the long run by leaving room for a new wave of much higher-density development (Peiser, 1989). Had room for infill development not been left, the new development would be at the periphery of the urban region and the total time and effort spent on commuting to the centre would be higher.

The development of land fixes its use for a long period and, thereby, prevents the property owner from taking advantage of possible future changes that could increase his or her options. The presence of uncertainty increases the value of having options and owners will delay development until prices rise to cover their perceived cost of the uncertainty. Owners have many different reasons for holding land, as pointed out by Brown et al. (1981). Some are keeping the land for their children and have no interest in making a profit. Some may place intrinsic value on the ownership itself regardless of the returns they can get for its sale or development. Others may have different discount rates or expectations of growth rates that affect their timing of the sale or of development. The variety of reasons for owning land implies that owners will sell or develop their land at different times. The result is the discontiguous pattern of suburban development that most city planners deplore. Keeping unused land within the city's perimeter increases infrastructure costs but it also leaves room for more intense development at a later time.

Since the development decision is not reversible, owners will wait until prices rise to cover their perceived cost of the irreversibility of the development decision (Capozza and Helsley, 1989) and the resulting land price gradient is illustrated in Figure 13.4. The base price is set by the residual value of the land just outside the periphery. The growth, uncertainty, and irreversibility premiums are added along with the cost of infrastructure to determine the market price of buildable lots at the periphery. The capitalized differential rent is added only inside the city. Constraints on development or increases in development costs at the edge of the city increase the price of all land inside the city. Zoning bylaws and differences in neighbourhood attributes create discontinuities in the rent and price profiles resulting in a gradient that is not as smooth as the one depicted in Figure 13.4.

The Changing Profile of Cities

Canadian metropolitan areas, historically characterized by monocentric development patterns and sharply falling density gradients, have dramatically changed over the last 30 years (see Grant and Filion, Chapter 18). Not only has their reach expanded, but an amalgam of demographic, labour market, and socio-economic shifts have resulted in what many believe to be fundamental changes in the spatial and social structure of Canadian cities (see Hutton, Chapter 7; Townshend and Walker, Chapter 8; Walks, Chapter 10). The return of higher-income households to the inner city is perhaps the most important change in the structure of cities in the last half-century (see Bain, Chapter 15). The changes are spectacularly visible in Vancouver and Toronto, where previously industrial lands and rail yards now are dotted with high-rise condominiums (Figure 13.6). Redevelopment of the inner city through a process called **gentrification** is attributed to the growth of smaller, non-family households, the emergence of a 'new middle class' of highly educated, quaternary-sector workers who reject the suburban lifestyle, and deindustrialization that left many inner cities with undervalued properties (Ley, 1996). Observers assign different relative weights to each of these explanatory factors. The incoming population tends to have specific preferences and values that translate into political lobbying for infrastructure and cultural facilities, which, when provided, further increase the value of inner-city land (Figure 13.7). In gentrification, the lower-income populations are displaced by the incoming, wealthier households. The lobbying and policies that facilitate inner-city investment, and support its upper- and middle-class residents,

Figure 13.6 Vancouver's redeveloped inner city. (Andrejs Skaburskis)

Figure 13.7 Use of nudity by the 'new middle class' to protest car culture. The ideologies of politically astute residents can shape transport and land-use policies with implications for property markets. (Andrejs Skaburskis)

have even been called 'revanchist' for what some describe as blatant attempts to exclude vulnerable populations (Smith, 1996).

To be sure, most population growth is still in the suburbs. Ever since the introduction of streetcars in the early 1900s, transportation improvements have continued to permit suburban development (see Perl and Kenworthy, Chapter 11). The rate of expansion increased dramatically in the 1950s with growth in prosperity, the development of mortgage financial institutions and instruments, and the construction of intra-urban highways. The neo-classical economic models, and their focus on the trade-offs between commute and housing costs, help explain the outward movement of the well-off and the filtering down of inner-city housing from higher- to lower-income households. The increasing income levels allow people to buy larger houses on larger lots, making them more sensitive to the per-square-metre price of lots and more willing to accept the longer commutes from the suburbs. Technological change and rising incomes increased automobile ownership, making jobs accessible from distant locations. One of the most important changes in the last half of the twentieth century was the changing employment prospects and interests of women that in part contributed to suburbanization. Not only did this increase the income of households with two income earners but it also increased their commute costs. Some companies moved their clerical offices out into the suburbs to be closer to their predominantly female workforce, helping to intensify the emerging centres that would define the new polycentric urban region (see Hutton, Chapter 7).

The spread of employment across the region increased the spread of the city. Surprising to many observers, however, was the resurgence of interest in the inner city by the well-off, leading to 'the return of the café society' as some have called this trend. The filtering process in housing used to pass housing down to lower-income people as it aged and as the higher-income households moved out to newer housing of higher quality. This process has reversed in most Canadian metropolitan areas since the early 1980s (see Walker and Carter, Chapter 20). Without exception, in all Canadian metropolitan areas, small and large, growing or staying still, the average rents in the older stock, which is located in the inner city, are higher than in the newer buildings (Skaburskis, 2006). This reversal reflects a revaluation of inner-city locations by higher-income households and it can be explained by at least two viewpoints: the neo-classical and the structuralist. Many believe, too, that the pursuit of more entrepreneurial urban policies is also partly responsible for the changes (see Allahwala, Boudreau, and Keil, Chapter 12).

Wheaton (1977) shows that the perceived cost of commute time has been increasing with household income, enough to change the way the commute-cost/housing-price trade-off is made and to make some higher-income people want to move back to the inner city. Demographic changes also contribute to the displacement of lower-income households by middle- and higher-income earners in gentrification. Average household sizes are declining in North America and Europe, and smaller households occupy less land and can benefit more by paying a higher per-square-foot price for land to reduce their commute costs. Employment insecurity makes central locations more attractive by reducing commute costs over the long run as job locations change. The very large increase in the proportion of one- and two-person households helps account for the increase in the demand for central locations. Two-worker households are more likely to locate near the urban centre to minimize joint commute costs when they work in different parts of the city. An inner-city location helps solve the co-location problem of well-educated, dual career couples and helps 'preserve the marriages of dual career households' (Costa and Kahn, 2000: 1289). Women professionals are more likely to find work in the downtown (Rose and Villeneuve, 1998).

The higher density of the inner city increases the options single people have in the marriage market. The delay in family formation since the 1950s has reduced young people's need for space and increased the time and money spent on leisure activities. Increases in the education and income of young people translate into a greater demand for the services and amenities generally found in the downtown, such as theatres, museums, cafés, and sushi bars. Glaeser, Kolko, and Saiz (2001), by observing that rents in cities increased faster than wages, conclude that demand for urban amenities increased faster than urban productivity. They suggest that people live in downtowns increasingly to consume, not to work. Increasing traffic congestion and gasoline costs also make inner-city locations more attractive.

A number of geographers see the changes in inner-city land markets as linked to shifts in the social structure brought about by global economic restructuring (Badcock, 1992; Maher, 1994). Global economic restructuring—at times coined as a transition from Fordist to post-Fordist regimes of accumulation—has polarized employment at two opposite extremes of the pay scale: white-collar professionals and low-level service workers. The two-tiered employment structure and a diminishing role of the **welfare state** under **neo-liberalism** have resulted in an ever-growing gap in income between those at the top and those at the bottom of the pay scale (Castells, 2002; Esping-Anderson, Assimakopoulou, and van Kersbergen, 1993). The 'social polarization' that is thought to follow from the new employment structure has been well documented, as well as criticisms surrounding the universality of the theory (Hamnett and Cross, 1998; Sassen, 1990, 1991). Social and economic bifurcation can be expected to change the structure of cities, especially through segregation and income inequality (Marcuse and van Kempen, 2000; Walks, 2001; see Walks, Chapter 10). Those at the top end of the income structure can out-compete others in the housing market, resulting in ever-escalating markets that fewer people can attain. The trends can help to explain the declining willingness of private developers to construct rental buildings that tend to be occupied by lower-income populations. Some observers have called this the 'dual' or the aforementioned 'revanchist' city (Mollenkopf and Castells, 1992; Smith, 1996). Some explanations of inner-city reinvestment draw on a Marxist perspective of investment cycles (Harvey, 1985; Smith, 1986). In the post-Fordist city, the move of capital away from the downtown resulted in inner-city decline, which, in turn, created the potential for inner-city **regeneration**. Redeveloping the districts and renovating the buildings in the inner city would restore the land values lost due to the flight of capital.

Changes in the structure of employment and emerging new technologies have prompted localities to devise policies to attract and expand their workforce. These policies focus on the development of sports venues, revitalizing waterfronts, holding urban farmers' markets, building marinas, and enacting urban design guidelines that are believed to generate the type of urban landscapes attractive to the so-called 'knowledge workers' or, to borrow Richard Florida's (2002) now much-used phrase, 'the creative class' of the **new economy** (Hall, 2006). Localities subsequently invest increasingly greater shares of public resources into the provision of cultural amenities, festivals, and '**urban renewal**' projects through '**place-making**' and 'place-marketing' strategies (Hackworth, 2007; Kipfer and Keil, 2002; see Bain, Chapter 15; Lynch and Ley, Chapter 19). Left with empty manufacturing buildings and polluted industrial sites, municipalities are devising strategies to attract cultural and creative activities seen now as essential to growth in the new economy (Figure 13.8).

The 'entrepreneurial' policies differ from previous 'managerial' policies that focused more heavily on providing public services such as housing (Harvey, 1989). The entrepreneurial policies change the amenity attributes and attractiveness of specific

Figure 13.8 Construction of cultural amenities in place of manufacturing in Montreal. (Markus Moos)

locations, shaping land markets differently than did the managerial policies that preceded them. Instead of investing in social infrastructure that promoted the development of stable neighbourhoods, the new policies harness the growth-generating potential of real estate markets by providing the type of infrastructure that enhances the value of land and location. Existing owners, investors, and the local tax base benefit from appreciating markets, while those wanting to enter the market or needing to pay escalating rents find it increasingly difficult to do so. Arguably, the new entrepreneurial policies cater to the consumption preferences of the workers in professional and managerial occupations, preferences not necessarily shared by those at the bottom of the income spectrum. The issue gets at fundamental ideological foundations that

separate entrepreneurial from managerial urban policies and at questions of whether government should help facilitate the operations of the market or whether it should help smooth its inequities.

In a market framework, land values necessarily differ by location and its uses and value are determined by those who have the means to pay for it. Valuing location in that manner may be efficient, but not necessarily equitable. Governments in Canada have traditionally intervened in land markets for reasons of equity and social justice when the inability to afford land and housing leads to a seriously disadvantaged social position. Whether recent increases in affordability problems and homelessness warrant renewed government programs is a question outside the scope of this chapter (see Walker and Carter, Chapter 20; Bourne and Walks, Chapter 25), but the economics of property markets clearly have broad consequences that warrant discussion about the relative role that markets and public policies ought to play in shaping our cities.

Review Questions

1. What factors—both internal and external to the city—determine the relative value of one location versus other possible locations for a home or business?
2. Why do we find expensive downtown land being held vacant or used for parking lots instead of being developed immediately at its highest and best use?

References

Alonso, W. 1964. *Location and Land Use: Toward a General Theory of Land Rent*. Cambridge, Mass.: Harvard University Press.

Arnott, R.J., and F. Lewis. 1979. 'The transition of land to urban use', *Journal of Political Economy* 87, 4: 161–9.

Badcock, B.A. 1992. 'Adelaide's heart transplant, 1970–88: 1. Creation, transfer, and capture of "value" within the built environment', *Environment and Planning A* 24: 215–41.

Blomley, N. 2004. *Unsettling the City: Urban Land and the Politics of Property*. New York: Routledge.

Brown, J.H., R.S. Phillips, and N. Roberts. 1981. 'Land markets at the urban fringe: New insights for policy makers', *Journal of the American Planning Association* 47, 2: 131–44.

Capozza, D., and R. Helsley. 1989. 'The fundamentals of land prices and urban growth', *Journal of Urban Economics* 26, 3: 295–306.

Castells, M. 2002. *The Information Age: Economy, Society and Culture, Volume 1: The Rise of the Network Society*. Malden, Mass.: Blackwell.

Costa, D., and M. Kahn. 2000. 'Power couples: Changes in the locational choice of the college educated, 1940–1990', *Quarterly Journal of Economics* 115, 4: 1287–315.

Esping-Anderson, G., Z. Assimakopoulou, and K. van Kersbergen. 1993. 'Trends in contemporary class structuration: A six-nation comparison', in G. Esping-Anderson, ed., *Changing Classes: Stratification and Mobility in Post-Industrial Societies*. London: Sage, 32–57.

Filion, P., and G. Gad. 2006. 'Urban and suburban downtowns: Trajectories of growth and decline', in T. Bunting and P. Filion, eds, *Canadian Cities in Transition: Local Through Global Perspectives*. Toronto: Oxford University Press, 171–91.

Florida, R. 2002. *The Rise of the Creative Class*. New York: Basic Books.

Glaeser, E., J. Kolko, and A. Saiz. 2001. 'Consumer city', *Journal of Economic Geography* 1: 27–50.

Hackworth, J. 2007. *The Neoliberal City: Governance, Ideology, and Development in American Urbanism*. Ithaca, NY: Cornell University Press.

Hall, C.M. 2006. 'Urban entrepreneurship, corporate interests and sports mega-events: The thin policies of competitiveness within the hard outcomes of neoliberalism', *Sociological Review* 54, 2: 59–70.

Hamnett, C., and D. Cross. 1998. 'Social polarisation and inequality in London: The earnings evidence 1979–1995', *Environment and Planning C* 16: 659–80.

Harvey, D. 1985. *The Urbanization of Capital*. Baltimore: Johns Hopkins University Press.

———. 1989. 'From managerialism to entrepreneurialism: The transformation in urban governance in late capitalism', *Geografiska Annaler* 71B, 1: 3–17.

Hotelling, H. 1931. 'The economics of exhaustible resources', *Journal of Political Economy* 39: 137–75.

Hulchanski, D., and M. Shapcott, eds. 2004. *Finding Room: Policy Options for a Canadian Rental Housing Strategy*.

Toronto: Centre for Urban and Community Studies Press.

Hurd, R. 1903. *Principles of City Land Values*. New York: Record and Guide.

Kipfer, S., and R. Keil. 2002. 'Toronto Inc? Planning the competitive city in the new Toronto', *Antipode*: 227–64.

Landcor. 2008. *Lessons from Expo 86 for the 2010 Winter Games*. The Landcor Report, Special Edition, 19 Aug. Vancouver: Landcor Data Corporation.

Ley, D. 1996. *The New Middle Class and the Remaking of the Central City*. Oxford: Oxford University Press.

———. and J. Tutchener. 2001. 'Immigration, globalization and house prices in Canada's gateway cities', *Housing Studies* 16: 199–223.

Maher, C. 1994. 'Housing prices and geographical scale: Australian cities in the 1980s', *Urban Studies* 31, 1: 5–27.

Marcuse, P., and R. van Kempen, eds. 2000. *Globalizing Cities: A New Spatial Order?* Oxford: Blackwell.

Mason, G. 2008. 'Just who is buying Vancouver's zillion-dollar condos?', *Globe and Mail*, 17 May. At: <www.theglobeandmail.com/servlet/story/GAM.20080517.BCMASON17/TPStory/TPComment>.

Mills, E. 1969. 'The value of urban land', in H. Perloff, ed., *The Quality of Urban Environment*. Baltimore: Resources for the Future, John Hopkins University Press.

Mollenkopf, J.H., and M. Castells, eds. 1991. *Dual City: Restructuring New York*. New York: Russell Sage Foundation.

Muth, R. 1969. *Cities and Housing*. Chicago: University of Chicago Press.

Peiser, R.B. 1989. 'Density and urban sprawl', *Land Economics* 65, 3: 194–204.

Ricardo, D. 1969 [1817]. *The Principals of Political Economy and Taxation*. London: Everyman's Library, J.M. Dent and Sons.

Richler, M. 1959. *The Apprenticeship of Duddy Kravitz*. Don Mills, Ontario: A. Deutsch.

Ronald, R. 2008. *The Ideology of Home Ownership: Homeowner Societies and the Role of Housing*. New York: Palgrave Macmillan.

Rose, D., and P. Villeneuve. 1998. 'Engendering class in the metropolitan city: Occupational pairings and income disparities among two-earner couples', *Urban Geography* 19, 2: 123–59.

Sassen, S. 1990. 'Economic restructuring and the American city', *Annual Review of Sociology* 16: 465–90.

———. 1991. *The Global City: New York, London, Tokyo*. Princeton, NJ: Princeton University Press.

Schultze, C. 1977. *The Public Use of Private Interest*. Washington: Brookings Institution.

Shoup, D. 1970. 'The optimal timing of urban land development', *Papers of the Regional Science Association* 75: 33–44.

Skaburskis, A. 2006. 'Filtering, city change and the supply of low-priced housing in Canada', *Urban Studies* 43, 3: 533–58.

Smith, A. 1970 [1776]. *The Wealth of Nations*. London: Everyman's Library, J.M. Dent and Sons.

Smith, N. 1986. 'Gentrification, the frontier, and the restructuring of urban space', in N. Smith, and P. Williams, eds, *Gentrification of the City*. Boston: Allen and Unwin, 15–34.

———. 1996. *The New Urban Frontier: Gentrification and the Revanchist City*. New York: Routledge.

Walks, A. 2001. 'The social ecology of the post-Fordist/ Global city? Economic restructuring and socio-spatial polarisation in the Toronto urban region', *Urban Studies* 38, 3: 407–47.

Wheaton, W.C. 1977. 'Income and urban residence: An analysis of consumer demand for location', *American Economic Review* 67, 4: 620–31.

Wicksell, K. 1935. *Lectures on Political Economy*. London: Routledge.

Wyly, E., M. Moos, D. Hammel, and E. Kabahizi. 2009. 'Cartographies of race and class: Mapping the class-monopoly rents of American sub-prime mortgage capital', *International Journal of Urban and Regional Research* 33, 2: 332–54.

Part III

'Placing' and Planning in the Twenty-First-Century City

Chapters in this section deal primarily with municipal policy-related concerns, 'planning' and 'placing', and more specifically with the recent role planning has assumed in the creation of the kinds of places believed to play a vital role in stimulating economic activity. In this sense, Part 3 is largely about the neo-liberal turn in urban governance and the economic role urban planning is given in this context. The chapters here describe and examine critically recent urban governance, development, planning, and lifestyle trends. The term 'placing' refers to efforts at designing or otherwise creating places that will 'look good' and 'feel good', thus attracting investments and middle- and upper-income customers with the wherewithal to generate economic activity.

Four planning-related chapters focus on contemporary issues that confront urban planners: Chapter 14 by Connelly and Roseland, on the new environmentalism; Chapter 18 by Grant and Filion, on attempts at changing land use; Chapter 23 by Gilliland, on the built environment and obesity; Chapter 24 by Blay-Palmer, on food systems. It might surprise some readers that we feature as topical planning concerns substantive work such as Gilliland's connection of community design with health and related problems, in particular obesity and other ailments that are induced by lack of physical activity, or Blay-Palmer's review of how food is delivered to and consumed within cities. It is indeed only in the last few years that these kinds of issues have been deemed 'legitimate' by younger or more forward-thinking members of the planning and related professions. On the other hand, while still reasonably new, the huge problem of environmental sustainability has been on the planner's radar screen for at least a couple of decades. In the case of environmentalism, what is new is its policy mandate. Chapter 14 underscores the importance that now firmly established environmental initiatives hold within contemporary professional planning circles. New inroads being made as environmental concerns are translated into policy planning are seen in large and small Canadian municipalities across the country. Ontario's 2006 *Growth Plan* policy statement and subsequent enactments such as the Greater Toronto Area Green Belt provide but one example. (The reader will also note that while important in and of itself as one initial step towards environmental remediation, the *Growth Plan* is also important in signifying a changing direction in municipal–provincial spheres regarding jurisdictional control over planning matters at the municipal and regional levels.)

Grant and Filion's chapter provides further discussion about other relatively new planning initiatives. Their focus is on recent strategies aimed at redressing the imbalance between suburban community design and movement patterns as discussed in earlier sections of this volume. Among other things, this chapter offers a critical view of 'New Urbanism' communities, concluding that, as developed to date, undue emphasis has been given to design and 'place' features of these communities at the expense of functional features such as trip destinations and modes of travel. As another relatively new development on the urban landscape, gated communities have been generally eschewed by planners due primarily to concerns about their exclusionary nature—this contrasts with New Urbanism design, which can be said to have generally aroused some level of enthusiasm within the profession. As developed in Canadian cities, however, both types of communities raise concern about movement patterns, as well as questions as to why people choose to live in them. Reasons for choosing these alternative community types seem to be more about distinction and difference (i.e., about 'placing'), and in the case of some gated communities about fear and refuge from the greater urban area outside (see also Cowen, Siciliano, and Smith, Chapter 17), as opposed to a choice for good community planning.

Inherent to discussions about 'placing' are three basic characteristics that the reader needs to have some knowledge of with regard to contemporary urban development. To begin, it is important to emphasize that the majority of so-called planned development today is carried out by the private sector with some influence from professional planning. Second, as passing familiarity with the established urban agenda for almost any Canadian city will attest, the latest and hottest trend is for highly designed projects reputed to 'sell places'. Finally, it follows that the overriding municipal preoccupation has been with creating attractive images and with the ultimate marketability of cities as competitive global places rather than with more 'bread and butter' issues like social housing, as was the case in the modern period (e.g., Florida, 2002, 2005).

By its title alone, Chapter 19, 'The Changing Meanings of Urban Places' by Lynch and Ley, clearly belongs to the debate about 'placing' cities, though there are also important connections back to recent planning initiatives. As discussed in Chapter 18, both New Urbanism and gated communities can be criticized for adhering more to principles of design and beliefs about what constitutes a 'good place' (as deemed desirable at least by selective segments of the real estate market) than to twenty-first-century concepts of 'good community planning'. Emphasis on placing also is pursued by Bain in Chapter 15, where the concept of the 'spectacular' is applied to raise awareness about the spectacular place-building that is the cornerstone of regeneration in the central- or inner-city zones of Canada's largest CMAs. Critics would argue that, so defined, 'placing' can be viewed as a strategic social construction of 'dreamscapes' and illusion. Proponents such as developers, however, will argue that placing is the very essence of what makes the twenty-first-century city distinctively different from the past and, most importantly, that placing has made metros like Vancouver and Toronto successful competitors for world city status, or, for that matter that, on a totally different scale, it is responsible for the appeal of smaller communities like Kelowna (and maybe even for the lack of appeal of slow- and no-growth places; see Donald and Hall, Chapter 16).

In a very different manner, overtones of placing also are found in the largely unplanned, mass social response that has made 'fear' synonymous with 'city'. New to Canadian cities in recent decades, fear is explored at length by Cowen, Siciliano, and Smith in Chapter 17. Whether we like it or not, today fear appears to be a significant force in urban landscapes. We even suspect that fear of a variety of features that are beyond the control of the average resident might have already become a new structuring parameter of large, contemporary Canadian cities. Fear, we are reminded, is a major reason why large numbers of households migrate from our biggest metropolitan areas to communities near their fringe—e.g., Port Coquitlam in the Greater Vancouver area or Milton just outside of Toronto. Fear is also responsible for all sorts of security systems such as surveillance cameras that are new to Canadian cities; fear may even be one of the reasons why too many Canadian urban dwellers, who have a choice, shun transit in favour of travelling in their own personal automobiles.

Other chapters in Part 3 identify planning concerns that, for the most part, have been raised in the recent past, that is, in the modern city, but these issues continue to represent problems in today's Canadian cities. Indeed, in many cases the 'neo-liberal turn' has been accused of making some of these problematic circumstances more acute than they ever were in the past—e.g., housing provision and homelessness. Certainly, it must be acknowledged that, like neo-liberalism itself, these concerns are ideological so not all urbanists should be expected to agree that they are in fact problematic. By and large, only scholars and policy-makers who affiliate themselves with centrist or left-of-centre political views or who are otherwise committed to social justice have their viewpoints represented in the four problem chapters. These concern: (1) the status of Aboriginal populations in cities; (2) the relative lack of attention given in urban policy matters to the plight of vulnerable older adults and dependent younger children and youth; (3) the problems of slow-growth and declining communities; and (4) the all-pervasive continuing concern with the provision of safe and affordable housing for all Canadians, involving sectors of society currently overlooked by the private marketplace and governments alike. Accordingly, these areas of urban affairs constitute a significant challenge to planners and municipal decision-makers. The reader will note that only selected and more salient problematic concerns are dealt with in Part 3. Some other problems, for example, those associated with racially different communities and other minority groups in the population, have been dealt with in previous chapters, or are simply excluded due to lack of space: for example, gay and feminist issues, or debate over education and the provision of schools. Finally, in order to highlight their relative importance, two of the most notable of problems inherited from the past that continue to plague twenty-first-century cities, home-lessness and urban sprawl, are dealt with in the concluding chapter, which alone comprises Part 4.

References

Florida, R.L. 2002. *The Rise of the Creative Class: And How It's Transforming Work, Leisure, Community and Everyday Life*. New York: Basic Books.

———. 2005. *Cities and the Creative Class*. New York: Routledge.

Ontario, Ministry of Public Infrastructure Renewal. 2006. *Growth Plan for the Greater Golden Horseshoe*. Toronto: Government of Ontario.

CHAPTER 14

Black Holes or White Knights? Cities and the Environment

Sean Connelly and Mark Roseland

Half of the world's people now live in cities. The activities of the world's urban residents contribute to a number of global environmental problems, from resource depletion, loss of bio-diversity, increasing greenhouse gas (GHG) emissions, increasing waste, and deterioration of water quality. These global concerns also have local impacts. In North America, buildings and urban infrastructure account for some 40 per cent of material consumption and a third of energy use. According to UN Food and Agriculture Organization estimates, the 16 million residents of New York City import 20,000 tonnes of food and export 10,000 tonnes of food waste per day (Kante, 2004).

A typical North American city of 100,000 inhabitants imports 200 tonnes of food, 1,000 tonnes of fuel, and 62,000 tonnes of water every day; it exports 100,000 tonnes of garbage and 40,000 tonnes of human waste each year (Morris, 1990). Indeed, it is these unsustainably 'developed' cities of the world that emit most of the world's GHG emissions, produce most of the world's solid and liquid wastes, consume most of the world's fossil fuels, emit the majority of ozone depleting compounds and toxic gases, and give economic incentive to the clearing of the world's forests and agricultural lands (UNEP, 1990). The world's cities take up just 2 per cent of the earth's surface, yet account for roughly 78 per cent of the carbon emissions from human activities, 76 per

cent of industrial wood use, and 60 per cent of the water tapped for use by people (UNEP, 2009). The environmental impact of cities stretches well beyond their borders. Cities require a land base much larger than that within their administrative boundaries to provide for the production of goods and services and the ecosystem services they require to absorb waste and pollution (Wacker-nagel and Rees, 1996; see Rees, Chapter 5).

The local and global environmental impacts of cities are a significant and pressing concern. Given these depressing facts, it is tempting to view cities as a source of global environmental problems—as black holes requiring ever-increasing resources and producing an ever-increasing amount of waste to meet societal demands. However, cities also offer reasons for hope. They provide many opportunities for improving the quality of life of citizens and provide fertile ground for producing innovations to help solve local and global environmental concerns. For example, Local Governments for Sustainability (also known as the International Council for Local Environmental Initiatives or ICLEI) represents a worldwide association of over 1,000 local governments committed to taking action to achieve tangible improvements in global sustainability through cumulative local action.

The City of Toronto currently emits 25 million tonnes of GHGs on an annual basis. But it is planning to reduce that amount by 80 per cent by 2050 through measures designed to tackle transportation, commercial, industrial, and residential emissions (Toronto, 2007). Since 1996, Vancouver has actively promoted alternative modes of transportation to driving and has seen a 44 per cent increase in walking, 180 per cent increase in bike trips, a 20 per cent increase in public transit use, and a 10 per cent decrease in vehicle trips (Vancouver, 2009).

Examples such as these demonstrate that cities can and do play a proactive role in addressing local and global environmental problems. We need our cities to serve as white knights that contribute to sustainability at local, national, and global levels. In this chapter, we argue that Canadian cities have demonstrated some success at sectoral initiatives to address environmental objectives, but for cities to act as a source of solutions to local and global environmental problems, much more attention needs to be given to the integration of these sector initiatives with community livability. Specifically, cities' thinking of environmental problems in terms of sustainable community development provides opportunities for engaging citizens and their governments around issues of environmental improvement, social progress, and economic enhancement.

In the next section we distinguish between two ways of conceptualizing environmental action in cities: ecological modernization and sustainable community development. The subsequent section examines two case studies exhibiting innovative approaches to sustainable community development, Toronto's Better Buildings Partnership and Surrey's East Clayton development. We conclude by revisiting our initial question, whether cities are black holes or white knights in the twenty-first-century project of living more sustainably on our planet.

Ecological Modernization or Sustainable Community Development?

While sustainable development is a relatively new term, the concept of ecological limits to human activity is not. It has been traced back to historical societies and traditional belief systems that recognized the relationship between humans and nature, and has been of concern through to Malthusian economic theories of limits to economic growth. More recently, social movements have focused on human-scale development, appropriate technology, green politics, bioregionalism, and social ecology (Mebratu, 1998; Roseland, 2001).

In the early twentieth century, the relationship between the environment and cities was primarily a local concern, focused on addressing the public health impacts of poor air or water quality and exposure to contaminants resulting from rapid industrialization within the city. Addressing environmental concerns relied on land-use zoning to separate industrial and residential areas and improved municipal infrastructure, such as sanitation and waste management that were designed to tame the negative by-products of the industrial city. The natural environment was identified as something external to the city, aside from manufactured landscapes of city parks designed to provide an oasis for urban residents (Hough, 2004).

At the same time, however, visionary planners such as Lewis Mumford were focused on a bottom-up transformation of the relationship between the city and the environment based on the recognition of the linkages between the biophysical environment, citizens, and cities (Friedmann, 1987). These concerns are more commonly recognized today through the concepts of sustainability and sustainable development.

The concept of sustainable development has achieved widespread recognition in the public, private, and NGO sectors. Yet, it has been interpreted in different and sometimes competing ways. Williams and Millington (2004) have characterized the diversity of interpretation along a spectrum from weak to strong sustainability, based on underlying concepts and worldviews of the relationships between the environment, economy, and society. The process of **ecological modernization** would be considered a weak sustainability approach, and is characterized by attention to technological solutions to environmental problems, the use of financial incentives, and 'greening' economic growth. Ecological modernization, on its own, is a reform-oriented rather than a transformative approach to environmental problems. It has been criticized for taking a symptomatic view of environmental problems by not challenging underlying process, structures, and values (such as our consumer-driven lifestyle), which create unsustainable communities—what William Rees (1995) refers to as 'staying the course' on the expansionist paradigm.

A strong sustainability approach is characterized by recognition of finite limits to growth and technological solutions to environmental constraints, and by greater attention to local *development* as opposed simply to growth. For example, a weak sustainability solution to addressing increasing emissions from private automobiles might be to provide financial incentives to citizens to encourage the purchase of hybrid cars. While such an approach likely would lead to an increase in hybrids on the roads, it does not necessarily address the vehicle miles travelled or reduce the need for cars in the first place. A strong sustainability approach would focus more on solutions that reduce the need to drive by providing housing closer to workplaces and making public transit, walking, and cycling more accessible. So what does this mean for cities and what role do cities play in advancing sustainability?

Examples from Canadian Cities

Canadian cities of all sizes have made commitments to move towards sustainability. There are numerous examples of cities that have been successful in implementing sustainability projects or initiatives. Each year the Federation of Canadian Municipalities (FCM), through its Sustainable Community Awards, recognizes the efforts of municipalities that have implemented innovative

sustainability projects in a variety of categories (Table 14.1).

Projects like these offer tangible improvements to the natural environment and in most cases represent significant cost savings to the municipality. On their own, however, they cannot transform our cities into more sustainable places. These examples demonstrate that some Canadian cities have been effective in making the connections between environment and economic concerns by relying on an ecological modernization framework of technological innovation and financial incentives to drive efficiency (Fisher and Freudenburg, 2001). While noteworthy, these examples represent the 'low-hanging fruit', those changes that are easiest to make because they do not challenge the status quo. They make financial sense and are focused on the relationships between the environment and the economy. However, they do not address the political conflicts among environment, economy, and equity goals that are at the heart of sustainable cities (Campbell, 1996).

Many Canadian cities have initiated sustainable development projects in such areas as green building programs, affordable housing, open space preservation, recycling, climate change, and smart growth, but they are occurring primarily on a project-by-project or issue-oriented basis, rather than being connected and integrated throughout municipal governments and their larger civic communities. In contrast to this ecological modernization approach to environmental problems, the framework of sustainable community development (SCD) illustrates a 'systems approach' that recognizes that cities are complex systems with multiple inter-relationships at different scales (rural/urban, local/national/global) and among different issue-based sectors (e.g., linkages of local air quality, transportation planning, and land-use densities). SCD for cities involves widespread community awareness-raising and integrated municipal involvement aimed at bringing about a shared understanding of what sustainability means and how to achieve

Table 14.1 FCM Sustainable Community Award Winners, 2008

Category	Winner
Buildings	Toronto's Arena Retrofit program was designed to reduce building operating costs and increase energy and water efficiencies. The program incorporates technological upgrades with an energy awareness program for staff, and has resulted in savings in operation costs of $1.25 million annually and, by reducing energy consumption, a reduction in CO_2 emissions of 4,600 tonnes.
Energy	The City of Saint John, New Brunswick, established a Municipal Energy Efficiency Program that has resulted in energy retrofits to over 50 buildings. The program has reduced GHG emissions by 17 per cent between 1996 and 2007 and generated a total cost savings of more than $5 million.
Planning	In Ontario, the City of Pickering's Sustainable Pickering project began in response to rapid population growth. The project is a city-wide initiative based on a 'lead by example' approach that engages citizens in a broad cross-section of sustainability initiatives. City council has since approved GHG reduction targets of 35 per cent per capita for the community and 50 per cent per capita for corporate entities by 2016.
Residential Development	The City of Kitchener, Ontario, used tax incremental financing and streamlined development processes as incentives for the redevelopment of an industrial brownfield site into a live-work community in the downtown core. The new development incorporates a diversity of housing styles that regenerate schools, parks, and community centres in the core.
Transportation	The Smart Commute program is an initiative of the Greater Toronto and Hamilton Area designed to link the trip-reduction strategies of Metrolinx, municipal governments, non-profit organizations, local universities, boards of trade, and chambers of commerce. More than 75 businesses have signed on, representing over 200,000 commuters. From 2004 to 2007, the project reduced the total number of vehicle kilometres travelled in the region by 75 million and GHG emissions by 17,400 tonnes.
Wastewater	For many years, the Saint Charles River in Quebec City suffered from sewage overflows and the natural ecosystem had been replaced by concrete channels. Since 1995, the Commission pour la mise en valeur du projet de dépollution et de renaturalisation has worked to implement major upgrades to wastewater treatment and naturalization of the river, creating 65,000 square metres of wildlife habitat in the city. A network of 14 retention ponds prevents six million cubic metres of wastewater annually from being discharged into the river.
Water	The Regional Municipality of Peel in Ontario is a rapidly growing region adjacent to Toronto. To address increasing demands for water supply and wastewater treatment, the Region developed a Water Efficiency Plan that includes a rebate program for water-saving products, a leak detection system, and an audit program. The program is expected to reduce peak demand by 10 per cent by 2015 and generate $10.4 million in capital cost savings.

Source: Adapted from FCM (2009).

it throughout all sectors of municipal government and the wider civic community (James and Lahti, 2004; Gruder et al., 2007).

Sustainable Community Development

The city is a complex system and you cannot have a sustainable city without all of the integrated parts being sustainable; therefore environmental problems in cities need to be thought of in an integrated manner. Integration is required both within the city and at larger scales. By raising the issue of scale, Mitlin and Satterthwaite (1996) remind us that the series of individual actions and decisions we make in our daily lives have repercussions for the sustainability of the bioregion and the globe. The central idea of bioregionalism is place. A bioregion literally means a life-territory. It is a place defined by its biota and topography, a region governed by nature instead of by human-made jurisdiction (Sale, 1985). Thinking of the city in bioregional terms helps to reorient our actions against the continuing destruction of natural systems, such as forests and rivers, and towards the renewal of natural systems based on a thorough knowledge of how they work, and the development of techniques appropriate to specific sites (Dodge, 1981; Thayer, 2003; Carr, 2004). The '**ecological footprint**' analysis developed by Wackernagel and Rees (1996) is a bioregional tool that considers the impact of cities on natural resources and ecosystems (see Rees, Chapter 5). Their work demonstrates that although some industrial cities may appear to be sustainable, they 'appropriate' carrying capacity not only from their own rural and resource regions but also from 'distant elsewheres', i.e., they import sustainability. Wackernagel and Rees (1996) estimate, for example, that the Vancouver region's ecological footprint is 19 times the actual land area, meaning that the resources consumed and wastes produced in the region far exceed the ecological carrying capacity.

Sustainability requires the transformation of our cities so that they are embedded in the patterns and processes of natural, sustainable ecosystems, achieving ecological regeneration, healthy communities, and viable economies within their bioregions. While conceptually this makes sense, the reality of sustainability in cities is chaotic and very much contextually specific, where sustainability initiatives can be sidetracked by conflict, power struggles, ambiguity, crisis, availability of resources, and a lack of leadership and commitment, among others. Sustainability outcomes will be dependent on what people living in communities do, through local projects and local conflicts, and how they link their communities to their local environment (Evans, 2002). Sustainability often arises out of particular local concerns for which there is enough sense of crisis to generate discussion about alternatives, and where concern over a particular issue is great enough to make the risks associated with designing alternatives more acceptable (Connelly et al., 2009). For example, some point to the successful public protest over the proposed highway through downtown Vancouver in the 1960s as the source of Vancouver's current reputation as one of the world's most livable cities (Berelowitz, 2005). As a result, Vancouver has never been cut off from its waterfront by freeways, and land-use planning and public infrastructure investments have been more focused on cycling and transit-oriented development than has been the case in many other North American cities.

The point is that sustainable solutions to environmental problems in cities can only be developed within those communities where the impacts are felt, given that the problems and solutions are contextually sensitive and result in different conflicts and complexities from place to place. Campbell (1996) clearly articulates the source of these conflicts and complexities as the competing priorities in cities between the economy, the environment, and social justice. These competing priorities are presented as points on a triangle and result in the all-too-familiar conflicts between environment and economy along one axis, growth and equity on the other, and environment and equity on the final axis. These conflicts form the basis of political debate at local, national, and international scales. Sustainability is presented as the heart of the triangle, where solutions to the conflicts among society, economy, and environment are resolved.

In both the sustainability literature and in practice, there has been a lack of emphasis placed on the relationships between social, cultural, economic, and equity issues and the environment, issues that in most communities remain politically sensitive (Pugh, 1996). Much emphasis in practice has been on reducing the environmental impacts of development, particularly in areas that lend themselves more readily to quantification, measurement, and win–win scenarios such as air quality, transportation, and waste management (Kenworthy, 2006; Mazza and Rydin, 1997). These types of initiatives tend to avoid political conflicts by focusing on technological solutions and efficiency arguments to promote change, rather than by addressing more conflictual social issues around equity and justice. The recycling movement that began in the 1980s, with the blue box program, is a good example. While promoted under the motto of 'reduce-reuse-recycle', the social and political consequences of addressing issues relating to reducing our patterns of over-consumption were avoided by squarely focusing on the technological fix of sorting recyclables.

If cities are to serve as white knights producing solutions to local and global environmental problems, they need to address problems in a more integrated manner. SCD applies the concept of sustainable development to the local or community level where the challenge is to integrate sustainable development principles, long-term planning processes, and specific community priorities. The community capital framework (Figure 14.1) is a useful way of framing the complexity of integrated development at the community scale and to describe the meaning and objectives of SCD.

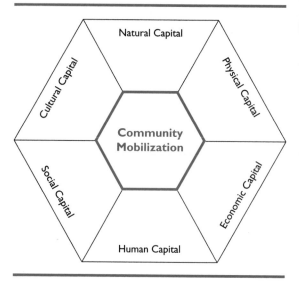

Figure 14.1 Community capital framework
Source: Roseland (2005).

The framework consists of six types of community capital (Table 14.2) that reflect the assets or resources in a particular community and, taken together, encourage discussion and acknowledgement of the linkages and interrelationships between them. It recognizes that the goal behind SCD is to adopt strategies, structures, and processes that engage citizens in their day-to-day activities and the structures of local government in the quantitative and qualitative improvement of all six forms of community capital. SCD is dependent on the active participation of citizens in democratic processes that identify sustainability values, visions, and actions.

Multi-stakeholder participation has emerged as a central aspect of developing sustainability solutions to community problems. Given the complexity of sustainability and the numerous interconnections between issues, it makes sense that business-as-usual, government-as-usual, and protesting-as-usual are insufficient approaches (Hemmati et al., 2002). The challenge, however, lies in developing processes that are effective at

Table 14.2 Types of Community Capital

Types of Capital	Characteristics
Natural capital	Natural (or environmental or ecological) capital consists of the biophysical resources, living systems, and life-support services of our planet. For example, vegetated surfaces are an important natural capital asset in urban areas that reduces surface run-off and water pollution.
Physical capital	Physical (or manufactured or produced) capital is the stock of material resources such as equipment, buildings, machinery, and other infrastructure that can be used to produce a flow of future income. For example, a city's public transportation infrastructure is an important physical capital asset that contributes to urban sustainability.
Economic capital	Economic (or financial) capital refers to the ways we allocate resources and make decisions about our material lives and includes cash, investments, and the monetary system. Unlike other capital types, its value is not intrinsic and derives, instead, from the human, physical, social, cultural, and natural capital it represents (e.g., via shares, stocks, cash).
Human capital	Human capital consists of health, knowledge, skills, motivations, competencies, and other attributes (such as emotional and spiritual capacity) embodied in individuals that facilitate the creation of personal, social, and economic well-being. For example, environmental literacy might be considered an important human capital asset for advancing urban sustainability.
Social capital	Social capital consists of relationships, networks, structures, and institutions that facilitate collective action and the shared knowledge, understandings, and patterns of interactions that a group of people bring to any productive activity. It includes families, communities, businesses, trade unions, voluntary organizations, legal/political systems, and educational and health institutions. Social capital is embodied in formal (e.g., government) and informal (e.g., social networks) structures, organizations, and institutions.
Cultural capital	Cultural capital is the product of shared experience through traditions, customs, values, heritage, identity, and history. Cultural capital is particularly important in Aboriginal communities and in other communities with a long and situated history.

Source: Adapted from Roseland (2005).

engaging a broad cross-section of citizens and organizations, each engaged in particular personal or sector issues, in meaningful discussion around the search for cross-cutting solutions based on principles of equity, accountability, transparency, and shared decision-making.

Neighbourhood councils have been used by a number of cities as an approach that provides a formal structure for public involvement at the neighbourhood level. For example, neighbourhood councils in Quebec City provide the basis for local citizens to engage with a multitude of issues at the local scale. The councils have balanced gender composition and are chaired by the non-voting local councillor, creating a direct link into political decision-making at the city level. These neighbourhood councils play an important role in defining and shaping discussions around values and visions for identifying and addressing neighbour-hood concerns (Bherer, 2002).

SCD requires local governments to lead by example in their day-to-day operations and to support initiatives that promote participation and build consensus around shared values focused on improving the quality of life in communities. Quebec City's Children's Council provides an interesting example. Each year, elementary schools in the city nominate a student to participate in the Children's Council that is focused on addressing a city-wide problem. The Council illustrates the importance of engagement in public life and provides City Council with a unique viewpoint and set of solutions to a particular problem. Through processes such as these, focused on community problem-solving, sustainability solutions will emerge that can make our cities and communities more resilient (Roseland, 2005).

Integrating SCD principles, planning processes, and community priorities remains a challenge for Canadian cities. The economic rationale of city-building is often emphasized, where the city's growth is seen as the main engine for economic development. Often, by comparison, the important cultural and social components of the urban sphere that help build community resiliency are neglected. The placement of environmental considerations within the list of municipal priorities remains a source of constant debate (e.g., City of Toronto's Roundtable on the Environment). Decision-making frameworks for incorporating environment and sustainability into municipal decision-making processes frequently are based on the land-use planning process, an area where municipal governments have direct control. There is no shortage of frameworks or tools for planning sustainable communities and for municipal decision-making (e.g., ICLEI, 1996; InfraGuide, 2003; James and Lahti, 2004; Ling et al., 2007; Robert et al., 2002; Roseland, 2005; Seymoar, 2004).

The 'natural step' framework is one example of a sustainability framework that incorporates high-level sustainability principles, planning processes, and community priorities. For example, it begins from the principles that we need to reduce our use of non-renewable resources and toxic substances, curtail the physical degradation of natural environments, and develop social capacity to meet our basic human needs. Based on these principles, community planning processes are used that engage citizens in visioning what their sustainable community might look like and in identifying some of the actions required to achieve that transformation and to address specific community priorities (James and Lahti, 2004).

In the next section, we examine two award-winning city cases to illustrate how they have dealt with the complex challenge of addressing environmental problems in a more integrated and sustainable way.

Innovative Approaches to Integration

Toronto's Better Buildings Partnership and Surrey's East Clayton development are two recent winners of the FCM Sustainable Community Awards. In each case, attempts were made to link a particular community concern over an environmental problem with broader and more integrated sustainability solutions. These examples serve to illustrate the potential impact that initiatives at the city level can have on advancing sustainability.

Toronto's Better Buildings Partnership
Toronto is Canada's most populated city with over 2.5 million people in 2006. Toronto's Better Buildings Partnership (BBP) provides a practical example of implementation related to the city's CO_2 emissions reduction goals. This program aims to decrease GHG emissions and improve urban air quality through energy-efficiency retrofits to buildings in the industrial-commercial-institutional building sector. The program, launched in 1996, provides comprehensive energy retrofits to private

and public buildings through lending schemes that allow building owners to pay back retrofit costs through efficiency gains. Building retrofits have included lighting efficiency upgrades, improvements to central heating and cooling systems, integrating alternative energy sources, to name a few examples. BBP has survived 11 years within the constraints of municipal financing and has contributed to improvements to over 600 buildings, resulting in a reduction of 200,500 tonnes of CO_2 annually, as well as $19 million in savings to building owners. As an example, the largest photovoltaic system in the country, at Exhibition Place, reduces GHG emissions by 115 tonnes per year and results in an annual savings of $10,000 in electricity costs (Figure 14.2).

The initial motivation for the BBP can be traced back to 1988 when the city hosted a conference on air quality and cities. At that time, Toronto was experiencing air-quality problems such as smog. This spurred the city to commit itself to a 20 per cent reduction in GHG emissions from 1988 levels, making it the first city to make such commitments. This goal served as the defining moment that spurred future and sustained commitment from the city.

Toronto's BBP exemplifies a city addressing community problems in an innovative manner. The municipal leadership was able to link specific community concerns over air quality with broad concerns about economic development during an economic recession by introducing a building

Figure 14.2 Solar roof at Toronto's Exhibition Place is one of the elements being used to achieve the goal of energy self-sufficiency by 2010. (Sean Connelly)

retrofit program. Over time, the BBP has improved the environmental and financial performance of buildings, created jobs in the 'green economy', and contributed to more efficient resource use in the city. The BBP has worked with Toronto Community Housing Corporation (the agency responsible for **social housing** in Toronto) to initiate building retrofits that protect low-income residents from rising energy costs and that engage residents in job-training, thus establishing linkages to social sustainability through the program. The BBP has succeeded in shifting the thinking around buildings and energy uses and has used its institutional capacity in this area to drive innovation and promote uptake of city demonstration projects more widely and in a more integrated way.

Surrey's East Clayton Neighbourhood Development

Surrey is British Columbia's second largest city, with a population in 2006 of approximately 394,000 people. The East Clayton development is a new planned neighbourhood designed to accommodate 13,000 residents at a density much higher than a conventional suburban development. The East Clayton Neighbourhood Plan arose from two sets of conditions: (1) the need to develop new urban areas in response to population growth, and (2) the need to balance development with the protection of agricultural land and salmon habitats. East Clayton is located in an area of Surrey that contains farmland and salmon habitat, and the City of Surrey had already received threats of lawsuits from farmers who contended that run-off from urban development would cause damage to their lowland farms. These factors combined to provide the motivation to explore sustainable neighbourhood design.

Guided by seven sustainability principles specific to the development,[1] the main priorities of the East Clayton project were to reduce urban stormwater run-off through on-site infiltration techniques, and to apply neo-traditional urban design considerations such as rear lanes, higher densities, live-work zoning, integration of commercial and business zones in the neighbourhood, and greenways (Figure 14.3). The increased density and reduction in paved surfaces resulted in a greater proportion of permeable surfaces, allowing for greater water infiltration. Greener stormwater infrastructure, such as retention ponds and swales that channelled run-off from the streets onto vegetated surfaces, was designed to address run-off concerns. The city created the Neighbourhood Concept Plan through engagement with local property owners, citizens, and city staff in a series of design charrettes with the goal of introducing sustainability measures to the new neighbourhood development through site design that went beyond the initial concern over stormwater management.

The East Clayton Neighbourhood Concept Plan and the subsequent development of the community have been successful in transforming the development standards of new residential communities throughout Surrey to integrate sustainability principles. The lessons learned from East Clayton about innovative green infrastructure and neighbourhood design all have been applied to a certain degree in new developments throughout the city. The East Clayton development has raised the bar for what can be expected from developers and the market has responded favourably to the mixture of housing types and affordability. While much of the success relates to the physical design of the neighbourhood, the development has contributed to an increased awareness of sustainability in Surrey. The city has recently adopted a sustainability charter that is intended to set long-term guidelines for future development based on sustainability principles.

Relationship to Community Capital

The Toronto and Surrey examples demonstrate value in addressing the environmental problems of cities in a more integrated way, and illustrate the potential benefits of local sustainability solutions

Figure 14.3 Live-work zoning in East Clayton as a strategy to provide a greater mix of uses in a neighbourhood. (Sean Connelly)

to global environmental problems. In each case, specific environmental problems were treated as catalysts for broader and more integrated sustainability solutions, improving the overall levels of community capital.

For example, the BBP began with widespread public concern over air quality, smog advisories, and GHG emissions, which were clear environmental or natural capital problems. The BBP also emerged, however, at a time when Toronto was undergoing an economic recession, which had greatly reduced the level of new property development and resulted in a loss of jobs in the construction industry, negatively impacting economic capital. By linking natural and economic capital concerns through an innovative program such as the BBP, the city was able to address other community capital issues as well.

The City of Toronto recognized that there was a need to intervene in society to advance SCD concerns, and that the city had a mandate to intervene. The BBP represents a major shift in the role and mandate of municipalities that challenged the status quo. The $23 million that made up the Toronto Atmospheric Fund endowment could instead have been used to reduce property taxes, support local business development, invest in roads or other infrastructure—actions more typically associated with local government mandates. It was recognized, however, that the city had relative strengths in terms of human and physical capital and the knowledge and capacity to intervene to address air-quality concerns through buildings. In addition, building retrofits were seen as a sector in which the city could lead by example and reduce the exposure of residents to rising energy costs, provide retraining and jobs for unemployed trade workers, and serve as an example for the broader community of the linkages between the environment, society, and the economy.

The success of the BBP has both contributed to and benefited from new social capital linkages between a diversity of organizations from small neighbourhood energy co-operatives, commercial property developers, environmental NGOs, and the city itself. These linkages have augmented the standing of air quality, energy, and climate change issues as key concerns for the City of Toronto.

Likewise, the East Clayton neighbourhood development emerged primarily as a result of concern over natural capital. Concern over urban stormwater run-off and the impact it would have on salmon bearing streams provided the rationale to explore alternative development options. Sustainability solutions were developed, however, that went beyond just addressing environmental problems associated with conventional suburban development. For example, low levels of social capital that often accompany conventional suburban developments were an issue; residents tend to be engaged in a range of communities that do not necessarily correspond to where they live. In these instances, a neighbourhood lacks a shared sense of cultural place, which is important to local civics and quality of life.

The focus of the East Clayton development was to transform the physical infrastructure and design for the neighbourhood to reduce the environmental impacts, and to encourage greater interaction among community members. Working with the development industry brought considerable economic capital to the project that provided incentives for community mobilization for the project.

If we examine the East Clayton development through the community capital framework, it is apparent that the city recognized the lack of social capital, which was identified as a limiting factor by charrette participants. Building social capital became a focus of many of the sustainability principles that guided the development. For example, the concepts of increased density and walkable neighbourhoods, inclusion of local commercial nodes, and a mixture of housing types were included to meet sustainability goals of reducing energy use, increasing local economic development and housing affordability, as well as providing more opportunities for neighbourhood resident interaction. The importance of creating opportunities for building social capital and the development of shared understandings of community were important in the development of a broader cultural shift towards sustainability in the city.

Examining both the Toronto and Surrey examples of sustainability initiatives through the community capital framework yields an appreciation for the need to move beyond the environment–economy axis and to think of environmental problems in a more integrated fashion to maximize each of the individual types of capital. In both the Toronto and Surrey examples, the importance of these linkages has been recognized. For example, the BBP has attempted to link retrofit projects to buildings operated by the Toronto Community Housing Corporation, thereby reducing low-income tenants' exposure to rising energy costs and providing training and job opportunities for residents. The East Clayton development was physically designed to increase opportunities for contact among neighbours as a first step in building social capital within the neighbourhood. In each example, the communities were able to use their strengths and capacity in one sectoral area as a catalyst for engaging in broader and more comprehensive initiatives for advancing **urban sustainability**.

While more initiatives are needed, there are examples of Canadian cities having addressed the environmental impacts on them in innovative ways. We must focus, however, on other initiatives as well, that advance sustainability and recognize that the environment is but one aspect of life. Initiatives for local food security, affordable housing, and the reduction of consumption patterns provide promising areas for addressing sustainability in a more comprehensive manner in Canadian cities (see Walker and Carter, Chapter 20; Blay-Palmer, Chapter 24). The challenge is no longer a lack of tangible examples to draw on, but how to mobilize citizens and their governments to identify priorities and implement strategies that are relevant locally.

Conclusion

Until citizens and their leaders become more engaged with strategies for reducing their ecological footprint, Canadian cities will be viewed as black holes that continue to appropriate ever-increasing quantities of resources and produce ever-increasing quantities of waste. The evidence from Canadian cities demonstrates numerous examples of innovative initiatives that seek to reconcile economic and environmental priorities. Cities still struggle, however, with engaging citizens in truly integrated efforts that address the economic, environmental, and social aspects of sustainability in their daily lives.

Attempts at advancing sustainability at the city level are limited by perceived mandate and jurisdictional issues and revenue constraints. In areas such as land-use planning and development, however, where greater degrees of local control exist, transformative change is possible based on groups of actors articulating a common vision through a collaborative process, identifying strategies, and taking collective action.

We began the chapter by questioning whether Canadian cities should be viewed as black holes or as white knights in terms of how they are contributing to local and global environmental problems. Are cities a source of environmental problems or can they be a source of sustainability solutions? As Newman, Beatley, and Boyer (2009) ask, are cities the problem or is the problem the way cities are organized? A pro-urban approach offers an optimistic scenario for the future, acknowledging that we have the capacity to change the form and function of cities and how they are organized to reduce their environmental footprints and to make them more sustainable. To change the way Canadian cities are organized requires shifts in the way citizens and their governments interact through planning processes, in the way that problems are identified, and in the way options are evaluated. Canadian cities that embrace sustainable community development will think of community problems in a more integrated manner, will recognize the complex economic, social, and environmental interactions within urban systems, and will identify key opportunities for change.

As we saw in the BBP and East Clayton examples, particular crises or problems in specific communities can effectively bring together citizen concerns, public awareness, and local government resources to address problems in a different and more integrated way. In each case, specific community priorities were articulated in a manner that was relevant for a broad cross-section of citizens and organizations. These stakeholders were then able to participate in processes that provided the opportunity for solutions to emerge that were more closely linked to sustainability principles. If cities are going to reach their potential as white knights for global sustainability and serve as solutions to local and global problems, they must organize themselves to construct visions and plans to rebuild their communities in a manner that addresses the social, economic, and environmental concerns of all citizens.

Review Questions

1. What is the difference between the concepts of 'ecological modernization' and 'sustainable community development' and how do the concepts relate to urban geography and planning?
2. Can cities be 'white knights' in our collective crusade to solve global environmental problems like climate change and natural habitat destruction, or are cities destined to remain 'black holes'?

Note

1. The sustainability principles included: increased density and walkable neighbourhoods; a mixture of housing types; promotion of social interaction through dwellings oriented more closely and interactively to the street; use of rear lanes for parking cars and for services at rear of dwellings; an interconnected street network and public transit to connect with surrounding region; narrow streets to reduce infrastructure burden; and preservation of the natural environment and promotion of natural drainage.

References

Berelowitz, L. 2005. *Dream City: Vancouver and the Global Imagination.* Vancouver: Douglas & McIntyre.

Bherer, L. 2002. 'Representation, expertise and participatory policies: The case of Quebec City's neighbourhood councils', paper presented at European Consortium for Political Research (ECPR). Turin, 22–7 Mar.

Campbell, S. 1996. 'Green cities, growing cities, just cities? Urban planning and the contradictions of sustainable development', *Journal of the American Planning Association* 62, 3: 296–312.

Carr, M. 2004. *Bioregionalism and Civil Society: Democratic Challenges to Corporate Globalism.* Vancouver: University of British Columbia Press.

Connelly, S., S. Markey, and M. Roseland. 2010. 'Strategic sustainability and community infrastructure', *Canadian Journal of Urban Research* 18, 1 (suppl.): 1–23.

Dodge, J. 1981. 'Living by life: Some bioregional theory and practice', *CoEvolution Quarterly* 32: 6–12.

Evans, P. 2002. *Livable Cities? Urban Struggles for Livelihood and Sustainability.* Berkeley: University of California Press.

Federation of Canadian Municipalities (FCM). 2009. 'FCM–CH2M Hill Sustainable Community Awards'. At: <www.collectivitesviables.fcm.ca/FCM-CH2M-Awards/db/awards.html>.

Fisher, D.R., and W.R. Freudenburg. 2001. 'Ecological modernization and its critics: Assessing the past and looking toward the future', *Society and Natural Resources* 14, 8: 701–9.

Friedmann, J. 1987. *Planning in the Public Domain: From Knowledge to Action.* Princeton, NJ: Princeton University Press.

Gruder, S., A. Haines, J. Hembd, L. MacKinnon, and J. Silberstein. 2007. *Toward a Sustainable Community: A Toolkit for Local Government.* Madison: University of Wisconsin Press.

Hemmati, M. 2002. *Multi-Stakeholder Processes for Governance and Sustainability: Beyond Deadlock and Conflict.* London: Earthscan.

Hough, M. 2004. *Cities and Natural Process: A Basis for Sustainability*, 2nd edn. London: Routledge.

ICLEI. 1996. *The Local Agenda 21 Planning Guide: An Introduction to Sustainable Development Planning.* Toronto.

InfraGuide. 2003. 'Planning and Defining Municipal Infrastructure Needs'. At: <www.infraguide.ca>.

James, S., and T. Lahti. 2004. *The Natural Step for Communities: How Cities and Towns Can Change to Sustainable Practices.* Gabriola, BC: New Society Publishers.

Kante, B. 2004. 'Local capacities for global agendas: Impact of cities on the global environment', World Urban Form: Dialogue on Urban Sustainability, Barcelona, 15 Sept.

Kenworthy, J.R. 2006. 'The eco-city: Ten key transport and planning dimensions for sustainable city development', *Environment and Urbanization* 18, 1: 67–85.

Ling, C., A. Dale, and K.S. Hanna. 2007. 'Integrated community sustainability planning tool'. At: <www.crcresearch.org/files-crcresearch/File/PlanningTool(1).pdf>.

Mazza, L., and Y. Rydin. 1997. 'Urban sustainability: Discourses, networks and policy tools', *Progress in Planning* 47, 1: 1–74.

Mebratu, D. 1998. 'Sustainability and sustainable development: Historical and conceptual review', *Environmental Impact Assessment Review* 18, 6: 493.

Mitlin, D., and D. Satterthwaite. 1996. 'Sustainable development and cities', in Pugh (1996: 23–62).

Morris, D. 1990. 'The Ecological City as a Self-Reliant City', in D. Gordon, ed., *Green Cities: Ecologically Sound*

Approaches to Urban Space. Montreal: Black Rose Books, 21–35.

Newman, P., T. Beatley, and H. Boyer. 2009. *Resilient Cities: Responding to Peak Oil and Climate Change*. Washington: Island Press.

Pugh, C.D.J. 1996. *Sustainability, the Environment and Urbanization*. London: Earthscan.

Rees, W.E. 1995. 'Achieving sustainability: Reform or transformation?', *Journal of Planning Literature* 9, 4: 343–61.

Robert, K., B. Schmidt-Bleek, J. Aloisi de Larderel, G. Basile, J. Leo Jansen, R. Kuehr, et al. 2002. 'Strategic sustainable development—Selection, design and synergies of applied tools', *Journal of Cleaner Production* 10, 3: 197–214.

Roseland, M. 2001. 'The eco-city approach to sustainable development in urban areas', in D. Devuyst, ed., *How Green Is the City? Sustainability Assessment and the Management of Urban Environments*. New York: Columbia University Press, 85–104.

————. 2005. *Toward Sustainable Communities: Resources for Citizens and Their Governments*, rev. edn. Gabriola Island, BC: New Society Publishers.

Sale, K. 1985. *Dwellers in the Land: The Bioregional Vision*. San Francisco: Sierra Club.

Seymoar, N.-K. 2004. *Planning for Long-term Urban Sustainability: A Guide to Frameworks and Tools*. Vancouver: +30 Network.

Thayer, R. 2003. *LifePlace: Bioregional Thought and Practice*. Berkeley: University of California Press.

Toronto, City of. 2007. *Change Is in the Air: Climate Change, Clean Air and Sustainable Energy Action Plan*. At: <www.toronto.ca/changeisintheair/pdf/clean_air_action_plan.pdf>.

United Nations Environment Program (UNEP). 1990. 'Call to a World Congress of Local Governments for a Sustainable Future', United Nations, New York, 5–8 Sept.

————, Division of Technology, Industry and Economics. 2009. At: <www.unep.or.jp/ietc/publications/freshwater/fms7/9.asp>. (25 Apr.)

Vancouver, City of. 2009. 'Greenest City: Quickstart Recommendations'. At: <http://vancouver.ca/greenestcity/PDF/greenestcity-quickstart.pdf>.

Wackernagel, M., and W. Rees. 1996. *Our Ecological Footprint: Reducing Human Impact on the Earth*. Gabriola Island, BC: New Society Publishers.

Williams, C.C., and A.C. Millington. 2004. 'The diverse and contested meanings of sustainable development', *Geographical Journal* 170, 2: 99–104.

CHAPTER 15

Re-Imaging, Re-Elevating, and Re-Placing the Urban: The Cultural Transformation of the Inner City in the Twenty-First Century

Alison Bain

Introduction

This chapter explores the new structural dynamics of metropolitan change at work in the inner city. As introductory chapters in this volume illustrate, it is the city's inner city that has shown most transformative change over the last quarter-century. In the last decade, in particular, with the apparent dominance of the new neo-liberal urban agenda, this transformation process has accelerated.

In this chapter, the prefix *re-* (meaning once more, again, or afresh) is used repeatedly in title and text to emphasize that cities are works in progress shaped by, and reflective of, prevailing political, economic, and social forces (Kyle, 2007). In the case of the inner city, this chapter demonstrates how cultural forces have transformed formerly industrialized and centralized portions of the city-scape into up-scale and fashionable post-industrial places in many Canadian metropolitan areas.

The reader should be aware that interpretations of this 'cultural turn' range from celebratory (Florida, 2002, 2005; Hall, 2003; Montgomery, 2008) to deeply and critically concerned (e.g., Kipfer and Keil, 2002; Ley, 2003; Walks, 2001). The current discussion leans towards the latter perspective. Fundamentally, this chapter asks what the contemporary built landscapes of core parts of Canada's largest urban centres reveal about who and what are privileged there. Conclusions are filtered through a social justice lens that reveals the vulnerable underbelly of spectacular but exclusive socio-spatial processes that (re)produce inequalities between people and places.

The argument asserted in this chapter is three-fold. First, that the twenty-first-century metropolitan quest for civic renewal has been dominated by a creative capital model of privately sponsored urban development (Allahwala, Boudreau, and Keil, Chapter 12). Second, that while culture can be used strategically to foster the development of place identities, high-profile cultural initiatives can exacerbate social divisions by creating consumption-driven urban environments with the potential for producing experiences that are elitist and exclusionary. Third, it proposes that a more socially inclusive urban future may be possible for Canadian cities, if urban decision-makers aim for a finer balance between the production of spectacle and the production of possibility in the informal spaces of everyday life.

Urban Development in Transition

When Canadian cities are examined, it becomes clear that each generation of citizens, decision-makers, and scholars has used their experiences, ideas, labour, and capital to intellectually and materially re-imagine and rebuild the urban fabric. Thus, '[b]elying the apparent inertia of concrete and steel, cities are dynamic entities' that have been dramatically transformed over the twentieth century. The following is a brief historical overview of urban development within a temporal framework of pre- and post-1945 and pre- and post-1975 (see also Bunting and Filion, Chapter 2; Filion and Bunting, Chapter 3).

In an earlier edition of this book, Ley and Frost (2006: ch. 11) document the historical development of the Canadian inner city as concept, region, and object of study. They present a fourfold typology of different kinds of districts: declining; stable; revitalized; and redeveloped. Although the inner city is at no time homogeneous, Ley and Frost's typology represents a useful transitional view of the inner city. They move from a discussion of the 1960s late-modern period, when deterioration and decline appeared to be all-pervasive across inner-city zones, to the early twenty-first century when inner-city redevelopment and **revitalization** became the celebrated characteristics of the new, creative city.

In Canada, the pre-1945 period has been referred to by Bunting and Filion, in Chapter 2, as an era of inner-city development driven by industrial production. At the close of the nineteenth century, municipal governments focused on building the basic infrastructure of cities: sidewalks, roads, bridges, sewer systems, and water filtration plants. In the early twentieth century, the role of municipal government expanded further to include the establishment of housing programs and parks and recreation facilities. Urban growth became focused on the central business district; industrial corridors radiated along waterways and railways, and high-density residential neighbourhoods segregated by income and ethnicity developed along public transit lines and roads close to retail and other educational, religious, and health-care institutions. With the creation of the **welfare state** in the 1940s, urban problems were framed as national problems, and federal and provincial governments became heavily involved in municipal affairs.

In the ensuing post-World War II period, 1945–75, metropolitan growth accelerated to include suburbs in its embrace. This was a time of massive government-led restructuring of urban space, underwritten by the increasing professionalization and bureaucratization of the urban planning process (Grant, 2006). At the same time, decline was becoming prevalent in many parts of the inner city. Central-city decline has been explained by the concept of 'urban filtering' whereby the construction of new homes for the affluent generated a 'chain' of household moves that 'filtered' homes downward through the real estate market to middle- and lower-income households (Bourne, 1981). More recently, central-city decline has been theorized by neo-Marxists such as Neil Smith (1996) as revanchist reinvestment of capital in property as a form of spatialized revenge by the privileged middle class against the poor and minorities who supposedly 'stole the inner city' from them. Either way, on the policy side, out of concern for central-city decline in the face of dramatic suburbanization, the public sector invested heavily in downtown redevelopment (Bourne, 1967). To improve the image of the core of many large Canadian cities, roads were widened, public transit was funded, and symbolic buildings were constructed. In Montreal, for example, during his nearly three-decade tenure as mayor, Jean Drapeau supported the construction of mega-projects that would increase the city's international profile and create highly visible urban landmarks: the Montreal Metro subway system; the Expo '67 site; and the 1976 Olympic Stadium (Germain and Rose, 2000; Lortie, 2004). In Montreal and other large agglomerations, the 1960–75 period was also a time of major private-sector investment in office buildings, plazas, underground shopping concourses, and downtown shopping malls, which together dramatically altered the built fabric and character of the central city (Filion and Gad, 2006; Gad, 1991).

The last quarter of the twentieth century was a period of economic restructuring and urban transition for Canadian cities, with an emphasis on suburban domination (Bourne, 1989). Loss of traditional industrial manufacturing and warehouse functions has reoriented downtowns towards recreational, cultural, educational, and residential functions. The 1970s witnessed the politicization of city spaces in Canadian urban centres (Blackwell, 2008). Citizen groups mobilized to preserve neighbourhoods and green spaces threatened by **urban renewal**, development projects, and highway expansions (Grant, 2006). While urban planning may have become more democratic and participatory through increased public consultation, the financial constraints wrought by the 1980s recession, the 1990s downloading of programs to municipalities (e.g., housing, welfare, transit), and the most recent global economic downturn have produced cash-strapped cities struggling to manage infrastructural decline. As cities are obliged to turn to the private sector for financial support, the participatory era of the 1970s has given way to an entrepreneurial era of urban development that yields authority to market forces (Allahwala, Boudreau, and Keil, Chapter 12). The twenty-first century marks a new era wherein cities are run more like businesses and local governments rely on risk-taking, inventiveness, self-promotion, and profit-maximization.

The Optics of the Twenty-First-Century Competitive City

The most recent distinctive and self-consciously entrepreneurial phase of urban development is characterized by 'flagship' strategies centred on profit, spectacle, and consumption (Hall, 2003; Kipfer and Keil, 2002; Logan and Molotch, 1987). As Lynch and Ley point out in Chapter 19, cities now are taking great care to use architectural form and urban design to promote a positive and high-quality image of place (Harvey, 1989; Zukin, 1995). This attention to **place-making** (see also Lynch and Ley, Chapter 19) creates a distinctive

development style that is especially notable in Canada's largest and fastest-growing metropolitan areas, where branding is applied not just to corporate products but also to whole cities in an effort to develop a distinctive civic image appealing to tourists, investors, and members of the creative class (Bradley et al., 2002; Evans, 2007). The goal of many municipalities is to replace perceptions of the city as a place of disinvestment, decay, crime, and poverty left behind by the post-World War II suburban transitions with images of growth, vitality, and prosperity (Avraham, 2004). Thus, at one level, increased concern about and enthusiasm for urban design can be understood as products of municipal boosterism. In a global economy, city image is of concern, particularly in the best-known and most central parts of the metropolitan region, because it influences where businesses locate, 'which, in turn, affects inward investment and the infrastructure of daily life for the city's inhabitants' (Tavernor, 2007: 159). But it is not just the built and physical environments that are central to imaging a city; the lifestyles and symbolic economies that exist in these spaces through recreation, leisure, and cultural activities also contribute to a city's image and perceived quality of life (Cronin and Hetherington, 2008; Zukin, 1995, 2004).

Culture, in particular, has become an instrument of economic development and 'urban spectacularisation, which serves both for real-estate speculation and for political propaganda' (Vaz and Jacques, 2007: 249). In addition to a bias towards the better-off in society, spectacle can have an inversely proportional relationship to popular participation: the more spectacular the interventions in urban revitalization, the less the participation of the population (Vaz and Jacques, 2007). Thus, who is included and who is excluded, who wins and who loses in the urban revitalization process are important considerations. Is this so-called 'circus' serving those in need of 'bread' and other necessities of life (Kipfer and Keil, 2002)? The assumed trickle-down mechanisms may not work. Poorer local residents may become further marginalized, their visions of less exclusive, corporate, or authoritarian urban futures may be sidelined or ignored by urban Elites (see Walks, Chapter 10).

Re-Imaging Canadian Cities through Spectacle

Over the past two decades, academics, urban planners, and municipal politicians across North America and Europe have enthusiastically favoured a flexible 'creative capital' model of urban development that privileges knowledge, creativity, and commodified difference as a means of civic renewal (Gertler et al., 2002; Markusen and King, 2003). Turning away from the more traditional primary and secondary sectors of the economy for support, this new model seeks instead to attract a broad 'creative class' of professionals in business, law, engineering, science, and health care who supposedly share similar approaches to complex problem-solving and a common work ethos that values individuality, difference, and merit (Florida, 2002, 2005; Vinodrai, Chapter 6). The iconic citizens of the new knowledge-based economy are presented as the 'storm-troopers' of city branding; they are said to be attracted to cities that offer 'the 3Ts': technology, talent, and tolerance. In a quest to lure members of the creative class, Canadian cities are remaking themselves as centres of arts and culture and marketing themselves as places that provide stimulation, diversity, and a richness of experience that inspire creativity and innovation. Nowhere is this more apparent in Canada than in Toronto, the country's largest census metropolitan area.

Since the turn of the twenty-first century, Toronto has witnessed an urban cultural building boom—what Stanwick and Flores (2007) describe as a 'design renaissance'—that seeks socially and architecturally to give cultural spaces new prominence in the city. Financial assistance from provincial and federal governments and wealthy private sponsors has supported 'starchitectural' makeovers of

high-culture institutions (e.g., the Royal Ontario Museum, the Gardiner Museum of Ceramic Art, the Four Seasons Centre for the Performing Arts, the National Ballet School, the Royal Conservatory of Music, and the Art Gallery of Ontario) by some of the world's high-profile architectural practitioners, including Frank Gehry, Daniel Libeskind, Lord Norman Foster, and Will Alsop. Daniel Libeskind's optimistic foreword to the book *Design City: Toronto* inspires authors Stanwick and Flores (2007: 8) to claim that: 'Toronto is now knee-deep in a phenomenal act of city (re)making. If a renaissance is about a rebirth and enlightenment, then Toronto's architectural renaissance is certainly on its way to maturity.' This trend of employing internationally renowned architects to create iconic architectural landmarks is not new. It began with the successful completion in 1997 of the franchised Guggenheim Museum in the industrial city of Bilbao, Spain, by the architect Frank Gehry (Marshall, 2001). The former Mayor of Winnipeg turned urban strategist, Glen Murray, who in February 2010 won a by-election in Toronto Centre to become a Liberal MLA in the Ontario legislature, has humorously referred to civic enthusiasts' mistaken belief that building iconic museums can be a shortcut to urban revitalization as the 'Irritable Bilbao Syndrome'. Instead of developing cultural planning strategies that use grassroots economic and cultural assets and resources for new uses, many urban decision-makers in Canada's largest and wealthiest cities are busy with iconic civic interventions, elevating buildings of undulating metal, glass prisms, and colourful pillars as branding projects. Such large-scale redevelopment initiatives, which often inadequately attend to the specificities of local context, have nevertheless become 'central icons in the scripting of the image of the future of the cities in which they are located' (Swyngedouw, 2007: 210).

The centrepiece of Toronto's cultural promotion package, initiated in the fall of 2005, was the $4 million, two-year, street-banner branding campaign—'T.O. live with culture'—and its accompanying website that listed arts events in the city. This marketing campaign was conceived of as a means of increasing local and tourist awareness of the extent of Toronto's cultural activities (e.g., corporate-sponsored events such as Scotiabank Nuit Blanche and L'Oréal Luminato). While such short-term events continue to draw crowds and have increased local cultural awareness, members of Toronto's arts communities have argued that city funding would be better spent on capital grants that would allow arts organizations and cultural workers to purchase permanent affordable work space, thus breaking the cycle that displaces them from gentrifying neighbourhoods (Barmak, 2008). Without such financial support, Toronto's street banners could become a misnomer: future cultural workers may not be able to afford to live downtown. Instead, the inner city may be left to developers to infill with soaring condominium towers.

Growing Sky High: The Condominium Boom

In large Canadian cities where significant inner-city redevelopment has occurred, the nineteenth-century, middle-class stigma once associated with living among the supposed deprivation, pollution, and pathologies of the working classes has lost some of its potency. As noted in the introduction to this chapter, Canadian urban geographer David Ley (1996a) has studied changes that have transformed the down-market 'modern' inner city into the highly prestigious 'postmodern' city. Ley has documented how, since the 1970s, the romanticized cosmopolitan character of downtown living has gained a cachet among the 'new middle class' (1988, 1993, 1996a, 1996b), persuasively arguing that the residential preferences, investment decisions, and cultural values of this new middle class have facilitated substantial gentrification and upgrading in many Canadian cities.

Unlike some American (e.g., Cleveland and Detroit) and European (e.g., Liverpool and Leipzig) 'shrinking city' counterparts that have witnessed urban population decline (www.shrinkingcities .com), large metropolises connected to Canada's urban system have not lost population and become hollowed out because of racial tensions, violence, or poor schools. In fact, some of urban Canada's most expensive real estate can be found in core-area neighbourhoods. Recognizing that land is a finite resource, many Canadian cities have sought to focus development on brownfield and infill sites, a process often referred to as 'intensification'. Brownfield sites are tracts of land occupied by a permanent structure that is now vacant, underused, or derelict and has the potential to be 'developed at densities appropriate to precious land near the economic heart of the metropolis' (Belmont, 2001: 368). Much of this intensification has taken the form of privately developed condominiums.

Condo-mania has a firm grip on Canadian cities across the country. Many urban planning departments are overrun with proposals for high-density residential living and must contend with the widespread process of condominium development as a major component of municipal comprehensive plans (Kern, 2007; Lehrer, 2008). Inevitably, such intensive residential development involves substantial physical and social change to the fabric of the city. Taking into consideration the potential long-term and wide-reaching consequences of converting formerly productive industrial, manufacturing, and office use to high-density, middle- and upper-class residential use, one must ask whether condominium infill is a socially and economically sustainable model of urban development.

Many potential benefits certainly may be associated with increased residential densities, for example, improved energy and land conservation; maximization of existing infrastructure; reduced reliance on the private automobile; improved economic capacity through the increased circulation of money in the local economy; an increased tax base; and the social and safety dimensions of a more vibrant street life. But drawbacks exist, also. It has been argued, for example, that intensification in the form of high-rise condominium towers that commonly cater to childless singles or couples from the middle and upper classes who seek maintenance-free living has created 'vertical gated communities' (Hwang, 2006; Townshend and Walker, Chapter 8). Condominium fees cover the costs of residents-only facilities (e.g., gym, swimming pool, spa, games room, movie theatre) and security systems, creating islands of wealth and privilege where residents may seldom interact with each other or with the people and spaces of the surrounding neighbourhood. People insulated from one another lose a sense of community and of commitment to a common social project. It is important, then, for planners to consider ways of fostering social interaction at the neighbourhood scale over the longer term, perhaps through the provision of accessible, inclusive, well-maintained public spaces (e.g., parks, parkettes, squares) that support a variety of uses and involve a range of different social groups. Intensified residential development can be a positive contribution to city life, if cities have the financial means and the political will to adequately invest in the provision and maintenance of public infrastructure.

Place-Making at the Water's Edge

Much condominium construction is concentrated along river-, lake-, or oceanfront property. The reclaiming of waterfront sites began in the 1980s and today these sites are considered by many to represent valuable urban amenities with a central role to play in contemporary city-making (Marshall, 2001). Waterfronts are complex places that have hosted a plurality of functions over time. Historically, the working part of the city was on the waterfront. The waterfront was where

industry located and ship and rail yards interfaced. Historically, urban waterfronts consisted of dirty, messy, and contaminated sites, and so were largely undervalued in the collective conscience. For much of the twentieth century, Canada's urban waterfronts were under-utilized parcels of land separated from the physical, social, and economic activity of the rest of the city by transportation corridors. Today, however, waterfronts have been revalued as city assets and efforts are underway to 're-capture' these areas by reconfiguring connections between the older, original city centre and the water's edge.

One of the most noteworthy examples of post-industrial city-making through waterfront redevelopment is Vancouver (Berelowitz, 2005; Punter, 2003). Since the 1990s, Vancouver has gained an international reputation for its high standard of urban planning and design practice, so much so that when North American architects and planners promote the idea of a high-residential density, high-public amenity central city they call it 'Vancouverism' (Boddy, 2005; also Lynch and Ley, Chapter 18). Located at the mouth of the Fraser River Valley, with the Coastal Mountains to the north and the Strait of Georgia to the west, the ocean is brought right into the heart of Vancouver. Together, the mountains and the ocean provide the raw material for the 'cult of the view', which, in conjunction with a series of overlapping official 'view corridors', has driven high-rise residential development in the inner city for the last 50 years (Berelowitz, 2005). Historic neighbourhoods in the inner city (e.g., Strathcona, Gastown, and Chinatown) remain largely intact as pedestrian-friendly, bicycle-friendly places because Vancouver has managed to avoid the worst of traditional North American urban renewal: highways, elevated and underground pedestrian systems, large shopping malls, and big-box retail (Perl and Kenworthy, Chapter 11). For example, within the city's municipal boundaries, no highways disrupt the regular street grid or cut the city off from its waterfront.

Since the 1970s and accelerating after Expo '86 and into the 2010 Winter Olympics, Vancouver has strategically transformed its waterfront from industrial and railway use to residential and recreational use. The construction of public pathways along the seawall has repositioned recreational activity in a continuous corridor along the inner-city waterfront. The downtown street grid and urban fabric have gradually been extended to the water's edge and filled in with tall, pale bluish-green glass condominium towers known locally as 'see-throughs' (Coupland, 2000). But the city has also faced criticism for privileging residential development over office development, thus jeopardizing its commercial land base, elevating land values, and creating more of a recreational resort than a working city.

With a backdrop of yacht clubs, glass towers, and cloud-shrouded mountain peaks a centrepiece of Vancouver's waterfront renewal is Granville Island. This former industrial site beneath the southern viaduct of the Granville Bridge was transformed in the 1980s into a commercial, arts, entertainment, and recreation destination. The foci of this complex are a public farmers' market and the Emily Carr Institute of Art and Design, combined with an eclectic mix of studios, galleries, theatres, restaurants, and specialty shops in converted warehouses. Granville Island has been characterized as a privatized public place of conspicuous consumption and play (Ley, 1996a). This island of wealth, investment, and pleasure stands in marked contrast to the poverty, homelessness, and struggles with addiction experienced by many residents of Vancouver's Downtown Eastside, perhaps Canada's poorest inner-city neighbourhood (Ley and Dobson, 2008; Smith, 2003).

Vancouver is arguably North America's most livable city. But it is also renowned for its high-cost central-city real estate, which only the very affluent can afford to purchase. As in most other fast-growing Canadian cities, the process of urban redevelopment and modernist planning practice has physically reorganized poverty and affluence in the city, often concentrating poor residents in neighbourhoods that are poorly serviced, far from employment opportunities, and physically separated from the rest of the city (Walks, 2001). The image of a new, affluent city of leisure seeks to hide such marginal spaces from view.

What Lies behind the Shimmering Façades and Visual Noise?

In a neo-liberal era of municipal financial constraint, where discourses of urban entrepreneurialism and competitiveness dominate and business elites hold the reigns of power in public–private partnerships, market forces are readily accommodated while social justice objectives that could help to meet the basic needs of the most marginal populations are all too easily abandoned (Miles, 2005). The large-scale cultural and residential redevelopment projects discussed in this chapter can create huge commercial gains for developers and landlords and increase the tax base of the city. However, '[w]hat this branding of commodified space also shares over time is the fact that little or none of the increased values accrue to the residents who have been decanted and displaced' (Evans, 2007: 198–200). Social displacements and dramatic inequities are particularly apparent in western Canada where the oil boom has accelerated urban growth.

The inflow of workers and money to cities such as Saskatoon, Calgary, and Vancouver has caused local housing markets to climb to record highs, thus minimizing the availability of affordable accommodation (Walks, Chapter 10). In numerous print media articles about Calgary, for example, journalists have documented how rental apartments have been converted into luxury condominiums and blocks of lower-end residences have been bulldozed to make room for condominium towers, many of which have yet to be completed because labour and material shortages have caused construction delays and cost overruns (Hutton, 2008). According to one journalist, in the last half-decade, Calgary has lost 220 modest downtown units, experienced rent increase of 67 per cent, and seen increases in homelessness of 264 per cent (MacGregor, 2008: A2). In an effort to speak to the complex social problem of homelessness the artists Ryan Nordlund and Ryan Schmidt created the Flowerbed Project for the exhibition *Out and Out in the Big City* at the Art Gallery of Calgary in 2001. The wooden flowerbed bench that these artists created on the Stephen Avenue pedestrian mall outwardly conformed to the aesthetic codes of the street, yet a door at the base opened to reveal an insulated and cushioned sleeping chamber (Jonsson, 2006). The temporary shelter was heated by the occupant's body heat, and belongings were stored beneath other benches in the planter. The Calgary police eventually padlocked the piece during non-gallery hours to prevent overnight use of the eminently practical sculpture. This intervention demonstrates just one way in which Canadian artists are seeking to challenge the social exclusions wrought by gentrification and the luxury condominium boom.

Artistic Interventions in the Gentrifying Urban Landscape

The story of inner-city neighbourhood change through gentrification and other kinds of upgrading is not a new one. Nor is it limited to the largest census metropolitan areas. It is national in scope, impacting cities diverse in population, age, size, and location. The following discussion focuses on the role of the arts and cultural workers in the transformation of one of Toronto's inner-city neighbourhoods in order to reflect on how the generalizations of urban development trajectories made at the outset of the chapter play out in the spaces of one neighbourhood and to suggest alternative approaches to urban change. For many readers, this case study will possibly also raise questions about displacement, social justice, equity, and balance. If there is a moral to the story, it comes in

the form of a question. Does inner-city revitalization that is driven almost entirely by private-sector profit-making threaten 'to kill the goose that lays the golden egg'?

Cities and city centres have long been celebrated by urban theorists as the places where the arts, culture, and creativity flourish. Cultural or creative activities tend to cluster in places with an 'urban edge: that is, a mix of old and new buildings, an active streetscape, mixed use, contemporary design, cafés and bars, nightclubs' (Montgomery, 2008: xviii). This has been the case in Toronto. Art districts have formed and reformed in different neighbourhoods in the city as artists have discovered the hidden beauty and potential for inexpensive quarters in gritty urban spaces ahead of developers and gentrifiers. For example, in the 1960s and 1970s, Toronto's Cabbagetown, one of the city's poorest neighbourhoods, witnessed the beginnings of gentrification that led to the widespread exodus of residents. Since then, gentrification has spread across most residential neighbourhoods in downtown Toronto. The process has exiled risk-taking and trend-setting households such as those of artists from the more centrally located Yorkville district to Queen Street West, then to King Street West, out to the Junction, over to Queen Street East, and away out to the inner suburbs of Etobicoke and Scarborough (Bain, 2003). This sequence of relocations closely parallels the 'stage theory' of gentrification (Gale, 1980).

'Stage-typologizing' seeks to explain who is involved in the gentrification process (Caulfield, 1994). The first stage includes 'marginal gentrifiers' (Rose, 1984), who are often non-family groups who rent. The second stage includes 'early gentrifiers', who may be first-time homeowners using sweat equity to renovate their homes. The third stage, 'gentrification proper', includes middle-class professional households and developers exploiting the 'rent gap' to maximize the greatest profit from the land. The 'advanced gentrification' stage includes major redevelopment projects funded by 'super-gentrifying' financiers who re-gentrify areas previously gentrified (Butler and Lees, 2006). One of the drawbacks of stage theory, however, is that it fails to concern itself with the location and welfare of displaced incumbents. Artists, often referred to as the storm-troopers of gentrification, are one such displaced group. While they are resourceful and adaptable, there may be limits as to how far into the suburbs they are willing to go to obtain large, bright, cheap, and accessible workspace.

In the central city, Queen Street West is the city's main arts axis. Extending westward from University Avenue, this street has the largest concentration of art galleries in the city and an extensively developed network of artist-run spaces, arts organizations, studios, and art-supportive cafés, bars, and restaurants. Since the 1950s this neighbourhood has been transformed from an economically depressed, largely immigrant, community in the light industrial garment district into a trendy 'new Soho'. The name was first applied by local 1970s journalists who noted similarities with the development of SoHo in New York. Both areas were vibrant city neighbourhoods of avant-garde arts activity with residential lofts in former warehouses, stores, and galleries. In New York in 1979, the Soho Merchants Association was formed and 'Welcome to Soho' signs appeared in store windows, incurring the wrath of some residents who saw it as an embarrassing derivative and a ploy to increase rents. Through the 1980s rents did increase and artists were displaced as developers converted warehouses into condominiums and offices.

The continued migration of artists exiled from Queen West proper has inspired the westward development of an eclectic retail mix of high-end independent and chain stores. But despite consumer hyperactivity, residual poverty is apparent in the mix of residents who call far Queen Street West home: the precariously housed, newcomers to Canada, psychiatric survivors, sex trade workers, methadone clinic users, and cultural workers (Slater, 2004). In 2003, with the city's support, the neighbourhood was officially re-branded by the local **business improvement area** (BIA) as the West Queen West Art and Design District. A website (westqueenwest.ca), pocket maps, and unique street signs and banners all have helped to reinforce the neighbourhood brand. Refurbished hotels with bars, such as The Cameron House, The Drake, and The Gladstone, host well-publicized cultural events and function as cultural lynchpins in the area. While The Drake is a relatively small player in the local real estate market, it has the symbolic cultural capital through artist-in-residency programs, music, lectures, performances, exhibitions, and sponsorship to make the arts community complicit in the processes of gentrification and displacement at work in the neighbourhood. The creation of these high-profile cultural destination points has helped to push up residential and commercial real estate values, leading one graffiti artist to make it onto the cover of *Fuse Magazine* by scrawling on the side of a Starbuck's coffee shop 'Drake, you pimp, it's all your fault!' (Blackwell, 2008).

Gentrification is hard at work on Queen Street West, physically restructuring the neighbourhood and displacing marginalized residents such as cultural workers, rooming-house residents, and refugees. Luxury condominiums are going up one after another, capitalizing on the trendy counter-culture first established by artists. The brochures given out at the sales centre for the Bohemian Embassy read: 'Join the ambassadors of hip on Queen Street West at Gladstone.' Such slick advertising campaigns co-opt the language of the arts and use cultural workers as bait to lure new buyers. These condominiums do not cater to real Bohemians, but to 'faux-hemians' (Kingwell, 2008) and 'neo-Bohemians' (Lloyd, 2006) who have corporate jobs, trust funds, and upscale tastes while maintaining superficial, self-declared Bohemian lifestyles.

Such change to the urban fabric has not gone unchallenged. For years, cultural workers have protested their forced evictions with demonstrations, petitions, mass e-mail campaigns, exhibitions, and urban interventions. Ute Lehrer (2008) has documented how Active 18, a coalition of local residents and business owners who work in the cultural industries, has opposed development proposals for the area that were out of scale with the neighbourhood and provided little in the way of mixed-use or mixed-income opportunities. Active 18 has successfully obtained some concessions from developers with respect to building heights, contributions for arts-related community improvement projects, and affordable housing. The non-profit organization Artscape has been heavily involved in cultural regeneration in Toronto since the mid-1980s. Functioning as a cultural intermediary between artists, developers, and municipal politicians, Artscape oversees the development and management of building conversions (e.g., the former police station at 1313 Queen Street West; the Distillery District; and the Wychwood Barns) into multi-tenant facilities that provide professional artists and arts organizations with affordable living, work, exhibition, and performance spaces (www .torontoartscape.on.ca). But many more such 'adaptive re-use' projects (Caulfield, 2005) need to be undertaken across the city, particularly in suburban neighbourhoods, if the exodus of artists to small-, mid-size, and rural areas is to be slowed. Without continued urban cultural activism, the city will continue to witness 'a massive art-brain-drain' (LeBlanc, 2008): cultural workers and creative industries will likely continue to be displaced in favour of the purchasing preferences of the middle class, and inner-city neighbourhoods likely will struggle to function productively as incubators for culture and entrepreneurship.

In the Queen Street West neighbourhood of Parkdale, some artists have worked collaboratively with BIAs, art councils, private sponsors, and boutique hotels to facilitate small-scale urban performance interventions to reanimate Queen Street West and to reveal complex layers, histories, and narratives about the neighbourhood. Heather McLean (2009) has focused on the work of the artist Darren

O'Donnell (2006) and his company, Mammalian Diving Reflex. O'Donnell's art, termed 'social acupuncture', uses social life transformed into art to make the world a more just and equitable place (Wilson, 2008). In his recent project, *Parkdale Public School vs. Queen West*, low-income schoolchildren from new immigrant families theatrically competed in six rounds of culinary, visual art, and music challenges at the Gladstone Hotel against adult 'artser' gentrifiers. This project found creative potential in social tensions and facilitated communication between social groups from the same neighbourhood usually divided by age, ethnicity, and class. As McLean (2009) notes, children, public schools, and public libraries are traditionally 'left out of the social imaginary of "hipster urbanism"', yet in this urban intervention a reversal of power gives young people the space and the opportunity to actively participate in the narration of their neighbourhood.

Such local initiatives by countercultural communities to shape and narrate urban space add 'cumulative texture' to cities in ways that cannot be reproduced by urban elites through macro-scale policy initiatives (Lloyd, 2005). Thus, it is important to credit local artists, and not just 'starchitects', with the power to re-image and re-place neighbourhoods within cities. The book *Utopia: Towards a New Toronto* (2006) is a product of Toronto's grassroots arts and culture scene. The book's contributors are an eclectic group of downtown artists, activists, journalists, historians, students, architects, musicians, writers, bartenders, and restaurateurs, who celebrate the reclamation of public space and the transformation of neighbourhoods (McBride and Wilcox, 2006: 11):

> increasingly, our homes are outside: in the streets, on patios, in bars, on concert stages, in bookstores, in parks . . . we are creating culture. We are re-claiming public space. We are transforming neighbourhoods. We are rediscovering or recovering history.

Thus urban change need not be large-scale and spectacular. Urban change can be small-scale, local, and ephemeral. A word of caution, though, is needed to consider who has the time, money, and energy to instigate such change. The activist cohort who have the privilege of creating humorous, ephemeral, and site-specific interventions that treat the central city as playground are primarily white, male, young, able-bodied cultural workers.

Just south of The Drake Hotel on Lisgar Street is a long blue warehouse complex slated for redevelopment and surrounded by the construction pits and showrooms of new condominium projects in the West Queen West neighbourhood. The entrance to Number 39 is a storefront façade constructed by the artists Dan Bergeron and Gabriel Reese for *A City Renewal Project* and corporately sponsored by Grolsch and Red Bull (torontoist.com/2008/11/fauxreel_specter_a_city_renewal_project.php). Their installation recreates a street inside the warehouse studio, replete with fake storefronts, original graffiti tags, and a TTC shelter and bus stop. The nearly life-sized black-and-white photographs of closed or derelict Toronto storefronts have been reconstructed out of condominium placards and accompanied by fake signage. The artifacts in this temporary urban archive testify to a city in transition. These artists are treating 'the city not as an arrangement of fixed sites but as a confluence of relationships that are constantly reworked' (Cronin, 2008: 66) and given new meaning through 'visual remixing' (Burnham, 2007). As Scott Burnham (2007: 183) explains, subversive 'creative interventions which remix the urban visual' serve to 'transform the static visuals of the city into a continuum of expression' that 'generate[s] personal relevance within deeply impersonal urban landscapes and reawaken[s] our relationship with urbanity.'

Such urban interventions as *Parkdale Public School vs. Queen West* and *A City Renewal Project* explore the boundaries between art, politics, and everyday life. They provide a unique method for using art to creatively, imaginatively, and critically engage with urban space in a way that fosters a personal connection to the city. Site-specific performances in everyday spaces create opportunities for participatory dialogue and engagement with diverse audiences. As David Pinder (2005) explains, artists and cultural practitioners who intervene in cities through creative practice have the ability to 'challenge norms about how urban space is framed and represented' and to disrupt routines of use, meaning, and value. In so doing, artists help to question and to re-envision who can claim a 'right to the city' (Staeheli et al., 2002: 198).

Conclusions and Improvisational Alternatives

To withstand waves of national and international competitive urbanism, Canadian cities have relied heavily on branding and spectacularization to reinvent and redevelop inner-city spaces. Across the country we see the creation of cultural megaprojects, luxury condominiums, hotel and retail districts, and festival waterfronts in redundant industrial areas. In the face of such apparent consensus on how the contemporary Canadian city should be re-imaged, re-elevated, and re-placed, one sees the danger of creating generic, homogenizing, single-thought cities where differences and conflict are neither tolerated nor valued and, most importantly, where 'the persons excluded from this spectacularisation process possess perhaps the key to its reversal, which would entail . . . popular participation . . . and deep acquaintance with urban spaces' (Vaz and Jacques, 2007: 250). This suggestion is persuasive and one for which the spatial dimension can be more productively teased out.

Popular participation and a deep, rather than a shallow, engagement with urban spaces can be encouraged through urban interventions that both produce and are a product of improvisational space. Improvisational space can be understood as space that is changeable, malleable, and affordable, that encourages spontaneous and intuitive activities, and that supports risk-taking (Bain, 2003). It supports small-scale experiential initiatives and local interventions in the urban fabric that allow for unexpected encounters with differences. Moreover, it can be occupied and used in ways that 'not only . . . disrupt the authoritative structures that govern them, but also… encourage a dialogue about the possibility for other forms of being and behaving' (Jonsson, 2006: 37). As in-between space with the potential for plurality and contradiction within it, improvisational space could be interpreted as a more socially sustainable alternative to spectacular space. Such localized 'spaces of hope' (Blomley, 2007) or 'Temporary Autonomous Zones' (Routledge, 1997) are the sites from which grassroots urban activism, optimism, and creativity that resist dominant ideologies and practices can emerge.

However, the opportunities for finding and securing improvisational space in cities are rapidly diminishing. Such space used to be found in lower-income, working-class, inner-city neighbourhoods. But cultural **commodification**, the condominium boom, and waterfront redevelopment projects have intensified residential development in the downtown core to meet middle- and upper-class consumptive and experiential practices, and in the process marginal social groups and cultural workers have been displaced. Neighbourhoods that remain affordable and can function as incubators for culture, entrepreneurship, and inclusive development are now located in the inner and outer suburbs, not in the central city. It is time to turn our collective practical and scholarly attention away from the repeated valorization of the inner city and develop a more nuanced appreciation of the changes occurring, and the creative possibilities awaiting, in the suburban periphery.

Review Questions

1. Why is the prefix 're-' (as in re-imaging, re-elevating, and re-placing) used with regard to inner-city change?

2. Explain the relationship between competitive city and inner city as 'spectacle'.

References

Avraham, E. 2004. 'Media strategies for improving an unfavorable city image', *Cities* 21, 6: 471–9.

Bain, A.L. 2003. 'Constructing contemporary artistic identities in Toronto neighbourhoods', *Canadian Geographer* 47, 3: 303–17.

———. 2006. 'Resisting the creation of forgotten places: Artistic production in Toronto neighbourhoods', *Canadian Geographer* 50, 4: 417–31.

———. 2009. 'Creative suburbs: Cultural "popcorn" pioneering in multi-purpose spaces', in Edensor et al. (2009).

Barmak, S. 2008. 'Lights! Camera! Get packing!', *Toronto Star*, 10 Aug., E1, E6.

Belmont, S. 2001. *Cities in Full: Recognizing and Realizing the Great Potential of Urban America*. Washington: American Planning Association Planners Press.

Berelowitz, L. 2005. *Dream City: Vancouver and the Global Imagination*. Vancouver: Douglas & McIntyre.

Blackwell, A. 2008. 'The gentrification of gentrification and other strategies of Toronto's creative class', *Fuse Magazine* 29, 1: 28–37.

Blomley, N. 2007. 'Critical geography: Anger and hope', *Progress in Human Geography* 31, 1: 53–65.

Boddy, T. 2005. 'Vancouverism and its discontents', *Vancouver Review*. At: <www.vancouverreview.com/past_articles/vancouverism.htm>.

Bonoguore, T. 2008. 'The downside of up', *Globe and Mail*, 19 July, M1, M4.

Bourne, L. 1967. *Private Redevelopment of the Central City: Spatial Processes of Structural Change in the City of Toronto*. Chicago: Chicago University Press.

———. 1981. *The Geography of Housing*. London: Edward Arnold.

———. 1989. 'Are new urban forms emerging? Empirical tests for Canadian urban areas', *Canadian Geographer* 33: 312–28.

Bradley, A., T. Hall, and M. Harrison. 2002. 'Selling cities: Promoting new images for meetings tourism', *Cities* 19, 1: 61–70.

Bunting, T., and P. Filion, eds. 2006. *Canadian Cities in Transition: Local Through Global Perspectives*. Toronto: Oxford University Press.

Burnham, S. 2007. 'The VJ of the everyday: Remixing the urban visual', in Marcus and Neumann (2007: 181–96).

Butler, T., and L. Lees. 2006. 'Super-gentrification in Barnsbury, London: Globalisation and gentrifying global elites at the neighbourhood level', *Transactions of the Institute of British Geographers* 31: 467–87.

Caulfield, J. 1994. *City Form and Everyday Life: Toronto's Gentrification and Critical Social Practice.* Toronto: University of Toronto Press.

———. 2005. 'Toronto: The form of the city', in H. Hiller, ed., *Urban Canada: Sociological Perspectives.* Toronto: Oxford University Press, 311–42.

Coupland, D. 2000. *City of Glass: Douglas Coupland's Vancouver.* Vancouver: Douglas & McIntyre.

Cronin, A.M. 2008. 'Urban space and entrepreneurial property relations: Resistance and the vernacular of outdoor advertising and graffiti', in Cronin and Hetherington (2008: 65–83).

——— and K. Hetherington, eds. 2008. *Consuming the Entrepreneurial City: Image, Memory, Spectacle.* London and New York: Routledge.

Davies, W.K.D. 1997. 'Sustainable development and urban policy: Hijacking the term in Calgary', *Geoforum* 43, 4: 359–69.

Edensor, T., D. Leslie, S. Millington, and N. Rantisi, eds. 2009. *Spaces of Vernacular Creativity: Rethinking the Cultural Economy.* London and New York: Routledge.

Evans, G. 2007. 'Branding the city of culture—The death of city planning?', in J. Monclus and M. Guardia, eds, *Culture, Urbanism and Planning.* Aldershot: Ashgate, 197–213.

Filion, P., and G. Gad. 2006. 'Urban and suburban downtowns: Trajectories of growth and decline', in Bunting and Filion (2006: 171–91).

Florida, R. 2002. *The Rise of the Creative Class: And How It's Transforming Work, Leisure, Community and Everyday Life.* New York: Basic Books.

———. 2005. *The Flight of the Creative Class.* New York: Harper Business.

Gad, G. 1991. 'Toronto's financial district', *Canadian Geographer* 35: 203–7.

Gale, D. 1980. 'Neighbourhood resettlement: Washington, DC', in S. Laska and D. Spain, eds, *Back to the City.* New York: Pergamon.

Germain, A., and D. Rose. 2000. *Montréal: The Quest for a Metropolis.* Toronto: John Wiley and Sons.

Gertler, M., R. Florida, G. Gates, and T. Vinodrai. 2002. *Competing on Creativity: Placing Ontario's Cities in a North American Context.* Toronto: Ontario Ministry of Enterprise, Opportunity, and Innovation.

Hall, T. 2003. 'Art and urban change: Public art in urban regeneration', in A. Blunt et al., eds, *Cultural Geography in Practice.* New York: Arnold, 221–34.

Harvey, D. 1989. 'From managerialism to entrepreneurialism: The transformation in urban governance in

late capitalism', *Geografiska Annaler: Series B, Human Geography* 71, 1: 3–17.

Huskins, B. 1999. '"Tale of two cities": Boosterism and the imagination of community during the visit of Prince of Wales to Saint John and Halifax in 1860', *Urban History Review* 28, 1: 31–46.

Hutton, David. 2008. 'Saskatoon grapples with boomtown pains', *Globe and Mail,* 3 July, B1, B4.

Hwang, I.S.Y. 2006. 'When does stacking become vertical sprawl?', *Transactions on Ecology and the Environment* 93: 283–92.

Jonsson, T. 2006. 'Space invaders: There goes the neighbourhood', *In/Site* 15: 37–9.

Kern, L. 2007. 'Reshaping the boundaries of public and private life: Gender, condominium development, and the neo-liberalization of urban living', *Urban Geography* 28: 657–81.

Kingwell, M. 2008. *Concrete Reveries: Consciousness and the City.* Toronto: Viking Canada.

Kipfer, S., and R. Keil. 2002. 'Toronto Inc? Planning the competitive city in the new Toronto', *Antipode* 34, 2: 227–64.

Kyle, W.J. 2007. 'Urbs Americana: A work in progress', in P. Swirski, ed., *All Roads Lead to the American City.* Hong Kong: Hong Kong University Press, 97–123.

LeBlanc, D. 2008. 'In constant exile, artists seek next colony', *Globe and Mail,* 21 Nov., G4.

Lehrer, U. 2008. 'Urban renaissance and resistance in Toronto', in L. Porter and K. Shaw, eds, *Whose Urban Renaissance? An International Comparison of Urban Regeneration Strategies.* London and New York: Routledge.

Ley, D. 1988. 'Social upgrading in six Canadian inner cities', *Canadian Geographer* 32: 31–45.

———. 1993. 'Past elites and present gentry: Neighbourhoods of privilege in Canadian cities', in L. Bourne and D. Ley, eds, *The Changing Social Geography of Canadian Cities.* Montreal and Kingston: McGill-Queen's University Press.

———. 1994. 'The Downtown Eastside: One hundred years of struggle', in S. Hasson and D. Ley, eds, *Neighbourhood Organizations and the Welfare State.* Toronto: University of Toronto Press.

———. 1996a. *The New Middle Class and the Remaking of the Central City.* Oxford: Oxford University Press.

———. 1996b. 'The new middle class in Canadian central cities', in J. Caulfield and L. Peake, eds, *City Lives and City Forms: Critical Research and Canadian Urbanism.* Toronto: University of Toronto Press.

———. 2003. 'Artists, aestheticisation and the field of gentrification', *Urban Studies* 40, 12: 2527–44.

———— and H. Frost. 2006. 'The inner city', in Bunting and Filion (2006: 192–210).

———— and C. Dobson. 2008. 'Are there limits to gentrification? The contexts of impeded gentrification in Vancouver', *Urban Studies* 45, 12: 2471–98.

Lloyd, R. 2006. *Neo-Bohemia: Art and Commerce in the Post-Industrial City*. New York: Routledge.

Logan, J., and H. Molotch. 1987. *Urban Fortunes: The Political Economy of Place*. Berkeley: University of California Press.

Lortie, A. 2004. *The 60s: Montreal Thinks Big*. Montreal: Canadian Centre for Architecture.

Lowes, M.D. 2002. *Indy Dreams and Urban Nightmares: Speed Merchants, Spectacle, and the Struggle over Public Space in the World-Class City*. Toronto: University of Toronto Press.

McBride, J., and A. Wilcox, eds. 2006. *uTOpia: Towards a New Toronto*. Toronto: Coach House.

Macgregor, R. 2008. 'The boom's ugly underside', *Globe and Mail*, 18 Feb., A2.

McLean, H. 2009. 'The politics of creative performance in public space', in Edensor et al. (2009).

Marcus, A., and D. Neumann, eds. 2007. *Visualizing the City*. London and New York: Routledge.

Markusen, A., and D. King. 2003. 'The artistic dividend: The hidden contributions of the arts to the regional economy', Project on Regional and Industrial Economics, Humphrey Institute, University of Minnesota, Minneapolis.

Marshall, R., ed. 2001. *Waterfronts in Post-Industrial Cities*. London and New York: Spon Press.

————. 2001a. 'Connection to the waterfront: Vancouver and Sydney', in Marshall (2001).

————. 2001b. 'Remaking the image of the city: Bilbao and Shanghai', in Marshall (2001).

Miles, S. 2005. 'Creativity, culture and urban development: Toronto examined', *disP—The Planning Review* 162, 3: 70–87.

Montgomery, J. 2008. *The New Wealth of Cities: City Dynamics and the Fifth Wave*. Aldershot: Ashgate.

Newman, H.K. 1999. *Southern Hospitality: Tourism and the Growth of Atlanta*. Tuscaloosa: University of Alabama Press.

O'Donnell, D. 2006. *Social Acupuncture: A Guide to Suicide, Performance and Utopia*. Toronto: Coach House.

Rose, D. 1984. 'Rethinking gentrification: Beyond the uneven development of Marxist urban theory', *Environment and Planning D: Society and Space* 2, 1: 47–74.

Routledge, P. 1997. 'Pollock Free State and the practice of postmodern politics', *Transactions of the Institute of British Geograhers* 22: 359–77.

Slater, T. 2004. 'Municipally managed gentrification in South Parkdale', *Canadian Geographer* 48, 3: 303–25.

Smith, H. 2003. 'Planning, policy and polarization in Vancouver's Downtown Eastside', *Tijdschrift voor Economische en Sociale Geografie* 94, 4: 496–509.

Smith, N. 1996. *The New Urban Frontier: Gentrification and the Revanchist City*. New York: Routledge.

Staehli, L., D. Mitchell, and K. Gibson. 2002. 'Conflicting rights to the city in New York's community gardens', *GeoJournal* 58: 197–205.

Stanwick, S., and J. Flores. 2007. *Design City: Toronto*. Chichester, West Sussex: John Wiley and Sons.

Swyngedouw, E. 2007. 'The state of the situation: Post-political cities', in P. Stouthuysen and J. Pille, eds, *The State of the City: The City Is the State*. Brussels: Brussels University Press, 203–27.

Tavernor, R. 2007. 'Composing London visually', in Marcus and Neumann (2007: 159–78).

Vaz, L.F., and P.B. Jacques. 2007. 'Contemporary urban spectacularisation', in J. Monclus and M. Guardia, eds, *Culture, Urbanism and Planning*. Aldershot: Ashgate, 241–53.

Walks, A. 2001. 'The social ecology of the post-Fordist/global city? Economic restructuring and socio-spatial polarization in the Toronto urban region', *Urban Studies* 38, 3: 407–47.

Wilson, C. 2008. 'Performance art that doesn't suck', *Toronto Life* 42, 12: 45–9.

Zukin, S. 1995. *The Cultures of Cities*. Cambridge, Mass.: Blackwell.

————. 2004. *Point of Purchase: How Shopping Changed American Culture*. New York: Routledge.

CHAPTER 16

Slow Growth and Decline in Canadian Cities

Betsy Donald and Heather M. Hall

One of the most striking findings from the 2006 Canadian census is that over 70 per cent of the Canadian urban system is either slowly growing (36.1 per cent) or declining (34.7 per cent). Yet, surprisingly little attention is paid to this phenomenon. Geographers, planners, and policy-makers have been involved in planning for decline *within* the city for several decades. This is not new. Research on intra-urban declivity has examined downtown and inner-city deterioration along with derelict brownfields, waterfronts, and industrial lands. What is new, however, is a heightened awareness and growing reality of rising disparities *between* cities—those that are rapidly growing and those that are not. This uneven inter-urban geography is confronting all developed countries to a greater or lesser degree and it is increasingly attracting attention across Canada, Germany, and the United States (Bontje, 2004; Lötscher, 2005; Bourne and Rose, 2001; Bourne and Simmons, 2003; Bunting and Filion, 2001; Downs, 1994; Hall, 2009; Hall and Hall, 2008; Leo and Anderson, 2006; Müller and Siedentop, 2004; Oswalt, 2005; Simmons, 2003; Popper and Popper, 2002; Rybczynski and Linneman, 1999). This literature shares a common view that if current demographic, economic, and policy trends persist, then the gap between cities that are rapidly growing and those that are not will become increasingly accentuated. The uneven inter-urban geography among Canadian cities will challenge the way we think about urban

development issues because the dominant discourse in Canada is about planning for growth, not decline. How we manage decline in many of our Canadian cities will become one of the most pressing urban policy issues of the twenty-first century.

This chapter is organized into four sections. The first section provides a typology of urban decline and includes a description of the broader demographic, economic, and policy trends producing this new and accentuated pattern of uneven growth. In the next section, we focus on the negative perceptions and 'psychology of failure' that often accompany academic or political labelling of a 'slow-growth' or 'declining city'. A discussion of the challenges and opportunities associated with different growth trajectories constitutes the third section. In the fourth section we focus on the geographic, planning, and policy implications for Canadian cities experiencing slow growth and decline. We conclude with a discussion of further challenges confronting urban areas in Canada.

Terminology, Trends, and Discourse

What Are 'Slow-Growth' and 'Declining' Cities?

In the Canadian urban geography literature, cities experiencing slower or declining rates of growth are referred to as 'slow-growth' or 'declining' (Bunting and Filion, 2001; Bourne and Simmons, 2003; Hall,

2009; Hall and Hall, 2008; Leo and Anderson, 2006; Simmons, 2003). In Europe and the United States, the term 'shrinking cities' is commonly used, along with 'stagnating cities' or 'urban areas in difficulty' (Atkinson, 2001; Bontje, 2004; ECOTEC Research and Consulting, 2007; Lötscher, 2005; Rybczynski and Linneman, 1999). While these different terms incorporate a number of trends, the common measurement for slow growth and decline is population change over time. No concise method exists for measuring this change, however, which produces a variety of results. For example, American policy analyst Anthony Downs classifies cities based on population change in a 10-year period as *rapidly declining* if they lost more than 4.9 per cent, *slowly declining* if they lost between 0.1 and 4.9 per cent, *slowly growing* if they gained less than 10 per cent, and *rapidly growing* if they gained more than 10 per cent. In Canada, five-year periods frequently are used, although this method only considers short-term trends and may not provide an accurate depiction of long-term trajectories of change (Bourne and Simmons, 2003; Bunting and Filion, 2001; Hall and Hall, 2008; Leo and Anderson, 2006; Simmons, 2003).

Greater Sudbury, Ontario, provides an excellent example of how these differences in measuring population change can create inconsistencies, which have implications for policy-makers. Using Anthony Downs' classification, the city has declined by 4.4 per cent between 1996 and 2006, placing it in the slowly declining category. But if the most commonly used measure in Canada is adopted—the five-year measure—between 2001 and 2006 the city experienced a 1.7 per cent growth in population (Statistics Canada, 2006, 2007). If we take a more strategic and long-term view of population change in Greater Sudbury, we discover that the city actually reached a peak population of 170,000 in 1971 compared to the present population of 157,857. It is not surprising that city officials favour statistics that indicate growth to maintain the development ethos that

underlies the culture of Canadian cities. The implications, however, for not having a standard 'decline' measurement across Canada means that all cities look for ways to promote and plan for growth at any cost, rather than openly contemplating the place-based realities they face.

Where Are Slow-Growth and Declining Cities?

Canadian declining cities are not alone. Similar patterns of slow growth and decline are emerging across Europe and the United States. Using population change over a five-year period, ECOTEC Research and Consulting (2007) found that 80 of the 258 Urban Audit cities experienced overall urban decline between 1996 and 2001. Declining cities were concentrated in eastern Germany and Central and Eastern European countries. Leo and Brown (2000) make reference to a number of slow-growth European cities—Vienna, Brussels, Copenhagen, Milan, Florence, and Rome. In the United States, Leo and Anderson (2006) reveal that 118 out of 280 metropolitan statistical areas were growing slowly or declining between 1990 and 2000. Meanwhile, in Canada 102 of 144 census agglomerations (CAs) and census metropolitan areas (CMAs) have experienced slow growth or decline between 1996 and 2006 (Table 16.1). As seen in Figure 16.1, this uneven pattern of growth in Canada is more pronounced in small and mid-size urban areas on the periphery. Prince Rupert, BC, had the largest decline, at 23.1 per cent, followed by Kitimat, BC (−19.3 per cent), and Elliot Lake, Ontario (−15.0 per cent). Meanwhile, most large city-regions like Toronto, Calgary, and Vancouver (with the exception of Winnipeg, Quebec City, and Montreal), as well as small and mid-size cities in oil-rich regions, grew rapidly.

In most developed countries, this uneven inter-urban geography is largely a result of several powerful demographic, economic, and policy trends (Bourne and Rose, 2001; Bourne and Simmons, 2003; Hall and Hall, 2008; Lötscher, 2005; Mäding,

Table 16.1 Trends in the Canadian Urban System, 1996–2006

	Large Urban Areas (>500,000)	Mid-Size Urban Areas (50,000–500,000)	Small Urban Areas (10,000–50,000)	All Urban Areas
Number of urban areas	9	50	85	144
Total pop. 1996	14,104,958	6,694,680	2,085,702	22,885,340*
Total pop. 2006	16,213,411	7,295,728	2,122,435	25,631,574*
Overall % pop. change, 1996–2006	14.9	9.0	1.8	12.0*
Number of slowly growing areas*	3	20	29	52
% slowly growing	33.3	40.0	34.1	36.1
Number of declining urban areas**	0	11	39	50
% declining	0	22.0	45.9	34.7
Lowest % pop. change	Winnipeg (3.4)	Cape Breton, NS (–10.1)	Prince Rupert, BC (–23.1)	Prince Rupert, BC (–23.1)
Highest % pop. change	Calgary (33.1)	Barrie, Ont. (49.2)	Okotoks, Alta (101.5)	Okotoks, Alta (101.5)

*The Canadian population grew by 5.4 per cent between 1996 and 2006, from 30,007,094 to 31,612,897.

**Slowly growing cities are defined as cities that grew by less than 10 per cent while declining cities are those that lost population between 1996 and 2006.

Source: Statistics Canada (2007).

2004; Müller, 2004; Müller and Siedentop, 2004; Oswalt, 2005; Popper and Popper, 2002). In Canada, we are in the final stages of the demographic transition in which lower fertility rates are resulting in the decreased role of natural increase for population growth (see Townshend and Walker, Chapter 8). As a result, immigration is the major source of demographic change and it is well documented that the majority of domestic and international migrants locate in large city-regions like Toronto and Vancouver, further accentuating the uneven population distribution between cities (Bourne and Rose, 2001; Bourne and Simmons, 2003; see Hoernig and Zhuang, Chapter 9). In some cities, especially on the Prairies, another major source of demographic change is the rising Aboriginal population (see Peters, Chapter 22). These demographic trends are expected to continue in the foreseeable future.

Economic restructuring and technological changes in Canada have led to fewer jobs in the resource and manufacturing sectors while job growth related to the knowledge economy has concentrated primarily in large city-regions as well as areas within their zone of influence (Barnes et al., 2000; Bourne and Simmons, 2003; Gertler, 2001; Norcliffe, 1994; see Vinodrai, Chapter

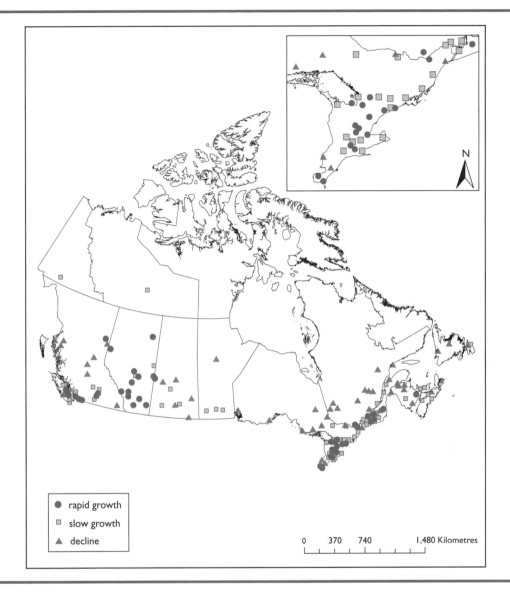

Figure 16.1 Pattern of growth and decline in the Canadian urban system, 1996–2006

6; Hutton, Chapter 7). In their work on Atlantic Canada and Quebec, Polèse and Shearmur (2006) argue that having a resource-dependent economic base is another precondition for decline. Resource dependency is an issue because the economies of many smaller peripheral cities lack the economic diversification to counteract the often volatile boom-bust cycle of a natural resources-based economy.

Other conditions contributing to decline relate to changing state policies and geography. In Canada, shifting national capital flows to the

United States, along with the changing role of government through trade liberalization and increased provincial responsibilities, have contributed to what Bourne and Simmons (2003) refer to as 'new fault lines' between cities that are rapidly growing and those that are not. Added to this is the shift away from state redistribution policies aimed at reducing spatial inequalities towards policies that stress competitiveness, innovation, and development from within urban areas (see Allahwala, Boudreau, and Keil, Chapter 12). In Canada, a peripheral location in the national or continental space economy or a location away from a major transportation axis or trade route are also cited as preconditions for decline (Polèse and Shearmur, 2006: 42). Thus, all of these macro social, economic, and political factors have accumulated to produce a new pattern of uneven growth, which has further accentuated existing patterns between a small number of cities and regions that are rapidly growing and a larger number of those that are not (Table 16.2).

Who Cares about 'Declining' Cities?

The literature on slow growth and decline is relatively small, although a number of themes are emerging. Most of the writing describes the causes, characteristics, and consequences of decline or slow growth (Bourne and Rose, 2001; Bourne and Simmons, 2003; Bunting and Filion, 2001; Leo and Anderson, 2006; Leo and Brown, 2000; Lötscher, 2005; Lötscher, Howest, and Basten, 2004; Müller, 2004; Müller and Siedentop, 2004; Oswalt, 2005; Polèse and Shearmur, 2006; Simmons, 2003). These scholars have initiated the discussion by providing detailed national accounts of the causes and by depicting the pattern of growth and decline. Meanwhile, a smaller literature examines the social, fiscal, economic, planning, and policy challenges

Table 16.2 Summary of Macro Trends Contributing to the Pattern of Uneven Growth

Trend	Description
Demographic	• Canada is in the final stages of the demographic transition • Canada is experiencing lower fertility rates and decreasing natural increase • Immigration to and within Canada is the major source of demographic growth • Demographic concentration is occurring in large Canadian city-regions
Economic	• Economic restructuring and technological changes in manufacturing and resource sectors • Shift to a more knowledge-intensive economy • Resource dependency and limits of (profitable) resource exploitation • Economic concentration is occurring in large Canadian city-regions
State policy	• Shifting national trade to the United States • Shifting Canadian capital flows • National trade liberalization and devolved responsibilities to lower levels of government • Shift away from state redistribution policies aimed at reducing spatial inequalities
Locational	• Peripheral location in the Canadian urban system • Location away from a major transportation axis or trade route • Outside the influence zone of a large city-region • Size: small and mid-size cities

associated with shrinkage or decline (Bontje, 2004; Franz, 2004; Hall, 2009; Mäding, 2004; Moss, 2008; Müller and Siedentop, 2004; Simmons and Bourne, 2007). New research explores the psychological dimension of open public discussion of the topic due to the growth-centric mentality that permeates civic politics and planning (Hall, 2009; Leo and Anderson, 2006; Oswalt, 2005; Popper and Popper, 2002; Seasons, 2007). A common thread running throughout all of this research is that the current demographic, economic, and policy trends most likely will persist and the spatial fallout of these trends will deepen the uneven inter-urban geography of the Canadian urban system.

Psychology of Failure and Political Sensitivity

How is the topic of decline perceived in the conduct of urban affairs? In the urban economic geography literature, a renewed interest in the qualities of 'place' has emerged as a key attribute of the knowledge-intensive economy (Donald and Morrow, 2003; Florida, 2002, 2005; see Vinodrai, Chapter 6; Lynch and Ley, Chapter 19). Most of this research has focused on large city-regions and has failed to consider the realities of slow-growth and declining cities. For example, creative economy variables like talent, tolerance, and technology are biased by design to favour large, core metropolitan areas (Figure 16.2). In Canada, it is not surprising that creative and high-tech industries are concentrating in and around large metropolitan areas like Toronto, Ottawa, and Vancouver and that these large city-regions are attracting greater proportions of 'talented' individuals who are highly educated (Beckstead and Gellatly, 2003; Gertler et al., 2002; Polèse et al., 2002). Meanwhile, smaller, peripheral and declining cities end up at the bottom, contributing not only to a reality of failure (because they rank at the bottom) but also to a psychology of failure for many small and mid-size urban areas (Gertler et al., 2002).

In a recent study on the social dynamics of economic performance in Kingston, Ontario, a slowly growing city, a key informant explained the role that this 'psychology of failure' is playing in economic governance (Lewis and Donald, 2010):

> If a place has a personality, it can also have a psychology. Just as individual psychologies reflect their experiences, their traumas, and upbringing and raising and so on, Kingston has had, one can argue, a psychology of loss, of failure. It didn't grow, it wasn't industrialized, it was bypassed by commerce, the military left, the capital left.

This psychology of failure and government malaise that often accompanies these slow-growth and declining cities, which tend to rank low on creativity-inspired indices, is a pressing issue that influences all aspects of decision-making from land-use planning to economic development.

In the local policy arena the issue of decline is rarely broached unless the discussion is focused on reversing the trend (Müller and Siedentop, 2004; Seasons, 2007). Seasons (2007: 6) refers to decline as the 'policy elephant in the living room' while Hall (2009) found that planners, economic developers, and policy-makers were uncomfortable discussing the topic at the local level. This discomfort is a direct result of the negative perceptions and psychology that accompany decline (Robinson, 1981). For example, Leo and Anderson (2006: 393) in their research on Winnipeg argue that there is a perception that 'any city that is not growing rapidly is being "left behind".' Hall (2007) discovered that local officials in Greater Sudbury associated decline with death, producing a negative image that would counteract investment and local economic development. One key informant suggested that 'if you're not growing, you're dying.' Put simply, cities view growth as expected, desirable, and necessary, while any trajectory other than growth is seen as failure. Moreover,

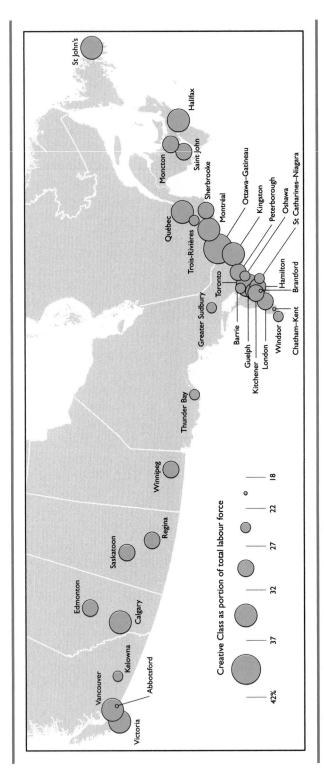

Figure 16.2 Pattern of the creative workforce

Source: Martin Prosperity Institute, based on 2006 Statistics Canada data.

anticipating decline is equivalent to accepting this failure.

This obsession with urban growth is firmly rooted in the politics of local economic development. From the early days of North American cities, demographic and economic growth and new urban development have been perceived essentially as ingredients for maintaining and improving urban quality of life. Harvey Molotch (1976) emphasized the entrenched attachment to growth in his now classic framing of the city as a 'growth machine'. Molotch (1976: 310) further stated that 'the clearest indication of success at growth is a constantly rising urban area population.' Growth is depicted as the magic potion that creates jobs, reinforces the tax base, and provides resources to solve social problems (Logan and Molotch, 1987). In the current economic context, a positive urban image is perceived to be a vital component for urban competitiveness and no city wants to be labelled as declining (Avraham, 2004).

Decline creates a policy dilemma for those cities experiencing it. Yet, if current trends persist, some cities will be unable to reverse their demographic situation and may have to make the best of what is perceived to be a non-ideal situation. Incessantly planning for future growth in its absence, however, may prove to be more costly in the long term. We argue that approaches reflecting place-based realities are essential because slow growth and no growth do not necessarily have to be viewed as problems if they are managed and planned for. As an example, Greater Sudbury has changed and improved dramatically over the last few decades without population growth, suggesting that city decline is different, but not deadly. Over the 35-year period of 1971–2006, as mentioned above, the population of Sudbury has trended downward, from about 170,000 to under 158,000. Despite this trend, the economic base has expanded so that Sudbury has become a regional service centre for northeastern Ontario for medical care, retail, tourism, government, and education. In 2005, the Northern Ontario School of Medicine was opened with a campus in Sudbury. This is the first medical school to be opened in Canada in over 30 years. When the city was rapidly growing during the 1960s the natural environment was barren and polluted due to smelting techniques used by the mining industry. Indeed, a widespread rumour claimed that US astronauts, in preparation for landing on the moon, trained on the supposedly similar barren terrain that Sudbury provided. While this rumour has since been dispelled, it gives some indication of the former state of the landscape. Since that time, key agents in the city have planted over eight million trees, earning a United Nations commendation among other distinctions. The Sudbury case depicts how communities can change and improve without demographic growth. As we discuss further in the following section, both declining *and* growing cities have unique opportunities and challenges.

Growth, Slow Growth, and Decline: Challenges and Opportunities

A common perception exists that rapid growth is the desideratum of urban development. There are challenges, however, associated with this type of development trajectory. Hall and Hall (2008) discuss how urban growth can lead to environmental concerns such as rural fringe development and the loss of prime agricultural land. Furthermore, challenges arise related to the provision and maintenance costs of infrastructure and municipal services as a city grows. Conflicts may arise between urban and rural land uses or other community social tensions, such as has occurred in Fort McMurray, located within the Regional Municipality of Wood Buffalo, Alberta. Between 1996 and 2006 the population of this region has increased by nearly 50 per cent, from 36,000 to 53,000 people (Statistics Canada, 2006). This rapid growth is a direct result of the region's proximity to the Athabasca oil sands and the booming oil and gas industry. Hoernig et al. (2005) discuss the social impacts of this rapid growth. High wages due to labour shortages in the oil sands development industry lead to shortages in other sectors that are unable to match the high oil wages. Other issues related to the quick demographic changes include a serious shortage of affordable housing, a high cost of living, homelessness, social and individual stresses, and, of course, environmental degradation on a massive scale. Thus, rapid growth is not a panacea but has challenges and opportunities that require realistic planning solutions.

On the positive side, large growing cities often offer demographic and cultural diversity as a result of their large immigrant populations, economic diversity in terms of employment opportunities and choices, and urban diversity, which includes more selection in dwelling types, such as lofts and condominiums (Gertler et al., 2002). In slow-growth or declining cities the populations tend to be relatively homogeneous, with a higher proportion of aging residents resulting from limited economic opportunities, which lead to youth out-migration and low immigrant settlement (Hanlon and Halseth, 2005; Simard and Simard, 2005; see Rosenberg and Wilson, Chapter 21). Declining cities can also become increasingly dependent on grants from provincial and federal governments as a result of their shrinking tax base. However, slow-growth and declining cities in peripheral locations often boast an abundance of landscapes rich in natural amenities: lakes and forests, better air quality, and less traffic congestion (Hall, 2009; Simard and Simard, 2005).

Leo and Anderson (2006) compare Vancouver, a fast-growing urban area, and Winnipeg, a slow-growing urban area. They found that housing costs are cheaper in Winnipeg, yet Vancouver has more ability to pay for its infrastructure and services. Vancouver also has sharper contrasts between the rich and poor. A remarkable characteristic of some

slow-growth cities is evident in Winnipeg, where dispersion has occurred despite its slow growth. This contrasts with Vancouver, a fast-growing city, which has had more success in limiting the outward growth of the region. This dispersion pattern in Winnipeg is related to decisions made by city officials who are more accepting of pro-development proposals in their desire to grow at any cost. Often these developments are unsuitable with regard to location, densities, and land uses and may place further pressure on services and infrastructure. In Greater Sudbury, Hall (2007) identifies a similar planning practice to 'take anything anywhere', in which decline is used as an opportunity to push through proposals. This trend has also occurred in Kingston, where a sprawl-like spatial pattern has emerged over the last 10 years without a significant population increase. This 'any growth is good growth' mentality becomes a problem, however, when city officials fail to see that more development also produces more pressure on municipal services and increases expenditures (Siegel, 2002).

Growth is an important component in municipal budgeting for the provision of infrastructure and services, presenting a challenge for cities in decline. Growth brings money into municipal accounts. In Canadian municipal budgeting the three main sources of revenue for municipalities are: user fees; unconditional or conditional grants; and taxes on the assessed value of property. Complicating municipal financing, moreover, has been the downloading in recent years of responsibilities from senior levels of government to municipalities without augmenting their capacity to secure new revenue streams (Bradford, 2002, 2004; FCM, 2006; Kitchen, 2002; Mintz and Roberts, 2006; Sancton, 2000, 2006; Vojnovic and Poel, 2000). For example, Mintz and Roberts (2006) describe how the Ontario government downloaded various social welfare programs, including social assistance, childcare, immigration services, **social housing**, and homelessness. Kitchen (2002: 5) argues that this downloading was the 'largest redistribution of services between a province and its municipalities ever witnessed in Canada', fuelling further a growth mentality because new property development increases the tax base and ultimately municipal revenue. Without substantial new growth the purse strings on the municipal budget become tighter and the challenge becomes how to deliver up-to-date and quality services and infrastructure without considerable new growth occurring.

One last opportunity associated with slow growth and decline is the option to manage land resources in a more efficient and forward-looking manner (Robinson, 1981). In rapidly growing cities, policies can be reactive; whereas in slow-growth and declining cities, the potential exists to become more proactive. Strategies like impact analyses and strategic assessments can be employed before making development decisions to determine how foreseeable changes will affect the city presently and over time. In reality, however, declining cities are not necessarily using that time to practise proactive planning. Thus, a variety of place-based opportunities and challenges are associated with rapid-growth, slow-growth, and declining communities (Table 16.3).

Implications for Urban Geography, Planning, and Policy

We argue that urban experts and urban policies have failed to consider the realities of slow-growth and declining cities. The 1950s and 1960s were a time of rapid economic and demographic growth, and during these decades 'most Canadian planning standards, norms and zoning practices were developed' (Wolfe, 1995: 55). Historically, planning has been tied to growth, and the current framework does not provide the proper tools for urban areas experiencing anything other than rapid growth at the urban scale.[1] Nevertheless, planners have experience managing areas in decline, such as inner cities, brownfields, and core areas, to increase

Table 16.3 Summary of Challenges and Opportunities

	Rapid-Growth Cities	Slow-Growth and Declining Cities
Opportunities	· Demographic and cultural diversity · Employment diversity · Variety in housing options and the built form · Higher tax base	· Lower cost of living · Housing affordability · Less traffic congestion · Better air quality · Natural amenities · Opportunities to be more proactive about decision-making
Challenges	· Sharper contrasts between rich and poor · Environmental concerns related to increased dispersion · Land-use conflicts · Social conflicts · Extreme pressure on infrastructure and services · Pollution (air quality) · Congestion	· Aging population · Homogeneous population · Youth out-migration · Limited economic opportunities · 'Any growth is good growth' mentality for urban development · Grant dependency · Less ability to pay for infrastructure and services

Source: Compiled by authors. For similar summary tables, see Bourne and Simmons (2007); Robinson (1981).

livability and attract investment. Most planning concepts, policies, and statutes, however, are devoid of the word 'decline', with common approaches like smart growth and the Places to Grow Act 2005 in Ontario focused on growth. A similar tale is told for the United States and Germany. Popper and Popper (2002) argue that US planning tools are designed to manage growth or a rising population. In Germany, municipal, regional, and state planning approaches are unprepared to meet the challenges of demographic decline (Bontje, 2004; Müller, 2004; Müller and Siedentop, 2004). More fundamentally, growth-centred solutions to no-growth scenarios are unrealistic and prevent the acceptance and management of decline as an acceptable trajectory. In the following sections we offer some solutions to ignite the debate in the search for alternatives.

We Need a New Decline Discourse

Canadian urbanists and policy-makers must turn their attention to the possibility that decline or slower growth will continue in some places and begin in others. Hall and Hall (2008: 14) argue that the Canadian urban literature 'repeats, and arguably reinforces, the message that quantitative [population] growth is the only legitimate response to decline' and that '[c]onsumers of this research are implicitly dared to concede defeat if they disagree with this perspective.' If we expect places to realistically plan for decline, then the topic can no longer be avoided (Bunting and Filion, 2001). Focus must now turn to how cities in decline can maintain quality of life when revenue sources depend on growth. This may require alternative financial tools and policies for slow-growth and declining cities. For example, many European cities rely much less on the property tax base for their revenue generation. In fact, Canadian municipalities have one of the highest dependencies on property taxes in the world. Alternative revenue sources more prominent in other jurisdictions include user fees, sales taxes, and income-based taxes. These alternative sources—especially the income tax—may diminish the pressure on cities to continuously search for and embrace land-development-related growth (see Slack, 2005, and Courchene, 2005, for discussions of fiscal constraints in Canadian cities). At a more fundamental level, the perceived notion that slow-growing or declining places are 'failing' in the metropolitan growth competition needs to change if we expect the political sensitivity of discussing the issue to diminish and become a more solutions-oriented conversation.

New Decline-Oriented Techniques: 'Smart Decline' and 'Qualitative Development'

Planning legislation and policy tools are ill-equipped to manage decline at the urban scale. Existing concepts like growth management, growth controls, and smart growth all imply substantial growth to manage. Müller and Siedentop (2004) call for a decline paradigm to establish more decline-oriented planning tools. Their suggestions include but are not limited to focusing on redevelopment, recycling land and buildings, adapting infrastructure, and strategic planning. Likewise, Popper and Popper (2002: 22) argue that we need a 'smart decline' strategy requiring us to think 'about who and what remains', and that this 'may entail reorganizing or eliminating some services and providing different ones.'

Based on previous research, we offer a number of alternative planning approaches to assist urban geographers, planners, and city officials (Hall, 2009). Our hope is to kick-start conversation on how to accept, manage, and plan for decline. We start with **qualitative development**, a concept used by Sudbury in the 1980s, which centred on the existing built form, promoting redevelopment, infilling, and conservation. These concepts are simple to implement and already essential to good planning. The challenge, however, is to forget about the growth mentality associated with quantitative development or expanding growth, services, and infrastructure in the city. An expansive urban boundary is *not* a metaphor of success. Other decline-oriented approaches involve realigned development where settlement boundaries and services actually are scaled back or reduced to reflect current expectations while maintaining and improving the existing built form. Controlled development would mean halting municipal infrastructure or service expansions to facilitate new growth, with developers paying the costs associated with development, and identifying spaces for redevelopment and infilling. A final decline-oriented technique that we offer is internal zoning, whereby a hierarchy of limited settlement areas are predetermined for development. Each zone or settlement area slated for urbanization or re-urbanization would be developed before another zone could be selected. Essentially, the key is not to overdevelop but to think strategically about where growth should occur within the city, which is something all cities should be doing. A summary of these approaches is provided in Table 16.4.

Fischler and Wolfe (2006: 348) describe efforts of the Montreal Metropolitan Community (MMC), which is planning for a smaller population by the year 2050. They explain that the MMC approach is to 'consolidate already urbanized areas, to rationalize regional infrastructure investment, and to arbitrate the inter-municipal competition for households and businesses by restricting development on greenfield sites'. Arguably, little risk befalls a city that plans for decline but experiences growth, other than negative image. If a city plans for growth and experiences decline, however, the risk is overbuilding and extending services and infrastructure that will be too costly to manage and upgrade in the future.

Moving Away from One-Size-Fits-All

In recent decades, government has been more concerned with large city-regions and their competition on a global scale (Donald, 2005). As Bürkner (2005) explains, state intervention policies that

Table 16.4 Summary of Decline-Oriented Techniques

Decline-Oriented Technique	Description
Decline-oriented planning/ decline paradigm	• Focus on redevelopment, recycling land and buildings, adapting infrastructure, and strategic planning
Smart decline	• Plan for who and what remains • Might involve reorganizing or eliminating some services and providing alternatives
Qualitative development	• A concept used by Sudbury, Ontario, in the 1980s • Focus on the existing built form, promoting redevelopment, infilling, and conservation
Realigned development	• Settlement boundaries and services are scaled back to reflect current expectations • Focus on maintaining and improving the existing built form
Controlled development	• No infrastructure or services expansion to facilitate new growth • Developers pay the costs associated with development • Identify spaces for redevelopment and infilling
Internal zoning	• Hierarchy of limited settlement areas are predetermined for development

Source: Compiled by authors from Müller and Siedentop (2004); Popper and Popper (2002); Hall (2009).

once aimed to *balance* spatial disparities between cities have shifted to *targeting* cities and regions that are already economically competitive or that are perceived to have potential. However, Atkinson (2001) cautions that this focus on competition may actually accentuate disparities and create more place-based winners and losers. In Ontario recently, most government policy has paid attention to the Greater Toronto Area and its surrounding region, exemplified through the City of Toronto Act, 2006 and the 2006 *Growth Plan for the Greater Golden Horseshoe*. The first provides the city with powers and legislative freedoms that other municipalities do not have, while the latter was designed to guide where and how growth should occur in order to promote healthy, strong, and prosperous communities in this large conurbation (Ontario Ministry of Municipal Affairs and Housing, 2009; Ontario Ministry of Public Infrastructure Renewal, 2006). More recently, a report published for the Ontario government by the Martin Prosperity Institute promotes the Greater Toronto Area to secure a future for Ontario centred on the creative economy (Florida and Martin, 2009).

We are not suggesting that Toronto and the broader Golden Horseshoe region are undeserving of attention. We are suggesting, however, that as the rift between have and have-not cities in the Canadian urban system increases, a new way of thinking will be needed to balance the needs of large, growing city-regions and their slow-growing and declining counterparts. This will require place-based, locally or regionally sensitive policies. It will require us to move beyond supply-side policies aimed at competitiveness that do little to manage spatial disparities between cities.

A similar spatial penchant for large city-regions is seen in much of the contemporary economic geography literature. The creative economy literature is biased by design to favour large cities like Toronto and Vancouver, while much of the literature on regional innovation systems and clusters is focused on economically advanced regions like Silicon Valley, Boston, Baden-Württemberg, and Toronto (Hall and Donald, 2009; see Vinodrai, Chapter 6). Unfortunately, this spatial bias creates a dualistic depiction that shows large cities, on the one hand, as creative and cosmopolitan economic drivers, and, on the other hand, many smaller peripheral places as economic failures. Hayter et al. (2003: 18) remind us that '[t]here is a whole world out there and not just a few cores or clusters.'

Does Population Growth Matter?

As described earlier in the chapter, growth is commonly measured in terms of population change. But is population growth everything? Is there a better way to measure a city's prosperity? Given the current economic realities and the expected demographic situation in many Canadian cities, we contend that these questions will require future attention. Population growth is important for municipal revenue generation but little is known about savings that could be made by planning for a smaller population. Rybczynski and Linneman (1999) believe that cities experiencing decline should not be trying to grow big again but finding out how to prosper as great, smaller cities. Other suggestions include the importance of planning for sustainability, liveability, and quality of life (Gonzáles, 2006). Furthermore, Molotch (1976: 328) foresees opportunities in the city that asks 'what it can do for its people rather than what it can do to attract more people'.

In Germany, Oswalt (2005: 13) describes a new slogan, 'shrinking as a new potential', to suggest that decline can result in a new compact urban core, which is an ideal European urban form. Meanwhile, the concept of Slow Cities or CittaSlow evolved out of the slow food movement in Italy. These two movements are responses to the fast-paced growth and cultural standardization most often associated with the broader trend of globalization. The slow cities movement encourages 'local, traditional cultures, a relaxed pace of life, and conviviality' (Knox, 2005: 6). The founding principles of CittaSlow include 'working towards calmer and less polluted physical environments, conserving local aesthetic traditions and fostering local crafts, produce, and cuisine' (ibid.). At the heart of the movement are the concepts of health, sustainability, and the environment (see Rees, Chapter 5; Connelly and Roseland, Chapter 14). Interestingly, participating cities are required to have populations of less than 50,000 (Knox, 2006). This model moves away from the obsession with rapid growth and instead encourages slow growth as the ideal trajectory for cities. Declining and slow-growing cities in Canada could draw on the CittaSlow movement for inspiring a new reality in which quality of life, quality of place, quality of employment opportunities, and sustainability become central to any policy-making, regardless of size. Thus, emphasizing local strengths and a *better* place, rather than a bigger place, could provide the opportunity for an alternative vision: one that ignores the obsession with growth and promotes quality (see Lynch and Ley, Chapter 19).

Conclusions: Moving Forward on a New Urban Agenda

For the last few years, Canadians have witnessed tremendous population growth in major urban centres from Toronto to Calgary to Vancouver. Many of the reasons behind this growth are linked more fundamentally to broader changes in the global economy and to increased flows of people, capital, and ideas from around the world to Canada (see Hall, Chapter 4). As a consequence, researchers and policy-makers interested in the Canadian urban system have become fascinated with documenting and promoting the dynamic, diverse, and creative Canadian urban centre. As detailed in this chapter, new research projects have explored the social dynamics of economic performance in these fast-growing cities and politicians have called for more explicit urban-based policies at the national level. However, this spatial bias accentuates inter-urban differences that are increasing across Canada.

As we have demonstrated, over 70 per cent of the Canadian urban system is either slowly growing (36.1 per cent) or declining (34.7 per cent). The questions we must ask are: How are we going to plan for Canadian cities in decline? What are the realistic urban scenarios and what are the effective policy tools at our disposal? Now is the time for urban scholars to turn more actively to insights from the environmental, green, sustainable, and quality-of-life literatures for ideas on how to plan for slow growth and decline (see Rees, Chapter 5; Connelly and Roseland, Chapter 14). Before doing this, however, we must first permit ourselves to face this reality and find new ways of constructively conversing about preparing for slow growth and decline in our Canadian cities. Only then can we enhance our collective capacity to respond to policy problems that are increasingly defined by the complex interdependence and spatially concentrated expression of the dynamic Canadian urban system. Coming to the aid of Canada's slow-growing or declining cities does not necessarily have to be a 'zero-sum' view of power relations between Canada's have and have-not places. Rather, it can be aligned with our ability and history as a nation to embed place-based policies that are sensitive to the realities of different regions across space and our reputation as a nation to deal with deeper economic unevenness with broader multi-scaled policy effects.

Acknowledgements

We would like to thank Richard Florida, Kevin Stolarick, and the Martin Prosperity Institute, along with Meric Gertler, David Wolfe, and the Innovation Systems Research Network. We are extremely grateful for statistical assistance from Jeff Moon, Maps, Data and Government Information Centre, Queen's University Library; for the GIS expertise of Greg McQuat, Department of Geography, Queen's University; and to Paulo Raposo, Martin Prosperity Institute.

Review Questions

1. What is a 'slow-growth' or 'declining' city, and where are they located in the Canadian urban system?

2. Is a 'qualitative development' strategy a good idea for municipalities in order to foster growth, or is it strictly a strategy for managing decline?

Note

1. For a history of Canadian planning, see Jill Grant, 'Shaped by Planning: The Canadian City Through Time', Chapter 18 in Bunting and Filion (2006), at: <www.oupcanada.com/bunting4e>.

References

Atkinson, R. 2001. 'The emerging "urban agenda" and the European spatial Development perspective: Towards an EU urban policy?', *European Planning Studies* 9, 3: 385–406.

Avraham, E. 2004. 'Media strategies for improving an unfavorable city image', *Cities* 21, 6: 471–9.

Barnes, T.J., et al. 2000. 'Canadian economic geography at the millennium', *Canadian Geographer* 44, 1: 4–24.

Beckstead, D., and G. Gellatly. 2003. 'The growth and development of new economy industries', *The Canadian Economy in Transition Series*. Statistics Canada Catalogue no. 11–622–MIE2003002.

Bontje, M. 2004. 'Facing the challenge of shrinking cities in East Germany: The case of Leipzig', *GeoJournal* 61: 13–21.

Bourne, L.S., and D. Rose. 2001. 'The changing face of Canada: The uneven geographies of population and social change', *Canadian Geographer* 45, 1: 105–19.

——— and J. Simmons. 2003. 'New fault lines? Recent trends in the Canadian urban system and their implications for planning and public policy', *Canadian Journal of Urban Research* 12, 1 (suppl.): 22–47.

Bradford, N. 2002. *Why Cities Matter: Policy Research Perspectives for Canada*. Ottawa: Canadian Policy Research Networks. At: <www.cprn.org/documents/29277_en.pdf>. (Apr. 2009)

———. 2004. *Place Matters and Multi-level Governance: Perspectives on a New Urban Policy Paradigm*. Ottawa:

Canadian Policy Research Networks. At: <www.cprn.org/documents/26856_fr.pdf>. (Apr. 2009)

Bunting, T., and P. Filion. 2001. 'Uneven cities: Addressing rising inequality in the twenty-first century', *Canadian Geographer* 45, 1: 126–31.

——— and ———, eds. 2006. *Canadian Cities in Transition: Local Through Global Perspectives*, 3rd edn. Toronto: Oxford University Press.

Bürkner, H.J. 2005. 'Polarization and Peripherization', in Oswalt (2005).

Courchene, T.J. 2005. 'Citistates and the state of cities: Political-economy and fiscal-federalism dimensions', *IRPP Working Paper Series* no. 2005–03.

Donald, B. 2005. 'The politics of local economic development in Canada's city-regions: new dependencies, new deals, and a new politics of scale', *Space and Polity* 9, 3: 261–81.

———. 2006. 'From growth machine to ideas machine: The new politics of local economic development in the knowledge-intensive city', in D.G. Tremblay and R. Tremblay, eds, *The Competitive City in the New Economy*. Montreal: University of Quebec Press, Political Economy Collection.

——— and D. Morrow, with A. Athanasiu. 2003. *Competing for Talent: Implications for Social and Cultural Policy in Canadian City-Regions*. Ottawa: Report prepared for Strategic Research and Analysis (SRA) Strategic Planning and Policy Coordination, Department of Canadian Heritage.

Downs, A. 1994. *New Visions for Metropolitan America*. Washington: Brookings Institution.

ECOTEC Research and Consulting Ltd. 2007. *State of European Cities Report: Adding Value to the European Urban Audit*. Birmingham: ECOTEC Research and Consulting Ltd. At: <http://ec.europa.eu/regional_policy/sources/docgener/studies/pdf/urban/stateofcities_2007.pdf>. (Dec. 2008)

Fischler, R., and J.M. Wolfe. 2006. 'Contemporary planning', in Bunting and Filion (2006).

Federation of Canadian Municipalities (FCM). 2006. *Building Prosperity from the Ground Up: Restoring Municipal Fiscal Imbalance*. Ottawa: FCM. At: <www.fpeim.ca/2006_FCM_Building_Prosperity_from_the_Ground_Up.pdf>. (Mar. 2009)

Florida, R. 2002. *The Rise of the Creative Class*. New York: Basic Books.

———. 2005. *Cities and the Creative Class*. New York: Routledge.

——— and R. Martin. 2009. *Ontario in the Creative Age*. Toronto: Martin Prosperity Institute. At: <http://martinprosperity.org/media/pdfs/MPI%20Ontario%20Report%202009%20v3.pdf>. (Feb. 2009)

Franz, P. 2004. 'Shrinking cities–shrinking economy? The case of East Germany', *German Journal of Urban Studies* 44, 1: on-line journal.

Gertler, M.S. 2001. 'Urban economy and society in Canada: Flows of people, capital and ideas', *ISUMA: The Canadian Journal of Policy Research* 2, 3: 119–30.

——— et al. 2002. *Competing on Creativity: Placing Ontario's Cities in North American Context*. Toronto: Ontario Ministry of Enterprise, Opportunity and Innovation and the Institute for Competitiveness and Prosperity. At: <www.competeprosper.ca/research/CompetingOn Creativity_061202.pdf>. (Jan. 2007)

Gonzáles, S. 2006. *The Northern Way: A Celebration or a Victim of the New City-Regional Government Policy?* ESRC/ DCLG Postgraduate Research Programme, Working Paper 28.

Grant, J. 2006. 'Shaped by planning: The Canadian city through time', in Bunting and Filion (2006).

Hall, H.M. 2007. 'Being Realistic about Planning in No Growth Communities: Challenges, Opportunities, and Foundations for a New Agenda', MA thesis, University of Waterloo.

———. 2009. 'Slow growth and decline in Greater Sudbury: Challenges, opportunities, and foundations for a new planning agenda', *Canadian Journal of Urban Research* 18, 1.

——— and B. Donald. 2009. 'Innovation and creativity on the periphery: Challenges and opportunities in northern Ontario'. Toronto: Martin Prosperity Institute Working Paper. At: <http://martinprosperity .org/media/pdfs/Innovation_and_creativity_on_the_ Periphery-H_Hall-B_Donald.pdf>. (Feb. 2009)

Hall, H., and P. Hall. 2008. 'Decline and no growth: Canada's forgotten urban interior', *Canadian Journal of Regional Studies* 31, 1: 1–18.

Hanlon, N., and G. Halseth. 2005. 'The greying of resource communities in northern British Columbia: Implications for health care delivery in already-underserviced communities', *Canadian Geographer* 49, 1: 1–24.

Hayter, R., T.J. Barnes, and M.J. Bradshaw. 2003. 'Relocating resource peripheries to the core of economic geography's theorizing: Rationale and agenda', *Area* 25, 1: 15–23.

Hoernig, H., et al. 2005. 'Planning for people: Integrating social issues and processes into planning practice', *Berkeley Planning Journal* 18: 35–55.

Kitchen, H.M. 2002. *Municipal Revenue and Expenditure Issues in Canada*. Canadian Tax Paper No. 107. Toronto: Canadian Tax Foundation.

Knox, P.L. 2005. 'Creating ordinary places: Slow cities in a fast world', *Journal of Urban Design* 10, 1: 1–11.

Leo, C., and K. Anderson. 2006. 'Being realistic about urban growth', in Bunting and Filion (2006).

——— and W. Brown. 2000. 'Slow growth and urban development policy', *Journal of Urban Affairs* 22, 2: 193–213.

Lewis, N., and B. Donald. 2010. 'A new rubric for "creative city" potential in Canada's smaller cities', *Urban Studies* 47, 1: 29–54.

Logan, W., and H. Molotch. 1987. *Urban Fortunes: The Political Economy of Place*. Berkeley: University of California Press.

Lötscher, L. 2005. 'Shrinking East German cities?', *Geographia Polonica* 78, 1: 79–98.

———, F. Howest, and L. Basten. 2004. 'Eisenhüttenstadt: Monitoring a shrinking German city', *Dela* 21: 361–70.

Mäding, H. 2004. 'Demographic change and local government finance—Trends and expectations', *German Journal of Urban Studies* 44, 1: on-line journal.

Mintz, J.M., and T. Roberts. 2006. *Running on Empty: A Proposal to Improve City Finances*. Toronto: C.D. Howe Institute.

Molotch, H.L. 1976. 'The city as a growth machine: Toward a political economy of place', *American Journal of Sociology* 82, 2: 309–32.

Müller, B. 2004. 'Demographic change and its consequences for cities—Introduction and overview', *German Journal of Urban Studies* 44, 1: on-line journal.

——— and S. Siedentop. 2004. 'Growth and shrinkage in Germany—Trends, perspectives and challenges for spatial planning and development', *German Journal of Urban Studies* 44, 1: on-line journal.

Norcliffe, G. 1994. 'Regional labour market adjustments in a period of structural transformation: An assessment of the Canadian case', *Canadian Geographer* 38, 1: 2–17.

Ontario Ministry of Municipal Affairs and Housing. 2009. City of Toronto Act, 2006. At: <www.mah.gov.on.ca/ Page343.aspx>. (Apr. 2009)

Ontario Ministry of Public Infrastructure Renewal. 2006. *Growth Plan for the Greater Golden Horseshoe*. At: <www .placestogrow.ca/images/pdfs/FPLAN-ENG-WEB-ALL.pdf>. (Apr. 2009)

Oswalt, P., ed. 2005. *Shrinking Cities: Volume 1—International Research*. Ostfildern-Ruit, Germany: Hatje Cantz.

———. 2005. 'Introduction', in Oswalt (2005).

Polèse, M., and R. Shearmur,. with P.M. Desjardins and M. Johnson. 2002. *The Periphery in the Knowledge Economy: The Spatial Dynamics of the Canadian Economy and the Future of Non-Metropolitan Regions in Quebec and the Atlantic Provinces*. Montreal and Moncton: Institut national de la recherche scientifique and the Canadian Institute for Research on Regional Development.

———— and ————. 2004. 'Why some regions will decline: A Canadian case study with thoughts on local development strategies', *Papers in Regional Science* 85, 1: 23–46.

Popper, D.E., and F.J. Popper. 2002. 'Small can be beautiful: Coming to terms with decline—Americans tend to think that places, once settled, stay settled', *Planning* 68, 7: 20–3.

Robinson, I.M. 1981. *Canadian Urban Growth Trends: Implications for a National Settlement Policy*. Vancouver: University of British Columbia Press.

Rybczynski, W., and P.D. Linneman. 1999. 'How to save our shrinking cities', *The Public Interest* 135: 30–44.

Sancton, A. 2000. 'Amalgamations, service realignment, and property taxes: Did the Harris government have a plan for Ontario's municipalities?', *Canadian Journal of Regional Science* 23, 1: 135–56.

————. 2006. 'City politics: Municipalities and multi-level governance', in Bunting and Filion (2006).

Seasons, M. 2007. 'Planning for uneven growth', *Plan Canada* 47, 2: 6.

Siegal, D. 2002. 'Urban finance at the turn of the century: Be careful what you wish for', in E.P. Fowler and D. Siegal, eds, *Urban Policy Issues: Canadian Perspectives*, 2nd edn. Toronto: Oxford University Press.

Simard, M., and C. Simard. 2005. 'Toward a culturalist city: A planning agenda for peripheral mid-size cities', *Canadian Journal of Urban Research* 14, 1: 38–56.

Simmons, J. 2003. *Cities in Decline: The Future of Urban Canada*. Toronto: Ryerson University, Centre for the Study of Commercial Activity.

———— and L.S. Bourne. 2007. 'Living with population growth and decline', *Plan Canada* 47, 2: 13–21.

Slack, Enid. 2005. 'Easing the fiscal restraints on Canadian cities', *Dialogues: Canada West Foundation* (Spring): 19–20.

Statistics Canada. 2006. 'Greater Sudbury', *Community Profiles 2006 Census*. Ottawa: Statistics Canada. At: <www12.statcan.gc.ca/census-recensement/2006/dp-pd/prof/92-591/index.cfm?Lang=E"http://www12.statcan.ca/census-recensement/2006/dp-pd/prof/92-591/index.cfm?Lang=E>. (Apr. 2009)

————. 2007. 'Census trends for census metropolitan areas and census agglomerations (table)', *2006 Census*. Ottawa: Statistics Canada Catalogue no. 92–596–XWE. At: <www12.statcan.ca/english/census06/data/trends/>. (Apr. 2009)

Vojnovic, I., and D. Poel. 2000. 'Provincial and municipal restructuring in Canada: Assessing expectations and outcomes', *Canadian Journal of Regional Science* 23, 1: 1–6.

Wolfe, J.M. 1995. 'Canada', in D. Lyddon, ed., *International Manual of Planning Practice*, 2nd edn. The Hague: International Society of City and Regional Planners.

CHAPTER 17

Fear, Insecurity, and the Canadian City

Deborah Cowen, Amy Siciliano, and Neil Smith

> No matter how many valuable functions the city has furthered, it has also served throughout most of its history as a container of organized violence and a transmitter of war. (Mumford, 1961)

On 4 December 2005, Vancouver's SkyTrain police became the first armed transit security force in Canadian history. Prompted by the perception of growing insecurity in and around the elevated light rail system, Vancouver's municipal transit authority, Translink, granted SkyTrain police the power to carry semi-automatic weapons along the 49.5-kilometre track. In 2007, Tasers were added to the SkyTrain police arsenal, while the power to carry guns was extended to police on select bus lines. More recently, in 2009, private security guards working on contract for Translink also were granted the power to carry guns.[1] These urban security initiatives are not without their critics. The Vancouver Bus Riders Union (BRU), a group of over 900 transit advocates, concluded that transit police are *the source* of insecurity for many transit-dependent women. Following extensive research on riders' experiences of the transit system, the BRU released a report that asserted, 'many women feel unsafe around SkyTrain police and have experienced harassment, intimidation and racial profiling.' If the city justifies its expansion of transit police authority as a protection of passengers, the BRU (2006) challenges this securitization, arguing that 'the real safety issues for women in transit are long waits at poorly lit bus stops, overcrowded buses, and the recent purchase of polluting, dirty diesel buses.' On the first day the SkyTrain police were armed the Canadian Press

(2005) reported that many riders 'didn't think the system was threatening or dangerous, but giving the officers more powers could make them feel safer.' One passenger, Robert Smith, echoed this intriguing logic when he said, 'I guess it makes me feel a little bit more secure although I didn't feel all that insecure' (Carmichael, 2005). Securitization may in fact *create* insecurity, heightening anxieties about fear and the city.

The arming of public transit guards in Vancouver suggests three crucial insights: the extent, contestation, and effects of security. First, this case highlights how public spaces like the transit system increasingly are defined as *insecure* and in need of *more security*. More and more, state and corporate initiatives in Canadian cities are oriented towards achieving 'security'. As Mariana Valverde (2009) points out, everyday issues become securitized: hunger is now a problem of 'food security', homelessness a problem of 'housing security'. Increased surveillance, new fencing and gating, the creation of 'secure areas', and the expansion of police, paramilitary, and even military power in cities increasingly are the norm.

Second, the case of the SkyTrain police and the response from the Bus Riders Union reveal the *contested nature* of securitization. Security projects impact people differently and social groups define the very meaning of security in dramatically varying ways. Thus, while security—rarely ever explained or defined—is presented as an objective public good, the BRU's challenge to the arming of transit police reveals the political nature of securitization. The BRU critique also points to specific ways that securitization reproduces established patterns of marginalization and stigmatization, in this case creating more insecurity for transit-dependent women. Finally, the BRU response highlights a broader shift underway in cities: the shift from *welfarist approaches* to *security responses* to social problems (on 'welfarism', see Bunting and Filion, Chapter 2; Allahwala, Boudreau, and Keil, Chapter 12).

Third and finally, this case reveals some complicated *effects of security*. If we move beyond the question of the extent of security in Canadian cities and its contestation, some very challenging questions about securitization emerge. The Sky-Train rider's response to the arming of guards suggests that efforts to secure the city can in fact produce *greater insecurity*. This transit rider admits that he did not feel insecure prior to these events, and yet he still describes feeling *more secure* once the transit police were armed. The paradox of securitization is that it can *create* insecurities that may not have existed before, and by doing so may also stimulate real demands for still more security initiatives. In a sense, securitization fuels itself. This raises a crucial point: it is nearly impossible to distinguish between 'real' and 'constructed' feelings of insecurity. Indeed, we take as our starting point that actual and perceived insecurities are impossibly entangled, while we contend that the relationship between the two demands careful investigation.

This chapter builds on these three themes: the extent, contestation, and effects of securitization. Everyday urban life presents genuine dangers and risk. Economic crisis, contaminated water supplies, and epidemics have created disruptions to everyday life in Canadian cities. However, the translation from actual risk to urban fear is complicated, requiring critical scrutiny. Fear is mobile, rarely reflecting actual risks that may arise in particular places. For instance, many women feel fear on city streets at night when the most likely place they will encounter violence is at home with men they already know. As we explore below, fear in public space is often expressed as a fear of racialized male strangers, an expression that takes up and reproduces racist stereotypes circulating in popular culture. Thus, fear may be motivated by actual risks, but, typically, fear is ordered by established social imaginaries. Fear sanctions the securitization of the city, but it tends to reproduce classed, gendered, and racialized relations of power. We further suggest that the particular kinds of uncertainty and risks generated by contemporary urban life in a globalized world are at once fuelling securitization of the city *and* pervasive insecurity. Securitization provides a promise of stability in a fundamentally precarious urban world, but at the same time generates its own rationale, namely, insecurity.

The chapter unfolds as follows. In the next section we provide a brief and selective historical sketch of the securitization of cities, highlighting the changing contexts for urban insecurity and tactics of securitization. We pose the complicated but unavoidable question 'how real is fear?' in order to explore the relationships between insecurity and securitization in contemporary Canadian cities. In the third section, to open some alternative ways of making sense of insecurity, we shift to look at different forms of urban fear governed under the rubric of security. We investigate forms of fear that are playing a profound role in reshaping urban space and **governance**: biohazards, terrorism, street crime, gendered violence, and security associated with property (such as the home). Each of these discussions addresses how actual risks are organized into security responses, revealing different aspects of the displacement of social and economic anxieties by fear and securitization. In a concluding section we reflect on this tango of fear and security in contemporary Canadian urbanism.

Social Anxieties and the Securitization of Urban Space

Concern for security in Canadian cities is not a new phenomenon but it has changed form over time. In fact, it could be said that cities have long been centres for security and defence. Taking a more global perspective, the city-states of Europe, Asia, and North Africa that predate nation-states were walled and heavily fortified, and were the primary space of sovereignty and protection (Graham, 2004a: 166; Mumford, 1961: 44). Our common-sense association of security issues with nation-states is a product of the modern system of nation-states and only 200–300 years old (Cowen and Smith, 2009; Singer, 2003). During the era of colonial settlement of North America, cities played crucial roles as administrative but also as defensive centres. With the consolidation of the states system, cities increasingly performed as spaces 'inside' sovereign national borders. With this shift, military and police authority were assigned separate spheres—the police kept order 'inside' while the military operated 'outside' the national borders (Giddens, 1985). Hence, domestic military deployments in cities seem shocking; they trespass the foundational mythologies of the nation-state and national territoriality. However, it is crucial to remember that in colonial and indeed in many post-colonial contexts, police and military authority remain thoroughly entangled in *paramilitary* forces. In Canada, the domestic deployment of military forces in response to Aboriginal land claim struggles, which have become increasingly urban affairs, reminds us of the persistent colonial politics of settler societies. The 1970 October Crisis in Quebec is another example of a deployment of military forces into city streets that was deeply startling to the public. The crisis unfolded when the Front de Libération du Québec, a revolutionary socialist movement that sought sovereignty for the province, kidnapped the British trade commissioner in Montreal and a provincial minister.

In response, the federal government declared a state of 'apprehended insurrection' under the War Measures Act, deployed the armed forces in city streets, and authorized arrests without charge. More than 450 people were detained. The 1970 deployment also raised questions about a persistent colonial relation between Québécois people and the Canadian state.

The Royal Canadian Mounted Police's changing form demonstrates this entanglement of 'inside' and 'outside' security and how security threats—both real and imagined—generate the perceived need for more security. In 1873, British imperialists established Canada's first police force, then called the North West Mounted Police, to defend colonial interests and sovereignty claims on the nation's western frontier. The 'taming' of the frontier saw the colonization of outside space transformed into the domestication of inside space. As the frontier became more densely settled, the role of the NWMP (which had 'Royal' added to its name in 1904 and later, in 1920, became the more generic Royal Canadian Mounted Police or RCMP), initially imagined to be a temporary force, shifted from a militaristic focus bringing colonial 'rule of law' in the West, to policing its rapidly expanding urban and rural settlements. The external mutated into the internal as colonialization proceeded.

Still, the eventual 'taming' of the western frontier brought the future of the federal police force into question until new fears arose. With the declaration of World War I and the passing of the War Measures Act in 1914, the Royal North West Mounted Police were granted 'special powers' to monitor and detain Austrian and German settlers, or 'enemy aliens'. Soon after the war, and on the heels of the Russian Revolution, perceived threats posed by 'radical and revolutionary' unions in Canada's rapidly industrializing cities again rejuvenated the force with the establishment of a secret security division designed to infiltrate union organizing. When 30,000+ workers walked off the job to protest poor working conditions during Winnipeg's General Strike in 1919, the RNWMP responded with the takeover of the city. Known as 'Bloody Saturday', the massive riot resulted in the death of one civilian and hundreds of others injured. Because municipal and provincial police forces actually were sympathetic to the workers— and with the military exhausted after World War I—the national government decided to reinvest in the federal police force, which otherwise might have been disbanded, to bridge instead the military–police-security divide. According to the RCMP's own history, the government 'looked to the Force's unique character as a semimilitary organization, whose discipline and traditions would make it a dependable arm of authority' (RCMP, 2009).

By the early 1970s, investigative journalists had exposed illegal 'disruption tactics' committed by the RCMP secret service in Vancouver and Montreal. The result was the formal separation of policing and security wings of the RCMP and the creation of the Canadian Security Intelligence Service (CSIS) (Security Intelligence Review Committee, 2005). Today, while CSIS is responsible for state intelligence, the passage of anti-terrorist legislation in Canada grants 'special powers' to police and orients them towards a more proactive, 'intelligence-led' model. The line between security, police, and military is once again blurring (Murphy, 2007). The expanded powers of surveillance and information-sharing within and between different security forces, especially after 2001, have led to unwarranted detentions, extraditions, and the violation of basic human rights of citizens and non-citizens. The case of Maher Arar, unlawfully deported by the US to several years of torture in Syria, all as a result of erroneous RCMP 'intelligence', serves as a vivid example of the consequences of inflated fears on public policy, policing, and security.

Since the mid-twentieth century cities have again become more explicitly entangled in traditional matters of security (Murphy, 2007). During World War II, cities in Europe and Asia became the targets of intentional air strikes. We can think of the devastation of cities like Coventry, London, Dresden, Hiroshima, and Nagasaki. Air strikes clearly had immediate impacts on targeted cities, but they also had lasting effects on urban planning practice. Density became associated with danger; urban concentration became a source of acute vulnerability to attack. Post-war city planning in North America was deeply impacted such that dispersal and decentralization became important planning principles (Farish, 2003; Light, 2003). Suburbanization can therefore be understood, at least in part, as a security response to the Cold War.

Security 'in' Cities

Security has been an issue not simply *of* cities but also *in* cities for a long time. Policing is often an important element in efforts to improve urban security, but as the BRU reminds us, *police can be the source of fear and insecurity* for many groups, particularly poor and racialized people, who may experience violence and intimidation associated with racial profiling and targeted policing. We will return to the question of policing shortly, but first, some extreme moments in urban insecurity demand scrutiny because of the way they have set new precedents and changed normal practice precisely at the boundaries of police and military authority. Starting in 1956, the US government declared a form of open warfare against a number of activist groups that posed a direct challenge to the social and economic status quo. Under the banner of COINTELPRO, for 'Counter Intelligence Program', the FBI engaged in a range of illegal and extreme practices, a form of domestic warfare that included the assassination of movement leaders against civil groups including the Black Panthers, women's rights groups, the American Indian Movement (AIM), and socialist and communist groups (Glick, 1989). Government response to riots in American cities in the 1960s also exceeded the boundaries of legal civilian law enforcement. National Guard forces were deployed into city streets, while the government initiated military planning for civilian unrest in US cities. A bitter irony: the riots of the 1960s were a direct response to racist police violence against African Americans. The US War on Drugs, initiated in 1971 under President Nixon, was largely responsible for the unprecedented rates of incarceration of black and Latino men. It also sanctioned a dramatic expansion of military and intelligence work inside the national borders and the militarization of policing, and in this way set new precedents in the blurring of police and military authority and jurisdiction (Burghardt, 2002).

In the wake of numerous acts of terrorism in urban settings, such as the 1995 Oklahoma City bombing, the 9/11 terrorist attacks of 2001 in the US, the 7/7 bombings of the London public transit system on 7 July 2005, and Israeli attacks on urban areas in Gaza and Fallujah, cities, globally, have become deliberate targets of terrorism by state and non-state actors. This urbanization of warfare has been particularly noticeable since the end of the Cold War and has enormous implications for violence against civilian populations and military preparations for urban warfare (Graham, 2004b). As one Australian Defence College report (Evans, 2007) suggests, globalization and urbanization processes are together responsible for the intensification of urban warfare. The report argues that military planners should take the work of human geographers very seriously. Quite extraordinarily, the report is premised on the argument that, from the perspective of military forces, urban warfare raises '*the problem of a civilian presence in cities*'. A curious problem indeed!

Globalizing 'Insecurity'?

Alongside the urbanization of the global population, warfare, too, is urbanizing. Yet, 'urban security' extends far beyond the traditional work of military and other security forces. Urban security practices have changed in response to events like anti-globalization protests, anti-war protests, union organizing, and public health crises. From Vancouver to Genoa and from Seattle to Quebec City, anti-globalization protests have provoked massive experimentation in urban security strategies. Disruption to trade summits has led governments to crack down on protests, sometimes banning activists entirely by creating special 'no-protest zones' or more permanent 'secure areas' where civic rights are severely circumscribed (Nemeth, 2009). And sometimes, world leaders organize their meetings far from the madding crowds of large cities to avoid urban insecurities and protests: the 2002 G8 summit, for example, was held at a secure and remote resort in Kananaskis, Alberta, and in February 2010 finance ministers and their entourages representing the G7 countries converged on Iqaluit, Nunavut. Security planning for world trade summits has extended co-operation and information-sharing practices between national governments, and between governments and private security companies. Paramilitary units proliferate in cities and military technologies are regularly imported into urban policing (Kraska, 2001; Rygiel, 2009).

To take the example of bio-security, recent processes of globalization are at the centre of hazards such as epidemics and pandemics. The intensified movement of people around the world—for instance, from cities in Canada to vacation destinations in the Caribbean, on business travel to East Asia, and to visit family in the Middle East—has accelerated dramatically, and this movement of people can facilitate the rapid spread of disease across large distances following very specific social geographies. The spread of the SARS virus, an example we explore in more detail below, reveals the certain relationship of epidemiology with patterns of human migration. Infectious diseases like SARS are managed through the lens of bio-security by local, regional, national, and international health authorities.

We can also think about the way globalization has created particular kinds of risks that make people less secure but may not register as 'security' issues. Indeed, the fact that terrorism receives extraordinary public attention does not mean it actually is the most pressing source of insecurity. The Worldwatch Institute reminds us that far more people die from car accidents (more than 43,000 a year in the US) or even from accidental falls (approximately 15,000) than from such terrorist acts as the events of 9/11 (2,975). This emphasis on particular types of threats suggests that we need to consider the issues that create social, economic, and biological precariousness but may not feature in debates about security. If we do, we would see that contemporary forms of precariousness are linked by the varied processes of globalization.

On the one hand, economic globalization increases **polarization** in Canadian cities. The movement of production to the global South has led to the decline of stable industrial employment for working-class populations in the North and the rise of an urban economy characterized by high-wage professions and low-wage service work. The proliferation of 'McJobs' has led to the expansion of the working poor while the **gentrification** of the city by the professional classes denies this group access to housing and services in the core. As we explore below, official responses to social polarization have relied increasingly on social discipline, policing, and even expanded prison sentencing in place of the welfarist approaches that dominated the post–World War II period (Friedmann and Wolff, 2006 [1982]).

The movement of production to the global South is paired with the movement of people from these developing countries to countries such as Canada in search of opportunities or to escape violence and instability. However, once in Canada, even highly skilled immigrant populations are concentrated in low-wage and precarious work (Vosko, 2000), such that social polarization and the racialization of poverty are two sides of the same Canadian coin (Galabuzi, 2006). The global South, now literally 'at home' in the cities of the global North, experiences the forms of security that were once more sharply delineated by national borders; they have been rescaled in the city, giving rise to new forms of *urban geopolitics* (Graham, 2004a). Of course, we also might identify the post-2008 financial crisis as a 'security problem' that emerged directly from the globalization of particular high-risk practices of finance and property capital. Even war is understood today as a means of policing globalization (Smith, 2004).

The brief sketch of changes in this chapter highlights pivotal events in the historical geography of urban securitization and, specifically, some of the socio-economic context for urban *insecurity*. However, insecurity cannot be *reduced* to social and economic fear. In fact, it may be impossible to identify definitively the source of insecurity, and such a quest can miss more important questions about the form and effects of specific *security projects* (see Valverde, 2009). Nevertheless, the relationship between *sources of fear and security responses* is critically important precisely because fear is at once real and manufactured. We have already seen in the effects on individuals by the Vancouver transit police that insecurity can be created by securitization. While fear is not a new phenomenon in cities, security projects have proliferated in recent years. Insecurity is not 'unreal'—fear stems from actual events and has powerful effects on urban life. Real threats exist, certainly; but fear also can be cultivated and channelled into particular kinds of responses.

Securing the City?

How do we make sense of the kinds of fears that define Canadian urbanism today, and what is the relationship between fear and insecurity? Through a series of brief vignettes we will explore the ways in which insecurity is animated by anxieties over changing social and economic geographies at multiple spatial scales, from the local to the trans-national. While urban fear can stem from a vast range of sources (communicable disease, gun violence, dark and empty streets, hunger) and can take an expansive number of forms when channelled into security projects (food security, community safety, bio-security, etc.), the major transformations provoking security responses stem from the changes in the population and economy of Canadian cities brought about by globalization. We examine five different issues that tend to be positioned for governing as problems of security. Little unites these issues; the disease of human bodies and the operation of a city's subway service are managed by different experts and institutions concerned with different sorts of threats to human health. Yet, with a serious threat of disruption to either one, for instance through epidemic outbreak or terrorist attack, they both may be considered under the rubric of 'security'. The move to govern such diverse problems as matters of 'security' is not benign.

As a number of scholars have argued, 'security' is a powerful but depoliticizing way of conceptualizing and governing social problems (Stasiulus and Ross, 2006).

Terror

Terror has become the issue that first comes to mind when we think about cities and security; however, an enormous dissonance exists between sources of terror and efforts to secure cities against it. The fear of terrorist attack has fuelled the widespread securitization of urban space lately. National borders are actively 're-scaled' within cities today (Graham, 2004a, 2004b). In other words, the threat of terrorism—typically categorized as a matter of national security—is being managed within urban space in a number of different forms. The physical securitization of public transportation systems in the name of counter-terrorism—the SkyTrain in Vancouver, for example, and other spaces and infrastructure devoted to the circulation of people

and goods—has been striking. Port spaces in Canadian cities are subject to a host of new security regulations largely inspired by the US-led 'War on Terror'. Concern for terrorism has provoked the Canadian federal government to deem ports 'secure areas', where access is heavily restricted by fencing, bollards, closed-circuit television cameras, and the extreme and possibly unconstitutional scrutiny of transportation workers' lives (Cowen, 2007; Cowen and Bunce, 2006). The creation of special secure areas has advanced rapidly in Canadian cities, although this does not compare to the US where as much as 40 per cent of urban land in Los Angeles is under special security regulations (Nemeth, 2009).

Concern for urban terror has prompted shifts in the organization of the security state. CSIS, the RCMP, and the military have seen their domestic authority extended under the 2001 Anti-Terrorism Act. Counter-terror work south of the border saw the Los Angeles police develop a plan to map Muslim communities in the city in an effort 'to identify at-risk communities' (Winton, Renaud, and Pringle, 2007). This plan eventually was dropped because of widespread outrage over such stark racial profiling; however, such targeted surveillance is shaping the work of Canadian police forces, too, even when it is not explicit public policy. The arrest of the 'Toronto 18' made an enormous media splash in 2006 when young Muslim men in Brampton were arrested on charges of 'homegrown terrorism', including a supposed plot to assassinate the Prime Minister. But the subsequent court hearings have done more to expose racial profiling and police incompetence than any terrorist activity in cities. In fact, as the trial became increasingly farcical, David Charters, an expert member of the Advisory Council on National Security, conceded that 'Canada's biggest case of alleged homegrown terrorism—the "Toronto 18"—is turning out to be little more than a bunch of "wanna-be jihadists".' To 'anyone the least bit familiar with security', he added, 'their so-called "plans" were scarcely credible' (*Ottawa Citizen*, 2008). Nonetheless, a number of these 'wanna-be jihadists' have been tried and convicted, with the so-called 'mastermind' receiving a life sentence for his part in the plot and offering an extensive apology in court for his role, stating that he deserved the 'absolute contempt' of Canadians (CBC News, 2010).

Whatever the ubiquitous talk about terror (Katz, 2001) and the fact that it has had enormous impacts on everyday life—particularly for people of colour—terrorism is statistically one of the least likely threats to life in Canadian cities. The discrepancy between terror talk and terror realities is vast. That the fight against terror is intimately related to the actual inspiration of terror finds no better evidence than in the US. A common thread links the major terrorist threats against the US in the last 10 years—Timothy McVeigh, the Oklahoma bomber, who killed 168 people in 2002; Al-Qaeda and Osama bin Laden; Saddam Hussein: all, at some point in their past, were employed by or financially aided and politically supported by the US military and security apparatus.

Crime

In 2007 Canada's national crime rate reached its lowest point in 30 years (Kormarnicki, 2008). Yet judging from media reports, political discourse, and recent changes to the Criminal Code that strengthen the security state, Canadians' fear of crime has never been greater. The perception of crime and violence is inextricably linked to the geographies of race and class. George Kelling and James Wilson's influential 1982 'broken windows' thesis illustrates this connection succinctly. It argues that minor issues of 'social disorder' lead to more serious forms of crime and violence if left unchecked. 'The citizen who fears the ill-smelling drunk, the rowdy teenager, or the importuning beggar is not merely expressing his distaste for unseemly behavior', they assert, but 'is also giving voice to a bit of folk wisdom that

happens to be a correct generalization—namely, that serious street crime flourishes in areas in which disorderly behavior goes unchecked. The unchecked panhandler is, in effect, the first broken window.' Kelling and Wilson argue that policing should address issues that are not quite criminal in order to pre-empt actual crime. This argument had dramatic impact in many cities, most notably New York but also Canadian cities like Windsor (Lippert, 2007), and indeed around the world, and it entailed the widespread criminalization of people (such as panhandlers) and activities (such as graffiti). 'Broken windows policing' extends the authority of the police well beyond the law, much as Kelling and Wilson intended: 'arresting a single drunk or a single vagrant who has harmed no identifiable person seems unjust, and in a sense it is.' Yet 'failing to do anything about a score of drunks or a hundred vagrants may destroy an entire community.' They acknowledge the danger of police use of 'discretionary powers', namely that they could be used to 'maintain the racial or ethnic purity of a neighborhood'.

While many critics have focused on the questionable logic of the broken windows thesis, the dangers of pre-emptive police power, and the racist and class violence that such a policy is premised upon, another element to broken windows policing has received less critical scrutiny in its original articulation: geographical targeting in police operations. Kelling and Wilson (1982) argue that additional police resources should be devoted to areas of the city that appear to be experiencing a breakdown of social order, evident to them in the physical deterioration of property, which they argue is 'vulnerable to criminal invasion'.

Thus, while broken windows policing has helped rationalize the spatial targeting of graffiti, loitering, panhandling, homelessness, and prostitution in marginalized urban areas, targeting has become increasingly influential in policing. The language of targeting comes from military practice, and as Kaplan (2006) argues, police 'targeting' mobilizes the same logic of 'precise positioning' and containment as the military—the definition of a social problem as one of location. This leads us directly to the question of gentrification, which the broken windows thesis was deployed to advance.

Property

As the 2008 economic crash suggests, the greatest threat to the 'security' of private property may stem from volatility in the global financial markets. However, growing fear regarding private property has fuelled the rapid expansion of the home security industry. The spread of gated communities exemplifies this manifestation of fear. Gated communities are now home to at least 9 million Americans, and 40 per cent of new homes in California are behind walls (Low, 2006). Lang and Danielson (1997) found that many people chose gated communities because they believe such places reduce risks, ranging from 'unwanted social exchange' to 'decline in property value'. Jill Grant has been studying Canada's gated communities for many years and suggests that gated communities have a complicated relationship to security. She writes, 'the gate is advertised as a security feature, but our observations indicate that gates mostly function to keep casual visitors and sight-seers out' (Grant, n.d.). She argues that there is a direct connection between socio-economic anxieties and security responses; gated communities are a reaction to 'urban problems that show no sign of easing'. They become an option when 'people feel they cannot rely on public regulations and political processes to protect their neighbourhoods from unwanted uses (or people)' (Grant, 2005). As a result, 'some find the option of voluntarily entering an exclusive community quite desirable.'

At the scale of the city, some distinct trends connect social and economic geography with security projects. Mike Davis (1990), a prolific writer on urban security, argues that the expansion of security is a response to growing social polarization. This sentiment is echoed by the World Urban Forum in its document *The Secure City*, produced for a Vancouver meeting in 2006, which argues that urban fear is a feature of the growing gap between 'haves' and 'have-nots' (World Urban Forum, 2006). In this case, the growing home security industry arms elite enclaves and citadel spaces (cf. Marcuse, 1997) against perceived threats from the poor. Indeed, social polarization in Canadian cities is generating more spatial segregation and the formation of 'citadel' spaces in gentrified downtown cores (Hulchanski, 2007). Citadel spaces are often fortified and create border zones within the city that become spaces of conflict. We can see the formation of citadel spaces in the construction of heavily securitized condominiums in gentrifying areas like Vancouver's Downtown Eastside, where security gates and surveillance cameras control access to residential and recreation spaces. But subtler forms of securitized elite spaces are becoming common in other Canadian cities. Toronto's gentrifying Cabbagetown neighbourhood has no visible signs of gating or cameras, yet the area is rife with conflicts over security. In 2007, out of concern for the protection of their property, middle-class Cabbagetown residents hired off-duty police to bring under surveillance the poor and working-class spaces in the neighbourhood.

Our point is not that existential and economic threats to Canadians' security are not real. For instance, the financial crisis has provoked a sweep of foreclosures. Rather, in the context of 30-year lows in crime rates, the question is why so much attention is paid to property security. Burglary occurs, even if at lower levels, but what can we learn from the fact that the language of burglary has been displaced by that of 'house invasions'? This rhetorical shift renders the household its own mini-state: traditionally, countries invade one another, households don't. Yet, efforts to secure property, such as by gating communities or by aggressively policing gentrifying neighbourhoods, channel economic fears that are pervasive into the targeting of victims of social inequality, most of whom do not pose a threat to their well-off neighbours.

Gendered Violence

Many women feel fear on city streets and public spaces. Indeed, the iconic image of a dangerous space may be a dark, empty city street. A perception of danger and insecurity centred on public spaces and strangers dominates popular imaginaries of gendered violence, even though the most likely place a woman will encounter violence is in the home (Bowman, 1993; Day, 1999, 2001; Valentine, 1992). While fear of public spaces may not accurately reflect the risks of gendered violence, it is nonetheless 'real' in the sense that it fuels concrete actions. Fear shapes the feelings people have about city spaces, the travel routes they take, and the demands they place on elected officials (Day, 2001). A 1976 transit safety audit by the Toronto Transit Commission illustrates this point perfectly; despite a very low crime rate, many women saw the subway as unsafe, causing them to limit 'their lives very dramatically by stopping their use of the public transit system altogether or at certain times, especially at night' (quoted in Shulz and Gilbert, 2003: 554).

The media play a crucial role in cultivating fear of violence in public. Media reporting paints particular urban areas as inherently dangerous, producing a lopsided fear of specific people and places. Poor and racialized communities are branded as dangerous spaces by the media (Day, 2001: 114), which rationalizes intensive and targeted policing of young racialized men in public spaces (Siciliano, 2007; Tanovich, 2006). Conversely, racialized ideologies of violence and the resulting targeted policing have a further impact, for they *create* fear and insecurity for racialized men (Day, 1999). Again, securitization can create fear, particularly for marginalized groups.

Fear has long been a central element of how women experience the city. The 'protection' of women has also served as a rationale for limiting women's access to specific spaces and institutions. It is not only racialized but also *class imaginaries* that shape gendered fear in the city and official responses through policing. Returning to the case of public transportation can reveal some of these dynamics. Concern for women's safety was accentuated with the opening of the first mass transportation systems in North America. In 1909, five years after the opening of New York's Interborough Rapid Transit, women's groups expressed safety concerns (Shulz and Gilbert, 2003). The Women's Municipal League proposed that the last car of every rush-hour train should be reserved exclusively for women to protect them from 'fearful crushes' and sexual aggression. However, as Shulz and Gilbert argue, 'a secondary purpose of [the] demand for segregated cars was less benevolent.' The Municipal League, 'representing the views of many upper-class women of the time, believed that some working-class women were willing participants in this subway rowdiness, and that creation of women-only cars would lead to more ladylike behaviour by those who needed such reforming.' In this case, class perspectives were plainly evident in the Women's Municipal League's efforts to protect working-class women *from themselves.*

These complicated politics of 'protection'—specifically, the use of 'women's security' as a rationale for social control of poor and racialized people—have important parallels at other spatial scales. International warfare is today premised on the need to *protect women's safety*, but this rationale helps produce the racial and gendered violence it claims to contest. Spivak (1985) famously described instances of gendered colonial violence as 'white men saving brown women from brown men'. The military mission in Afghanistan, where Canada plays a leading role, provides parallels. Today, the language of 'urban security' frames debates about women's safety in the city, and this framing supports more securitized and even militarized responses to fear.

Biohazards

Biohazards cover a wide range of events and processes, from bio-terror to oil spills, from the results of climate change to a broad array of everyday environmental hazards. Here we focus on the very real threats from communicable disease as they affect urban life. The dense concentration of populations and the rapid circulation of people between cities worldwide enable the geographical transmission of disease (Ali and Keil, 2006). Outbreaks of infectious disease are not new to urban life; plague, cholera, tuberculosis, HIV, and many other contagious diseases all have had an impact on our urban history. What distinguishes the present situation is that infectious disease is being brought under the rubric of security planning and management. Bio-security has become the official framework through which governments, NGOs, and corporations conceptualize epidemics and emergency response. However, bio-security approaches often displace effective prevention efforts, such as investment in public health, by resorting instead to expensive security initiatives that do little to mitigate contagion (Fearnley, 2008; Hooker and Ali, 2006).

The 2003 outbreak of Severe Acute Respiratory Syndrome (SARS) provides a perfect example. SARS is understood to have emerged in southern China. It spread around the world impacting a series of globally networked cities, including Toronto. Encountering an unprepared public health system, SARS had a big impact on Toronto. As a result, the World Health Organization issued a warning against any but essential travel to Canada's largest city. Federal authorities adopted stricter monitoring of international travellers, and almost 30,000 people in the city were examined for quarantine. In total, 44 people died as a result of the outbreak in Toronto.

The SARS outbreak took place less than three years after another environmental tragedy: the 2000 Walkerton, Ontario, E. coli outbreak, in which several people died and many became ill after provincial cuts to water inspection services contributed to undetected E. coli contamination of the town's drinking water. There are important political parallels between Walkerton's water and Toronto's SARS outbreak beyond loss of life from infectious disease. As Hooker and Ali (2006) argue, Walkerton 'received a great deal of national and international media coverage and the accompanying inquiry commission identified many neoliberal structural influences common to the SARS outbreak.' Like Walkerton, responsibility for the strained response to SARS can be placed squarely on the retrenchment of investment in public health. Three major commissions of inquiry into the SARS outbreak led by different levels of government concurred that the response failed because of underfunding of health infrastructure. All three recommended renewed public health investment (Sanford and Ali, 2005).

As with other 'security issues', the SARS outbreak reveals how human welfare approaches have been reframed and even displaced by the lens of security. Managing SARS as a security issue contributed to a 'geopolitical' response to the outbreak. Containing SARS became an occasion for racism against Chinese and other East Asian people, with many incidents of official and unofficial anti-Asian sentiment reported daily. Popular racism became so salient that Ontario's commissioner of public health was compelled to speak out, stating that 'It is both wrong and prejudicial to fear or shun any or all people in the Asian community based on the assumption that they must have SARS.' People avoided Chinese businesses and racial slurs were reported on the streets, but as Hung (2004) writes, official responses also were characterized by racism: 'Many immigration officers wore face masks when attending audits for Chinese migrants applying for citizenship, though all eligible applicants must have resided in Canada for at least two years before the interview.'

Conclusion

On 13 October 2007, soon after SkyTrain police obtained Tasers, officers in the Vancouver airport encountered a newly arrived visitor from Poland. Robert Dziekanski spoke no English, was held alone in a security area for nearly nine hours, seemed disoriented and combative, and was Tasered by RCMP officers, after which he soon was declared dead. The Dziekanski case exemplifies much that is problematical about the securitization of cities. On the one hand, Dziekanski's arrival in Vancouver had everything to do with new global patterns of migration—in his case, a post-Berlin Wall world. The case became an important news item in Poland and around the world and a major embarrassment in Canada. On the other hand, it was a totally local event in the airport terminal in Vancouver. The geography of fear and security in Canadian cities can be studied as a question of scale: what happens in Canadian cities partakes in a global story of movement and flows, and what happens globally may land locally. We have seen that traditional matters of national security are rescaled in the city. Indeed, global patterns of movement are directly connected to the securitization of Canadian cities, which are increasingly managed as insecure border spaces.

Three conclusions stand out. First, securitization often entails a displacement of very real social and economic fears, refocused against surrogate social targets. Second, attempts at securitization breed their own very real experiences of insecurity. Third, the geography of fear and insecurity is central to both of these shifts insofar as the sources of insecurity are routinely displaced by implanting that insecurity in a specific place—identified by class or race or national or other forms of social difference—whether at global or local scales. The question of fear and insecurity in the Canadian city is only going to intensify, and critical responses to this intensification need to develop apace.

Review Questions

1. What groups are most targeted by urban security measures? How are they affected by these measures?
2. Why does the public demand additional urban security?
3. What are the devices used to enhance urban security?

Note

1. In addition to arming police and security guards in their effort to secure the system for the 2010 Olympics, Translink spent upward of $23 million on the 'physical hardening' of the system, including new cameras and video software, lights, 'smart fences' able to detect human motion, chemical sensors for Sky-Train and the West Coast Express, and counter-terrorism training for transit police (Keast, 2008).

References

Ali, S.H., and R. Keil. 2006. 'Global cities and the spread of infectious disease: The case of severe acute respiratory syndrome (SARS) in Toronto, Canada', *Urban Studies* 43: 1–19.

Bowman, C. 1993. 'Street harassment and the informal ghettoization of women', *Harvard Law Review* 106: 517–80.

Burghardt, T. 2002. *Police State America: US Military 'Civil Disturbance' Planning*. Toronto: Arm the Spirit Press.

Bus Riders Union (BRU). 2006. 'Skytrain police pose threat to bus riders', press release. At: <http://bru.vcn.bc.ca>. (May 2009)

Canadian Press. 2005. 'They have guns, will travel; Vancouver becomes first city in Canada to specially train, arm transit police force', *Kitchener-Waterloo Record*, 6 Dec.

Carmichael, A. 2005. 'Vancouver arms transit cops', *Globe and Mail*, 5 Dec.

CBC News. 2010. 'Toronto 18 co-leader apologizes to Canadians', 14 Jan. At: <www.cbc.ca/canada/toronto/story/2010/01/14/amara-trial.html>.

Cowen, D. 2007. 'Struggling with "security": National security and labour in the ports', *Just Labour* 10: 30–44.

———— and S. Bunce. 2006. 'Competitive cities and secure nations: Conflict and convergence in urban waterfront agendas after 9/11', *International Journal of Urban and Regional Research* 30: 427–39.

——— and N. Smith. 2009. 'After geopolitics? From the geopolitical social to geoeconomics', *Antipode* 41: 22–48.

Davis, M. 1990. *City of Quartz: Excavating the Future of Los Angeles.* London: Verso.

Day, K. 1999. 'Embassies and sanctuaries: Race and women's fear and welcome in privatized public space', *Environment and Planning D: Society and Space* 17: 307–28.

———. 2001. 'Constructing masculinity and women's fear in public space in Irvine, California', *Gender, Place and Culture* 8: 109–27.

Evans, M. 2007. *City without Joy: Urban Military Operations into the 21st Century.* Canberra, Australia: Australian Defence College, Occasional Paper No. 2.

Farish, M. 2003. 'Disaster and decentralization: American cities and the Cold War', *Cultural Geographies* 10: 125–48.

Fearnley, L. 2008. 'Signals come and go: Syndromic surveillance and styles of biosecurity', *Environment and Planning A* 40: 1615–32.

Friedman, J., and G. Wolff. 2006 [1982]. 'World city formation: An agenda for research and action', in R. Keil and N. Brenner, eds, *The Global Cities Reader.* New York: Routledge.

Galabuzzi, G. 2006. *Canada's Economic Apartheid: The Social Exclusion of Racialized Groups in the New Century.* Toronto: Canadian Scholars' Press.

Giddens, A. 1985. *The Nation-state and Violence.* Cambridge: Polity.

Glick, B. 1989. *War at Home: Covert Action against U.S. Activists and What We Can Do About It.* Cambridge, Mass.: South End Press.

Graham, S. 2004a. 'Postmortem city: Towards an urban geopolitics', *City* 8: 165–96.

———. 2004b. 'Introduction: Cities, warfare and states of emergency', in S. Graham, ed., *Cities, War and Terrorism.* Oxford: Blackwell.

Grant, J. 2005. 'The function of the gates: The social construction of security in gated developments', *Town Planning Review* 76: 339–61.

———. n.d. 'Gated Communities in Canada', project website. At: <http://gated.architectureandplanning.dal.ca/welcome.htm>. (May 2009)

Hooker, C., and S. Ali. 2006. 'SARS and security: Public health in the "new normal"', paper presented at the annual meeting of the American Sociological Association, Montreal.

Hulchanski, D. 2007. *The Three Cities within Toronto: Income Polarization among Toronto's Neighbourhoods, 1970–2000.*

Toronto: University of Toronto, Centre for Urban and Community Studies, Research Bulletin No. 40.

Hung, H.-F. 2004. 'The politics of SARS: Containing the perils of globalization by more globalization', *Asian Perspective* 28: 19–44.

Kaplan, C. 2006. 'Precision targets: GPS and the militarization of U.S. consumer identity', *American Quarterly* 58: 693–714.

Katz, C. 2001. 'Vagabond capitalism and the necessity for social reproduction', *Antipode* 33: 709–28.

Keast, K. 2008. 'B.C. Transit security and 2010', *Vancouver Sun*, 13 Aug.

Kelling, G., and J. Wilson. 1982. 'Broken windows', *Atlantic Monthly* (Mar.): 29–38. At: <www.theatlantic.com/magazine/archive/1982/03/broken-windows/4465/>. (May 2009)

Kormarnicki, J. 2008. 'Crime rate hits 30-year low, Statscan says', *Globe and Mail*, 17 July.

Kraska, P. 2001. *Militarizing the American Criminal Justice System: The Changing Roles of the Armed Forces and the Police.* Boston, Mass.: Northeastern University Press.

Lang, R., and K. Danielson. 1997. 'Gated communities in America: Walling out the world?', *Housing Policy Debate* 8: 867–99.

Light, J. 2003. *From Warfare to Welfare: Defense Intellectuals and Urban Problems in Cold War America.* Baltimore: Johns Hopkins University Press.

Lippert, R. 2007. 'Urban revitalization, security and knowledge transfer: The Case of broken windows and kiddie bars', *Canadian Journal of Law and Society* 22: 29–54.

Low, S. 2006. *Behind the Gates: Life, Security, and the Pursuit of Happiness in Fortress America.* New York: Routledge.

Marcuse, P. 1997. 'The enclave, the citadel, and the ghetto', *Urban Affairs Review* 33: 228–64.

Mumford, L. 1961. *The City in History: Its Origins, Its Transformations, and Its Prospects.* New York: Harcourt, Brace and World.

Murphy, C. 2007. 'Securitizing Canadian policing: A new policing paradigm for the post 9-11 security state?', *Canadian Journal of Sociology* 32: 449–73.

Nemeth, J. 2009. 'The closed city: Downtown security zones and the loss of public space', paper presented to the Association of American Geographers annual meeting, Las Vegas.

Ottawa Citizen. 2008. 'Toronto 18 terror suspects posed little danger: analyst', 6 Mar. At: <www.canada.com/topics/news/story.html?id=7e6e2029-447a-4181-bc52-d5ed06044bc2>. (May 2009)

Royal Canadian Mounted Police (RCMP). 2009. *An Interactive History 1873–1973.* At: <www.rcmphistory.ca/>. (May 2009)

Rygiel, K. 2009. 'The securitized citizen', in E. Isin, ed, *Recasting the Social in Citizenship*. Toronto: University of Toronto Press.

Sanford, S., and S.H. Ali. 2005. 'The new public health hegemony: Response to severe acute respiratory syndrome (SARS) in Toronto', *Social Theory and Health* 3:105–25.

Security Intelligence Review Committee. 2005. *Reflections: Twenty Years of Independent External Review of Security Intelligence in Canada*. Ottawa: Government of Canada.

Shulz, D., and S. Gilbert, S. 2003. *Women and Transit Security*. At: <www.fhwa.dot. gov/ohim/womens/chap30.pdf>. (May 2009)

Siciliano, A. 2007. 'The cultural politics of control: The "year of the gun" in Toronto', paper presented to the Association of American Geographers annual meeting, San Francisco.

Singer, P. 2003. *Corporate Warriors: The Rise of the Privatized Military Industry*. Ithaca, NY: Cornell University Press.

Smith, N. 2004. *Endgame of Globalization*. New York: Routledge.

Spivak, G. 1985. 'Can the subaltern speak? Speculations on widow-sacrifice', *Wedge* 7–8 (Winter–Spring): 120–30.

Stasiulis, D., and D. Ross, D. 2006. 'Security, flexible sovereignty, and the perils of multiple citizenship', *Citizenship Studies* 10: 329–48.

Tanovich, D. 2006. *The Colour of Justice*. Toronto: Irwin Law.

Valentine, G. 1992. 'Images of danger: Women's source of information about the spatial distribution of male violence', *Area* 24: 22–36.

Valverde, M. 2009. 'Questions of security', unpublished paper available from author, at: <m.valverde@utoronto.ca>.

Vosko, L. 2000. *Temporary Work: The Gendered Rise of a Precarious Employment Relationship*. Toronto: University of Toronto Press.

Winton, R., J.-P. Renaud, and P. Pringle. 2007. 'LAPD to build data on Muslim areas—Anti-terrorism unit wants to identify sites "at risk" for extremism', *Los Angeles Times*, 9 Nov.

World Urban Forum. 2006. *The Secure City*. Vancouver: Vancouver Working Group Discussion Papers. At: <www.cd.gov.bc.ca/LGD/intergov_relations/library/wuf_the_secure_city.pdf>. (May 2009)

CHAPTER 18

Emerging Urban Forms in the Canadian City

Jill L. Grant and Pierre Filion

How have cities changed over the last two or three decades and what factors have shaped the transformation? Chapter 3 described the 1950s as a period of urban transition as development accommodated widespread automobile use and consequently reshaped the city. By the 1970s optimism and faith in modernism gradually gave way to concerns about the environment and the quality of urban life. Critics like Jane Jacobs (1961) argued that the form of the city mattered to the health and well-being of residents. New theories of sustainable development, New Urbanism, and smart growth gradually gained adherents and began to influence public policy and market trends. Although conventional patterns of car-oriented development persisted in many parts of Canada, and new ones emerged, the last decades have produced innovative urban forms, some of which respond to critiques of post-war urban form. In this chapter we survey the Canadian urban environment to identify and explain the innovative urban forms that developed over the last decades.

We begin by discussing the key processes and factors that have affected the emergence of new physical forms in the Canadian city in recent experience. Then, we explain some ways in which government policy has responded to changing economic and cultural processes to frame the legal and regulatory contexts within which urban development occurs. Finally, we describe the new urban forms emerging as a result of policy and market dynamics. The new physical forms include multi-functional developments like mixed nodes and complete communities, and segregated patterns like auto-oriented pods and private communities.[1]

Factors Contributing to Urban Change

We might expect components of the city that are sold or rented to behave like other consumer products, which are the object of constant efforts at innovation in response to changing circumstances or simply to cravings for novelty (Baudrillard, 1994; McDonald and McMillen, 2007). Cities are inherently stable because innovations must jibe with infrastructure networks and dominant technologies while conforming to regulations such as building codes, zoning bylaws, and infrastructure standards. Moreover, the home purchasing process itself induces conservatism (Lipson, 2006; Steacy, 1987). A home is not just a commodity but an investment. Accordingly, the cautious homebuyer considers the potential preferences of the subsequent purchaser who will determine the resale value of the house. Lending institutions shy away from unusual house designs, preferring proven styles. Buzzelli and Harris (2006) suggest that local builders tend to avoid innovations that may increase their risks, and Harris (2004) documented the role of the Canada Mortgage and Housing Corporation in standardizing suburban development after the 1940s.

These circumstances account for a rate of change in the city that lags behind most consumer products. While consumers typically replace computers every three or four years, for instance, houses last many decades, even centuries. In cities, circumstances favour urban stability, not innovation. Yet major forces of change can challenge the built-in inertia of cities.

Over the last decades, economic transformations have had profound impacts on cities. The decline in manufacturing changed demand for land in urban areas while altering employment structure. Rising consumption of services represents a related economic change with powerful urban reverberations (Gallouj, 2002; Industry Canada, 1996). The mushrooming of restaurants, gyms, health centres, and countless other categories of service establishments reveals the effects of the transition. In many cities, rising levels of service consumption contribute to the growing appeal of 'revitalized' residential areas in the core and the inner city (Meligrana and Skaburskis, 2005; Skaburskis and Moos, 2008). Deepening income polarization resulting from the loss of middle-class jobs and reduced transfer payments is having a dramatic effect on the social geography of Canadian cities, as the number of middle-class neighbourhoods dwindles (Apparicio, Séguin, and Leloup, 2007; Hackworth, 2006; Hulchanski, 2007; Keil, 2002; Walks, 2001; Walks and Bourne, 2006; Walks, Chapter 10).

In the competitive context of urban development in high-growth areas, some developers and builders responded by targeting projects at particular niche markets. New urban forms and practices have emerged as developers have looked for ways of reducing development costs, achieving economies of scale, controlling the chain of production, and improving the marketability of projects. To reduce the costs of urban infrastructure, some developers turn to building private streets and services sold to occupants through condominium corporations. These various economic factors contribute to new patterns of social and spatial sorting in urban landscapes.

Although technological innovations—like the bulldozer and the automobile—significantly affected the development of new urban forms in the early and mid-twentieth century, technology figured less prominently among sources of new urban forms over the last decades. The recent past has not seen the rapid spread of any new technology that could have had such a dramatic impact. For instance, we find little evidence that computer and Internet use are transforming dominant spatial patterns. Construction techniques have remained remarkably consistent over the last decades. Cities have stayed car-dependent, despite public transit holding its own in large centres. Some industrial and retailing innovations of an organizational nature are, however, having an impact on the city. We might argue that 'just-in-time' inventory systems via truck delivery are beginning to render some spaces surplus in the city, as warehouses become open to transformation to new uses. Also, the shift in retailing to auto-oriented big-box vendors has influenced the distribution and uses of commercial spaces in many cities.

Changing cultural values and practices have had significant effects on **urban form** over the last two or three decades. Demographic shifts induced some forms of urban innovation. Over the period under consideration, several demographic trends had a major impact on housing (Bourne and Rose, 2001; Foot with Stoffman, 1996; Isaacs et al., 2007). First, the demographic structure changed as birth rates fell and longevity increased, leading to aging of the population. Second, household size dropped steadily as more people chose to live alone and divorce rates increased. Third, immigration levels increased with more immigrants coming from Asia and other parts of the world accustomed to high-density living. These trends influenced options for recent forms of residential development. For instance, households with no children or with older members constitute a significant proportion of the target market for gated communities (Maxwell, 2004), one of the urban forms coming to maturity in the study period. The popularity of condominium developments during the last two decades owes a debt to the growth in small and immigrant households.

Consumer preferences reflect cultural values. While many Canadians continue to look for homes in car-oriented conventional suburbs, the last few decades have shown that a segment of the population shifted its priorities. Urbanism—a commitment to participating in urban life—inspired a resurgence of interest in downtown and inner-city living (Meligrana and Skaburskis, 2005; Skaburskis, 2006). These consumers want specialty services like coffee shops; they seek access to public transportation. They willingly give up residential space for urban amenities. They drive the growth in the development of mixed nodes. Other households appreciate elements of traditional towns and provide a market for New Urbanism communities. Households seeking privacy and security may purchase homes in private or gated enclaves. The greater diversity in values associated with residential choices supports the trend towards niche markets and specialization in urban forms.

In the immediate post-war period, consumers and governments seemed to believe that energy and materials were endless, technology could solve any problem, and the future promised unlimited opportunities. For many Canadians, that optimism has succumbed to concerns about pollution, climate change, dwindling fossil fuels, urban gridlock, and health risks from urban living. Cultural values associated with environmentalism—commitment to environmental protection and appreciation—have increasingly influenced public policy and local planning. Environmentalism supports efforts to limit land consumption and to reduce reliance on the automobile. Recent trends to protect waterways, wetlands, and forested areas reflect, in part, the shift to these values. At the same time, however, Canadians have proven remarkably resistant to the idea of giving up their cars.

Public Policy Responses

Governments respond to prevailing economic and cultural trends. As early as the 1970s cities like Toronto, Vancouver, and Ottawa were adopting policies promoting urban intensification and environmental protection. Those policies took some time—and changing economic conditions—to influence urban outcomes. Over the last 35 years governments have strengthened policies and improved implementation mechanisms. In part, this reflects a shift in dominant philosophies of what it takes to make cities and economies thrive. Whereas at one time economic development strategies recommended land-use segregation, decongestion, and garden suburbs, contemporary development strategies advocate density and a mix in land use. Spatial planning at a large scale—including amalgamated city regions—is making a comeback as cities seek to improve their competitiveness in a global market.

Laws, policies, and regulations have influenced urban form outcomes in various ways. For instance, one legal innovation had a profound effect on urban form outcomes over the last decades. In the 1970s provincial governments across Canada adopted legislation to permit condominium tenure—a form of property ownership that allows people to own units within shared property managed by a corporation of which they are members. By the 1980s condominium tenure had become the predominant form of high-density residential development (Hulchanski, 1988). High housing costs and declining household sizes confirmed the trend during the 1990s and 2000s. By 2006 almost 11 per cent of Canadian households lived in condominiums: 40 per cent of those units were townhouses (CMHC, 2008). With almost 32 per cent of housing starts in 2007 in condominium tenure, the potential for private residential enclaves—whether gated communities in the suburbs, private courts in redeveloped urban **brownfield sites**, or high-rise towers near transit stations—was growing rapidly.

The rise of neo-liberal philosophies in the 1980s and 1990s encouraged governments to change their ways of managing urban development (see Allahwala, Boudreau, and Kiel, Chapter 12). Many municipalities transferred responsibility for building urban infrastructure to developers, who then passed on the costs to purchasers. To reduce the cost of roads in large development sites zoned for multi-family housing, developers often created bare-land condominium projects with attached or free-standing homes sharing private access streets. The condominium projects offered the additional attraction to municipalities that residents would remain responsible for road maintenance, snow clearing, and garbage collection. In an era when municipalities were struggling to meet their financial obligations while keeping property taxes as low as possible, private communities proved attractive.

By the 1990s plans and policies in most Canadian cities promoted greater densities, compact form, multi-functionality, mixed housing, environmental protection, quality urban design, and transportation options to promote the values of efficiency, competitiveness, and diversity. Demonstration projects and master planned developments provided opportunities to introduce innovative planning and design practices, like complete communities. Responding to economic and demographic pressures, and changing values on the part of the public, governments adopted plans that called for urban intensification and reduced dependence on the car (e.g., Metro Vancouver, 2007; Ontario, 2006). At the same time, however, market forces continued to fight for space for auto-oriented development forms that met the needs of business.

New Forms Appearing

In response to the factors driving urban change we see several important emerging urban forms in Canada. The forms differ according to three criteria:

- the extent to which they mix a range of uses or tend towards reproducing the post-war segregated pattern of uses;
- the degree to which commercial or residential uses dominate the mix of uses;
- the location within the city (urban core or suburban fringe).

Figure 18.1 illustrates the new forms emerging. New forms cluster in two general types: predominantly mixed use (left side of graphic), and predominantly segregated use (right side). Public policy and planning agendas advocate a mixed-use agenda; public investment in infrastructure in urban areas experiencing rapid growth promotes mixed use urban forms. The new, segregated use patterns reflect the persistent strength of economic and market forces even in the face of policy efforts to shift development paradigms away from auto-dependency.

Within the general types are further categories of new forms that differ according to the types of uses that dominate within them, and depending on where they are in the city. Mixed-use projects include mixed nodes and complete communities. Segregated use forms include auto-oriented pods and private communities.

Mixed-Use Forms

Governments seek to integrate uses by enabling **mixed-use development**. Mixed-use forms reflect the philosophy of Jane Jacobs (1961) and the growing influence of sustainable development, New Urbanism, and smart-growth principles in urban planning. Mixed-use projects are most common in high-growth areas seeking to facilitate higher density land use and transportation options.

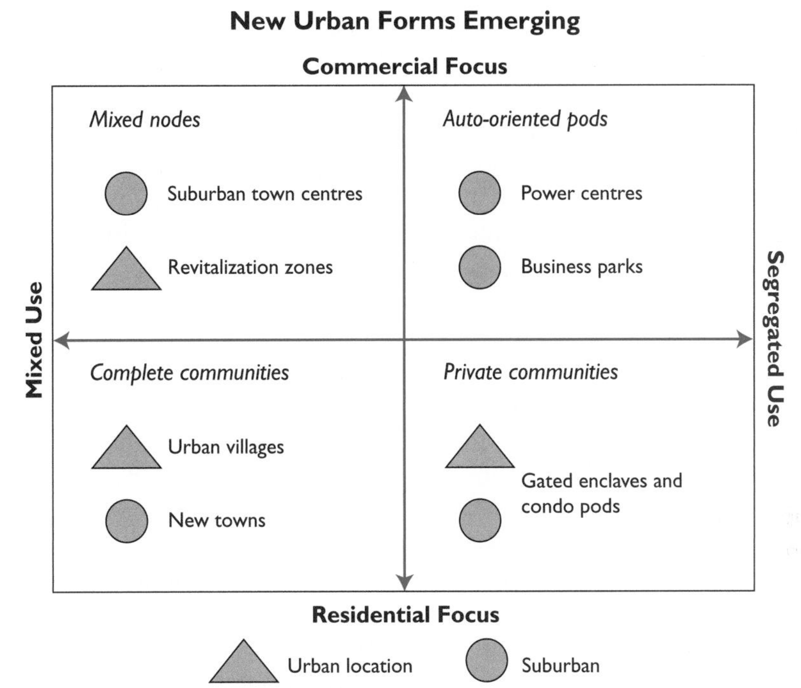

New Urban Forms Emerging

Commercial Focus

Figure 18.1 Emerging new urban forms
New urban forms vary by the degree of integration of uses, by the type of use that dominates the mix, and by location in the city.

In some cases the projects reuse parts of cities hollowed out by the decline in manufacturing. Most projects combine residential and commercial uses, and sometimes add civic uses. Uses may be vertically integrated (within the building) and/or horizontally mixed (within the block). In mixed nodes, commercial uses represent the more common use, with residential being ancillary. In complete communities, residential uses tend to occupy more territory, but commercial and other uses are planned into the mix to provide the range of services needed for daily living.

Mixed Nodes

In this chapter, nodes are understood as medium- to high-density multi-functional developments, usually with planned access to public transportation networks. Plans see nodes as transit-oriented developments linked to each other by high-quality public transportation networks, an objective that is, however, generally unmet. The planning and implementation of nodes stems from the adherence of governments since the 1970s to urban intensification in an effort to manage sprawl and accommodate more residents and uses within the existing urban fabric. The public sector thus played a key role in the launching of nodes.

We consider two types of nodes; the first version being the suburban node, variously labelled suburban town centre, sub-centre, suburban downtown, or regional city centre. These nodes purport to transpose in the suburbs the dynamics found in successful traditional downtowns, characterized by intense pedestrian-based interaction between their diverse land uses. The second type of nodal development we examine pertains to the premier node of major metropolitan regions, the downtown area. We demonstrate how the density and diversity of the downtown were enhanced recently by the large-scale redevelopment of abandoned or under-utilized industrial, commercial, or institutional sites. These efforts enriched downtowns with a substantial increment in residential, recreational, and cultural uses.

Suburban Town Centres

Suburban town centres are planned intensification nodes located in large suburbs on the periphery of major city regions. Local authorities attempt to mimic the dynamics of traditional downtowns, typically by including retail, office space, public-sector institutions and services, and housing uses. In order to encourage public transit patronage, nodes may include rail transit stations and /or assume the role of bus service hubs. Their multi-functionality and density distinguish these nodes from the suburban low-density and segregated land-use norm.

The nodal approach is a keystone of metropolitan region planning across North America, often linked to the ideal of transit-oriented development. Calgary and Edmonton promote transit-oriented development around certain stations of their light rail systems, and Ottawa does the same for stations on its Transitway, a system of roads for the exclusive use of buses (Calgary, 2008; Edmonton, 2009; Ottawa, 2007). At present, transit-oriented development remains in its initial stages in Canada, notwithstanding the fact that suburban town centres can be perceived as large transit-oriented developments.

Virtually all North American urban regions that engage in metropolitan scale planning adopt some version of the nodal model used by Atlanta, Boston, Chicago, Los Angeles, Miami, Minneapolis–St Paul, Montreal, Portland (Oregon), San Francisco, Seattle, Toronto, Vancouver, and Washington, DC. Toronto and Vancouver, where regional administrations have pursued a nodal strategy for over 25 years, are among those North American metropolitan regions where nodal development is most advanced (Davis and Perkins, 1993; GVRD, 1994; Metro Toronto, 1981; Metro Vancouver, 2008; Ontario, 2006; Raad, 2006).

Toronto features three major suburban nodes—North York City Centre, Scarborough Town Centre, and Mississauga City Centre—as well as many smaller ones. An additional major node, Markham City Centre, is under development. In the Vancouver metropolitan region eight major nodes or Regional Town Centres are at different stages of development: Lonsdale, Richmond City Centre, Metrotown, New Westminster, Coquitlam Town Centre, Surrey Centre, Maple Ridge Town Centre, and Langley Town Centre. In Montreal, the current metropolitan plan continues earlier traditions of proposing a nodal structure (CMM, 2005), but the embryonic developments at Carrefour Laval, Fairview in Pointe-Claire, and Galleries d'Anjou reveal the challenges of implementing the plan (see Figure 18.2).

How well do the suburban centres function? Toronto's major nodes meet some of their objectives, but not to the extent anticipated in planning documents (Filion, 2007). Over the last 15 years a near-stalling of high-density office development threatened their multi-functionality. Meanwhile, residential development, in the form of high-rise condominiums, boomed between 2000 and 2009, causing an imbalance between an expanding residential function and a stagnating employment base. Moreover, while suburban nodes offer more transit options than other suburban areas, workers and residents remain dependent on the automobile. Nodes suffer from Canadians' enduring reliance on driving for shopping, and from poor access to public transit in the suburban environments that constitute local catchment areas. Much of the space in nodes serves the automobile, with deleterious effects on walking: for example, office workers within Scarborough Town Centre and Mississauga

Figure 18.2 Mississauga City Centre. Like many suburban nodes, Mississauga City Centre developed around a regional mall and its parking area. (Pierre Filion)

City Centre do not patronize local restaurants as much as they could if large expanses of surface parking did not separate offices from restaurants (Filion, 2001). The best-laid plans to induce mixed use and to reduce auto-oriented development patterns continue to face significant challenges in practice.

Downtown Revitalization Zones

The traditional post-war image of the downtown entailed a central business district (CBD) occupied exclusively by offices, retailing, and ancillary activities. While zoning contributed to land-use specialization, market forces drove the process. Until recently, office buildings, retail uses, and hotels outbid other activities and thus collectively took over downtowns. Middle- and upper-income households migrated to the suburbs to uniformly residential districts (Murphy, 1972). Downtown Calgary comes close to retaining the conventional CBD model, although residential development has recently dotted its edges. The financial districts of Toronto and Montreal (where the highest office buildings are found) remain highly specialized (Gad, 1991). Other areas and other cities have moved increasingly to a nodal model of mixed uses downtown.

Planning attempts to depart from the CBD model began in Toronto in the 1970s as the city responded to the infrastructure overload caused by a massive wave of office development. Authorities attributed worsening traffic and transit congestion to the growing concentration of downtown employment. Planners argued that housing more residents downtown would reduce in-bound commuting. Downtown residents could walk to work and other core area activities. The presence of housing would assure that the downtown would remain lively in the evenings rather than emptying after office hours (Toronto, 1974). Developers received bonuses allowing them to build higher and larger buildings if they included housing components in their projects. The mid- to late 1980s witnessed a surge in downtown residential development, followed by massive condominium construction in the 2000s. Railway lands left fallow for decades now feature condominium developments.

Vancouver also actively encouraged high-density housing development in its core area. It first allowed high-rise residential redevelopment in the West End, an amenity-rich neighbourhood between Stanley Park and downtown. High-rise apartment construction proceeded apace in the 1960s and 1970s. A second high-rise residential boom occurred in the 1990s with redevelopment of the Expo '86 site and of other waterfront land adjacent to downtown (Berelowitz, 2005; Punter, 2003). New high-rise residential buildings spread through the downtown. With the most rapid growth of downtown housing in Canada, Vancouver registers a downtown population nearly as large as that of Toronto, a metropolitan region that is considerably bigger (Table 18.1). It developed its own style of high-density residential development, referred to as 'Vancouverism': narrow glass towers of condominium units sit on podiums that provide three- or four-storey façades at street level. Ground-level uses include townhouses, stores, cafés, and restaurants. Public policy requires that developers provide affordable housing and family units in downtown condominium structures. In some cases new downtown schools have been integrated in the developments. Downtown Vancouver arguably offers one of the most livable high-density residential formulas (Punter, 2003) (see Figure 18.3).

Downtown residential intensification strategies have largely achieved their planning objectives in the largest cities. Downtown nodes have become livelier with a wide range of activities (including supermarkets) that cater to the growing resident population. Also, transportation objectives have been met. In downtown Toronto, 35.6 per cent of home-based journeys are made on foot, more than combined auto-driver and auto-passenger modal shares (29.2 per cent). Fostering

Table 18.1 Downtown Population in Large Canadian Census Metropolitan Areas (2006 population of more than 1 million), 1971–2006* with Percentages of Total Population

CMA	1971	1986	2001	2006	1971–86	1986–2001	2001–6	1971–2006
Toronto	55,795	69,235	98,584	111,442	24.1%	42.4%	13%	99.7%
Montreal	84,150	63,705	72,234	77,786	–24.3%	13.4%	7.7%	–7.6%
Vancouver	47,170	50,030	82,895	101,569	6.1%	65.7%	22.5%	115.3%
Ottawa	72,965	55,380	59,626	58,094	–24.1%	7.7%	–2.6%	–20.4%
Edmonton	22,275	18,360	19,109	23,547	17.6%	4.1%	23.2%	5.7%
Calgary	33,665	30,555	33,866	36,651	–9.2%	10.8%	8.2%	8.9%

*Includes census tracts mostly located within 2 km of peak value intersection.

Source: Compiled by authors from 1971, 1986, 2001, 2006 Statistics Canada census data.

Figure 18.3 High-rise condominiums, Vancouver. Vancouver has experienced massive downtown residential growth, much of it in the form of high-rise condominium buildings. (Pierre Filion)

downtown residential concentrations thus offers a clear environmental benefit in reduced vehicle use. Lively downtowns attract young professionals and the employers seeking their services: some indicators suggest that the kinds of businesses and workers occupying downtown offices differ from those selecting suburban highway-related locations (Gad, 1985; Ley, 1996).

Urban nodes are not without problems. Despite Vancouver's attempts to promote social diversity, its downtown has become mainly an upper- and middle-income enclave due to the high cost of most condo units and the replacement of low-income housing with new condominium towers. Threats to social diversity are even more serious in Toronto where little effort was made to vary the units in new condominium developments (see Bain, Chapter 15). In addition, the growing presence of condominium developments may limit downtown functional diversity as they crowd out office buildings (Boddy, 2005; McCullough, 2005). Nodal development has not alleviated social polarization nor can it guarantee continuing diversity of uses in the context of market processes that may favour specialization.

As shown in Table 18.1, other large Canadian metropolitan regions have not experienced anything close to the downtown residential intensification that took place in Toronto and Vancouver. These findings suggest that large-scale, condominium-based downtown intensification requires conditions not replicated in all urban areas. These include rich downtown amenities (entertainment and culture in Toronto and natural features in Vancouver), sustained demographic and economic growth, high real estate values due to land scarcity, and accommodating planning interventions. Mid-size cities have similarly struggled to try to reverse processes hollowing their cores. While cities like London, Ontario, and Winnipeg have had little success in revitalizing their centres, others, such as Halifax and Kingston, have had some success in luring new residents downtown (Burayidi, 2001; Filion et al., 2004).

Complete Communities

While nodes reflect attempts at creating or enhancing poles of attraction within the urban environment, complete communities are new areas envisioned as complete small towns or urban villages. In contrast to conventional development trends that generated homogeneously residential suburbs, complete communities plan for a mix of uses, housing, and densities. Most involve suburban locations, but some entail comprehensive redevelopment of brownfield sites. The new complete communities reflect largely interchangeable principles associated with the contemporary development paradigms of New Urbanism, sustainable development, and smart growth. The complete community creates a miniature urban structure with a town centre of commercial and office uses, sets high design standards, implements a mix of housing types, creates a connected street network (often supplemented with rear service lanes), and includes a system of open and public spaces. These are large projects intended to house hundreds to thousands of homes and the services residents require to meet their daily needs. Despite the effort to be 'complete', however, most plan for limited employment opportunities within the development.

In some cases, like Cornell in Markham (in the Toronto metropolitan region), provincial and municipal governments played a significant role in assisting to plan and develop these new projects. In other cases, like Calgary and Surrey (to the east of Vancouver), municipal policy adopted the principles promoting complete communities advocated by project developers. In the last several years, Ontario made smart growth principles provincial policy for the Greater Golden Horseshoe area. While most of the development and building of new communities remains in the hands of the private sector, various levels of government set a policy context that promotes complete communities as a desired form of urban development.

Since the mid-1990s, New Urbanism principles have influenced residential development in many parts of Canada (Grant, 2003). New Urbanism originated in the US with the development of projects such as Seaside, Florida, and Kentlands, Maryland. Its proponents advocate building complete and compact communities with a mix of uses and housing types, pedestrian-oriented streets, and an attractive public realm (Duany, Plater-Zyberk, and Speck, 2000). Many of the projects feature traditional architectural and urban design qualities. A recent study identified 42 New Urbanist communities in Canada, mostly in rapidly growing suburban areas in Ontario, Alberta, and British Columbia (Grant and Bohdanow, 2008).

In Ontario the provincial government set the policy context for promoting New Urbanism in the early 1990s by arranging for design competitions and providing a large parcel in Markham for a demonstration project. Cornell, designed by Duany Plater-Zyberk Associates, soon revealed the potential of New Urbanism to support environmental policies and objectives to protect sensitive areas (Gordon and Tamminga, 2002).

The East Clayton community in Surrey established an ambitious environmental agenda in its construction. Designers employed sustainable development principles to guide the site and landscape design. Green infrastructure manages storm water on the site and park designations protect waterways from development (Condon, 2003). New Urbanism principles influenced the urban design: small lots, traditional architectural designs, and rear service lanes are common features throughout the project.

Developers began work on McKenzie Towne in southeast Calgary in the mid-1990s. Creating a New Urbanism community was a private-sector marketing strategy to give the project a strong identity in a region where many developments constructed artificial lakes to anchor subdivisions. Despite a promising start, however, McKenzie Towne struggled in the marketplace. The distance of the development from Calgary city centre and the continuing lure of large suburban houses with attached garages made small homes on small lots a challenge to sell. The developer, Carma, returned to a conventional development concept after building three phases. Another Calgary new community, Garrison Woods, proved more successful in the market. Its development was launched in the late 1990s, when Canada Lands began planning to transform a former military base in Calgary into New Urbanism communities. The commercial centre, built adjacent to a successful commercial district and existing transportation networks, attracted a range of uses. The project sold out two years ahead of schedule. Property values increased rapidly, though, pricing homes out of the range of the average Calgary household.

Complete communities ideally contain a robust mix of uses and housing types, and enjoy good access to transportation options. An attractive and well-connected street network encourages walking. Each community includes a town centre with civic, commercial, office, and residential uses. Businesses in the town centres in developments like McKenzie Towne and Cornell struggled to survive on the limited trade available within the developments. In these early applications of the New Urbanism model, the mixed-use area was

located centrally in the development, but later developments moved the commercial district close to major arterials to take advantage of larger markets (Figure 18.4). While many of the projects have some access to public transportation, few are well-sited for rapid transit to major employment centres. The mix of housing types provided has not resulted in the degree of socio-economic diversity that planners hoped to achieve in designing 'complete' communities. Although the projects attract a range of household types (from single persons to multi-generational families), they tend to be relatively homogeneous in income levels. McKenzie Towne and Cornell offered reasonably affordable housing in their early years of development, but escalation of prices in the Toronto and Calgary markets soon priced them out of range of many households. Apart from a small number of well-known projects, few of the Canadian New Urbanism communities have lived up to the potential of the concept. None of the projects is fully 'complete' in terms of the original design.

Segregated Use Forms

While government policy over the last decades has generally encouraged the mixing of uses, market pressures continue to promote conventional forms of segregated use—as in suburban residential pods—and to develop new forms of separation. The new segregated use forms reflect developers' continuing efforts to find niche markets to appeal to consumers and to raise returns on development. The forms presume that most users will have access to private automobiles.

Although we classify these forms as segregated, we recognize that some projects may involve a small admixture of uses other than the dominant ones. The projects are occurring in cities of varying sizes across Canada. While the forms reflect market dynamics at work, they occur because government policies have made space for them. New practices in retail and commercial developments

Figure 18.4 The commercial area of Cornell (Markham, Ontario). The amount and range of retailing is very limited in New Urbanism developments, as seen here in Cornell's commercial area. (Pierre Filion)

may take advantage of land that local governments previously designated for future industrial use.

Auto-Oriented Pods

In the 1960s through the 1980s many cities designated, zoned, or developed land for industrial use. The decline in manufacturing in Canada over the last few decades meant a slowing demand for industrial land. Many communities were oversupplied in land they originally hoped would house industrial uses. Looking for alternative uses for the surplus land, many accepted proposals for big-box retail outlets or for business parks.

Power Centres

Most mid- and large-sized cities now feature one or more **power centres**: concentrations of big-box

and factory outlet retail stores, entertainment and recreational facilities, and hotels. The new power centres generally locate on the urban periphery near major highway transportation corridors and junctions. They feature large parking areas, sometimes shared by several retailers. By contrast with the suburban nodes promoted by public policy and investment, power centres continue the tradition of auto-oriented land-use segregation. They become significant shopping destinations, catering to consumers looking for bargains and willing to drive some distance to find what they want (Lorch, 2005).

The power centre reflects a new form in its scale and particular mix of commercial uses that some see as the latest innovation in retailing. The pattern and location of retail uses in cities is constantly changing in response to economic conditions and consumer behaviour (Zukin, 2005). A century ago residential neighbourhoods enjoyed a wide mix of uses, including local grocers and barbers. By the early twenty-first century, the neighbourhood mom-and-pop corner store struggled to survive in the face of competition from chain outlets on arterial streets. The shopping malls of the post-war economic boom looked frayed around the edges: some closed or converted to discount outlets or call centres; others renovated in the hopes of staving off redundancy. The pattern of commercial uses transformed over the decades not only in the largest urban centres, but even in smaller settlements. Consumers looking for bargains and variety drove the success of the new power centres.

Most of the power centres have appeared in suburban or urban fringe locations. For instance, in Halifax, the Bayer's Lake big-box centre developed in the 1980s on land prepared for industrial uses. With industry and warehouses in the region preferring to buy land in Burnside Industrial Park in Dartmouth (across the harbour), Halifax council decided instead to permit a big-box retailer to open in Bayer's Lake. Within a decade the area had become a big-box power centre subject to considerable traffic congestion. In the early 2000s, now amalgamated as a Regional Municipality with Dartmouth and the county, Halifax approved a new power centre at Dartmouth Crossing: the new site dwarfs other regional shopping facilities and serves as a tourism destination for the Maritimes region. In 2009, the project developers applied to Halifax Regional Municipality for a zoning change to permit them to add a residential component to the project. Most Canadian power centres have limited residential uses, but many 'leisure centres' in American cities have added condominium projects.

While the power centres in Canada generally are in the suburbs, developers have shown some interest in proposing them in central places, even in Toronto. The City of Toronto opposed a recent effort to transform a former film studio in Leslieville to a power centre, citing policy aimed at protecting the area for employment uses (Vincent, 2009). Even in contexts where municipal policy clearly discourages such auto-oriented commercial concentrations, developers continue to promote segregated development models.

Business Parks

The growth in business parks—sometimes called office parks or research parks—similarly involves a new use for lands previously held for industrial activities. Searching since the 1980s for less expensive land with ample parking for employees, office developers have increasingly looked to the periphery of urban regions to create up-scale business parks (Lang, 2003). These projects do not include warehousing or heavy manufacturing facilities. Many are based around natural or environmental amenities; the business parks house growing concentrations of office uses, accompanied by business-oriented retail and service uses. Landscaping and design standards in business parks may screen parking lots from view, but these developments generally are of low density and auto-oriented. The campus configuration of business parks and their easy car access by a wide suburban labour pool have proved to be especially popular with businesses. The stalling of office development in downtowns and suburban nodes is in part a consequence of the success of business parks.

Private Communities

While government policy in many larger cities advocates complete communities, market factors are creating new types of developments in cities of varying sizes. The popularity of condominium tenure underpins the building of private communities. Private communities involve clusters of buildings in areas zoned for medium- to high-density housing with land held in common ownership. Sometimes units share recreational amenities, but the communities are not complete because they rarely contain commercial or civic uses. Many of these residential districts have private streets, and some are gated. Private enclaves are appearing in both urban cores and suburban areas.

For most of the twentieth century, open-concept community designs were popular in Canada. Post-war subdivisions often avoided fences or hedges, preferring unbounded lawns that facilitated views and easy access. By the 1980s, though, gated communities—developments restricting public access to internal streets—began to appear in parts of British Columbia. The popularity of gated communities in the US at this time arguably inspired their development in Canada. The trend accelerated by the late 1990s, with some development companies exploiting what they recognized as a niche market for affluent retirees choosing to move into condominium townhouse projects. By 2003, Canada had thousands of private communities and over 300 gated enclaves (Grant, Greene, and Maxwell, 2004). Two-thirds of Canadian gated projects are in British Columbia: provincial laws in BC and Alberta make it possible for developers to target projects for seniors, a population that finds private enclaves attractive.

Municipal authorities eager to reduce the costs of providing services such as garbage collection, snow removal, and road maintenance have proven sympathetic to private townhouse condominium developments and have directly or indirectly facilitated gating. In high-growth areas, medium-density townhouse projects have become a common development form providing more affordable housing options. Many such projects employ private streets built by developers and managed by condominium corporations. Those private streets, classified as 'shared access driveways' by some municipalities, can easily be gated to restrict access when residents decide the cost of doing so is worthwhile.

Most enclaves occur in suburban areas, but some appear in urban districts where sites are redeveloped for high-end residential use. The smallest private enclaves have only a few homes around a cul-de-sac, but the largest have more than a thousand units. Most enclaves have fewer than 100 units and feature modest clubhouses or pool facilities for the exclusive use of residents. Many involve small households of high socio-economic status seeking privacy, exclusivity, and a degree of security. Large gated projects sometimes include recreational amenities like private golf courses, lakes, or marinas and may house social elites. Few Canadian gated communities employ guards.

Critics suggest that private communities exacerbate social and spatial fragmentation in suburban landscapes and encourage automobile use (Grant and Curran, 2007). Certainly the growth in gated communities parallels advancing social polarization in Canadian society and exacerbates the spatial fragmentation that already characterizes suburban areas. Gated developments like Swan Lake in Markham, with over 1,000 homes planned, force pedestrians, cyclists, and drivers to go around a large enclosed area. As an urban form, the private enclave constitutes a space marked for privacy and social distance.

Planning policies that promote urban connectivity, social integration, and mixed uses have not proven effective in preventing the proliferation of private and gated communities. Developers continue to respond to consumer demands for privacy. Having permitted ground-oriented condominium development originally as a way to address the need for more affordable housing for Canadians, governments now find themselves unable to prevent an urban form that generates some significant social and spatial concerns (see Figure 18.5).

Conclusion

The urban forms that have appeared over the last decades reflect different strategies that local governments have employed to address the fiscal and environmental challenges they face. The desire to promote urban efficiency has led to plans for new

Figure 18.5 A gated community in Calgary. (J.L. Grant)

community forms that are denser, mixed in use, and accessible by transit. The most populous cities with rapid growth and high rates of immigration have seen some progress in making new forms of mixed-use nodes and complete communities happen, both in their core areas and in suburban town centres. Cities with low rates of growth or with population decline have often adopted similar ideals and prescriptions in their plans, but show few examples of successful mixed-use projects. While intensification suits the interests of government and of land developers in high-growth areas, the dynamics to promote higher urban densities do not exist in many regions. Urban patterns in the largest cities are changing in ways different from those in smaller settlements.

Local governments anxious to find resources to provide services and programs continue to respond to the imperative to find uses for land they have available for development. They need development to provide property tax revenues. They rely on development strategies that reduce the costs of delivering infrastructure. These pressures lead governments—especially those in slow-growth and smaller cities—to make compromises with development interests that result in segregated land-use forms like auto-oriented pods and private communities. As long as consumers shop in big-box stores and buy homes in private communities, developers will produce segregated land-use forms. Rising gasoline prices may eventually limit such sprawl, but in the interim it continues unabated.

Our account of new urban forms has a strong transportation dimension that points to an ongoing predominance of the car and its impact on land use. Among our examples were attempts at creating multi-functional environments in order to abate dependence on the automobile. Predictably, these efforts have met with more success in transit- and walking-oriented core areas than in car-dependent suburbs. Meanwhile, despite their inconsistency with prevailing planning objectives, segregated uses enjoy strong growth potential by virtue of their adaptation to high levels of car use.

Review Questions

1. What emerging trends are influencing public policy responses in cities? What form are these responses taking?
2. What urban forms are surfacing in the contemporary city? Are these new urban forms likely to alter the course of urban development?
3. Among the new urban forms described in the chapter, which are likely to most influence future urban patterns?

Note

1. To place the issues discussed in the chapter within the broader context of contemporary Canadian planning, see Raphaël Fischler and Jeanne M. Wolfe, 'Contemporary Planning', Chapter 19 in Bunting and Filion (2006), at: <www.oupcanada.com/bunting4e>.

References

Apparicio, P., A.-M. Séguin, and X. Leloup. 2007. 'Modélisation spatiale de la pauvreté à Montréal: Apports méthodologiques de la régression géographiquement pondérée', *Canadian Geographer* 51: 412–27.

Baudrillard, J. 1994. *Simulacra and Simulation*. Ann Arbor: University of Michigan Press.

Berelowitz, L. 2005. *Dream City: Vancouver and the Global Imagination*. Vancouver: Douglas &McIntyre.

Boddy, T. 2005. '"Downtown" a fool's paradise? Council policy has encouraged the construction of new condos but not new office towers', *Vancouver Sun*, 10 Aug., B2.

Bourne, L.S., and D. Rose. 2001. 'The changing face of Canada: The uneven geographies of population and social change', *Canadian Geographer* 45: 105–19.

Burayidi, M., ed. 2001. *Downtowns: Revitalizing the Centers of Small Communities*. New York: Routledge.

Buzzelli, M., and R. Harris. 2006. 'Cities as the industrial districts of housebuilding', *International Journal of Urban and Regional Research* 30, 4: 894–917.

Calgary, City of. 2008. *Transit Oriented Development (TOD): Development around Rapid Transit Stations*. Calgary: City of Calgary, Mar. At: <www.calgary.ca/portal/server.pt/gateway/PTARGS_0_2_527219_0_0_18/Transit+Oriented+Development+.htm>.

Canada Mortgage and Housing Corporation (CMHC). 2008. 'Changing patterns in homeownership and shelter costs in Canada: Information from the 2006 census', *National Housing Research Committee Newsletter* (Fall): 8.

Communauté Métropolitaine de Montréal (CMM). 2005. *Cap sur le monde: Pour une région métropolitaine de Montréal attractive—Projet de schéma métropolitain d'aménagement et de développement*. Montréal: CMM.

Condon, P. 2003. 'Green municipal engineering for sustainable communities', *Municipal Engineer* 156: 3–10.

Davis, H.C., and R.A. Perkins. 1993. 'The promotion of metropolitan polynucleation: Lessons to be learned from the Vancouver and Melbourne experience', *Canadian Journal of Urban Research* 1: 16–38.

Duany, A., E. Plater-Zyberk, and J. Speck. 2000. *Suburban Nation: The Rise of Sprawl and the Decline of the American Dream*. New York: North Point Press.

Edmonton, City of. 2009. *Stadium Station Transit Oriented Development*. Edmonton: City of Edmonton. Mar. At: <www.edmonton.ca/city_government/planning_development/stadium-station-transit-oriented-development.aspx>.

Filion, P. 2001. 'Suburban mixed-use centres and urban dispersion: What difference do they make', *Environment and Planning A* 33: 141–60.

———. 2007. *The Urban Growth Centre Strategy in the Greater Golden Horseshoe: Lessons from Downtowns, Nodes and Corridors*. Toronto: Neptis Foundation.

———, H. Hoernig, T. Bunting, and G. Sands. 2004. 'The successful few: Healthy downtowns of small metropolitan regions', *Journal of the American Planning Association* 70: 328–43.

Foot, D.K., with D. Stoffman. 1996. *Boom, Bust and Echo: Profiting from the Demographic Shift in the New Millennium*. Toronto: Macfarlane Walter and Ross.

Gad, G. 1991. 'Toronto's financial district', *Canadian Geographer* 35: 203–7.

Gallouj, F. 2002. *Innovation in the Service Economy: The New Wealth of Nations*. Cheltenham, UK: Edward Elgar.

Gordon, D.L.A., and K. Tamminga. 2002. 'Large-scale traditional neighbourhood development and pre-emptive ecosystem planning: The Markham experience 1989–2001', *Journal of Urban Design* 7, 2: 41–54.

Grant, J. 2003. 'Exploring the influence of new urbanism in community planning practice', *Journal of Architectural and Planning Research* 20, 3: 234–53.

———— and S. Bohdanow. 2008. 'New urbanism communities in Canada: A survey', *Journal of Urbanism* 1, 2: 111–30.

———— and A. Curran. 2007. 'Privatised suburbia: The planning implications of private roads', *Environment and Planning B: Planning and Design* 34, 4: 740–54.

————, K. Greene, and D.K. Maxwell. 2004. 'The planning and policy implications of gated communities', *Canadian Journal of Urban Research, Canadian Planning and Policy* 13, 1 (suppl.): 70–88.

Greater Vancouver Regional District (GVRD). 1994. *Livable Region Strategic Plan*. Burnaby, BC: GVRD.

Hackworth, J. 2006. *The Neoliberal City: Governance, Ideology, and Development in American Urbanism*. Ithaca, NY: Cornell University Press.

Harris, R. 2004. *Creeping Conformity: How Canada Became Suburban, 1900–1960*. Toronto: University of Toronto Press.

Hulchanski, J.D. 1988. 'The evolution of property rights and housing tenure in postwar Canada: Implications for housing policy', *Urban Law and Policy* 9: 1350–56.

————. 2007. *The Three Cities within Toronto: Income Polarization among Toronto's Neighbourhoods, 1970–2000*. Toronto: Centre for Urban and Community Studies, University of Toronto (Research Bulletin 41). At: <www.urbancentre.utoronto.ca/pdfs/research bulletins/CUCSRB41_Hulchanski_Three_Cities_ Toronto.pdf>. (Mar. 2009)

Industry Canada. 1996. *Canada's Service Economy*. Ottawa: Industry Canada.

Isaacs, B., G. Miller, G. Harris, and I. Ferguson. 2007. 'Bracing for the demographic tsunami: How will seniors fare in an aging society', *Plan Canada* 47, 4: 20–1.

Jacobs, J. 1961. *The Death and Life of Great American Cities*. New York: Random House.

Keil, R. 2002. '"Common-sense" liberalism: Progressive conservative urbanism in Toronto, Canada', *Antipode* 34: 541–601.

Lang, R.E. 2003. *Edgeless Cities: Exploring the Elusive Metropolis*. Washington: Brookings Institution Press.

Lipson, B.D. 2006. *The Art of the Real Estate Deal*. Toronto: Thomson Carswell.

Lorch, B. 2005. 'Auto-dependent induced shopping: Exploring the relationship between power centre morphology and consumer spatial behaviour', *Canadian Journal of Urban Research* 14, 2: 364–83.

McCullough, M. 2005. 'Office space suffers lack of demand: Real Estate 1 Cheapest condos now worth more per square foot than the priciest office space', *Vancouver Sun*, 30 Apr., G3.

McDonald, J.F., and P. McMillen. 2007. *Urban Economics and Real Estate: Theory and Policy*. Malden, Mass.: Blackwell.

Maxwell, D.K. 2004. 'Gated communities: Selling the good life', *Plan Canada* 44, 4: 20–2.

Meligrana, J., and A. Skaburskis. 2005. 'Extent, location and profiles of continuing gentrification in Canadian metropolitan areas, 1981–2001', *Urban Studies* 42: 1569–92.

Metro Toronto. 1981. *Official Plan for the Urban Structure*. Toronto: Metro Toronto.

Metro Vancouver. 2007. *Choosing a Sustainable Future for Metro Vancouver: Options for Metro Vancouver's Growth Management Strategy*. Burnaby, BC: Metro Vancouver. At: <www.metrovancouver.org/about/publications/ Publications/RGS_Options.pdf>. (Mar. 2009)

————. 2008. *Our Livable Region 2040: Metro Vancouver Growth Strategy (Preliminary Draft)*. Burnaby, BC: Metro Vancouver.

Murphy, R.E. 1972. *The Central Business District*. Chicago: Aldine-Atherton.

Ontario, Ministry of Public Infrastructure Renewal. 2006. *Growth Plan for the Greater Golden Horseshoe*. Toronto: Government of Ontario.

Ottawa, City of. 2007. *Transit-oriented Development Guidelines*. Ottawa: City of Ottawa. At: <www.ottawa.ca/ residents/planning/design_plan_guidelines/completed/ transit/index_en.html>.

Punter, J. 2003. *The Vancouver Achievement: Urban Planning and Design*. Vancouver: University of British Columbia Press.

Raad, T. 2006. 'Turning transit stations into transit villages: Examples from Greater Vancouver', *Plan Canada* 46, 2: 25–8.

Skaburskis, A. 2006. 'Filtering, city change and the supply of low-priced housing in Canada', *Urban Studies* 43: 533–58.

——— and M. Moos. 2008. 'The redistribution of residential property values in Montreal, Toronto, and Vancouver: Examining neoclassical and Marxist views on changing investment patterns', *Environment and Planning A* 40: 905–27.

Steacy, R. 1987. *Canadian Real Estate*. Toronto: Stoddart.

Toronto, City of, Planning Board. 1974. *Core Area Task Force: Report and Recommendations*. Toronto: City of Toronto, Planning Board.

Vincent, D. 2009. 'OMB rejects big box plans in Leslieville', *Toronto Star*, 5 Mar. At: <www.thestar.com/news/gta/article/596848>.

Walks, R.A. 2001. 'The social ecology of the post-Fordist/global city? Economic restructuring and socio-spatial polarization in the Toronto urban region', *Urban Studies* 38: 407–47.

——— and L.S. Bourne. 2006. 'Ghettos in Canada's cities? Racial segregation, ethnic enclaves and poverty concentration in Canadian urban areas', *Canadian Geographer* 50: 273–97.

Zukin, S. 2005. *Point of Purchase: How Shopping Changed American Culture*. New York: Routledge.

CHAPTER 19

The Changing Meanings of Urban Places

Nicholas Lynch and David Ley

Urban life in Canada, now more than ever, is shaped by the complex connections and relationships found in the many local places in which we live, work, consume, and socialize. In many ways, talking about Canadian cities has become a way of talking about Canadian places: how we make them, how we define them, how we change them, and how they change us. In recent decades, intentional design and planning of distinct urban places has become increasingly important. Buildings, waterfronts, streetscapes, and entire neighbourhoods are meticulously enhanced to make them more attractive and to offer new forms of urban sociability and interaction to workers, residents, and visitors. But the less-planned forms, functions, and meanings of contemporary urban places have their own special histories and politics as well.

In this chapter we present an interpretation of the evolving meanings of places in the diversifying Canadian city. We begin by considering place as a key geographic concept and how it is related to history and identity. In the following sections we briefly examine the development of Canadian urban place(s) through four distinct historic contexts: the emergence of place in Canadian urbanism up to the 1960s; the use of place to reflect and promote **multiculturalism** and social diversity; the role of **place-making** in the production of new landscapes centred on leisure, conviviality,

and aesthetics; and the contemporary rise of global urban places and their role in reshaping an increasingly unequal urbanity.

History, Identity, and the Politics of Making Meaningful Place

The materiality of place, or place as physical location, is one starting point for understanding the concept of place. But digging deeper, the idea of place as mere location does not suffice (Cresswell, 2004). Places are locations infused by a variety of meanings consolidated through the spatial routines of everyday life, or inherited from tradition, or internalized through the power of persuasion. While the meaning of places is shared, they vary in the understanding of different social groups. Meanings may be in conflict: for example, the instrumental purposes of a developer for a block of old houses might provoke the preservationist ethos of local residents. Places, both as mental and as physical constructs, are also fabricated in the imagination of novelists, artists, journalists, and travellers.

Places mean more than bounded localities or the collection of physical structures; as Doreen Massey (1992: 11) has argued, they are 'formed in part out of the particular set of social relations which interact at a particular location'. While space comprises the two- or three-dimensional environment in which objects and events are located, place is regarded as a space invested with human meaning and understanding—it is a 'humanized space', or a space that possesses 'personality' (Ley, 1977; Pocock, 1981; Relph, 1976; Tuan, 1974). The pluralism of meanings renders places 'multi-coded' as individuals, communities, and governments 'read and write different languages in the built environment' (Goss, 1988: 398). Not only do personal and collective experiences, and cultural and socio-economic status, impact how we approach places, but they guide the various motives for enacting place, whether they be to 'disseminate propaganda; to reveal the politics of context; to perpetuate tradition; to instill beliefs and values; [or] to rebel against these patterns'

(Wortham-Galvin, 2008: 32). Consequently, the varied meanings brought to a place are not passive and ahistorical. The contemporary face of Kingston's Market Square, the oldest operational public market in Canada, represents a specific place formed by many generations of cultural, economic, political, and social interaction and transformation (see Figure 19.1). Shifting from an Aboriginal meeting place, to the heart of the city, to a zone of discard, and back again to a symbolic place of congress, the Market Square speaks both to the role of historical roots in embedding a legacy to place and the power of institutional action in reconstructing place meaning (Sack, 1993).

Of course, the recently revitalized Market Square combines past memories and present

Figure 19.1 Displayed in front of Kingston's city hall is the Market Square, a revitalized place and now the focus for numerous public events. (Pierre Filion)

experiences, as 'old' elements are fused with the 'new' to forge a tourist-historic symbol sympathetic with the surrounding heritage landscape. History not only narrates past memories but also builds, in the present, a sense of distinction and authenticity. This relationship between history, memory, and place has been subject to much debate (Hobsbawm and Ranger, 1983; Lowenthal, 1985; Tunbridge and Ashworth, 1996). For many, public places offer a means of anchoring either individual pasts or collective memories through symbolic sites such as war memorials or heritage landmarks that act to forge the cohesion and identity of a group. Like flags and anthems, such places tell a particular (hi)story of a people, a state of belonging, a definition of the 'citizen'. Such meanings reflect the memories and projects of popular culture; currently, the *Ottawa Citizen* has urged the establishment of a memorial to ice hockey, 'Canada's game', in the National Capital Region as a project of symbolic nation-building. Yet, cultural and historical narratives sutured into place by those in power are often partial and distorting, carrying with them a discourse that defines the past worth remembering (Lowenthal, 1985). So place-making can become a crucial tool for governments and planning authorities to fix and legitimize national or local agendas and identities.

The connections between place and identity have been noted repeatedly (Agnew and Duncan, 1989; Ley, 1983; Relph, 1976). As memory and identity are localized in place to preserve or reproduce specific events and ideas though time, so the 'sense of place' also can evoke or prompt certain attitudes and actions, for example belonging or attachment, from the scale of the home to that of the nation. Yi-Fu Tuan (1974) has used the term '**topophilia**' or the 'love of place' to explain the human emotions that develop from individuals' or communities' shared experiences in or of a favoured place, including security, belonging, independence, fulfilment, and spirituality. One's home, school, neighbourhood, or place of worship are physically and emotionally fused, places whose meanings and authenticity circulate from relationships between culture, time, and locale (Chastain, 2006: 6). But besides topophilia we must also acknowledge 'topophobia', an aversion to places of stigma, fear, or insecurity that repress and repel, eliciting distressing memories and destructive practices (Ley, 1983; Wacquant, 2007).

Places are complex and contingent building blocks of urban life, physically and socially constructed features of the urban environment, with continuously circulating meanings among different groups reflecting history and supporting identities. With this in mind, we move on to explore four historic contexts of place-making in the Canadian city.

Emerging Cities, Emerging Places (1900–60s)

The earliest urban places in Canada were 'empire outposts', garrison communities whose main functions were to establish control points for imperial expansion, resource extraction, and trade (Marshall, 1994: 40). Quebec City, Kingston, and Halifax were originally 'planted' in the eighteenth century as representative markers of British and French territories. Their form and function reflected rational European development with carefully designed streets, markets, and public spaces. Reflecting a hierarchical social structure, the early elites lived and concentrated their activities in the town centre, while the lower classes and small ethnic communities were forced to populate the margins (Stelter and Artibise, 1986). Such planning not only 'reflected the triumph of political authority over the landscape', for new settlements required masterful engineering in the harsh Canadian environment, but also positioned the expanding town centres as a central symbol of elite British or French power (Grant, 2006: 321). While imperial British landscapes and land uses in Halifax were granted a central location that is preserved still, the established black population was marginalized in Africville, a settlement later demolished (Barber, 2009; Clairmont and Magill, 1974). In larger cities, the presence of a dominant British idiom continued well into the twentieth century in the elite districts of Westmount (Montreal), Rosedale and Forest Hill (Toronto), Shaughnessy (Vancouver), and their aspiring followers as places of privileged homeowners, resolutely defended by ratepayers' associations (Ley, 1993).

The industrializing city meant something different for working-class newcomers flooding Canada's urban centres. From 1881 to 1921, a demographic surge of 'revolutionary' proportions led to the national urban population racing from 1.1 million to 4.3 million (Rutherford, 1984). The new inhabitants, primarily European, worked in factories, warehouses, and labouring sites, and found housing in expanding working-class districts. While explosive population growth symbolized economic opportunity, early absorption into mainstream society usually was based on arriving with adequate capital, assured employment, and proficiency in the English language. American and British immigrants generally fared well, but for many without social and economic capital the new metropolis could be a place of deprivation and oppression. Rampant epidemics, poverty, racism, and a lack of public services were commonplace, magnified for many immigrant families packed into overcrowded inner-city neighbourhoods. By 1915, there was acute separation into sub-communities in Winnipeg and Regina based on ethnicity, religion, and class, creating pronounced 'divided cities' (Artibise, 1984). In Winnipeg, the clean and spacious living conditions in the Anglo-Canadian wards to the south contrasted with the mainly Slavic and Jewish immigrant wards in the North End, which were crowded and lacked consistent access to heat and bathing facilities (Hiebert, 1991). Similarly, in Regina's 'Germantown', an area of 33 city blocks, some 700 Eastern European immigrant families lived in shacks consisting of little more than three

rooms located on tiny, 25-foot lots (Warren and Carlisle, 2005: 36). Like other immigrant enclaves at the time, the lack of sufficient ventilation and sanitation also meant that these places usually were the hardest hit by influenza and endured the highest rates of infant mortality.

The prevailing sense of a 'divided city' was perpetuated by the socio-cultural 'othering' that shaped the daily lives of many immigrants. The dominance of Anglo- or Franco-Canadian groups marked a fundamental distinction between these and the various 'others' who lived, worked, and socialized in urban places. Vancouver's Chinatown was a distinctly racialized place, produced both physically and discursively as a separate outcast territory in the heart of the city. In a pernicious conjunction of place and identity, Chinatown, the place, effectively distanced the 'exotic' Chinese people from mainstream society. It was considered a place that was '"their" domain, "their" home away from home, "their" doing and "their" evil' (Anderson, 1991: 106). However, such spatial and social compartmentalization of the city could not be sustained. Amid the construction of new syncretic places, the 'Little Italys', 'Japantowns', and 'Chinatowns' of the city continually interpenetrated, creeping beyond their expected physical and social boundaries. The tarnished meaning of the inner city and its working-class and polyglot neighbourhoods redefined the urban as a threat to 'respectable' values, places to withdraw from to more congenial streetcar, and later automobile, suburbs.

An unflattering image of the industrial city became a powerful tool in legitimizing what was perceived to be a more progressive urbanism defined by the post-1945 'technical fix': a corporate-style model of **urban renewal** and freeway development, the high modernist epitome of Le Corbusier's freeway, high-rise city (Purdy, 2003, 2005; Sewell, 1993). Public-sector intervention in 'slums', like the sites of Regent Park in Toronto and Habitations Jeanne-Mance in Montreal, rested on 'normalizing' the built environment through new design formulas that would create spatially and socially 'ordered' places, with both public and private developers mass-producing standardized neighbourhoods. In Regent Park, modern planning practices were borrowed from Le Corbusier's '**Tower in the Park**' concept. Wholesale clearance of this working-class community made way for minimalist, high-density apartment complexes and multi-family townhouse clusters, closed streets and cul-de-sacs, large central open spaces, and the separation of pedestrian and vehicle traffic. Upon completion, Regent Park was praised as an innovative approach to housing through its sanitized aesthetic and its clean modernist styles, leading many other cities to implant the formula in their own 'problem' areas.

Other visions of the 'good city' informed by the rationale of high modernist planning were carried to suburban development and design. Pervasively inculcated by Le Corbusier's machine-age aesthetic extolling the automobile, homes and services were predicated on widespread car ownership. Park Royal shopping centre, opened in 1950 in the suburb of West Vancouver, is said to be Canada's first automobile-based mall, but soon the serried ranks of acres of parked vehicles became common across the country. No longer small and exclusive, the modern suburb was built for the masses during the post-1945 economic and baby boom and was made available to the affluent working and middle classes, presenting a 'refuge' for undefiled community and domesticity. Precedent-setting was the corporate suburb of Don Mills, outside Toronto. Modern design permeated all facets of its development, offering the middle-class homebuyer a slice of what was to become the 'Canadian dream' (Sewell, 1977). Its creator, Mackling Hancock, applied Ebenezer Howard's garden city principles through sprawling wide lots and ranch-style houses sited on crescents or cul-de-sacs. Individual neighbourhood units were defined through their clustered orientation

and by the interspersed greenways and parks laid out between street sections. Soon after its construction, Don Mills became the icon of Canadian suburbia. It not only represented the values of the post-war middle class in search of conformity, privacy, and progress, but also reproduced these values in the making of the modern urban landscape.

Correcting, Connecting, and Celebrating Urban Places (1960s–80s)

By the mid-1960s, perceptions of the Canadian city were changing. While visions of 'good' city form and the ideals of suburban life still were entrenched in planning and political circles, they no longer enjoyed a monopoly as place-making solutions. The city was the stage for the decade's critical social movements, the podium for mounting reactions against political and social conservatism and the technocratic practices of the state. A new sense of cultural pluralism was unfolding. Alongside Quebec's demands for self-expression during the 'Quiet Revolution', Western European immigrant arrivals began to fall dramatically, and by the 1970s gave way to growing immigration from Asia, Latin America, and the Caribbean. To manage this rapidly shifting character of the nation, the 'cultural mosaic' metaphor became the basis for Canada's multicultural policy (Qadeer, 1997). Indeed, as new families from Hong Kong, India, Jamaica, and elsewhere settled in major cities, they accentuated the multilingual and multicultural identities of urban places and also championed new policies for compatible urban spaces, public services, and political and social representation. In the same vein, feminists and gay activists urged that modern, male-dominated urban places had enjoyed excessive supremacy in shaping the daily lives of others. This 'cultural turn' increasingly challenged the socio-spatial divisions of the metropolis, destabilizing dominant visions of urban places as supportive of masculine, heterosexual, and white culture and lifestyles.

Urban communities similarly questioned the previously unassailable wisdom of planning experts and their abstract modern designs. The massive and demoralizing concentration of poverty characteristic in urban renewal projects, the effacing of uniquely diverse places like Toronto's Bohemian Yorkville or Montreal's Chinatown, and the revolts against massive infrastructure like Vancouver's Strathcona and Toronto's Spadina expressways became new focal points for reformers seeking alternatives to an impersonal, placeless, and formulaic modernity. As John Sewell (1993: 198) put it, 'there was no mistaking the intention or the result of the reform wave that gained a foothold in the city . . . the day of destroying the existing city to replace it with a new Jerusalem was over.' In this volatile climate, the urban philosophy of the American cum Canadian social planner Jane Jacobs (1961) became increasingly relevant. In her massively influential book, *The Death and Life of Great American Cities*, Jacobs argued that good communities and healthy societies are located in complex and creative metropolises, not in the rebuilt and fragmented modernist city. Age, density, and diversity—all interconnected elements—provide the real units for urban living, challenging the sense of homogeneity, abstraction, and oppression creeping into modern city spaces. She anticipated a postmodern vision, connecting and celebrating what was being demolished and suppressed: old buildings, city parks, social and land-use diversity, meeting places, pedestrian travel and public transportation, ethnic neighbourhoods. Jacobs's insistence on the need to plan 'people places' inspired neighbourhood movements comprising ethnic and working-class residents and especially the professional middle class. Her conceptual legacy created new urban landscapes, priming both a wave of heritage protection and the beginnings of inner-city **gentrification** in the early 1970s (Caulfield, 1994; Ley, 1996).

Together, the concerted efforts of community activists and reform-minded municipal and federal officials eventually redirected urban policy away from modern renewal and design solutions. By 1969 a federal task force reported sternly on urban renewal's public housing projects, calling them 'ghettos of the poor' and places of 'stigma' and 'alienation' (Ley and Frost, 2006: 203). During the 1970s federal policy shifted decisively away from large-scale clearance and rebuilding to more incremental policies of housing rehabilitation and neighbourhood enhancement; whereas in 1970 all federal monies were committed to renewal, by 1979 the ratio of rehabilitation and improvement funds to renewal was of the order of 50 to one (Ley and Frost, 2006: 203).

While large public and private development projects housing hundreds and thousands of residents continued to be undertaken, two important examples, Toronto's St Lawrence project and Vancouver's False Creek South, incorporated innovative design philosophies. Sewell (1993: 191) noted that St Lawrence was planned 'with the underlying assumption that it would become much like a traditional Toronto neighbourhood', in contrast to private-sector St Jamestown or public-sector Regent Park. Mirroring postmodern ideology, Vancouver's False Creek South was heralded as 'a delightful and humane living environment for people of different ages, lifestyles, and economic backgrounds . . . [where, it was hoped,] most of the residents of this community will have the opportunity to express their identity as well' (cited in Ley, 1987: 46). Both neighbourhoods were built with explicit notions of place-making and community-building in mind. They incorporated mixed uses and a mix of tenures to foster local vitality and a sense of community through diversity. Constructed on previously derelict industrial lands, both developments sought to minimize residential displacement and to fit in with their surrounding urban landscapes (Hulchanski, 1984).

Consultation with residents also meant that buildings, although designed at medium to high densities, were at a 'human scale', avoiding the super-blocks common at the time through the construction of three-storey row houses and modestly sized apartment buildings. False Creek also was among the first projects that helped to form the design strategy now known as 'Vancouverism', a design innovation being simulated in cities from Dubai to Dallas, and offering a replacement for the dense Euclidean grid of 'Manhattanism'. Instead, inner-city density is promoted through the construction of widely separated residential towers intermixed with low-rise multi-functional buildings (Punter, 2003). This innovation, illustrated in its early form in False Creek, connected multi-unit residential and retail uses to adjoining public spaces, parks, and streetscape/waterfront edges, helping to overcome previous concerns over higher-density living.

Such initiatives ushered in a postmodern strategy for place-making, one that continually brought attention to the revitalizing capacity of the urban neighbourhood. As a result, they emphasized the defence and enhancement of the neighbourhood as a prevailing solution to reanimate the 'placeless' urban scene, to encourage local heritage, cultural diversity, and security, and, to revitalize ailing volunteerism and social support networks fundamental to fostering active communities. Ethnocultural districts were no longer perceived as simply the deprived territories of Canada's immigrant families. Rather, stable ethnic communities were encouraged to express identity in the built environment.

In Montreal, the expression of distinct cultural communities has been marked by religious contrasts. Mile End, a neighbourhood adjoining Outremont and the Plateau Mont-Royal, has been a key Jewish enclave and an inspiration for artists and novelists including Mordecai Richler (*St. Urbain's Horseman*, *The Apprenticeship of Duddy Kravitz*). In Richler's time, Mile End's diversity included Lithuanian, Romanian, and Russian Jews, Greeks, Portuguese, and residual English and French populations. Richler chronicled Mile End's rich tapestry, exposing its intimate connections to uniquely defined places like Wilensky's, Jack and Moe's barbershop, or Schacter's Cigar and Soda. Mile End became a central influence for the identity of many residents, including Richler (2007: 19) himself: 'I do feel forever rooted in Montreal's St Urbain Street. That was my time, my place, and I have elected myself to get it right.'

By the mid-1960s, however, many upwardly mobile Jewish residents had moved on, leaving behind a strengthening Hasidic and Orthodox Jewish diaspora. This highly conservative group is committed to shaping a particular ethno-religious identity (Germain and Rose, 2000). Concentrated in Mile End and Lower Outremont, Hasidic groups founded *shtiebls* (synagogues or 'prayer rooms'), *mikvahs* (ritual baths), and religious schools, along with retail and kosher groceries (Germain and Gagnon, 2003). On the streets, this distinct religious culture was marked by Yiddish, both spoken and legible on retail signs and posters, and by the visibility of Hasidic men wearing black coats, top hats, side curls, and beards, while women wore hair scarves, wigs, and conservative, long dresses. The community's coexistence with other minorities and an increasingly gentrifying cohort has not been without tension. Claims for new and expanding Hasidic synagogues frequently have met opposition (Germain and Gagnon, 2003). The marked difference of the orthodox Jewish community, embodied in culture and in the built environment, from other religious and secular identities has sparked a contest between an intellectual acceptance of minority religious values and their material expression in public space.

Such identity politics in constructing urban places reoccurs in the making and meaning of gay villages. The expression of difference based on sexuality did not initially receive the same social and political legitimacy. By and large, the central gay enclaves in Montreal (rue Ste-Catherine Est), Vancouver (Davie Street), and Toronto (Church and Wellesley Village) were originally founded from relatively quiet and inconspicuous beginnings in the more private confines of bars, clubs, and communal houses. More explicit appropriation of urban areas developed primarily from the desire of growing gay communities to find a larger space of sexual acceptance. The rainbow flags, charged street atmosphere, and eclectic storefronts that demarcate today's gay villages have evolved from complex contestations over sexual identities and their varied expression. According to Catherine Nash (2006: 2) Toronto's gay village, arguably Canada's largest, has 'experienced a precarious existence as simultaneously a location of community and individual freedom and as material proof, for some, of Canada's descent into moral and social decay'. By 1980, however, the 'gay ghetto', by then a visible residential and commercial village, was envisaged as the 'rightful home of a minority group and as the foundation for political and economic strength and community building' (Nash, 2006: 13). Gay villages also provide a central place for political protest and celebration (especially with the popularity of gay pride parades), and for inventing and enacting alternative urban identities (Podmore, 2006).

Place and the New Urbanism (1980s–)

Change in the Canadian city from the late 1960s— political and planning reform, multiculturalism, and the preservation of social and cultural diversity in the neighbourhood movement—established the groundwork for a new kind of urbanism. A less welcome trend was growing public deficits that beset senior governments and inspired both the rise of public–private partnerships and a more entrepreneurial public sector aiming to capture the attention (and dollars) of baby boom professionals, members of the 'creative class', and urban tourists seeking vital central-city places. Indeed, these key economic groups were increasingly drawn not only by centralized professional-managerial employment and gentrifying opportunities, but also by the cultural diversity and service-rich amenities

concentrated 'downtown' (see Bain, Chapter 15). State and private entrepreneurs began remaking urban places to promote a sense of conviviality incorporating urban livability permeated with an affluent lifestyle and stylized, cosmopolitan leisure.

Among the most important places for this new emphasis are waterfronts and parks embellished from the remnants of past enterprises or built anew as convivial landscapes aimed at re-imaging the postmodern city. Plans for waterfront reclamation from Victoria to St John's have included ambitious designs combining older functions such as naval bases and port facilities with new spaces like recreational seawalls, specialty shopping quays, and waterside residential and hotel development (Hoyle, 1992; Norcliffe, Bassett, and Hoare, 1996). Along Halifax harbour and the Bedford basin, state-led redevelopment has created a central cultural and tourism district, a vital economic engine for the city and region. On the renovated waterfront pedestrians now make use of the three-km Harbourwalk and visit rehabilitated warehouses accommodating maritime museums and Atlantic crafts. In Vancouver, waterfront **revitalization** includes high-rise condominiums and hotels seeking mountain and ocean views, replacing deserted docks, silos, and railway lines. The city's networks of seawalls, beaches, waterside parks, and marinas significantly mark the leisure culture that is now emblematic of west coast living. Yet, even in non-coastal cities like Montreal and Ottawa where the Lachine and Rideau canals were once vital industrial arteries, federal and provincial reclamation plans have created heritage sites and waterfront trails that lure cyclists, boaters, pedestrians, and skaters.

Other significant redevelopments in reclaimed spaces are festival marketplaces and pedestrian malls, specialized shopping districts refashioned for inner-city leisure and tourist use. Following the lead of James Rouse, creator of Boston's Faneuil Hall and South Street Seaport in New York, designs epitomize a postmodern ethic: the absence of an 'anchor' department store, a mélange of unique boutiques and distinct shopping environments, a mix of tenants and land uses, an assortment of attractions, and an explicit link to local and regional heritage. These multifarious places are for a clientele that habitually rejects the standardization (of products *and* design) in the shopping centres and strip malls that dot the suburban landscape. They are dynamic consumption places, offering tasteful goods in a fashionable setting.

Among Canada's popular festival marketplaces are Market Square, St John's, Byward Market, Ottawa, The Forks, Winnipeg, and Prince Arthur Street, Montreal. But the most successful is Vancouver's Granville Island, creatively guided by the Canada Mortgage and Housing Corporation to be Vancouver's waterfront 'oasis in the architectural

desert . . . a perfect antidote to the massive boxes' of the downtown core (Kemble, 1980: 1627) (see Figure 19.2). Central to its design and planning was a strategy emphasizing a diversity of uses and users, and a comprehensive heritage policy. The variety of tenants on this relatively small island (actually a peninsula) is exceptional: a public market, craft shops, art studios, fine dining, a boutique hotel, theatres, the city's art college, marinas, a community centre, a micro-brewery and even a concrete-making plant. Setting the scene is a collage of scavenged and recycled industrial structures, materials, and designs. Corrugated tin and exposed wood, large original doorways, vaulted ceilings, exposed piping, and multi-pane industrial glazing compose the design palette. The area features irregular interstitial public spaces between

Figure 19.2 A unique post-industrial place, Vancouver's Granville Island offers visitors a diverse range of services and experiences. (Nicholas Lynch)

buildings, meandering footpaths, platforms for local buskers, and a general lack of sidewalks (aside from the enveloping seawall), allowing pedestrians to walk alongside slow-moving cars on old industrial arteries and conserved rail lines. For 30 years, these place-making strategies have been successful in creating a distinct sense of place, whose sustainability is an important and long-standing issue in the changing dynamics of place design (Oldenburg, 1999; Relph, 1976).

In Ottawa's Byward Market, located close to Parliament Hill, decades of inner-city revitalization have consistently altered the meaning of this previous zone of discard. Efforts by the National Capital Commission (2004) to celebrate the 1967 centennial with a ceremonial 'Mile of History' along Sussex Drive sparked a flurry of commercial upgrading in the Market area, with the reuse of old warehouses and low-grade retail spaces for small independent outlets, and with arts and craft vendors selling their wares to the growing middle-class, tourist, and public service populations. Distinct francophone identities established in French schools and shops were displaced, together with the city's red-light district and numerous homeless shelters (Tunbridge, 2001). This commercial gentrification was enhanced in the early 1980s through an attempt to develop a tourist-historic district and a ceremonial route called Confederation Boulevard to enhance the legibility of the city centre and make explicit route linkages for tourists visiting national museums, Parliament Hill, and other nearby places of interest (Ashworth and Tunbridge, 2000; Tunbridge, 2001). More recently, the Peace-keeping Monument and the American embassy have completed an 'official' vista displaying both a sense of national memorializing and international connection, a backdrop to the specialty restaurants and national festivals in the Market quarter. Proceeding apace, residential infill, new condominiums, and especially the mixed-use redevelopment of the Daly Building site (Ottawa's first department store) on Rideau Street have further marked the Byward Market as a growth pole for property reinvestment.

In suburban communities, the preoccupation with designing attractive and centralized destinations has led some municipalities, in concert with development corporations, to construct suburban downtowns more reminiscent of nearby central cities. In suburban Scarborough (Toronto) and St Foy (Quebec City) expansive shopping centres incorporate office buildings, theatres, hotels, restaurants, and apartment complexes in an attempt to promote more concentrated growth. Throughout the Greater Toronto Area, a number of such malls have become the nuclei for large suburban business centres. Mississauga's Square One Shopping Centre and Scarborough's Town Centre are foci for intensified retailing but also hubs for regional corporate headquarters and high-rise residential developments. Such suburban nodes are commonly built as master-planned superblocks, offering not only parking facilities but traffic-free inner pedestrian zones connecting users to transit systems and other public facilities. Their aim, according to Edward Relph, 'seems to be to make something similar to the pedestrianized centrum of a European new town', thereby evoking a distinct 'sense of suburban community identity' (Relph, 1991: 421–3). Mixed uses, increased densities, transit options, and, in some cases, the development of elaborate public buildings, notably city halls, are designed to create both physical and symbolic landscapes akin to more dynamic urban cores. Moreover, with the recent diversification of ethnoburbs like Richmond or Markham, suburban downtowns are increasingly designed to be reflective of local cultures (Bourne, 1996; Evenden and Walker, 1993). In Richmond, a predominantly Asian suburb of Vancouver, a cultural retrofit of grand proportions is visible in large shopping facilities like the Aberdeen, Yaohan, and Empire Centres that offer both distinct ethnic retailing and cultural amenities catering primarily to Chinese-Canadian consumers. While there is a risk of 'marginalized ethnic enclaves, which can diminish community cohesiveness' (quoted in Kom, 2009), such ethnically themed shopping centres can also give other communities exposure and access to immigrant cultures (Qadeer, 1999; Zhuang, 2008).

The emphasis on suburban downtowns is not the only contemporary suburban place-making strategy. Neo-traditional housing styles have become commonplace to implant memory and identity into new subdivisions, while a number of developments have proven more ambitious, employing New Urbanism in their planning and design (Grant and Bohdanow, 2008). New Urbanist philosophies unite historicist neo-traditional architecture with systematic planning that emphasizes many of the village-like elements previously admired by Jane Jacobs, including mixed land use, pedestrian-friendly street networks, reduced vehicular traffic, demographic heterogeneity, and 'eyes on the street' design components, such as front porches. McKenzie Towne in Calgary, a pioneering development, was constructed as a series of villages individually distinguished by unique architectural elements (Townshend, 2006). The conscious construction of a 'sense of place' incorporates pedestrian trails, recreational areas, and outdoor community spaces highlighted by character-laden village squares and gazebos (Townshend, 2006). Moreover, in McKenzie Towne and other New Urbanist landscapes (e.g., Cornell in Markham), environmental sustainability and, latterly, personal fitness are top priorities. Densities are up to twice those of conventional suburbs, and usually include public transit networks and some local employment in retail and other sectors, reducing car travel (Skaburskis, 2006).

However, regardless of the strides made by New Urbanism and its popularity in the housing market, its capacity to contain urban sprawl has been overstated, while its effect on limiting car use remains unproven (see Grant and Filion, Chapter 18; Grant and Bohdanow, 2008; Skaburskis, 2006). For some, neo-traditional place-making attracts a rather limited sense of diversity, offering nostalgic places primarily for those who share the specific culture and history that are recreated (Lehrer and Milgrom, 1996). Nevertheless, in terms of its ambitious and multiple objectives to address sustainability, affordability, community, and health and fitness, no other competing proposal to New Urbanism is presently on offer.

The Rise of Global Places (1990s–)

Toronto, Vancouver, and Montreal, more than other Canadian cities, are described as 'international', 'global', 'cosmopolitan', or 'world class'. Over the last 20 years, a preoccupation with global markets and the hegemony of pre-eminent world cities (New York, London, Tokyo) have led many municipalities to seek new ways to remain competitive before global as well as national consumers (see Hall, Chapter 4). In these 'place wars' (Haider, 1992) to attract investment and position places as commodities, even the earlier and costly urban design projects intended to fix identities and stimulate the consumption and leisure lifestyles are being re-imaged internationally to display, above all, diversity, multiculturalism, and entertainment. Goaded by imagineering consultants, place has become the *local* tool by which public and private interests seek to attract *global* assets, including corporate headquarters, transportation and other infrastructure, tourists, and skilled national and foreign workers.

In this fervour to remain competitive and construct a 'cutting-edge' global profile, the repertoire of strategies involves urban marketing, branding, and benchmarking (e.g., slogans, icons, quality-of-life surveys), hallmark events, iconic architecture, and urban mega-development projects (Hall, 2004; Kavaratzis, 2007; Sklair, 2006). Some authors speak of global 'imagineering' to describe the Disney-like practice of merging 'imagination' with 'engineering' (Paul, 2004; Short, 1999; Yeoh, 2005).

However, the manipulation of the urban environment and the promotion of a global sense of place not only represent an economic tactic to lure global resources but also a political tool as urban elites often promote their own visions of 'global status and connectivity' while downplaying or ignoring the problems and realities of local city life (Paul, 2004).

In Vancouver, global imagineering is particularly intense. Frequently represented as a 'gateway' for the flow of Asian immigrants especially, Vancouver also is seen as a staging-ground for international development and investment (Ley and Murphy, 2000). Mega-events have been crucial in this repositioning. Expo '86 repackaged Vancouver from a 'sleepy provincial town' to a city with a global presence (Mitchell, 2004: 43), showcasing for visitors its 'arts, architecture, its "world class" dining and shopping, and its immigrant energy' (Olds, 2001; Whitson, 2004: 1221). City officials and business elites have positioned Vancouver with an even wider global image through promotional campaigns for the 2010 Winter Olympics. As McCallum et al. (2005: 24) demonstrate, branding packages from the outset of the Olympic bid projected the city as embodying 'pristine urban nature, multicultural social harmony, and a vibrant local culture of sport', while recent marketing slogans prefiguring the games include BC's new motto: 'The Best Place on Earth'. The Olympics, and its promotional fare, thus involved a nuanced 're-imaging' of the city through what David Whitson (2004: 1223) argues is a set of practices that 'build', 'signal', and 'transform' a global character both as display for the world and also 'on show' for its citizens.

These hallmark events have been the focus of and the 'trigger' for new rounds of substantial public and private investment (Olds, 2001). Earlier legacies from Expo '86 include the Canada Place convention centre and the spine of the SkyTrain rapid transit system. Private spinoffs include Pacific Place, built on the old Expo lands, and among the largest and most globally integrated downtown projects in North America. On land purchased by the Hong Kong tycoon, Li Ka-shing, Pacific Place is distinguished by its massive scale (over 80 hectares), high-rise luxury condominiums, enveloping seawall, and adjacent leisure districts including two sports stadiums, Chinatown, and an emergent downtown arts precinct. On the ground, the development was envisioned with neighbourhood identity and livability in mind, with carefully orchestrated views, public space, and pedestrian-scale streets to tame the impact of the towers on the sidewalk (Vancouver, 2003).

Critics of Pacific Place's high-rise lifestyle point out that such private mega-developments too often cater to domestic and transnational capitalist elites and ignore local issues like affordable housing and participatory policy-making (McCann, 2002). Indeed, Expo's intent was to graft the ailing regional industrial economy into the global economic powerhouse of Asia Pacific (Ley, 2010). Expo '86 was the loss leader to induce investment and attract wealthy immigrants, and Pacific Place is part of a massive flow of Asian capital into Vancouver's property market. From 1988 to 1997 the liquid assets of business immigrants alone to Vancouver amounted to an estimated $35–$40 billion, creating a predictable bubble in the regional property market (Ley, 2010). The promotion of such cosmopolitan consumption has some critics calling Vancouver's downtown a residential 'resort' ushering in new rounds of urban gentrification and the displacement of the remaining low-income communities (Boddy, 2006). In the City of Toronto considerable publicity has been given to the identification of income polarization and the creation of 'three cities', an inner ring of privilege, a central transitional ring, and an outer ring where real income decline has occurred in the old suburbs. The outcome, writes Hulchanski (2007), is no longer a city of neighbourhoods, but a city of disparities. The cheerful jingoes of the convivial **global city** often conceal a reality of growing inequality among its parts.

Smaller-scale approaches to global imagineering often involve the production of iconic architectural projects tied to cultural programming. Art galleries, museums, and libraries are globally branded to promote a sense of the 'cultural city' (Evans, 2003; Sklair, 2006). The recent popularity and success of 'cultural flagships' like the Bilbao Guggenheim have many cities seeking a similar formula: 'Eye-popping architecture + cultural attractions = more tourists' (Rybczynski in Hannigan, 2003: 254). In Toronto, this formula has been applied recently to the Royal Ontario Museum (ROM), which has been given an architectural facelift of sculptured pyramidal glass designed by 'superstar' architect Daniel Liebeskind (see Figure 19.3). The explosive crystal-shaped facade imposes a spectacular, even jarring, quality to the local streetscape. The global iconicity of both design and architect are intended to reinforce the

cultural–entertainment primacy of Toronto and its Yorkville district. Adding to Toronto's 'cultural renaissance' are changes to the Art Gallery of Ontario (AGO) and the Ontario College of Art and Design (OCAD). Similarly inspired by the pursuit of a global signature, they are employing, respectively, the asymmetric designs of Frank Gehry and the playful motifs of Will Alsop. In Ottawa, the national museums play the same role, while Vancouver's architectural signature quickly is becoming its coliseum-themed public library (Lees, 2001).

Although the entanglement of public and private sectors has expanded the range and feasibility of globally attractive destination features, their achievements have involved relinquishing some 'public' authority over public spaces in the city so that calculated decisions over who belongs and how they can act falls more into the hands of private interests. In the United States and the

Figure 19.3 The recent facelift of the Royal Ontario Museum in Toronto epitomizes the current trend of iconic architecture in building a sense of global place. (Nicholas Lynch)

United Kingdom, the movement to safeguard public spaces and landmark projects through public and private policing, the implementation of surveillance technologies, target hardening, and gated communities has been at the forefront of urban analysis and debate, even before its apogee after the events of 9/11 (Blakely and Snyder, 1997; Davis, 1990; Fyfe, 1999; see Cowen, Siciliano, and Smith, Chapter 17). Carefully crafted street fixtures (e.g., bum-proof benches and mobile barriers) have been widely used to protect urban places from 'unwanted' users. The risk is the tolerance of 'paranoid urbanism' that systematically eradicates diversity in urban places by threatening a wider variety of users not considered 'legitimate' patrons (Flusty, 2001; Mitchell, 1997). Though more muted, such features have entered the Canadian urban landscape. In Montreal and Toronto open-street closed-circuit television systems and groups of private security guards have been increasingly deployed along dense commercial corridors (including rue St Denis in Montreal and Yonge Street and Dundas Square in Toronto) to promote a sense of security for shoppers and residents (Walby, 2006). In many cases, this securitization stems from a desire to contain the 'transgression' of poverty and crime from adjacent lower-income areas. In Vancouver, the potential extension of the Downtown Eastside into neighbouring areas, including tourist-historic Gastown or Chinatown, represents a perceived threat not only to the viability of these tourist districts but also to Vancouver's global image. With the 2010 Winter Olympics, the Downtown Eastside became a target of intensified police activity and surveillance. As early as 2006 the mayor launched 'Project Civil City', attempting to cleanse the city of 'street disorder' and reprimand individuals who did not fit a blemish-free image. Activists claim that the project's mandate disproportionately affected the city's most desperate citizens and pushed homelessness and drug-dealing into other areas without addressing root causes. In addition, acute development pressures, strengthened with the hype of the Olympics, placed the city's affordable housing stock into serious decline and left some of its low-income residents with little alternative to living on the streets.

The election of a new mayor and council in 2008 led to the scrapping of Project Civil City and to a more enlightened view of combatting homelessness with new **social housing**. But the stark confrontation of a temporary global spectacle and grinding everyday poverty draws attention to the growing inequality among urban places that is a structural feature of the global city (Hamnett, 2003; Hulchanski, 2007). It highlights the shifting priorities of the state from welfare manager to neo-liberal entrepreneur (Harvey, 1989).

Conclusion

Like all systems of classification, the four-fold division we have used in this chapter is both real and fabricated. Different themes overlap and cannot be neatly partitioned, and the dynamism of urban life and urban landscapes makes any categorization incomplete and unfinished. Two global issues in particular are likely to introduce significant new pressures on Canada's urban places. Global environmental change adds a chronic new condition to the attentions of mayors and planning directors. Some cities have responded vigorously, including Vancouver's 'ecodensity' policy that promises to contribute significant new inputs to place-making. A second pressure is the acute global recession that began in 2008. Dramatic public intervention suggests to some a new **Keynesianism** with the state back in its managerial driving seat; certainly, infrastructure spending as economic stimulus could profoundly shape cities. But another interpretation sees, instead, an intensification of hyper-competition with the state re-established as the dominant entrepreneur, and public investment applied even more resolutely to wealth generation rather than redistribution. These global issues are not independent, as construction of Vancouver's Olympic Village—the showcase of its sustainability program—ran afoul of its New York financial lenders, leaving the city vulnerable to financial exposure, which is the classic nightmare of public–private partnerships. Once again, shifting and interdependent historical contexts in combination with the agency of city agendas will re-chart the identity of urban places.

Review Questions

1. How does the landscape of your city reflect its history?
2. How are the emerging multicultural mosaic and emerging values reflected in urban place?
3. How does present urban space differ from previous urban forms?
4. Why is urban branding important in a globalizing age?

References

Agnew, J.A., and J.S. Duncan. 1989. 'Introduction', in J.A. Agnew and J.S. Duncan, eds, *The Power of Place: Bringing Together Geographical and Sociological Imaginations*. Boston, Mass.: Unwin Hyman.

Anderson, K. 1991. *Vancouver's Chinatown: Racial Discourse in Canada, 1875–1980*. Montreal and Kingston: McGill-Queen's University Press.

Artibise, A. 1984. 'Divided city: The immigrant in Winnipeg society, 1874–1921', in G. Stetler and A. Artibise, eds, *The Canadian City: Essays in Urban and Social History*. Montreal and Kingston: McGill-Queen's University Press.

Ashworth, G., and J. Tunbridge. 2000. *The Tourist-Historic City: Retrospect and Prospect of Managing the Heritage City*. Oxford: Pergamon.

Barber, L. 2009. 'Landscapes of conflict: The politics of heritage in Halifax, Nova Scotia', unpublished paper, Department of Geography, University of British Columbia.

Blakely, E., and M. Snyder. 1997. *Fortress America: Gated Communities in the United States*. Washington: Brookings Institution Press.

Boddy, T. 2006. 'New Urbanism: The Vancouver model', *Places* 16, 2: 14–22.

Bourne, L. 1996. 'Reinventing the suburbs: Old myths and new realities', *Progress in Planning* 46, 3: 163–84.

Bunting, T., and P. Filion, eds. 2006. *Canadian Cities in Transition: Local Through Global Perspectives*. Toronto: Oxford University Press.

Caulfield, J. 1994. *City Form and Everyday Life: Toronto's Gentrification and Critical Social Practice*. Toronto: University of Toronto Press.

Chastain, T. 2006. 'Forming place informing practice', *Places* 123: 6.

Clairmont, D., and D. Magill. 1974. *Africville*. Toronto: McClelland & Stewart.

Cresswell, T. 2004. *A Short Introduction to Place*. Malden, Mass.: Blackwell.

Davis, M. 1990. *City of Quartz: Excavating the Future in Los Angeles*. London: Verso.

Evans, G. 2003. 'Hard branding the cultural city: From Prado to Prada', *International Journal of Urban and Regional Research* 27: 417–40.

Evenden, L., and G. Walker. 1993. 'From periphery to centre: The changing geography of the suburbs', in L.S. Bourne and D. Ley, eds, *The Changing Social Geography of Canadian Cities*. Montreal and Kingston: McGill-Queen's University Press.

Flusty, S. 2001. 'The banality of interdiction: Surveillance, control and the displacement of diversity', *International Journal of Urban and Regional Research* 25: 658–64.

Fyfe, N. ed. 1999. *Images of the Street: Planning, Identity, and Control of Public Space*. London: Routledge

Germain, A., and D. Rose. 2000. *Montreal: The Quest for a Metropolis*. Chichester: Wiley.

——— and J. Gagnon. 2003. 'Minority places of worship and zoning dilemmas in Montreal', *Planning Theory and Practice* 4: 295–318.

Goss, J. 1988. 'The built environment in social theory: Towards an architectural geography', *Professional Geographer* 40: 392–403.

Grant J. 2006. 'Shaped by planning: The Canadian city through time', in Bunting and Filion (2006).

——— and S. Bohdanow. 2008. 'New urbanism developments in Canada: A survey', *Journal of Urbanism: International Research on Placemaking and Urban Sustainability* 12: 109–27.

Haider, D. 1992. 'Place wars: New realities of the 1990s', *Economic Development Quarterly* 6: 127–34.

Hall, C.M. 2004. 'Urban entrepreneurship, corporate interests and sports mega-events: The thin policies of competitiveness within the hard outcomes of neoliberalism', *Sociological Review* 54, 2: 59–70.

Hamnett, C. 2003. *Unequal City: London in the Global Arena*. London: Routledge.

Hannigan, J. 2003. 'Symposium on branding, the entertainment economy and urban place building: Introduction', *International Journal of Urban and Regional Research* 27: 352–60.

Harvey, D. 1989. 'From managerialism to entrepreneurialism: The transformation in urban governance in late capitalism', *Geografiska Annaler* 71B: 3–17.

Hiebert, D. 1991. 'Class, ethnicity and residential structure: The social geography of Winnipeg, 1901–1921', *Journal of Historical Geography* 17: 56–86.

Hobsbawm, E., and T. Ranger, eds. 1983. *The Invention of Tradition*. Cambridge: Cambridge University Press.

Hoyle, B.S. 1992. 'Waterfront redevelopment in Canadian port cities: Some viewpoints on issues involved', *Maritime Policy and Management* 19: 279–95.

Hulchanski, D. 1984. *St. Lawrence and False Creek: A Review of the Planning and Development of Two New Inner City Neighbourhoods*. Vancouver: School of Community and Regional Planning, University of British Columbia (Planning Papers 10).

———. 2007. *The Three Cities within Toronto: Income Polarization among Toronto's Neighbourhoods, 1970–2000*. Toronto: University of Toronto, Centre for Urban and Community Studies.

Jacobs, J. 1961. *The Death and Life of Great American Cities*. New York: Random House.

Kavaratzis, M. 2007. 'City marketing: The past, the present and some unresolved issues', *Geography Compass* 1: 695–712.

Kemble, R. 1980. 'Granville Island', *Canadian Architect* 25: 1627.

Kom, J. 2009. 'Call to limit ethnic malls angers Asians; Consultant warns city enclaves can become ghettos', *Calgary Herald*, 6 Mar., B3.

Lehrer, U., and R. Milgrom. 1996. 'New suburbanism: Countersprawl or repackaging the product', *Capitalism, Nature, Socialism* 7, 26: 49–64.

Lees, L. 2001. 'Towards a critical geography of architecture: The case of an ersatz colosseum', *Cultural Geographies* 81: 51–86.

Ley, D. 1977. 'Social geography and the taken for granted world', *Transactions of the Institute of British Geographers* 24: 498–512.

———. 1983. A *Social Geography of the City*. New York: Harper and Row.

———. 1987. 'Styles of the times: Liberal and neo-conservative landscapes in inner Vancouver, 1968–1986', *Journal of Historical Geography* 13: 40–56.

———. 1993. 'Past elites and present gentry: Neighbourhoods of privilege in the inner city', in L. Bourne and D. Ley, eds, *The Changing Social Geography of Canadian Cities*. Montreal and Kingston: McGill Queen's University Press.

———. 1996. *The New Middle Class and the Remaking of the Inner City*. Oxford: Oxford University Press.

———. 2010. *Millionaire Migrants: Trans-Pacific Life Lines*. Oxford: Blackwell-Wiley.

——— and H. Frost. 2006. 'The inner city', in Bunting and Filion (2006).

——— and P. Murphy. 2000. 'Immigration in gateway cities: Sydney and Vancouver in comparative perspectives', *Progress in Planning* 55: 119–94.

Lowenthal, D. 1985. *The Past Is a Foreign Country*. Cambridge: Cambridge University Press.

McCallum, K., A. Spencer, and E. Wyly. 2005. 'The city as an image-creation machine: A critical analysis of Vancouver's Olympic bid', *Association of Pacific Coast Geographers Yearbook* 67: 24–46.

McCann, E. 2002. 'The cultural politics of local economic development: Meaning-making, place-making, and the urban policy process', *Geoforum* 33: 385–98.

Marshall, J. 1994. 'Population growth in Canadian metropolises: 1901–86', in F. Frisken, ed., *The Changing Canadian Metropolis*. Toronto: Canadian Urban Institute.

Massey, D. 1992. 'A place called home?', *New Formations* 17: 3–27.

Mitchell, D. 1997. 'The annihilation of space by law: The roots and implications of anti-homeless laws in the United States', *Antipode* 29: 303–35.

Mitchell, K. 2004. *Crossing the Neoliberal Line: Pacific Rim and the Metropolis*. Philadelphia: Temple University Press.

Nash, C. 2006. 'Toronto's gay village 1969–1982: Plotting the politics of gay identity', *Canadian Geographer* 50: 1–16.

National Capital Commission. 2004. *Sparks Street Mall Vocation Study—Phase 2 Vision: Strategic Recommendations*. At: <www.canadascapital.gc.ca/data/2/rec_docs/1382_Phase2EngFinalSep04.pdf>. (May 2009)

Norcliffe, G., K. Bassett, and T. Hoare. 1996. 'The emergence of postmodernism on the urban waterfront: Geographical perspectives on changing relationships', *Journal of Transportation Geography* 4: 123–34.

Olds, K. 2001. *Globalization and Urban Change: Capital, Culture, and Pacific Rim Mega-projects*. Oxford: Oxford University Press.

Oldenburg, R. 1999. *The Great Good Place: Cafes, Coffee Shops, Bookstores, Bars, Hair Salons, and Other Hangouts at the Heart of a Community*. New York: Marlowe.

Osborne, B. 1996. 'Texts of place: A secret landscape hidden behind the everyday', *GeoJournal* 38: 29–39.

Paul, D. 2004. 'World cities as hegemonic projects: The politics of global imagineering in Montreal', *Political Geography* 23: 571–96.

Punter, J. 2003. *The Vancouver Achievement: Urban Planning and Design*. Vancouver: University of British Columbia Press.

Pocock, D. 1981. 'Place and the novelist', *Transactions of the Institute of British Geographers* 6: 337–47.

Podmore., J. 2006. 'Gone "underground"? Lesbian visibility and the consolidation of queer space in Montréal', *Social and Cultural Geography* 7: 595–625.

Purdy, S. 2003. '"Ripped off" by the system: Housing policy, poverty and territorial stigmatization in Regent Park housing project, 1951–1991', *Labour/Le Travail* 50, 2: 45–108.

———. 2005. 'Framing Regent Park: The National Film Board of Canada and the construction of outcast spaces', *Media, Culture and Society* 27: 523–49.

Qadeer, M. 1997. 'Pluralistic planning for multicultural cities: The Canadian practice', *Journal of the American Planning Association* 63: 481–94.

———. 1999. *The Bases of Chinese and South Asian Merchants: Entrepreneurship and Ethnic Enclaves*. Toronto: Joint Centre of Excellence for Research on Immigration and Settlement (Working Paper 11).

Richler, M. 2007. *Mordecai Richler Was Here: Selected Writings*. Toronto: Madison Press.

Relph, E. 1976. *Place and Placelessness*. London: Pion.

———. 1991. 'Suburban downtowns of the Greater Toronto Area', *Canadian Geographer* 35: 421–5.

Rutherford, P. 1984. 'Tomorrow's metropolis revisited: A critical assessment of urban reform in Canada', in G. Stelter and A. Artibise, eds, *The Canadian City: Essays in Urban and Social History*. Montreal and Kingston: McGill-Queen's University Press.

Sack R. 1993. 'The power of place and space', *Geographical Review* 83: 326–9.

Sewell, J. 1977. 'The suburbs', *City Magazine* 2, 6: 19–55.

———. 1993. *The Shape of the City: Toronto Struggles with Urban Planning*. Toronto: University of Toronto Press.

Short, J.R. 1999. 'Urban imagineers: Boosterism and the representation of cities', in A. Jonas and D. Wilson, eds, *The Urban Growth Machine*. Albany, NY: SUNY Press.

Skaburskis, A. 2006. 'New urbanism and sprawl: A Toronto case study', *Journal of Planning Education and Research* 25: 233–48.

Sklair, L. 2006. 'Iconic architecture and capitalist globalization', *City* 10, 1: 21–47.

Stelter, G., and A. Artibise. 1986. 'Introduction', in G. Stelter, and A. Artibise, eds, *Power and Place: Canadian Urban Development in the North American Context*. Vancouver: University of British Columbia Press.

Townshend, I.J. 2006. 'From public neighbourhoods to multi-tier private neighbourhoods: The evolving ecology of neighbourhood privatization in Calgary', *GeoJournal* 66: 103–20.

Tuan, Y.-F. 1974. 'Space and place: Humanistic perspective', *Progress in Geography* 6: 211–52.

Tunbridge, J. 2001. 'Ottawa's Byward Market: A festive bone of contention?', *Canadian Geographer* 42: 356–70.

——— and G.J. Ashworth. 1996. *Dissonant Heritage: The Management of the Past as a Resource in Conflict*. New York: J. Wiley and Sons.

Vancouver, City of. 2003. *'False Creek North', Vancouver's New Neighbourhoods—Achievements in Planning and Urban Design*. At: <vancouver.ca/commsvcs/current planning/urbandesign/br2pdf/falsecreek.pdf>. (May 2009)

Wacquant, L. 2007. *Urban Outcasts: A Comparative Sociology of Advanced Marginality*. Cambridge: Polity Press

Walby, K. 2006. 'Little England? The rise of open-street closed circuit televisions surveillance in Canada', *Surveillance and Society* 1: 29–51.

Warren, J., and K. Carlisle. 2005. *On the Side of the People: A History of Labour in Saskatchewan*. Regina: Coteau Books.

Whitson, D. 2004. 'Bringing the world to Canada: "The periphery of the centre"', *Third World Quarterly* 25: 1215–32.

Wortham-Galvin, B.D. 2008. 'Mythologies of placemaking', *Places* 20: 32–9.

Yeoh, B.S.A. 2005. 'The global cultural city? Spatial imagineering and politics in the multicultural marketplaces of south-east Asia', *Urban Studies* 42: 945–58.

Zhuang, Z.C. 2008. *Ethnic Retailing and the Role of Municipal Planning*. Toronto: Joint Centre of Excellence for Research on Immigration and Settlement, Working Paper.

CHAPTER 20

At Home in the City: Housing and Neighbourhood Transformation

Ryan Walker and Tom Carter

Housing is the single largest user of space in Canadian cities, consuming approximately 51 per cent of all urban land (Hodge and Gordon, 2008). Having a secure, affordable, personalized home in a location that fits the pattern of a person's everyday life is important to Canadians. In urban areas, housing can be one of the greatest points of public policy debate, whether centred on neighbourhood design and residential density, building for energy efficiency, or issues of social equity such as access to affordable and adequate housing for everyone. Mostly, people want a good home that they can afford and there is evidence to suggest that without one, people's circumstances relating to health, education, and income security are adversely affected. We argue throughout this chapter that there is a need for stronger public policy in the housing sector, not just to address the housing needs of Canadians but to strengthen the economic, social, and environmental sectors, all of which are recognized to be intertwined. Ironically, the last few decades have witnessed a time of unprecedented affluence in Canada yet, for the first time in our history, there have been rising rates of homelessness. That said, this chapter does not deal with homelessness per se. Rather, the topic is featured in Chapter 25 by Larry Bourne and Alan Walks, where it is used to illustrate the complex, contradictory, and unexpected challenges of the postmodern city.

The next section develops a profile of the characteristics of housing in Canadian cities using Statistics Canada data. Like the health and education systems in Canada, we have a housing system and some of the key players in the system are discussed. The role of housing and home in people's lives is examined in a subsequent section, as are social housing and the changing roles of the state and civil society actors in providing it. Housing is a central focus of neighbourhood revitalization schemes, and revitalization with resident displacement versus initiatives that centre on the residents already in place is examined. Finally, the twenty-first-century city will increasingly be centred on its environmental footprint (see Rees, Chapter 5). No discussion of urban environment is complete without consideration of its largest consumer of land, the housing we live in.[1]

Housing Characteristics in Canadian Cities

The stock of housing in Canada is about 13.2 million units (Statistics Canada, 2008a). Single family homes on individual lots consume the greatest amount of residential land and dominate the housing inventory, although as city size increases so does the proportion of higher-density residential dwellings. Fifty-five per cent of all residential dwellings in Canada in 2006 were single detached units. This proportion fell to 47 per cent in the larger cities (census metropolitan areas or CMAs) but rose to 63 per cent in smaller cities (census agglomerations or CAs). Semi-detached and row housing constituted 11 per cent of all dwellings in Canada, 12 per cent in CMAs, and 9 per cent in CAs. Residential units in apartments of less than five storeys make up 51 per cent of the total housing stock in Montreal but only 8 per cent in Toronto, where 27 per cent of the stock is in high-rise units compared to only 8 per cent in Montreal (Statistics Canada, 2008b). Toronto is Canada's high-rise city, Montreal the low-rise city. Only Vancouver comes even close to these two cities in the discussion of high-density housing accommodation.

Over the past couple of decades levels of homeownership in Canadian cities have risen. Currently 68 per cent of all privately occupied dwellings are owned, up from 60 per cent in 1971 (Statistics Canada, 2008a). Owner-occupied dwellings rise as high as 81 per cent in Barrie, while in Montreal, long a city with a renters' lifestyle, the ownership rate is 53 per cent, the lowest of Canadian CMAs (Statistics Canada, 2008c). Contributing to this increase in ownership has been the growth in the number of households that own condominiums. The level of condominium ownership increased 31 per cent between 2001 and 2006. Condo ownership constitutes about 11 per cent of the housing market, whereas only 4 per cent of households lived in condominiums in 1981 (Canada Mortgage and Housing Corporation, 2008).

Paralleling the rise in ownership has been a decline in the proportion of renters, from 40 per cent in 1971 to 32 per cent in 2006. The inventory of rental units has been declining, dropping from 3,908,000 in 2001 to 3,880,490 in 2006 (Statistics Canada, 2008a). The conversion of rental units to condominiums, demolitions, and losses through other means has exceeded new construction. These factors have contributed to tight supply and low vacancy rates in Canada's major cities for a number of years, averaging 2.7 per cent in 2008. The availability of rentals is below the 3 per cent considered an acceptable level that gives consumers sufficient choice in the market and provides rental developers and owners with the revenue necessary to operate and make a profit. Vacancy rates, using 2008 data, are less than 1 per cent in many cities: Quebec City, Regina, and Winnipeg, for example. Rates are high (above 5 per cent) in a few centres like St Catharines–Niagara, Abbotsford, and Windsor. Average monthly rents (for two bedrooms) range between $1,000 and $1,200 a month in larger or rapidly growing cities such as Vancouver, Calgary, Toronto, Edmonton, and Victoria but are closer to $500–$600 in smaller or **slow-growth cities** (Canada Mortgage and Housing Corporation, 2008). Renters in most Canadian cities in recent years have faced low vacancy rates, declining inventories of rental units and rising rents.

The housing stock in Canadian cities is relatively new, mainly because much of it was built following World War II to accommodate the baby boom and the boomers as they moved into their own homes during the 1970s and 1980s. Only 12 per cent of dwellings in cities were built prior to 1946 (Statistics Canada, 2008a). Being relatively new, most units are in good condition, with only 6 per cent requiring major repairs—mostly units built before 1929. Little new construction occurred during the Depression and World War II. Canadians invest heavily in the renovation of their dwellings and money spent on renovations generally approximates money invested in the purchase of new homes. In 2007, for example, $49.5 billion was spent on renovation. The amount invested in renovation was only marginally exceeded by investment on new construction of $52 billion. In 2008, $21.3 billion was spent on home renovation in the 10 largest cities alone and 40 per cent of homeowners in these centres undertook some form of renovation, spending on average $12,600 each (Canada Mortgage and Housing Corporation, 2009).

Demand for housing continues to be high in expanding Canadian cities. Almost 10 per cent of the total housing stock was built in the five-year period 2001–6. In the cities of Barrie, Kelowna, and Calgary, demand was so significant that almost 20 per cent of the total housing stock in these cities was built during this five-year period (Canada Mortgage and Housing Corporation, 2008). Housing demand in Canada's cities is currently being driven by a number of demographic and economic factors. The rapid influx of immigrants to address labour shortages and strong growth in the Aboriginal population in some cities have increased the demand for housing. Changing living arrangements and household structures, smaller families, and an aging population with more elderly single individuals have resulted in declining household size. Between 1951 and 2001, one-person households rose from 7.4 per cent of Canadian households to 25.8 per cent (Walks, 2006). With more unattached individuals (e.g., seniors and younger people), empty-nesters, lone parents, and overall declines in household size, more units are required to accommodate the same number of people. This means that even in centres of very slow growth a demand continues for housing and growth in residential land development.

Despite the generally positive housing scenario for Canadian cities and the fact that most Canadians are well-housed, not everyone has experienced positive housing outcomes. Forty-two per cent of urban renters spend more than 30 per cent of their income on shelter, leaving many with too little disposable income to cover the cost of other essentials like food, clothing, health care, and education. In urban communities approximately 14 per cent of households are in *core need*. This means that they spend in excess of 30 per cent of their income on housing, their units are too small for the household size and composition, or their homes need major repairs. Some households experience all three of these problems (Canada Mortgage and Housing Corporation, 2008). Governments have provided little relief for such households as few new affordable housing units have been constructed over the last two to three decades. The proportion of housing units

that are part of the social housing sector (whose costs are subsidized by government resulting in below-market rents) is small and declining. Approximately 630,000 units are in this sector, the majority in the major metropolitan centres. They constitute approximately 5 per cent of the Canadian housing stock, the smallest percentage among Western nations with the exception of the US (Carter, 2008).

The Canadian Housing System

The Canadian housing system consists of many actors and an intricate set of relationships between these actors that is dictated by both legislation and the development and construction process. James McKellar (1994) and Paul Knox and Linda McCarthy (2005: ch. 11) put these actors into three groups: government, the private sector, and the community or non-profit sector.

All three levels of government are major actors and the government focus is mainly, but not entirely, on the regulatory aspects of the housing system. The federal government has tremendous influence in the housing sector via control of interest rates, economic policy (which it often uses to stimulate housing activity), immigration policy (as it has overall control of the number of new arrivals, which affects housing demand), and legislation that controls the banking system and mortgage lending practices. Responsibilities of provinces and municipalities may vary by provincial jurisdiction but, in general, provinces control building code regulations, general standards for construction, and land-use policies. Municipalities, because of their control of the regulatory environment, influence aspects of housing development such as lot regulations, maintenance and occupancy bylaws, subdivision design, the land development process, and health and safety aspects of dwellings.

Within the private sector land developers develop residential land, generally in suburban locations, for builders who buy lots from the developers for construction of residential units designed by architects and built by a variety of sub-trades ranging from those that do the concrete work, to framers that build the structure, to trades that stucco, paint, shingle, and provide cabinetry—just to name a few. In many cases, developers also assume the role of builders. Professional planners plan subdivisions that may be landscaped by landscape architects. Renovators are a growing sector of the housing industry. Specialized subsectors, such as historic renovators who restore, preserve, or rehabilitate older properties, have grown up within this industry.

The real estate sector handles the marketing and sale of units, both new and existing. It is made up of appraisers who determine value (market or replacement), realtors who sell units, and lawyers who legalize sales and leasing transactions. In the property management sector, property managers handle rental and leasing of units and provide general property management functions. Lenders within the financial sector provide mortgage funds for developers, builders, and purchasers of homes, and the insurance industry provides mortgage and home insurance.

Community and non-profit organizations, a sometimes forgotten sector of the industry, play a significant role in the provision of affordable housing. Charitable, Aboriginal, neighbourhood, religious, and co-operative organizations are strong components of this sector. These organizations play various roles in the provision of housing, ranging from raising funds, participating in the planning and design of projects, handling project and property management functions, and assuming outright ownership of social and affordable housing projects. Few homes get built without the involvement of a significant number of these actors during the planning, development, construction, and marketing process.

The Role of Housing and Home in People's Lives

Four important dimensions of housing in people's lives include the *financial, psychological/social, locational,* and *physical* aspects of housing (Figure 20.1). Housing is an asset with *exchange value*. It is a tradable commodity for homeowners and constitutes the largest financial expenditure and asset for many throughout much of their adult lives. The equity many people have in their house on retirement constitutes wealth, which helps finance their later years that may be characterized by failing health.

Housing is also 'home' and has *use value*. When it is secure, adequate, and affordable, it augments quality of life by giving households continuity, privacy, a domain of control, space for self-expression, and a place for maintaining social networks (Després, 1991; Harris, 2006; Saunders, 1989; Smith, 1994). The home is the most important setting for the development of personal values and social reproduction and it represents the physical space most intimately linked with one's sense of self (Bratt, 2002; Harris, 2006; Imrie, 2004). A sense of *home*, which embodies all of these elements that go beyond shelter from the elements, is very important because of the extent to which people construct their lives and deal with their existential questions through homes that are stable, secure, and personalized (Dunn, 1998).

Housing cannot easily be separated from the neighbourhood characteristics that affect how people feel about their home, the safety of their environment, and their residential satisfaction in general. Location of the home also determines how well households are connected to jobs, transportation, and many other services. Isolation from services, for example, can be a significant barrier for high-need households and those who depend on public transportation. Location may also facilitate or hinder networking with friends and family and weaken or strengthen the social support network that is so important to people.

Adequate and affordable housing is a right of **citizenship**, although it remains unavailable for too many Canadians. The Universal Declaration of Human Rights (1948) and the International Covenant on Economic, Social and Cultural

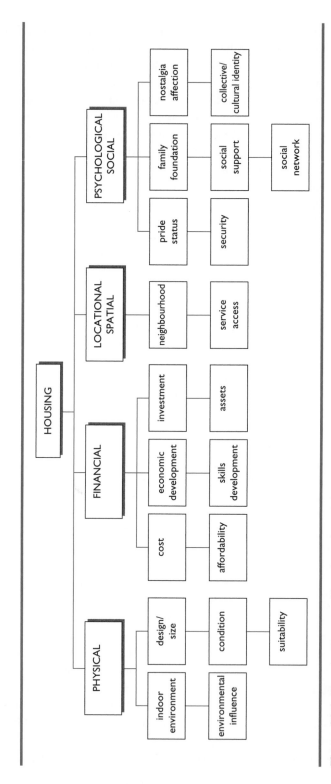

Figure 20.1 The significance of housing to people's lives

Sources: Carter and Polevychok (2004), adapted from Dunn (2003); Platt (1996); Porteous and Smith (2001).

Rights (1976) articulate substantive rights to hous-ing, including legal security of tenure, availability of services, affordability, habitability, accessibility, location, and cultural adequacy (Shapcott, 2008). It is important to bear in mind, and too often is over-looked, that in the period since the end of World War II we agreed as Western capitalist nations that we would not unduly hinder the accumulation of personal wealth, but that we would secure basic minimum standards for all. In wealthy Western countries such as Canada one of these standards was entitlement to a home that is affordable and adequate.

It is common to break public policy debate into discrete sectors, like housing, health, and edu-cation. Increasingly, though, housing scholars are concerning themselves more with important link-ages between sectors, and it is clear why. Housing is such a central part of people's ability to enjoy quality of life that it is no wonder that having, or not having, adequate and affordable housing is linked to outcomes in health policy, educational attainment, and employment. As Dowell Myers (2008) put it:

> Housing fits in the middle of everything. It is physical design, it is community economic development, it is social development, it is important to health and educational outcomes, it can be a poverty reduction tool, and it is an investment, a wealth creator and a generator of economic development. It is both an individ-ual and public good.

Tom Carter and Chesya Polevychok have argued convincingly that housing policy is good social and public policy from at least two perspec-tives (Carter and Polevychok, 2004; Carter et al., 2009). The first is that adequate and affordable housing is a necessary household good, in and of itself. The second is that it facilitates the success of initiatives taken in other policy sectors like health, education, and income security.

Affordable housing is an integral component of income security. Reasonable housing costs leave families with more disposable income to secure the other elements of a healthy and happy life. High housing costs 'trap' many in perpetual pov-erty, driving some to actually become homeless on the street or to become part of the hidden home-less population, couch-surfing or doubling- (or tripling-) up households in a single dwelling.

Living in poor-quality housing has impacts on physical and mental health. Poor physical design, overcrowding, poor state of repair, or forced relocation all significantly impact health and well-being (Kearns et al., 1991). On a socio-economic level, the domestic environment at home, monthly expenditure on housing and shelter-related costs, exercise of control, and security of tenure are sig-nificant predictors of physical and mental health (Dunn, 2000). Anxiety and depression increase with housing instability or poor-quality housing. People who actually are homeless, a phenom-enon strongly correlated with the discontinuation of federal social housing programs in the 1990s, experience the worst health consequences; indeed, their mortality rates have been found to be as much as 10 times greater than for people who are properly housed (Guirguis-Younger et al., 2003). Because other types of support, such as for men-tal health, addiction, or physical health issues, are mostly ineffective for homeless people, given that being homeless causes such stress and exacerbation of health problems, it is as much as five times more expensive to shelter (in emergency shelters) and treat homeless people than it would be simply to house them properly (De Jong, 2008). In Canada we have in the past decade come to focus quite heavily on shelters for the homeless as a policy initiative in itself, perhaps too easily forgetting that we need to focus on housing and homes.

The impact of poor housing circumstances on educational outcomes is significant. Children living in overcrowded housing have poorer educational outcomes, and those who are poorly housed or homeless have more problems with anxiety, depression, behavioural problems, and low educational achievement (Lezubski et al., 2000; Mueller and Tighe, 2007; Mullins et al., 2001). Researchers in the US found that residential mobility as a child had a greater impact on high school graduation than either poverty or welfare dependency (Mueller and Tighe, 2007). Residential mobility connects directly with the security of tenure, affordability, and adequacy of housing. Low-income families in adequate and affordable social housing who pay rent geared to their income can put their children on a more level playing field as far as life chances are concerned than if they are required to move repeatedly because of rising rents or inadequate housing.

Social Housing and the Changing Roles of State and Civil Society

As housing is good social policy in its own right and as a basis for success in other policy sectors, why have governments so badly neglected this policy field? One of the fundamental roles of the state is to intervene when there is market failure, especially in areas of significant public interest like the housing sector. Housing scholars agree that the only real way to address the shortage of adequate and affordable housing for those who cannot satisfy their needs in the diminishing private rental market or the high-priced homeownership market is for the state to build it (Moore and Skaburskis, 2004; Walker, 2008; Walks, 2006). Experience since the 1970s shows that it is most effective for the state to meet its building targets by channelling adequate and dependable resources through community-based social housing organizations for building and operating units at below-market rent to those who need assistance.

The way the state intervenes in the housing market has changed significantly since the 1960s. During the period when the federal government led the way in building a public, non-profit, and co-operative 'social' housing sector (i.e., 1964–93),

the state was a leader in program planning, implementation, and long-term funding (Harris, 2006; Skelton, 2000). In the period after 1993, the state virtually stopped building new social housing. The consequences were noticeable soon afterward, most palpably in the drastic rise in rates of homelessness (Walks, 2006; see Bourne and Walks, Chapter 25).

The need continues, as it always will, for sustained state intervention in housing to serve those who cannot make it in the private market. From the 1960s through the early 1990s the governments of Canada played a role, although from the mid-1980s to 1993 there was a steady diminution of annual unit targets, from 25,000 units a year in 1983 to zero in 1993 (Hulchanski, 2002). A total of 4,450 units were built between 1994 and 1998 (Conference Board of Canada, 2009). Following this period, governments, particularly the federal government, have undertaken measures to placate demands for new social housing investment through small short-term investments like the Affordable Housing Initiative, launched in 2001, and recently a suite of affordable housing trusts; but governments have mostly abandoned the notion that social housing should stand as a pillar of the welfare state alongside health, education, and income security.

One way of theorizing the shift away from state-led social (housing) programs concerned with providing a strong foundation for common citizenship is presented by Anthony Giddens and, in Canada, adapted by Jane Jenson and Denis Saint-Martin (Giddens, 1998; Jenson and Saint-Martin, 2003). They argue that in response to neo-liberal critiques of the welfare state, social democratic governments have moved away from attending to social rights such as affordable and adequate housing for all citizens. **Social cohesion**, understood as the strength of social bonds in society, replaces social *rights* as the goal of policy- and decision-makers. A focus on keeping people engaged in active citizenship, producing their own welfare, is seen as a way to increase social cohesion. Giddens conceptualizes this turn as a new relationship between a 'social investment state' and an 'active civil society', where both state and civil society actors are said to be better able to adapt to shifting economic conditions and to change priorities and policy directions quickly and strategically in response to market and social forces.

The earlier focus on collective social rights is being replaced with the primacy of individual (including corporate) consumption rights substantiated in the marketplace. Personal 'freedom' appears to be realized through the consumption of goods, services, and private assets of personal choosing, with minimal state interference. The flipside of this type of 'freedom' is that individuals also bear responsibility for their own welfare, investing in their own education, housing, income security, and health. Shortfalls in personal development are increasingly seen as the result of individual failings and a lack of 'entrepreneurial virtues', not as a shortfall of our collective political community and its apparatus, the state (Harvey, 2005). This shift in attitude is particularly troubling when seen against the backdrop of the growing gap between the richest and poorest Canadians (Green and Milligan, 2007; Yalnizyan, 1998). We are either witnessing a rapid rise of individual failing in Canadian society, or evidence that there is a need for a stronger state to correct market inadequacies, such as in the housing sector.

Housing and Neighbourhood Revitalization

To thrive, cities need people from diverse cultural and socio-economic backgrounds, ages, and genders. People of diverse backgrounds all play their respective roles in sustaining the economic, social, and cultural life of the city and all have claims to quality of life in the city. Those with the greatest financial capital get to set the terms of urban development, largely through their consumption choices and political influence. If they wish to live in the suburbs, then they will. If they wish to

follow the cultural pioneers of our cities, such as artists, into older neighbourhoods with a bit more diversity and 'edge' to them, they may do so. As those with resources choose to shift their housing and neighbourhood preferences to different parts of the city, those with less financial capital must move somewhere else. Whether this seems right or not, and whether it places enough value on the stability and use value of a home for residents who have perhaps been in a neighbourhood for decades, is a political question that can be addressed through public policy. It is possible for citizens or their governments to create policy and programs to safeguard people's claim to a home, both in terms of the housing unit and of the neighbourhood. We explore briefly some of the issues around **revitalization** by looking at gentrification versus measures initiated by those already residing in older areas. Both groups—the higher-income gentrifiers and the lower-income residents of an area subject to gentrification—often draw on the policy and financial mechanisms of public authorities.

Gentrification: Revitalization with Displacement of People

Gentrification is a revitalization phenomenon where middle-class professional households move into inner-city districts and renovate older housing stock or redevelop lot by lot (Ley and Frost, 2006). The process, prominent from the 1970s onward, is stimulated not only by lower housing costs but also by lifestyle preferences for inner-city living in 'character neighbourhoods'. On the positive side, it needs to be acknowledged that gentrification certainly preserves the older housing stock and extends the longevity of the oldest housing units at very little cost to the state; on the negative side, gentrification disrupts the process whereby older properties filter down to those with lower incomes, thus fostering lower-cost ownership and rental opportunities (Bourne, 1981). Instead, gentrification generally has been shown to have just the opposite effect of 'filtering up' (Ley and Frost, 2006).

This change of trajectory has been driven by the amenity values of these neighbourhoods and the financial capacity of would-be suburban households to increase the property values through major renovation or redevelopment. Perhaps most importantly, gentrification has represented a cultural shift, from a modern to postmodern urban identity linked to reclaiming the distinctive or 'authentic' inner city by an increasingly professional-managerial middle-class for lifestyle reasons (Ley, 1996). The most dramatic impact of gentrification is on the housing market, but in many cases local businesses and public amenities are transformed considerably as well on account of new middle-class demand for cosmopolitan place-making ventures like galleries, theatres, diverse restaurants, and interactive waterfronts (Bunting and Rutherford, 2006). In some Canadian cities, particularly those with strong cultural and knowledge economies, the cycle of gentrification—or the reclamation of the inner city by the post-Fordist middle class—may have already pretty much run its course and given way to booms in condo and loft developments in inner-city locations (Bunting and Rutherford, 2006; see Bain, Chapter 15).

David Ley and Heather Frost (2006) discuss instances where housing prices have doubled in as short a period as two years, such as in Don Vale, Toronto (1979–81) and Fairview Slopes, Vancouver (1980–1). Displacement of lower-income households can be in the thousands when property values and rents rise, and as purpose-built rental housing becomes subject to renovation for up-market rental or conversion to individually titled condominiums, or when lower-cost rental units simply are torn down and redeveloped. In Vancouver roughly 7,500 rental units were demolished from 1973 to 1981 and in Toronto roughly 18,000 units were lost to conversions between 1976 and 1985 (Ley and Frost, 2006). In Saskatoon roughly 10 per cent or just under 2,000 units of purpose-built rental housing were converted to condominiums during 2007–8, and many of these were in inner-city neighbourhoods undergoing gentrification. In instances where conversions and redevelopment occur, we have an acute point of tension between the *use value* of a home to current residents and the *exchange value* of properties to landlords and developers. In Saskatoon, for example, it was common to see residents of secure, adequate, and affordable purpose-built rental housing, who had been living in a unit for 10–40 years, served notices to vacate their homes in order for conversion of the building to condominiums for sale. In the public realm, this becomes a debate over use and exchange values and how to balance these through policy and statute.

The profiles of gentrifying census tracts show remarkable increases in young adult households, large reductions in household size, a dramatic increase in population with university education, and increased rates of household mobility (Meligrana and Skaburskis, 2005). While no inner-city area is fully immune from gentrification, certain characteristics decrease the likelihood of gentrification or slow it down. David Ley and Cory Dobson (2008) propose that the absence of architectural distinction, including heritage and aesthetic features, in the housing stock can impede gentrification, as can the lack of a nearby environmental amenity (e.g., waterfront, leafy streets, a mature park) or significant cultural institutions (e.g., art gallery, theatre) and poor access to downtown. Operating industrial sites and active train and truck lines are disincentives to gentrification. Similarly, public or social housing developments in any perceptible concentration can create a sense of deep poverty in an area and deter middle-class would-be gentrifiers and investment by the business community. Social housing also effectively takes property out of the private market. While low-income residents in market rental housing can be displaced as neighbourhood property values rise, low-income residents of social housing enjoy greater security of tenure.

With the enduring dominance of a neo-liberal policy paradigm in urban affairs generally and in housing specifically, it is unusual to see active government intervention to improve areas of the city with the current residents in mind. For example, while we might see tax breaks and other financial incentives given to property owners to improve privately their properties in older areas (e.g., a number of cities have enterprise zones that do this, encompassing older neighbourhoods where buildings and infrastructure are in need of reinvestment), we are less likely to witness initiatives that lead to property improvements while taking measures to prevent the displacement of incumbent residents. The Residential Rehabilitation Assistance Program, which has been delivered with some consistency in Canada since 1974, is a notable exception, where funds can be made available to homeowners and landlords to physically improve their housing without necessarily causing resident displacement.

Ley and Dobson (2008) discuss the Vancouver Agreement—a partnership between all three orders of governments—that runs until 2010 and aims to revitalize the inner city without displacing low-income residents. Targeting the Downtown Eastside, Canada's poorest urban area by standard socio-economic indicators, the Vancouver Agreement aims to maintain the level of affordable housing by building new social housing while permitting some new housing development for middle-class households. This area of Vancouver has many of the architectural and aesthetic features, accessibility to downtown, and environmental and cultural amenities that would make it ideal for gentrification (Ley and Dobson, 2008). But it also has a deep poverty culture, thousands of units of public and social housing, and a politicized public that sees the area's improvement tied to the rights of residents to a safe and affordable place to live. The Vancouver Agreement thus departs from the more abstract notion of physical improvement (minus the people) that exists in some other cities or other parts of Vancouver.

Gentrification is based on a class-distinguished 'other' coming into a neighbourhood, most often in the inner city. The gentrifying households displace those who benefited from the filtering down process and the prior preference of middle-class households for larger homes and lots in the suburbs. The gentrification phenomenon can be contrasted with neighbourhood revitalization directed at improving the lives of those people already there (Redfern, 2003).

Revitalization without Displacement

In economically depressed neighbourhoods, improving access to affordable, good-quality housing can boost the health, quality of life, and future prospects of local residents. It can also stimulate community economic development and physical as well as social neighbourhood renewal. The West Broadway Development Corporation in Winnipeg was formed in the early 1990s by residents of the community concerned about neighbourhood decline and worsening crime and security problems in their community (West Broadway Development Corporation, 2009). The 50–60-member group that initiated the West Broadway Development Corporation began a place-specific process of 'urban citizenship' (Walker, 2006). Local residents and businesses got together and established a common vision for the neighbourhood, created a housing plan and engaged different levels of government to help implement programs that would fulfill the vision held by incumbent residents. Urban citizenship of this kind, where new state–society relations are forged to address priorities articulated by a community occurs among communities rooted in-place who feel they have a right to be where they are and to engage the state in future plans that satisfy local aspirations (Holston, 1995, 2001).

Although an active civil society has been central to the approaches taken in West Broadway, the state side of the urban citizenship equation has been pivotal to the sustainability of the revitalization effort. The province and the City of Winnipeg have made program funding available for many of the initiatives and have provided money for staff time and organizational capacity-building. Funding through Manitoba's Neighbourhoods Alive Program, which focuses on the revitalization of inner-city neighbourhoods, has been crucial to the success of the initiatives in West Broadway. Housing investment in the area over the last three decades has improved housing quality, provided more affordable rental accommodation, helped increase the level of homeownership, and improved the quality and safety of older rooming houses.

The West Broadway Development Corporation attempted to remove some of the neighbourhood land from the private market to prevent gentrification and preserve affordability for incumbent residents. In the late 1990s West Broadway introduced a Community Land Trust Program. By developing a program under which the community absorbed the cost of the land and held the land in trust, it hoped to withdraw the land from market pressures to make ownership more affordable for present and future residents. Although this program attracted initial support from governments and interest from community residents, it was discontinued. Among the reasons behind the demise of the land trust was the fact that those who bought homes under the program did not accept forsaking capital gains on the land, in part because this would prevent them from moving to a new owner-occupied home in the future.

West Broadway today faces a new challenge: gentrification. The success of its initiatives and the appealing features of the neighbourhood, such as proximity to downtown, to the University of Winnipeg, to the legislature, and to both natural and built amenities have combined to make this area attractive to wealthier households. Because they realize the potential to attract higher-income tenants in a revitalized neighbourhood, rental property owners are fixing up their properties and increasing rents. This gentrification is reducing the inventory of affordable housing in the area and displacing some of the lower-income tenants. Other people are also starting to buy and fix up homes in this area. Rising property values make it even more difficult to make affordable housing programs work. As low-income people are displaced from West Broadway, what alternatives do they have for affordable housing elsewhere? Government social housing policy can make the difference because, despite all the efforts and positive work of residents and community-based organizations, they alone cannot augment or preserve the stock of affordable housing without policy direction and resources from senior and local governments.

Housing, Environment, and Neighbourhood Transformation

Housing plays an important role in addressing environmental issues like climate change and non-renewable energy consumption. One way of thinking about how housing relates to the environment is by concentrating on the housing unit itself. A second way is by considering the impact of residential subdivision design on **urban form** and the environmental impact of a city (see Grant and Filion, Chapter 18). The age of housing stock, its size and construction standards, and home systems including heating and air conditioning all relate to the environmental impact of a particular housing unit. Subdivision design and density, location relative to daily activity patterns, and the relationship between housing and neighbourhood design, on the one hand, and transportation, on the other, determine how housing contributes to the environmental impact of urban form. We need to be aware of the environmental impact of housing when developing new residential areas, but also when transforming existing homes and their neighbourhoods. Of particular environmental significance is the renovation and redevelopment work that takes place in pre-World War II areas. In Saskatoon, for example, there is enough of a market for environmentally advanced, affordable housing to create a space for a new development company,

Shift Development Inc., to focus on designing and building energy-efficient infill homes in mostly inner-city (pre–World War II) neighbourhoods (Figure 20.2).

Moving from single detached homes to semi-detached and multiple-unit structures reduces home energy demand on account of having less wall and roof area per housing unit that interfaces with the outside causing heat loss. Home construction techniques that place wider distances between boards in framing, leaving more continuous space for insulation and composing less of the exterior building interface with wood and nails that transfer cold into the home, improve environmental performance. Minimizing window area on north-facing walls, and maximizing area on the south face, aids passive solar heating. Triple-glazed windows, higher levels of insulation in walls and ceilings, and improved vapour barriers to reduce heat loss are techniques that should immediately become industry standards.

Home size and the consumption of materials and energy are directly proportionate. We could stop increasing the size of new homes and go back to smaller units, which use fewer resources to build and less energy to heat and cool. Canada-wide, the average house size in 1945 was just over 800 square feet. By 1975 this had increased to 1,075, while today the average is just over 1,700 square feet (Canadian Home Builders, 2002). This increase in average house size occurred despite the fact that households have fewer people.

Building at higher densities reduces land consumption and the length of municipal infrastructure networks, and makes public transportation more viable. Nevertheless, we still see suburban

Figure 20.2 The first infill Shift home (right), Riversdale, Saskatoon. (Curtis Olson)

expansion at low densities (e.g., 3–5 units per acre). Some cities are introducing new standards for neighbourhood design, but arguably are not moving as quickly as they should given all that is known about the inefficiency of low-density development and the need for an interactive mix of uses planned around efficient public transportation and designed so as to promote alternative modes such as walking and cycling (Harris, 2006). Taking Saskatoon as an example, in August 2009 City Council voted to accept a two-year-old report by planning and engineering consultants, with non-binding recommendations for new neighbourhood and design standards centred on smart-growth principles (Saskatoon, 2007).

Saskatoon's Land Branch has developed a new subdivision, Evergreen Neighbourhood, which registers the city's highest (projected) residential density (8.6 units per acre), has positioned 60 per cent of front or back lot lines to maximize solar orientation, has created village and district commercial and service centres, linear parks, and a naturalized storm-water management system, and has adopted a marginally higher standard of transit-oriented development than other neighbourhoods. While Evergreen is a local good-news story for Saskatoon, one must wonder why such slow and hesitant movement towards environmental sustainability in neighbourhood design persists when these ideas have been around for many years now. Why, too, are provincial building codes not more ambitious?

Cities can use residential development as a way to transform neighbourhoods and minimize environmental impact. For example, cities can select high-value areas of land in new residential subdivisions and reserve them for builders certified by green building standards (e.g., Energy Star, R-2000, LEED for Homes in Canada), providing an incentive to raise the standard of building practices. Cities can also use zoning. Vancouver's EcoDensity Charter 2008, which includes rezoning for green building standards, is perhaps the most ambitious in

the country. Ontario's Places to Grow Act of 2005 is another of Canada's most ambitious standards for regulating the impact of residential development on the regional environmental impact of urban form. Permitting the construction of laneway, garden, and garage housing on existing residential lots is a tool that is becoming more common, and the cities of Vancouver and Saskatoon have travelled this route recently.

Conclusion

Where are we going as a society with respect to social and environmental objectives in housing? Considering the many years that a residential subdivision and homes built in it will remain on our urban landscape, to what extent can we permit even one single additional unit of housing to be built to an environmental standard that is not in tune with our commitments to address climate change, energy efficiency, and conservation of water and other materials, and to use less land to house more people? Accepting that housing fits in the middle of everything, as Myers (2008) has put it, we have argued in this chapter that there is a need for stronger public policy in the housing sector to smooth out areas where the market is failing. Public-sector responses we recommend include the production of housing for low-income renters and finding ways to protect the use value of homes for low-income residents. We also recommend an accelerated implementation of state-of-the-art environmental design for housing units and residential areas in the certainty that our housing consumption patterns are contributing to what is the most pressing issue of the twenty-first century, the severe degradation of our life-supporting environment.

Some of the housing challenges twenty-first-century cities face, drawing from our discussion, are:

- the declining rental inventory and the marginalization of households in the rental sector;

- addressing the environmental imperatives in home-building;
- sustainable approaches to residential subdivision design;
- strengthening housing policy so that it plays a supporting role to other policy sectors like health and education;
- social housing for the very low-income and homeless;
- effective use of housing for neighbourhood revitalization;
- protecting the affordable stock in gentrifying neighbourhoods;
- building for an aging and diversifying population.

These complex issues will not be resolved through private decision-making in the marketplace alone. There continues to be a role for ambitious public policy in this sector, which is so central to our individual personal and collective societal development.

Review Questions

1. How is adequate and affordable housing a precursor to good outcomes in other key aspects of urban life, such as education, employment, health, and income security?
2. What role does housing play in addressing such environmental issues as climate change, the degradation of natural systems, and non-renewable energy consumption?

Note

1. For a discussion of the production process of housing and the different uses made of the home, see Richard Harris, 'Housing: Dreams, Responsibilities, and Consequences', Chapter 15 in Bunting and Filion (2006), at: <www.oupcanada.com/bunting4e>.

References

Bourne, L.S. 1981. *The Geography of Housing*. New York: Edward Arnold.

Bratt, R. 2002. 'Housing and family well-being', *Housing Studies* 17: 13–26.

Bunting, T., and P. Filion, eds. 2006. *Canadian Cities in Transition: Local Through Global Perspectives*, 3rd edn. Toronto: Oxford University Press.

———— and T. Rutherford. 2006. 'Transitions in an era of globalization and world city growth', in Bunting and Filion (2006).

Canada Mortgage and Housing Corporation. 2008. *Canadian Housing Observer*. Ottawa: Canada Mortgage and Housing Corporation.

————. 2009. *Renovation and Home Purchase Report*. Ottawa: Canada Mortgage and Housing Corporation.

Canadian Home Builders. 2002. *Pulse: Manitoba and Saskatchewan*. Winnipeg: Manitoba Homebuilders Association.

Carter, T. 2008. *Canadian Policy Paper: Tri-Country Conference, Canadian Policy Update*. Ottawa: Canadian Housing and Renewal Association.

———— and C. Polevychok. 2004. *Housing Is Good Social Policy*. Ottawa: Canadian Policy Research Networks.

————, ————, A. Friesen, J. Osborne, A. Gunn, and C. Wolfe. 2009. *Housing for Manitobans: The Next Ten Years*. Winnipeg: University of Winnipeg.

Conference Board of Canada. 2009. *Building from the Ground Up: Enhancing Affordable Housing in Canada*. Ottawa: Conference Board of Canada.

De Jong, I. 2008. 'Streets to homes in Toronto', a presentation to housing and homelessness stakeholders in Saskatoon.

Després, C. 1991. 'The meaning of home: Literature review and directions for future research and theoretical development', *Journal of Architectural and Planning Research* 8: 96–115.

Dunn, J. 1998. 'Social Inequality, Population Health and Housing: Towards a Social Geography of Health', Ph.D. thesis, Simon Fraser University.

————. 2000. 'Housing and health inequalities: Review and prospects for research', *Housing Studies* 15: 341–66.

————. 2003. 'A needs, gaps and opportunities assessment for research', *Housing as a Socio-Economic Determinant of Health*. Ottawa: Canadian Institute of Health Research.

Giddens, A. 1998. *The Third Way: The Renewal of Social Democracy*. Cambridge: Polity Press.

Green, D.A., and K. Milligan. 2007. *Canada's Rich and Poor: Moving in Opposite Directions*. Toronto: Canadian Centre for Policy Alternatives.

Guirguis-Younger, M., V. Runnels, and T. Aubry. 2003. *A Study of the Deaths of Persons who are Homeless in Ottawa—A Social and Health Investigation*. Ottawa: Saint Paul University and the Centre for Research on Community Services.

Harris, R. 2006. 'Housing: Dreams, responsibilities, and consequences', in Bunting and Filion (2006).

Harvey, D. 2005. *A Brief History of Neoliberalism*. Toronto: Oxford University Press.

Hodge, G., and D.L.A. Gordon. 2008. *Planning Canadian Communities: An Introduction to the Principles, Practice, and Participants*, 5th edn. Toronto: Thomson Nelson.

Holston, J. 1995. 'Spaces of insurgent citizenship', *Planning Theory* 13: 35–51.

———. 2001. 'Urban citizenship and globalization', in A.J. Scott, ed., *Global City-Regions: Trends, Theory, Policy*, Toronto: Oxford University Press.

Hulchanski, J.D. 2002. *Housing Policy for Tomorrow's Cities*. Ottawa: Canadian Policy Research Networks.

Imrie, R. 2004. 'Disability, embodiment and the meaning of the home', *Housing Studies* 19: 745–63.

Jenson, J., and D. Saint-Martin. 2003. 'New routes to social cohesion? Citizenship and the social investment state', *Canadian Journal of Sociology* 28: 77–99.

Kearns, R., C. Smith, and M. Abbott. 1991. 'Another day in paradise? Life on the margins in urban New Zealand', *Social Science and Medicine* 33: 369–79.

Knox, P., and L. McCarthy. 2005. *Urbanization: An Introduction to Urban Geography*. Upper Saddle River, NJ: Pearson Prentice-Hall.

Ley, D. 1996. *The New Middle Class and the Remaking of the Central City*. Toronto: Oxford University Press.

——— and C. Dobson. 2008. 'Are there limits to gentrification? The contexts of impeded gentrification in Vancouver', *Urban Studies* 45: 2471–98.

——— and H. Frost. 2006. 'The inner city', in Bunting and Filion (2006).

Lezubski, D., J. Silver, and E. Black. 2000. 'High and rising: The growth of poverty in Winnipeg', in J. Silver, ed., *Solutions That Work: Fighting Poverty in Winnipeg*. Winnipeg and Halifax: Canadian Centre for Policy Alternatives—Manitoba and Fernwood Publishing.

McKellar, J. 1994. *The Canadian Housing System in the 1990s*. Ottawa: Canada Mortgage and Housing Corporation.

Meligrana, J., and A. Skaburskis. 2005. 'Extent, location and profiles of continuing gentrification in Canadian metropolitan areas, 1981–2001', *Urban Studies* 42: 1569–92.

Moore, E., and A. Skaburskis. 2004. 'Canada's increasing housing affordability burdens', *Housing Studies* 19: 395–413.

Mueller, E.J., and J.R. Tighe. 2007. 'Making the case for affordable housing: Connecting housing with health and education outcomes', *Journal of Planning Literature* 21: 371–85.

Mullins, P., J. Western, and B. Broadbent. 2001. *The Links between Housing and Nine Key Socio-Cultural Factors: A Review of the Evidence*. Brisbane: Australian Housing and Urban Research Institute.

Myers, D. 2008. *Failed Urban Policy: Tear Down HUD*. At: <www.planetizen.com>. (28 Aug. 2008)

Platt, K. 1996. 'Places of experience and the experience of place', in L.S. Rouner, ed., *The Longing for Home*. South Bend, Ind.: University of Notre Dame Press.

Porteous, J.D., and S.E. Smith. 2001. *Domicide: The Global Destruction of Home*. Montreal and Kingston: McGill-Queen's University Press.

Redfern, P.A. 2003. 'What makes gentrification "gentrification"?', *Urban Studies* 40: 2351–66.

Saunders, P. 1989. 'The meaning of "home" in contemporary English culture', *Housing Studies* 4: 177–92.

Saskatoon, City of. 2007. *New Neighbourhood Design and Development Standards: Final Report, Part I, Summary of Recommendations*. Saskatoon: City of Saskatoon.

Shapcott, M. 2008. 'The right to housing and the right to the city', presentation at the Planners Network Conference, Winnipeg.

Skelton, I. 2000. 'Cooperative and nonprofit housing in Winnipeg: Toward a re-engagement of the provision infrastructure', *Canadian Journal of Urban Research* 9: 177–96.

Smith, S. 1994. 'The essential qualities of a home', *Journal of Environmental Psychology* 14: 31–46.

Statistics Canada. 2008a. *Population and Dwelling Counts for Canada and the Provinces*. Ottawa: Queen's Printer.

———. 2008b. *Population and Dwelling Counts for Census Metropolitan Areas*. Ottawa: Queen's Printer.

———. 2008c. *Population and Dwelling Counts for Census Agglomerations*. Ottawa: Queen's Printer.

Walker, R.C. 2006. 'Searching for Aboriginal/Indigenous self-determination: Urban citizenship in the Winnipeg low-cost-housing sector, Canada', *Environment and Planning A* 38: 2345–63.

———. 2008. 'The political challenge of implementing social housing policy in Saskatchewan', in H. Leeson, ed., *Saskatchewan Politics: Crowding the Centre*. Regina: Canadian Plains Research Centre Press.

Walks, R.A. 2006. 'Homelessness, housing affordability, and the new poverty', in Bunting and Filion (2006).

West Broadway Development Corporation. 2009. 'About Us'. At: <www.westbroadway.mb.ca/>.

Yalnizyan, A. 1998. *The Growing Gap: A Report on Growing Inequality between the Rich and Poor in Canada*. Toronto: Centre for Social Justice.

CHAPTER 21

Younger Cities, Older Cities, and Cities in the Balance: Spaces and Places of the Younger and Older Population

Mark W. Rosenberg and Dana H. Wilson

Larry Bourne (2000) describes visions of the city as 'entrepreneurial', 'equitable', 'compact', 'energy-efficient/wired', 'humane', 'affordable', 'sustainable/green', 'empowered and privatized', 'safe', and 'efficient'. In our vision, we want to think about Canada's cities as 'younger', 'older', and hanging 'in the balance'. These three terms are not for categorizing Canadian cities according to the years in which they were founded. We use these terms as a basis for focusing on two demographic groups, the younger population (the population aged 15 to 24) and the older population (the population aged 65 and over), who are under-researched, under-considered, or even ignored in the urban geography, planning, and policy-making of our changing twenty-first-century Canadian cities.

Not only can we think about Canada's largest cities as younger, older, or in the balance demographically, but we can also think about places within cities as foci for the younger population, such as Queen Street West in Toronto, or for the older population, such as gated communities in Kelowna, or in an uneasy balance between an older population and a younger population, such as the 'student ghetto' adjacent to Queen's University in Kingston. Cities and places also take on different meanings over time, from season to season and from day to night. City places are dynamic, gendered, and racialized. Places within cities are often intersections of all age groups, sometimes in

co-operation with each other and sometimes contested among some or all of the groups. Gore Park in the centre of Hamilton, for example, was once a classic Victorian-style park and at the centre of that city's civic life. Today, it remains a hub of activity during the day but is deserted at night, considered unsafe by older people, younger women, and visible minorities.

In the next part of the chapter, we examine Canada's largest urban areas, the census metropolitan areas (CMAs), to make the case for thinking about cities as younger, older, and in the balance, suggesting some of the salient reasons why we might think about Canadian cities this way. We then discuss places of the younger population, followed by a parallel discussion of places of the older population. We conclude by considering what thinking about Canadian cities as younger, older, and in the balance might mean for urban geography, planning, and public policy as the twenty-first century continues.

Defining Cities That Are Younger, Older, and in the Balance

We readily acknowledge that defining the younger and older populations as between 15 and 24 years old and aged 65 and over is arbitrary. In the case of the younger population, however, this is the approximate time in the **life course** when people make decisions as to whether they will remain in school, seek employment, and leave the parental home. Defining the older population as aged 65 and over follows long-established conventions linked to retirement and eligibility for federal and provincial government transfers (e.g., Old Age Security and Canada/Quebec Pension Plan payments). Nonetheless, concepts like continuing education, full-time and part-time employment, leaving the parental home, and retirement all have become much more fluid concepts in recent years.

Myriad changes are occurring in Canada's CMAs. In terms of overall growth and decline, most CMAs had positive growth between 1996 and 2006, but some CMAs, such as Saint John, Saguenay, Greater Sudbury, and Thunder Bay, did lose population (see Donald and Hall, Chapter 16). Among the CMAs that grew, the rates of growth varied from just over 49 per cent (Barrie) to barely 0.5 per cent in the case of Regina. Six CMAs over the 1996–2006 period experienced declines in the proportion of their younger population. Among those CMAs where the younger population grew, the percentage changes were from a low of 3.79 per cent (Moncton) to a high of over 64 per cent (Barrie). In contrast, every CMA had positive growth in older populations, and with only three exceptions growth was in double digits, ranging from approximately 11 to over 50 per cent.

To classify CMAs as younger, older, or in balance, the percentage of the older population is divided by the percentage of the younger population. A value between 0.0 and 0.89 defines a younger city, a value between 0.9 and 1.09 defines a city in balance, and a value of 1.1 and above defines an older city. Table 21.1 lists the CMAs based on their ratios of older to younger populations.

Younger Cities

The distribution of youth among urban centres in Canada continues to change as youth gravitate to places offering opportunities for work, education, and desirable cultural and lifestyle amenities. The

Table 21.1 Cities Younger, Older, and In the Balance

CMAs	Ratio
Younger Cities	
St John's (Newfoundland and Labrador)	0.79
Halifax (Nova Scotia)	0.86
Ottawa–Gatineau (Quebec part)	0.75
Ottawa–Gatineau (Ontario part)	0.88
Oshawa (Ontario)	0.83
Toronto (Ontario)	0.89
Kitchener (Ontario)	0.80
Guelph (Ontario)	0.86
Barrie (Ontario)	0.84
Regina (Saskatchewan)	0.85
Saskatoon (Saskatchewan)	0.74
Calgary (Alberta)	0.65
Edmonton (Alberta)	0.72
Older Cities	
Saguenay (Quebec)	1.17
Quebec City (Quebec)	1.14
Trois-Rivières (Quebec)	1.34
Peterborough (Ontario)	1.24
Hamilton (Ontario)	1.14
St Catharines–Niagara (Ontario)	1.36
Greater Sudbury (Ontario)	1.12
Thunder Bay (Ontario)	1.21
Kelowna (British Columbia)	1.48
Victoria (British Columbia)	1.38
Cities in the Balance	
Moncton (New Brunswick)	1.02
Saint John (New Brunswick)	1.06
Sherbrooke (Quebec)	1.03
Montreal (Quebec)	1.08
Kingston (Ontario)	1.09
Brantford (Ontario)	1.09
London (Ontario)	0.95
Windsor (Ontario)	0.96
Winnipeg (Manitoba)	0.98
Abbotsford (British Columbia)	0.91
Vancouver (British Columbia)	0.95

Source: Adapted from Statistics Canada (2006).

younger cities of Canada are the magnets attracting people seeking the employment and social opportunities characteristic of these places. Within this cluster, other demographic forces at work are regionally specific. For example, there can be no doubt that immigration fuels not only the growth of Toronto, but also the CMAs and regional municipalities surrounding it (e.g., Barrie, Oshawa). In western Canada, the general movement of young people out of rural areas and small towns and the growth of the Aboriginal youth cohort have made prairie cities like Regina and Saskatoon younger. In addition, Calgary, Edmonton, and more recently St John's have drawn young people from all across Canada as a result of the economic prosperity and employment opportunities generated by the recent oil boom. The prairie cities of Edmonton, Calgary, and Saskatoon have recently become three of the youngest in Canada with median ages of 36.4, 35.7, and 35.8 years, respectively (Statistics Canada, 2009; Wyman, 2008). Other cities where youth have begun weighing in and have tipped the scales in their favour include Ottawa–Gatineau and St John's.

Older Cities

Much has been written about the aging of Canada's population and how it is geographically differentiated (Moore and Pacey, 2003; Moore and Rosenberg, 1997; Moore et al., 2000). With the general decline in fertility rates and the tendency towards aging-in-place, Canada is growing older everywhere. Among the older cities, two subgroups can be identified. In the first group, with the exception of Peterborough, are the CMAs in Ontario and Quebec that were the mining and manufacturing centres of Canada that spurred much of Canadian economic growth through the 1950s, 1960s, and into the 1970s. These places also were the first victims of industrial restructuring and free trade in the 1980s and 1990s (see Vinodrai, Chapter 6; Hutton, Chapter 7). Many of the cities in this subgroup are among the slowest-growing

CMAs in Canada, with growing older populations aging-in-place (see Donald and Hall, Chapter 16). In contrast, the second subgroup—which includes Peterborough, Kelowna, and Victoria—is attracting retirees because of the quality of life and amenities associated with these cities, in addition to the population that is aging-in-place.

Cities in the Balance

Cities in the balance share some of the characteristics of younger and older cities. For example, a relatively large flow of immigrants normally is associated with younger cities, and the decline of manufacturing normally is associated with older cities. Both of these characteristics are evident in the case of Montreal. In Winnipeg, growth in the Aboriginal youth population, a characteristic of younger cities on the prairies, combines with the decline in manufacturing associated with older cities. Some of the cities we find hanging in the balance—like Sherbrooke, Kingston, and London—combine a younger population associated with post-secondary educational institutions in a city with an older population attracted by the quality of life, amenities, and health-care services available. Over the next 10–20 years many of the cities in this group likely will become older cities, while a few that are highly attractive to younger people and immigrants, such as Vancouver and Abbotsford, more likely will take on profiles of younger cities; but this prediction is open to debate.

Youthful Places

Youth Trends and Growing Pains in Canada

Recent trends indicate that many young Canadians are extending their 'youthful' years and delaying the transition into adulthood by living with their parents longer, completing more years of education, getting married later in life, delaying child-bearing, and having fewer (if any) children than their parents and previous generations did (Bushnik and Garner, 2008; Clark, 2008; see Townshend and Walker, Chapter 8).[1] Canadian youth are more diverse ethnically than previously, with increasing proportions of new residents from outside Canada in CMAs like Vancouver, Toronto, and Montreal (see Hoernig and Zhuang, Chapter 9). Canada has more immigrant youth than ever before. More Aboriginal peoples in urban centres, and their higher fertility rates, have led in recent years to growing Aboriginal youth populations in many CMAs, especially in western Canada (see Peters, Chapter 22). Canadian youth in general are more inclined to migrate for employment and education opportunities than in previous generations, with patterns of youth mobility facilitated by advancements in communication technology and transportation, and reflecting shifts in regional and global economies (Statistics Canada, 2008).

Young people are taking advantage of economically prosperous CMAs, such as Calgary and Edmonton, which have benefited from recent oil-based resource booms. The influx of youth to CMAs is a sign of the economic growth and prosperity of those centres. It also forewarns that local infrastructure, services, and amenities must keep pace to meet the basic needs and the consumption desires of larger youth populations. Edmonton initially struggled to meet the diverse needs of youth as the city swelled with a large younger population arriving in the city for work (Diotte, 2008). Those youth wanting to start families were faced with crowded schools and hospitals, rising housing costs, and a lack of amenities for children and young people in new suburbs while the city worked to keep up with the growing young population. Many transient youth migrating for work also lacked the resources and social support to help them become established in their new environment. Despite acquiring relatively high disposable incomes, the lifestyles of transient workers and lack of connectivity to social services placed many younger workers at increased risk of workplace injury, engaging in risky behaviours (binge partying), and contracting sexually transmitted infections (STIs) (Goldenbert et al., 2008). Socio-economic disparities and variable civic amenities have contributed to the marginalization and exclusion of many youth, particularly those not directly benefiting from the economic prosperity in growing youthful cities. Increased youth poverty and homelessness have resulted as the cost of living rises in urban centres.

The mix of youth from diverse cultural and ethnic backgrounds also has led to considerable discrimination, racism, and violence in some cities. Gang membership among male youth has become a growing concern in urban centres in Canada, although gang activity and violence among females has been increasingly noted in cities in western Canada (Astwood Strategy Corp., 2002). While a lively debate ensues about the actual magnitude of youth involvement in criminal and gang activity, as opposed to the perceived magnitude created by media attention in recent decades, the marginalization and disengagement of youth are a legitimate and growing concern. Factors attracting youth into gang membership often mirror the fundamental requirements of youth, including but not limited to excitement and stimulation, protection, financial security, and a sense of belonging. Racism and discrimination are also factors contributing to violence among youth groups in urban centres. Racial discrimination probably plays a central role, for example, in Saskatoon's incidence of high numbers of Aboriginal youth involved in physical and gun violence (Public Safety Canada, 2007). Higher numbers of low-income immigrant (non-white) and Aboriginal youth in urban centres have created challenges for integration among communities, often exacerbated by living in families with high rates of unemployment and below the poverty line.

Youth Places in Urban Centres

Young people spend a great deal of time in home, school, work, and leisure environments arranged and controlled by adult guardians or supervisors.

With so much of their lives spatially and temporally structured, youth during free time often seek out autonomous places that provide them with a sense of belonging and an opportunity to express themselves and explore their developing identities (Dumont and Provost, 1999; Harris et al., 2002; He et al., 2004; Malbon, 1998; Roth and Brooks-Gunn, 2003; Simpson et al., 2006; Valentine et al., 1998). While many consider youth to be a carefree stage in life, it is in fact a critical and often stressful transitional period between childhood and adulthood when the skills needed to assume the roles and responsibilities of adulthood are developed. Youth is marked by important life events and experiences that can have lifelong (and often cumulative) implications, including completing or leaving school, joining the workforce, engaging in risky behaviours (e.g., drug and alcohol consumption, tobacco use, gambling), instigating or being a victim of violent or delinquent acts, sexual encounters, bearing children, or civic involvement (Bradizza and Stasiewicz, 1999; Bronfenbrenner, 1986; Brooks-Gunn, 2001; Call et al., 2002; Cook et al., 2002; Harris et al., 2002; He et al., 2004; Hedberg et al., 1999; Kershaw et al., 2005; Morrongiello and Dawber, 2004; Richter, 2006; Roth and Brooks-Gunn, 2003; Wilson and Ross, 2009). Hanging around and chatting with friends in safe and welcoming places may be central to identity formation for youth, although this type of behaviour often holds little value from a civic planning perspective and may not be encouraged by the adult majority (Nairn et al., 2003) (Figure 21.1).

The scarcity of financial resources available to civic authorities, privatization of previously public infrastructure (e.g., recreation facilities like ice hockey arenas), and new building developments or redevelopment in urban areas have resulted in a declining quantity and quality of public places and recreational or leisure programs available to youth in many cities (Breitbart, 1998; Katz, 1998; Matthews et al., 2000; Thomas, 2005; Valentine, 2008; Valentine et al., 1998). Declining fertility rates and smaller family sizes have worked further to marginalize youth in urban centres, socially and spatially, through a reduction in resources targeting youth in favour of services and amenities for the aging baby boomer generation. For example, playgrounds provide amenities for children (swings and slides) and guardians (benches and picnic tables), but not necessarily for youth (basketball courts, skateboard facilities) (Driskell et al., 2008). In fact, teens are often considered too old or rowdy for public places like park playgrounds, and are often discouraged from using playground equipment and amenities. On the flip side, (younger) youth are often considered too young, irresponsible, or immature to frequent adult places like concert halls, bars, galleries, and nightclubs, or can do so only with adult accompaniment. As a result of the age-specific spatial ordering of many cities, few public places are reserved exclusively for youth, where they might feel safe and welcome to spend time in an unstructured manner.

Shifts to dual-income families and single-parent households have given youth more time, on average, without parental or guardian supervision. Youth today thus typically have more time without supervision, fewer places where they are welcome to spend leisure time, and a longer transitional period before they are considered responsible adults. Many public and semi-private places in Canadian urban centres are not suitable or sufficient for youth. Often urban youth have few designated and desirable options for a 'third place' away from home and school, where they can feel safe and welcome and engage in meaningful interactions with peers or mentors. With few designated places for them, youth are often left to claim places in the public fringes to serve their needs. While public spaces like sidewalk networks, stairwells, alleyways, parking lots, and transit stations are designed for the passage of people between places, youth often attempt to claim certain portions of these spaces to facilitate socializing and bonding (Thomas, 2005; Valentine et al., 1998).

Figure 21.1 Four youth sitting and chatting in Kensington Market (Toronto) on a weekday afternoon. Kensington Market is famous for its eclectic shops, countercultural values, and large immigrant and artist population. Kensington Market is a place where radical ideas and images are embraced and a place where many youth may feel accepted as they chart their course to adulthood. Note unsolicited graffiti on storefronts is a common sight in the Market area. See www.kensington-market.ca. (Dana Wilson)

Social exchanges that occur in public spaces and meanings taken from these relations can also make youth feel more or less excluded (Nairn et al., 2003). For example, while libraries or public pools may be places where youth should feel welcome, tension arising from relations and exchanges with supervisors (e.g., librarian, staff, lifeguard) or other users of the facility (e.g., adults, alone or with younger children) such as real or perceived disapproval, discrimination, or condemnatory attitudes may make them unwelcoming to youth or even places of exclusion (Nairn et al., 2003; Valentine et al., 1998). Female and minority youth or those

in the younger years of the 15–24-year-old cohort may be even more excluded from public places and instead rely on semi-private locations for a safe alternative place to spend leisure time. Females often have greater restrictions on the places they can occupy due to safety issues, gendered ideals, and expectations of feminine behaviours and activities (Collins and Kearns, 2001; Thomas, 2005; Valentine et al., 1998). The level of (in)security at particular places or in places at certain times, such as parks in the evening, may further limit access by female youth, youth of minority groups, or those who are younger (Nayak, 2003; Valentine et al.,

1998; see Cowen, Siciliano, and Smith, Chapter 17). Walking unaccompanied in a particular place or at a certain time of night, for example, may not be feasible for certain youth groups (Bauder, 2001).

Shopping malls, coffee shops, and fast-food restaurants that are well-lit and supervised (by security guards) have become popular locations for youth (Matthews et al., 2000; Nairn, 2003). Semi-private places—gyms, bars, movie theatres, Internet cafés, and shopping centres—also are popular, but these often have a user fee associated with them. Places linked with consumerist activities effectively marginalize economically disadvantaged youth from frequenting them, and youth must instead rely on spending time in places without a pre-arranged or consumerist program. Minority youth, who are on average more likely to be poor, thus face even greater challenges to carving out safe and desirable places in their urban environments where they can engage in meaningful activities and have quality interactions with peers or mentors.

Fewer youth-friendly opportunities for constructive behaviours and activities at third places have resulted in greater time spent inside the home. Many youth are spending more time at home than ever before and much of this time is spent watching TV, playing video games, or using the Internet (see Gilliland, Chapter 23). The rise in popularity and accessibility of information and communication technologies in private dwellings has reduced the reliance on alternative public places for informal social exchanges (Liu, 2009). The combination of scant public space and potentially infinite virtual spaces for youth (through home computers, laptops, mobile devices) has resulted in many youth spending little real time developing social skills and negotiating peer culture outside of school. Instead, they may rely on virtual environments and realities provided through the Internet and (reality) TV for meaningful exchange and leisure with local and remote members of their social networks (Katz, 1998). Social networking websites like Facebook and Twitter have become tremendously popular among youthful populations where much socializing now is done virtually and is available 24 hours a day.

Contesting Youth in Urban Places

The presence of youth in the public fringes or marginal places of cities is often a source of tension for other (adult) groups who share some stake in those places. Parents, adults, and police may frown on youth using public places as spaces of leisure and consider their presence disruptive and unnerving. In spite of our knowledge that youth are vulnerable and impressionable, they are often subject to negative stereotypes in relation to urban places, such as being irresponsible or delinquent (Collins and Kearns, 2001; Nayak, 2003; Pain, 2001; Valentine et al., 1998). Much attention paid to youth in recent years has been negative, focusing on their presence in urban places and the problems associated with youthful populations. Youth groups are often problematized through the creation of negative stereotypes and moral panics (working to generate representations of delinquent youth) around issues like graffiti, loitering, vandalism, violence and crime, delinquency and truancy, teen pregnancy, and drug use and other risk-taking behaviours (Lucas, 1998).

Traditional and less progressive initiatives typically consider youth as a problematic population and their presence in public places as an undesirable feature of urban landscapes. For example, some municipalities and provinces have explored the possibility of imposing curfews on young people as a way to keep them out of such public places as streets and parks between certain hours. Curfews are generally imposed in the hope of reducing mischief and the occurrence of youth crime and violence. In the counties of Middlesex and Lambton, the Ontario Provincial Police have invoked a curfew for the summer months, when youth are most active outdoors, with the aim of reducing vandalism and criminal activity. In contrast to case-by-case municipal bylaws, this curfew falls under the Provincial Child and Family Services Act, which states that: '[b]etween the hours of midnight and 6 a.m., if a child apparently or actually under 16 years of age [is] in a public place without a person over 18 years (authorized by their parents), police may apprehend that child without warrant as a child in need of protection' (Ontario, Child and Family Services, 2009). This provincial Act can be invoked by local police departments if deemed necessary or desirable.

Another example of regulating the temporal and spatial places of youth is the City of Saskatoon's Bylaw No. 2954, 'A bylaw of The City of Saskatoon to restrict improper use of streets, lanes, parks and City property', which states that:

> No person or persons shall stand in groups or sit or lounge on a public street in front of a licensed premises, restaurant, poolroom, boarding house, hotel or place of public accommodation or place of public entertainment so as to cause any obstruction to the free use of the street, or on the step or approach to any premises or dwelling open to a street whereby the public are subjected to disturbance or annoyance. (Saskatoon, 2004)

This bylaw not only authorizes the removal of homeless persons and panhandlers, but also effectively serves as another tool to manage youth, signalling that they are not welcome to frequent or linger in the proximity of commercial areas except strictly on a purposeful consumerist basis.

Curfews or other bylaws and devices designed to limit the use of places by youth are controversial and considered discriminatory by many since they prohibit (or penalize) all youth from public places based solely on their age (Collins and Kearns, 2001). Implicit in these approaches to addressing youth are that they are problematic and their presence (particularly in public places) should be limited to avoid naturally occurring tendencies towards deviant behaviour. One example of an initiative to control youth movement and gathering in public spaces is called 'the mosquito', a 'teen repellent device' that has been tested or installed in some Canadian CMAs, including Vancouver, Edmonton, and Montreal. The electronic device emits high-pitched ultrasonic noise supposedly audible—and intolerable—only to young people. The device is marketed for use in areas where youth congregate but are viewed as undesirable or troublesome. A distributor of the device, No Loitering Technology (2007), describes its reputed value to retailers:

> The Mosquito ultrasonic teenage deterrent is the solution to the eternal problem of unwanted gatherings of teenagers in shopping malls, around shops, in parks and schools after hours and anywhere else they are causing problems. The presence of anti-social teenagers discourages genuine customers from coming into stores, affecting your turnover and profits. With loitering comes vandalism.

Obviously a controversial issue, the device has been contested by youth and human rights activists, who claim it is discriminatory and violates human rights legislation. Concerns over the safety of the device have also been raised (CBC News, 2009).

Yet, the problem is not with youth and inherent deviant behaviours, but instead that civic authorities are not well equipped to engage youth meaningfully in public places. Youth are often left to spend time in areas of the city where few spaces are designed with them in mind. Youth lacking opportunities and resources to be engaged meaningfully in civic and social systems may turn to deviant behaviours partly to contest their marginalization in public places and civic processes (Valentine, 2008; Vanderbeck, 2007). The degree to which street youth, gangs, and youth-related criminal activity in an urban centre pose persistent and real problems to the mainstream urban culture may go some way as a useful measure for assessing the gap between services and amenities provided by civic authorities for youth and the actual needs and desires of youth.

Youth homelessness, gang membership, and crime are often a culmination of other complex and interrelated social and health inequalities, including family disruption and inadequate parental support, absence of positive role models, academic difficulties, physical or emotional abuse, sexual exploitation, risky behaviours, and racial or other discrimination (particularly for Aboriginal youth) (Gaetz, 2004; Haley et al., 2004; Public Safety Canada, 2007; Saewyc et al., 2006; Weber et al., 2002). These issues have become more pressing in urban centres with high proportions of young people and it has become apparent that urban centres must provide the resources to support youth adequately and engage them meaningfully. Judicial or disciplinary intervention and greater controls on youth behaviour are not likely to be sufficient to get at the root causes of youth violence and disengagement. We propose that greater youth involvement and programming will provide more realistic and effective solutions to address many of the root causes of youth violence and criminal activity, including youth homelessness and social marginalization.

Contemporary Approaches to Engaging, Attracting, and Retaining Youth in Urban Centres

Young people have been provided with few opportunities to engage in civic planning processes that ultimately affect their future economic, social, and environmental well-being. While some traditional impressions of youth and youth interventions have not been overwhelmingly positive or progressive, more recent initiatives have been directed at the positive attributes of youth and the identification of a series of constructive arrangements of youth places within urban design (Breitbart, 1998). Many CMAs in Canada have begun actively working to engage youth in civic issues as a way to 'get it right' in efforts to attract new and retain existing youth populations as part of their economic growth strategies. The changing face of youth has challenged

urban geographers and city planners to find ways to include reconstructed and evolving notions of culture, difference, place, productivity, recreation, neighbourhood, and commerce into their theories, strategies, and programming.

Some cities have begun taking stock of what key services and amenities their centres have to offer youth and how they measure up against those offered in other Canadian urban areas. For example, Edmonton's Next Gen Task Force was launched by Mayor Stephen Mandel in 2005 as a way to engage youth in civic issues and identify ways to retain and attract young, talented, educated, and energetic citizens for the city's future sustainability (Diotte, 2006; Next Gen, 2007). The FUSION Halifax project and Urban Playground in Saskatoon are two other examples of initiatives created to improve the marketability and livability of cities as progressive and competitive places for young entrepreneurs and workers. Both Halifax and Saskatoon retained Next Generation Consulting (NGC) to assess the challenges and advantages for attracting and retaining youth. NGC, featuring leading urban youth specialist Rebecca Ryan, focuses on seven areas of assessment to understand a city's potential for attracting skilled young professionals: (1) vitality (environmental quality and health-promoting features); (2) earning (job growth and diversity potential, income level); (3) learning (quality education opportunities, library use, Wi-Fi); (4) social capital (social accessibility, tolerance, safety); (5) cost of lifestyle (cost of living); (6) after hours (activities and venues for weeknights and weekends); and (7) getting around town (public transit, **walkability**, commute times) (Next Generation Consulting, 2009). These seven areas of interest are becoming common features that cities are evaluating and promoting as they begin to prioritize the needs of youth in their civic planning.

Natural place amenities such as water bodies, mountain ranges, or mild climates are geographical features that may appeal to a young and active population. Constructed amenities like pedestrian-friendly streetscapes, art galleries, libraries, bike lanes, light-rail transit systems, and urban Wi-Fi zones are examples of contemporary civic initiatives designed to create unique and vibrant streetscapes that accommodate a variety of uses and appeal to modern and youthful lifestyles (Florida, 2002; see Lynch and Ley, Chapter 19) (Figure 21.2). A Wi-Fi zone in a park setting or downtown area, for example, can provide an alternative site for productivity, recreation, and leisure for youth and fits well with technologically savvy young

Figure 21.2 Example of planned mural artwork (in place of graffiti) in Kensington Market (Toronto), created by local youth artists between 17 and 20 years of age. This artwork was funded by the City of Toronto and co-ordinated by the Harbourfront Community Centre's Mural Art Project. See www.harbourfrontcc.ca. (Dana Wilson)

populations. Table 21.2 provides some examples of amenities and initiatives in the five youngest CMAs in Canada (by proportion of youth aged 15–24 years) to involve youth in the civic planning process, engage them in their local environment, and encourage them to assist in developing ideas and programs to address their needs and aspirations.

Places of the Older Population

Recent Population Trends in Urban Centres among the Older Population of Canada

The places of the older population are more difficult to define because the 'baby boom' generation is now starting to retire and most older people age in place. Among the larger CMAs, a typical pattern has evolved where, increasingly, the central parts of the CMA are dominated by the older population, while faster-growing suburban regions surrounding the CMA are dominated by working-age and younger populations. Nowhere is this spatial pattern better illustrated than in the case of Toronto and the surrounding regional municipalities (Table 21.3). For the Toronto CMA, the percentage of the population aged 65 and over is 11.87 per cent. In the City of Toronto, already 14.12 per cent of the population was aged 65 and over in 2006, contrasting with its surrounding regional municipalities where the percentages were below that of the CMA. At the census tract level, the percentage of the population aged 65 and over in the City of Toronto can be well above 14 per cent.

Age-Friendly Cities, Age-Friendly Communities, and Naturally Occurring Retirement Communities

From the global to the local levels, public agencies have called for the creation of **age-friendly cities** and **age-friendly communities** to account for the particular needs of older people who will make up a quarter or more of the population in their home communities by between 2031 and 2051 (Saanich, n.d.; PHAC, n.d.; WHO, 2007). The call for

Table 21.2 Examples of Youth Amenities and Initiatives in Five Youthful CMAs

	Calgary	Edmonton	Saskatoon	Ottawa–Gatineau	St John's
Total population	1,079,310	1,034,945	233,923	1,130,761	181,113
% pop. change (1996–2006)	23.87	16.65	6.36	11.42	3.90
Total youth pop. (000s)	155.29	159.36	39.01	155.50	26.29
Youth as % of total	14.39%	15.40%	16.68%	13.61%	14.51%
Provincial median age	35.7	35.7	37.9	41.0	42.5
Youth gang members (000s)	0.22	0.22	1.34	0.29	0
Youth criminal offences (2006/7)	8,016	8,016	5,165	25,102	738
Civic park hours	500h–2300h	500h–2300h	500h–000h	500h–2300h	700h–2300h
Downtown Wi-Fi	N	Y	Y	N	N
Designated bike lanes	N	Y	Y	Y	N*
Youth engagement or retention initiatives	Calgary Urban Vibe <www.calgary urbanvibe.ca> Child & Youth Friendly Calgary <www.cyfc.ca> Mayor's Youth Council <www.youthare awesome.com>	Next Generation Task Force <www.edmonton nextgen.ca> City of Edmonton Youth Web Resource <www.yedmonton.ca> City of Edmonton Youth Council <www.ceyc.ca>	Child and Youth Friendly Saskatoon <www.childfriendly saskatoon.sk.ca> Urban Playground <www.urban playground.ca>	Ottawa Youth Commission <www.ottawayouth commission.ca> Ottawa Youth Zone <www.ottawa.ca/ residents/yzj>	Futures in Newfoundland and Labrador's Youth <www.finaly.ca> Youth Retention and Attraction Strategy <www.lmiworks. nl.ca/yras> Provincial Youth Council <www.finaly. ca/provincial-youth-council>

*Projects for bike lines announced.

Source: Adapted from Statistics Canada (1996, 2006); Astwood Strategy Corporation (2003); Public Safety Canada (2007); official websites for each of the five cities.

Table 21.3 Comparing Age Groups in the Greater Toronto Area

	Total Population	Age 15–24		Age 65 and over	
		Population	%	Population	%
Toronto (CMA)	5,113,145	683,940	13.38	607,025	11.87
Toronto (GTA)	5,555,908	741,260	13.34	664,885	11.97
Toronto (City)	2,503,280	318,650	12.73	353,445	14.12
York (Regional Municipality)	892,710	126,475	14.17	91,920	10.30
Halton (Regional Municipality)	439,255	54,310	12.36	54,840	12.48
Peel (Regional Municipality)	1,159,405	163,170	14.07	104,525	9.02
Durham (Regional Municipality)	561,258	78,655	14.01	60,155	10.72

Source: Adapted from Statistics Canada (2006).

Table 21.4 Age-Friendly Vision of Communities for Older Populations

Aspect of Vision	Characteristics
Participation	Positive images of older persons Accessible and useful information Accessible public and private transportation Inclusive opportunities for civic, cultural, educational, and voluntary engagement Barrier-free and enabling interior and exterior spaces
Health	Places and programs for active leisure and socialization Activities, programs, and information to promote health and social and spiritual well-being Social support and outreach Accessible and appropriate services Good air/water quality
Security and independence	Appropriate, accessible, affordable housing Accessible home-safety designs and products Hazard-free streets and buildings Safe roadways and signage for drivers and pedestrians Safe, accessible, and affordable public transportation Services to assist with household chores and home maintenance Supports for caregivers Accessible stores, banks, and professional services Supportive neighbourhoods Safety from abuse and criminal victimization Public information and appropriate training Emergency plans and disaster recovery Appropriate and accessible employment opportunities Flexible work practices

Source: Adapted from WHO (2007).

age-friendly cities and communities is a top-down approach asking local governments to incorporate a particular vision of the older population in civic affairs along three dimensions: participation, health, and security and independence (Table 21.4).

Others have conceptualized areas where the older population is concentrated as naturally occurring retirement communities (NORCs). NORCs are described as:

> neighbourhoods or buildings in which a large segment of the residents are older adults. In general, they are not purpose-built senior housing or retirement communities and were neither designed nor intended to meet the particular health and social service needs and wants of the elderly. Most commonly, they are places where community residents have either aged in place, having lived in their own homes over several decades, or are the result of the significant migrations of older adults into the same housing constructs or neighbourhoods, where they intend to spend the rest of their lives. (United Jewish Communities, 2009)

What age-friendly communities and naturally occurring retirement communities have in common is the shared recognition that areas within the city where the older population concentrates should also be places that both positively enhance the participation of older people and eliminate the barriers that reduce participation in civic life (for more on NORCs, see Masotti, 2009). For example, positive enhancements can be as simple as placing seating on each city block so that older people who are limited in their mobility can always find a place to rest, enhancing their ability to get out and walk. The elimination of barriers can be as simple as ensuring that all buses in a transit system can 'kneel' to allow easier access.

Implicit also in the concepts of age-friendly cities, communities, and naturally occurring retirement communities is the assumption that the older population wants to live in demographically heterogeneous communities (i.e., communities that have mixed aged groups from the youngest to the oldest). This assumption is challenged by the growth of age-segregated places, such as gated retirement communities and seniors-only apartment and condominium developments (see Townshend and Walker, Chapter 8). These places are marketed to older people with many of the same claims as age-friendly communities and NORCs, but with emphasis towards services specifically for them and claims of enhanced physical security found only by living with other older people (see Grant, 2005; Laws, 1993, 1995).

These contrasting views of age-friendly communities versus age-segregated communities reflect a parallel debate in the delivery of services for the older population, and particularly those who are disabled or in poor health. Should all services be made accessible to all age groups regardless of health status? Or should special services be provided as an alternative for those who cannot access services as they are regularly provided? Nowhere has this challenge been played out as obviously as in the provision of public transportation in CMAs. In most if not all CMAs, special transit services are provided for the frail elderly and disabled, and most transit authorities are making their regular services more accessible with adaptations such as 'kneeling' buses. Special transit services are not the same as regular transit services, however, and having some kneeling buses and some subway stations equipped with elevators does not create the same accessibility for older people as it does for the rest of the population. While the technology exists to make all public transit services, buildings, and the built environment accessible, related policy issues are who pays and what constitutes accessibility.

Contested Places and Intergenerational Equity

There are few if any public places (e.g., streets, parks) or even semi-public places (e.g., shopping malls) where the mixing of age groups does not occur. Normally, these are shared places with few or no problems in how they are shared. Shared places, however, can easily become contested if a group tries to dominate the place or if the use of the place changes between day and night. For example, urban parks are generally shared places among the younger and older populations during the day. At night, they are often depicted as unsafe and to be avoided by older people for fear of attack by younger people. Whether such fears are justified or not they are socially constructed out of inter-generational differences.

Intergenerational differences have spurred a more general debate about intergenerational equity (see Gee and Guttman, 2000; McDaniel, 2000). Looking at Canadian cities, the debate about equity is a debate about how cities should be planned in the future as an increasing proportion of the population is aged 65 and over. Will an older population demand that more public money fund services that will enhance their lives, at the expense of less public money being spent on services to enhance the lives of younger people in the city. For example, in older cities where the younger population is declining and the older population is growing proportionately, should schools be closed and seniors' centres opened? Even if the older population does not make such demands, will local politicians direct more public money towards enhancing the lives of older citizens to curry their electoral favour? Should public spending be proportional to the age groups in a community or should public spending be based on need (see McDaniel, 2000)?

The above types of scenarios, predicated on the growth of the older population, have been labelled 'apocalyptic demography' (Gee and Gutman, 2000). To date, the evidence contradicts arguments that the older population will make demands for more services at the expense of the younger population. Whether cities are younger, older, or in the balance, the challenge will be to create all-ages-friendly cities at a time when local governments are asked to provide more services without a commensurate growth in funding from either their own sources (e.g., local property taxes) or from higher levels of government.

An All-Ages-Friendly City

At the beginning of the chapter, we divided Canada's largest cities into younger, older, and those hanging in the balance. Obviously, all age groups are represented in any large city. How might we define an all-ages-friendly city? We have emphasized that places in the city can provide positive experiences for different age groups either by enhancing the landscape (e.g., adding skateboard areas for youth or benches for the older population in parks) (Figure 21.3). Or, through the failure to think about differences among younger and older people, barriers can be produced (e.g., broken or uneven sidewalks or poor lighting) that limit people's engagement with the places where they live. An all-ages-friendly city needs to take into account how places can be improved for all age groups, where separation is required and where integration must be enhanced. The other challenge is how to do this throughout the day and night and for all seasons. A prerequisite for creating an all-ages-friendly city is to engage different age groups in civic planning and to plan according to changing needs, as some population groups grow and others decline, while not falling into the trap of dividing resources based on the proportions of various age groups. Finally, age cannot be the only metric for creating an all-ages-friendly city. Gender, race, ethnicity, poverty, health, and other socio-economic factors must be taken into account.

Conclusion

Thinking about Canadian cities as younger, older, and in the balance is a useful way of considering the challenges faced by urban geographers, planners, and policy-makers as the demographic trends of the first half of the twenty-first century

Figure 21.3 Two young males skateboarding at the skate park in Mississauga's Civic Centre. One of seven multi-use ramp locations opened beginning in 2003 in Mississauga, this skate park features mural artwork created by youth, seating, use of indoor washrooms daily until 10 p.m. and a portable washroom onsite. See www.mississauga.ca/portal/residents/skatepark. (Dana Wilson)

transform the urban landscape. Many cities will become older and face the dual challenge of how to attract a younger population while trying to support an older population in an age-friendly way. Urban centres in Canada have long realized that a critical component of the workforce is an influx of new, often young, people, and that retaining a stable and healthy youth population contributes to overall future economic sustainability and prosperity. Urban centres both with and without large youth populations have identified the need to retain and reclaim local youth (i.e., those present and those who may have moved away) as well as attract young migrant and immigrant populations from elsewhere. Young people have traditionally

been provided with few opportunities to engage in civic processes that ultimately affect their future economic, social, cultural, and environmental well-being. While youth have been largely missing from traditional research in human geography and urban planning, future economic success and stability for urban centres depend on their younger population. As increasing numbers of youth take up residence in the smaller number of 'younger cities', and other urban areas feel the impact of dwindling numbers of youth, it is becoming apparent that youth have been excluded to varying degrees from urban development processes in different places. Some cities have been more or less youth-friendly than others. Communities and urban centres need to recognize the importance of public places and programs that focus on youth.

For the older population and 'older cities' of Canada, the challenge for urban geographers, planners, and policy-makers is to understand the complex forces required to attract and retain younger people while adapting and creating a built environment and services that are age-friendly. This challenge is made more complex by the declining fiscal resources of local governments and the growing responsibility they have for services downloaded upon them from higher levels of government. As cities continue to move towards greater involvement of youth in their civic processes, we believe a better balance can be struck between the needs of youth and of adult and older populations and how best to negotiate and plan for their exclusive and shared spaces in urban settings. The ultimate challenge will be to create age-friendly cities for all ages regardless of whether a city is younger, older, or in the balance. While more initiatives focus on the older or younger populations exclusively, we believe that much overlap occurs in the core values, key place-making themes, and services that both the older and younger populations require and desire. The challenge is to provide nuanced services and infrastructure that can appeal to younger populations, older populations, and those communities that hang in the balance.

Review Questions

1. What can municipal authorities and civic organizations do to attract and retain youth to their cities?

2. What would an all-ages-friendly city look like and, from your personal perspective, how would it feel to live in one?

Note

1. A notable exception, however, is the case of Aboriginal women in Canada, who are more likely to have several children and begin giving birth at a younger age, often between 15–24 years of age (Clark, 2008).

References

Astwood Strategy Corporation. 2003. *Canadian Police Survey on Youth Gangs*. Ottawa: Solicitor General of Canada, Catalogue no. PS4–4/2002.

Bauder, H. 2001. 'Agency, place, scale: Representations of inner-city youth identities', *Tijdschrift voor Economische en Sociale Geografie* 92, 3: 279–90.

Bourne, L. 2000. 'Urban Canada in transition to the twenty-first century: Trends, issues, and visions', in Bunting and Filion (2006).

Bradizza, C., and P. Stasiewicz. 1999. 'Introduction', special issue on 'Addictions in Special Populations', *Addictive Behaviors* 24, 6: 737–40.

Breitbart, M. 1998. 'Dana's mystical tunnel', in Skelton and Valentine (1998: 305–27).

Bronfenbrenner, U. 1986. 'Ecology of the family as a context for human-development: Research perspectives', *Developmental Psychology* 22, 6: 723–42.

Brooks-Gunn, J. 2001. 'Children in families in communities: Risk and intervention in the Bronfenbrenner tradition', in P. Moen, G. Elder, and K. Lüscher, eds, *Examining Lives in Context: Perspectives on the Ecology of Human Development*. New York: American Psychological Association.

Bunting, T., and P. Filion, eds. 2006. *Canadian Cities in Transition: Local Through Global Perspectives*. Toronto: Oxford University Press.

Bushnik, T., and R. Garner. 2008. *The Children of Older First-time Mothers in Canada: Their Health and Development*. Ottawa: Statistics Canada, Children and Youth Research Paper Series, Catalogue no. 89–599–M–005.

Call, K., A. Riedel, K. Hein, V. McLoyd, A. Petersen, and M. Kipke. 2002. 'Adolescent health and well-being in the twenty-first century: A global perspective', *Journal of Research on Adolescence* 12, 1: 69–98.

CBC News. 2009. 'Quebec convenience store chain tests "Mosquito" to get teen loiterers to buzz off', 30 Mar. At: <www.cbc.ca/canada/montreal/story/2009/03/30/mtl-mosquito-loitering-0330.html>.

Clark, W. 2008. 'Delayed transitions of young adults', *Canadian Social Trends* no. 84. Statistics Canada, Catalogue no. 11–008.

Collins, D., and R. Kearns. 2001. 'Under curfew and under siege? Legal geographies of young people', *Geoforum* 32: 389–403.

Cook, T., M. Herman, M. Phillips, and R. Settersten. 2002. 'Some ways in which neighborhoods, nuclear families, friendship groups, and schools jointly affect changes in early adolescent development', *Child Development* 73, 4:1283–309.

Diotte, K. 2006. 'Youth task force a breath of fresh air', *Edmonton Sun*, 24 June.

Driskell, D., C. Fox, and N. Kudva. 2008. 'Growing up in the new New York: Youth space, citizenship, and community change in a hyperglobal city', *Environment and Planning A* 40: 2831–44.

Dumont, M., and M. Provost. 1999. 'Resilience in adolescents: Protective role of social support, coping strategies, self-esteem, and social activities on experience of stress and depression', *Journal of Youth and Adolescence* 28, 3: 343–63.

Fengshu, L. 2009. 'It is not merely about life on the screen: Urban Chinese youth and the Internet café', *Journal of Youth Studies* 12, 2: 167–84.

Fitzpatrick, M. 2008. 'Half of Canada's Aboriginal population under 27 years of age', CanWest News, 16 Jan.

Florida, R. 2002. *The Rise of the Creative Class*. New York: Basic Books.

Gaetz, S. 2004. 'Safe streets for whom? Homeless youth, social exclusion, and criminal victimization', *Canadian Journal of Criminology and Criminal Justice* (July): 423–55.

Gee, E., and G. Gutman, eds. 2000. *The Overselling of Population Aging: Apocalyptic Demography, Intergenerational Challenges, and Social Policy*. Toronto: Oxford University Press.

Goldenberg, S., J. Shoveller, A. Ostry, and M. Koehoorn. 2008. 'Youth sexual behavior in a boomtown: Implications for the control of sexually transmitted infections', *Sexually Transmitted Diseases* 84, 3: 220–3.

Grant, J. 2005. 'Planning responses to gated communities in Canada', *Housing Studies* 20: 273–85.

Haley, N., E. Roy, P. Leclerc, J.-F. Boudreau, and J.-F. Boivin. 2005. 'HIV risk profile of male street youth involved in survival sex', *Sex Transmitted Infections* 80: 526–30.

Harris, K., G. Duncan, and J. Boisjoly. 2002. 'Evaluating the role of "nothing to lose" attitudes on risky behavior in adolescence', *Social Forces* 80, 3: 1005–39.

He, K., E. Kramer, R. Houser, V. Chomitz, and K. Hacker. 2004. 'Defining and understanding healthy lifestyle choices for adolescents', *Journal of Adolescent Health* 4, 35: 26–33.

Hedberg, V., C. Bracken, and C. Stashwick. 1999. 'Long-term consequences of adolescent health behaviours: Implications for adolescent health services', *Adolescent Medicine* 10, 1:137–51.

Katz, C. 1998. 'Disintegrating developments: Global economic restructuring and the eroding ecologies of youth' in Skelton and Valentine (1998: 130–44).

Kershaw, P., L. Irwin, K. Trafford, and C. Hertzman. 2005. *The BC Atlas of Child Development*. Victoria: Human Early Learning Partnership and Western Geographical Press.

Laws, G. 1993. '"The land of old age": Society's changing attitudes towards urban built environments for old elderly people', *Annals, Association of American Geographers* 83: 672–93.

Laws, G. 1995. 'Embodiment and emplacement: Identities, representation and landscape in Sun City retirement communities', *International Journal of Aging & Human Development* 40: 253–80.

Lucas, T. 1998. 'Youth gangs and moral panics in Santa Cruz, California', in Skelton and Valentine (1998: 145–60).

McDaniel, S. 2000. '"What did you ever do for me?" Intergenerational linkages in a restructuring Canada', in Gee and Gutman (2000: 130–52).

Malbon, B. 1998. 'The Club', in Skelton and Valentine (1998: 266–86).

Massotti, P., R. Fick, A. Johnson-Masotti, and S. MacLeod. 1995. 'Healthy naturally occurring retirement communities: A low-cost approach to facilitating healthy aging', *American Journal of Public Health* 96: 1–7.

Matthews, H., M. Taylor, K. Sherwood, F. Tucker, and M. Limb. 2000. 'Growing up in the countryside: Children and the rural idyll', *Journal of Rural Studies* 16:141–53.

Moore, E., D. McGuinness, M. Pacey, and M. Rosenberg. 2000. *Geographic Dimensions of Aging: The Canadian Experience, 1991–1996*. Hamilton, Ont.: Social and Economic Dimensions of an Aging Population, Research Paper 23.

———— and M. Pacey. 2003. *Geographic Dimensions of Aging in Canada, 1991–2001*. Hamilton, Ont.: Social and Economic Dimensions of an Aging Population, Research Paper 97.

———— and M. Rosenberg, with D. McGuinness. 1997. *Growing Old in Canada: Demographic and Geographic Perspectives*. Ottawa and Toronto: Statistics Canada and ITP Nelson.

Morrongiello, B., and T. Dawber. 2004. 'Identifying factors that relate to children's risk-taking decisions', *Canadian Journal of Behavioural Science* 36, 4: 255–66.

Nairn, K., R. Panelli, ad J. McCormack. 2003. 'Destabilizing dualisms: Young people's experiences of rural and urban environments', *Childhood* 10, 1: 9–42.

Nayak, A. 2003. 'Through children's eyes': Childhood, place and the fear of crime', *Geoforum* 34: 303–15.

Next Gen. 2007. *A Gathering of Energy: The Voice of Edmonton's Next Generation*, Edmonton Next Gen Final Report. At: <www.edmontonnextgen.ca/finalreport/index.html>. (July 2009)

Next Generation Consulting. 2009. 'Attracting and retaining talent to Saskatoon—Discover, dream & design: Findings & recommendations'. At: <www.urbanplayground.ca>. (Aug. 2009)

No Loitering Technology. 2007. 'Noloitering.ca: Home of the Mosquito Ultra Sonic Teen Deterrent!', C.S. Environmental. At: <www.noloitering.ca>.

Ontario, Child and Family Services. 2009. Child and Family Services Act R.S.O. 1990, Chapter C.11 Section 79(5). At: <www.e-laws.gov.on.ca/html/statutes/english/elaws_statutes_90c11_e.htm>. (July 2009)

Pain, R., 2001. 'Gender, race, age and fear in the city', *Urban Studies* 38, 5–6: 899–913.

Public Health Agency of Canada (PHAC). n.d. Global Age-Friendly Cities Project. At: <www.phac-aspc.gc.ca/seniors-aines/pubs/age_friendly/>. (18 Sept. 2009)

Public Safety Canada. 2007. *Building the Evidence: Youth Gangs*. National Crime Prevention Centre. Ottawa, Catalogue no. 2007–TG–1.

Richter L. 2006. 'Studying adolescence', *Science* 312: 1902–5.

Roth, J., and J. Brooks-Gunn. 2003. 'Youth development programs: Risk, prevention and policy', *Journal of Adolescent Health* 32: 170–82.

Saanich, District of. n.d. 'District of Saanich, British Columbia, Canada', *World Health Organization's Global Age-Friendly Cities Project*. At: <www.gov.saanich.bc.ca/municipal/docs/pdfs/who_web_final.pdf>. (18 Sept. 2009)

Saewyc, E., C. Skay, S. Skay, S. Pettingell, E. Reis, L. Bearinger, M. Resnick, A. Murphy, and L. Combs. 2002. 'Hazards of stigma: The sexual and physical abuse of gay, lesbian, and bisexual adolescents in the United States and Canada', *Child Welfare* 2, (Mar.–Apr.): 195–213.

Saskatoon, City of. 2004. 'Bylaw No. 2954, A bylaw of The City of Saskatoon to restrict improper use of streets, lanes, parks and City property'. At: <www.saskatoon.ca/DEPARTMENTS/City%20Clerks%20Office/Documents/bylaws/2954.pdf>. (July 2009)

Shaienks, D., and T. Gluszynski. 2009. *Education and Labour Market Transitions in Young Adulthood*. Ottawa: Statistics Canada, Culture, Tourism and the Centre for Education Statistics Research Papers, Catalogue no. 81–595–M, no. 075.

Simpson, K., I. Janssen, W. Boyce, and W. Pickett. 2006. 'Risk taking and recurrent health symptoms in Canadian adolescents', *Preventive Medicine* 43: 46–51.

Skelton, T., and G. Valentine, eds. 1998. *Cool Places: Geographies of Youth Cultures*. London: Routledge.

Statistics Canada. 1996. *1996 Community Profiles*. At: <www12.statcan.ca/english/profil/PlaceSearchForm1.cfm?LANG=E>.

———. 2006. *2006 Community Profiles*. At: <www12.statcan.gc.ca/census-recensement/2006/dp-pd/prof/92-591/index.cfm?Lang=E>.

———. 2008. *Annual Demographic Estimates: Census Metropolitan Areas, Economic Regions and Census Divisions, Age and Sex 2002 to 2007*. Ottawa: Statistics Canada, Demography Division, Catalogue no. 91–214–X.

Thomas, M. 2005. 'Girls, consumption space and the contradictions of hanging out in the city', *Social & Cultural Geography* 6, 4: 587–605.

Totten, M. 2008. *Promising Practices for Addressing Youth Involvement in Gangs*. In support of the Strategy, Preventing Youth Gang Violence in BC: A Comprehensive and Coordinated Provincial Action Plan. Victoria: Ministry of Public Safety and Solicitor General of British Columbia.

United Jewish Communities. 2009. 'All about NORCs'. At: <www.norcs.com/page.aspx?id=119552>. (18 Sept. 2009)

Valentine, G. 2008. 'Living with difference: Reflections on geographies of encounter', *Progress in Human Geography* 32: 321–35.

———, T. Skelton, and D. Chambers. 1998. 'An introduction to youth and youth cultures', in Skelton and Valentine (1998: 1–33).

Vanderbeck, R. 2007. 'Intergenerational geographies: Age relations, segregation and re-engagements', *Geography Compass* 1, 2: 200–21.

Weber, A., J.-F. Boivin, L. Blais, N. Haley, and E. Roy. 2002. 'HIV risk profile and prostitution among female street youths', *Journal of Urban Health: Bulletin of the New York Academy of Medicine* 79, 4.

Wilson, D., and N. Ross. 2009. 'Place, gender and the appeal of video lottery terminal gambling: Unpacking a focus group study of Montreal youth', *Geojournal*.

World Health Organization (WHO). 2007. *Global Age-Friendly Cities: A Guide*. New York: WHO.

Wyman, D. 2008. 'From lagging to leading: Newfoundland and Saskatchewan dig into the resource boom', *Canadian Economic Observer* (May). Catalogue no. 11–010–XIB.

CHAPTER 22

Aboriginal People in Canadian Cities

Evelyn J. Peters

Perspectives of Indigenous[1] realities rarely focus on life lived in major metropolitan centres. Instead, the tendency is to frame rural and remote locations as central to the survival of Indigenous cultures and societies. However, recent censuses show that in many countries Indigenous populations are now mainly urban populations. In 2006, 84 per cent of Maori in New Zealand, 76.1 per cent of Indigenous Australians, and approximately 60 per cent of Native Americans lived in cities (Meredith, 2008; National Urban Indian Family Coalition, 2007; Taylor, 2009). Del Popolo et al. (2007), examining the 2000 round of censuses for 10 countries in Latin America, found lower Indigenous urbanization rates of about 40 per cent, but indicated that this rate had increased rapidly in the past decade. Increasingly, then, the Indigenous experience is an urban one.

According to the 2006 Canadian census, 53.2 per cent of Aboriginal people live in urban areas (Statistics Canada, 2008a). Aboriginal populations comprise the largest minority group in many prairie cities, and their social and economic conditions are central to the future of these cities. While researchers and popular accounts document Aboriginal urbanization and their 'migration' to urban areas, it is important to emphasize that many Aboriginal people are travelling within their traditional territories, and that the urban centres they migrate to are often built on traditional Aboriginal settlement places. In fact, the earlier absence of Aboriginal people in urban areas was the result of explicit policies to remove them from growing urban centres (Wilson and Peters, 2005).

Despite the importance of urban life to contemporary Aboriginal people and their importance in many Canadian cities, very little work in geography or in other disciplines explores these realities. In their 2000 review of over 200 geographic papers, books, and dissertations on Aboriginal peoples in the United States and Canada published in the previous decade, Rundstrom et al. (2000) cited only two papers that focused on urban situations. In a review for *The Canadian Geographer* in the same year, I similarly noted the paucity of attention to urban Aboriginal peoples by Canadian geographers (Peters, 2000a: 52–3). While a number of papers focusing on urban Aboriginal issues have appeared since then, they are still in the minority compared to work on Aboriginal people in history or in areas that are remote from the metropolitan centres of Canadian society. Without recognizing the realities of urban Aboriginal peoples, it is difficult to understand how many cities, particularly those on the prairies and in the North, work. Many Aboriginal people arrive in cities expecting their histories and their status as Aboriginal people to make a difference to their access to institutions and services. In this way, Aboriginal people are not like other urban residents. These characteristics pose unique challenges to policy and to urban theory.

The following sections examine three main aspects of urban Aboriginal life and their associated challenges to public policy. The first section examines assumptions about Aboriginal peoples' abandonment of reserves and rural communities and subsequent concentration in cities. Then the issues raised by an emphasis on Aboriginal cultures and rights to self-government in the city are explored. Finally, the challenges associated with the socio-economic diversity of urban Aboriginal communities are considered.

An explanation about terminology is in order at the outset. According to Section 35.1 of the Constitution Act, 1982, Aboriginal peoples are the 'Indian, Inuit and Métis peoples of Canada'. Most contemporary writers use 'First Nations' instead of 'Indian', although many official government policy documents and statistics still use the colonial terms 'North American Indian' and 'registered Indian'. The former refers to all individuals who identify themselves as such in censuses and other situations, while the latter refers to First Nations peoples who are registered pursuant to the Indian Act. It is important to recognize that these general terms homogenize particular Aboriginal identities and cultures. Many First Nations, for example, identify with their particular cultural groups, for example Mi'kmaq, Cree, Algonquin. The general term 'Inuit' includes a variety of different cultural groups. The term 'Métis' also has different and contested meanings. Sometimes it refers to descendants of First Nations and European people regardless of geographical location or historical period, and sometimes it refers specifically to descendants of the Métis Nation that emerged in the Prairie provinces, was centred on the Red River settlement in Manitoba, and was a consequence of relations between Aboriginal women and European fur traders.

Urbanization Processes

Expectations that Aboriginal people would leave their reserve and rural communities and move to urban areas have a long history in Canada. Until the mid-twentieth century, 'Indian' policy assumed that reserve lands would be abandoned as First Nations peoples assimilated (Tobias, 1983). Urbanization was seen as a partial solution for reserve and rural poverty, and the Department of Indian Affairs organized a relocation program in 1956 designed to assist First Nations to move to urban areas (Peters, 2002). Urbanization rates began to increase in 1950 and climbed sharply in the 1970s and 1980s (Kalbach, 1987: 102). Since then the proportion of Aboriginal people living in cities has increased for each census period. Table 22.1 describes changes in urbanization between 1996 and 2006 for individuals who identified themselves as Aboriginal in response to census questions. Between 1996 and 2006 the proportion of the Aboriginal population living on reserves and in rural areas declined from 53.2 to 46.8 per cent; concomitantly, the proportion living in urban areas increased from 46.8 to 53.2 per cent.

Urbanization rates vary for different legal categories of Aboriginal peoples. Métis people are most highly urbanized, with almost 70 per cent living in urban areas, as defined by Statistics Canada. While this is lower than the Canadian urbanization rate, which is above 80 per cent, it is substantially higher than the rate for other Aboriginal groups. The Inuit have the lowest proportion living in cities. Urbanization for First Nations increased from 40 per cent to almost 45 per cent between 1996 and 2006. However, the proportion of registered Indians (First Nations people registered under the Indian Act) living in cities remained relatively constant at close to 40 per cent. It is also important to note that for several census periods a number of reserves have refused to participate in census taking, with the result that the number and proportion of First Nations people, and especially

Table 22.1 Urbanization Patterns for Different Groups of Aboriginal People, 1996–2006

	Total Aboriginal Identity*	Métis**	First Nations	Inuit	Registered Indian
Total population, 1996	1,101,960	204,115	529,040	40,220	488,040
Total population, 2006	1,172,790	389,780	698,025	50,480	623,780
On reserve, 1996 (%)	32.8	1.5	47.4		46.0
On reserve, 2006 (%)	26.3	1.1	43.1	0.9	48.1
Rural, non-reserve, 1996 (%)	20.4	31.4	12.8		13.0
Rural, non-reserve, 2006 (%)	20.5	29.5	12.2	61.5	11.3
Urban, 1996 (%)	46.8	67.1	40.0	28.0	41.0
Urban, 2006 (%)	53.2	69.4	44.7	37.6	40.6

*The total Aboriginal identity population includes persons who reported more than one Aboriginal identity group and those who reported being a registered Indian and/or band member without reporting an Aboriginal identity.

**The counts for Métis, First Nations, and Inuit were based on single responses to census questions about Aboriginal ancestry. Some individuals identifying as Aboriginal claimed more than one Aboriginal identity.

Sources: Statistics Canada (2003, 2006).

registered Indians, living on reserves are underestimated. Norris and Clatworthy (2003: 54) suggested that if unenumerated reserve populations had been included in 1996, approximately 60 per cent of registered Indians would live on reserves.

Aboriginal urbanization rates cannot be explained solely by population measures such as fertility, mortality, and migration (Guimond, 2003). Legislation passed in 1985 allowed for the reinstatement of First Nations people who had lost their status through a variety of processes, and this accounts for part of the increase. However, another component of the increase is the result of individuals who did not identify as Aboriginal in previous census years now choosing to do so. Researchers have documented a similar phenomenon in the US, identifying as contributing factors US ethnic polities that embraced ethnic pride and Indian activism (Nagel, 1995). Siggner (2003) suggests that shifting attitudes towards Aboriginal peoples in Canada are important in changing patterns of self-identification. US researchers suggest that urban residents are more likely to reclaim their Indian identities, and similar processes may be occurring in Canadian cities.

An examination of absolute numbers shows that, despite increasing urbanization rates, the number of First Nations and Métis people in rural and reserve areas increased substantially between 1996 and 2006. In other words, reserves and rural Métis communities are not being depopulated as urban Aboriginal populations grow. Migration data from 1986 to 2001 show a substantial movement back and forth between cities and reserves and rural areas (Norris and Clatworthy, 2009). While some migrants may return to reserve and rural communities because of problems with urban life, researchers found that these communities of origin remain important for individuals (Wilson and Peters, 2005). Many Aboriginal people emphasize ties to the land as a continuing element of their cultural identity, and migration may be one reflection of these ties (Todd, 2000–1). Migration back to rural and reserve communities may also represent an attempt to maintain vital and purposeful community relationships.

The continuing importance of reserves and rural communities and the migration between them and urban areas has some important policy implications. Clearly, early assumptions about the depopulation of reserves and rural Métis communities have not materialized. The continuing connections many urban Aboriginal people have with these rural communities raise questions about the appropriate scale of policy interventions. A recent study of First Nations homelessness in Prince Albert, Saskatchewan, found that First Nations homelessness in urban areas was linked to the condition of reserve housing (Peters and Robillard, 2007). Initiatives only in cities or only in rural/reserve areas may not address significant factors at work among urban Aboriginal people. Migration back and forth also raises the importance of interface mechanisms that connect urban and rural populations. The appropriate geographies for addressing issues facing urban Aboriginal people may extend beyond urban boundaries.

Aboriginal Cultures and Rights in Urban Areas

A long history in Western thought sees urban and Aboriginal cultures as incompatible (Berkhoffer, 1979). Early writing about Aboriginal migrants to urban areas reflected ideas that Aboriginal cultures were an impediment to successful adjustment to urban society. As a result, services to Aboriginal migrants emphasized integration (Peters, 2002). Ideas about the incompatibility of urban and Aboriginal cultures have not dissolved. Presenters to the Urban Roundtable of the Royal Commission on Aboriginal Peoples (RCAP) talked about cities as 'an environment that is usually indifferent and often hostile to Aboriginal cultures' (RCAP, 1993: 2). An important challenge

for policy-makers is to recognize that vibrant Aboriginal cultures can contribute positively, not only to Aboriginal communities in cities, but also to non-Aboriginal communities. Researchers have found that skilled workers are attracted to particular cities both for economic and for social reasons. Cultural diversity was one of the elements that made cities attractive (Bradford, 2002). Aboriginal cultures have the potential to be part of this cultural diversity, contributing not only to elements such as art, dance, theatre, music, food, and media, but new perspectives on **governance**, a greater depth to urban histories, and different approaches to environmental issues and educational practices (Walker, 2008a).

Presenters to the public hearings of the RCAP saw vibrant urban Aboriginal cultures as important prerequisites for Aboriginal peoples' success in cities (Peters, 2000b). Presenters argued that Aboriginal cultures worked against alienation, provided important values for urban residents, and helped to build strong urban Aboriginal communities. Native studies professor David Newhouse (2000) argues that urbanization and the reinforcement of cultural identities are occurring simultaneously. Recognizing that Aboriginal cultures in urban areas are not simply transplanted non-urban cultures, Newhouse suggests that urban Aboriginal people are reformulating Western institutions and practices to support Aboriginal cultures and identities, so that Aboriginal people can survive as distinct people in contemporary societies. In this way urban Aboriginal people are participating in, to use Marshall Sahlins's phrase, the 'indigenization of modernity' (Sahlins, 2000: 495).

The Indigenous right to self-determination is an important component of Aboriginal identities, and as a result Aboriginal people distinguish themselves from other minority cultural groups, rejecting approaches based on **multiculturalism** (Johnson, 2008; Walker, 2008b). Increasingly, Aboriginal people in urban areas strive to express this right by attempting to define a role in policy-making and by determining and delivering programs and services to urban populations. The Canadian government has recognized the Aboriginal right to self-government as an inherent right of all Aboriginal people (Canada, 1997). However practical and political challenges face the implementation of the right to self-government in urban areas.

Not all urban Aboriginal people are represented by existing Aboriginal political bodies, and consequently they are denied a strong political voice in this arena. This lack of representation derives both from geographic variations in organizational structures and from differential access to these organizations for different categories of Aboriginal people. The Congress of Aboriginal Peoples (CAP), formerly the Native Council of Canada, has defined itself as the voice of off-reserve Aboriginal people in Canada since 1971. CAP currently has very limited support in the prairies and Ontario. The Assembly of First Nations' (AFN) stated interest in urban issues is relatively recent, sparked in part by assumption of responsibility for off-reserve member services by some First Nations governments and tribal councils. At present, though, the AFN has little involvement in urban First Nations issues. The Métis National Council, which is working to affirm Métis rights more generally, also does not have a strong presence in urban areas. Provincial First Nations and Métis organizations similarly do not have a well-developed focus on urban Aboriginal issues.

At the level of particular cities, however, some political organizations have emerged with a particular focus on urban Aboriginal issues. Arrangements are extremely variable geographically. One approach is an urban-focused organization that represents all Aboriginal people (First Nations, Métis, and Inuit) in the city, for example, the Aboriginal Council of Winnipeg (ACW). Founded in 1990 with the union of the Urban Indian Association and the Winnipeg Council of Treaty and Status Indians, it serves as a political and advocacy voice representing the interests of the Aboriginal community of Winnipeg regardless of their legal status. While there have been attempts to establish similar organizations in other cities, none appears to have been as stable as the ACW. In some cities, individual First Nations have set up urban offices to represent and provide services to their members in the city. In a few cities, separate organizations represent and provide services for urban First Nations and Métis people. For example, in Saskatoon, political representation for First Nations is provided by the Saskatoon Tribal Council, which also delivers a wide range of services, some to First Nations and some to all Aboriginal people living in Saskatoon. The Central Urban Métis Federation Inc. (CUMFI) is a Métis local of the Métis Nation Saskatchewan that provides a political voice as well as programs and services to Métis people in Saskatoon. Finally, the location of some reserves within city boundaries of a few urban areas means that members of those First Nations have political representation within the city.

In most cities, though, urban Aboriginal people do not have political representation. Even where there are urban-focused organizations such as the ones in Saskatoon, they do not provide a voice for all urban Aboriginal people, leaving out First Nations people who are not registered or who do not belong to First Nations that are represented on the tribal council. Moreover, as Walker (2003, 2006) has described, urban Aboriginal people, like many other urban residents, often do not participate in non-Aboriginal community organizations and consultations. This, in addition to their uneven access to political representative bodies, means that they often do not have a direct voice in public policy-making.

Most urban Aboriginal people experience some level of self-government through the emergence of a variety of urban Aboriginal organizations that deliver programs and services in a wide range of policy sectors. The emergence of many urban Aboriginal organizations marked the increase in migration to the city around 1950. Like ethnic groups in Canadian cities, Aboriginal people worked to develop formal and informal institutions to meet the need of migrants (Ouart, 2009). Unlike ethnic organizations, though, these institutions are often viewed as steps towards self-government. The Royal Commission suggested that the development of urban Aboriginal organizations created meaningful levels of control over some of the issues that affect urban Aboriginal residents' everyday lives (RCAP, 1996a: 584). Morgan's (2006: 373–4) description of the ambiguous status of urban Aboriginal organizations in Australia is relevant to the Canadian situation. Heavily dependent on state bureaucracies, they are viewed as symbolic spaces of self-determination:

> Indigenous leaders often construct their organisations as sequestered spaces, reclamations of pre-colonial decision-making processes. In doing this, the formal legal relationship of organisations to the state is buried amidst the cut and thrust of Aboriginal politics. The traditional social and political life of clan groups is reinvented as community politics in the era of self-determination.

Aboriginal-controlled social services generally have greater scope in delivering programs that incorporate Aboriginal principles, beliefs, and traditions, they create important employment opportunities for urban Aboriginal residents, and they result in significant economic benefits for Aboriginal communities (Hylton, 1999: 85–6).

Heather Howard-Bobiwash's (2003: 567) account of the strategies of Aboriginal women in Toronto between 1950 and 1975 describes how Aboriginal women in the emergent 'Native "middle class"' did not 'equate their relative economic success with assimilation', but rather 'utilized their class mobility to support the development of Native community organizations and promote positive pride in Native cultural identities' in the city. Similar to the emergence of 'pan-tribal' urban organizations in US cities (Straus and Valentino, 2001), small numbers and the nature of federal policies meant that early organizations in Canadian cities were 'pan-Aboriginal', in other words, they did not distinguish by legal status or different Aboriginal cultures. The emphasis was on programs to facilitate integration and address poverty (Peters, 2002). While some cultural programming was supported as a way of providing a familiar milieu for urban Aboriginal residents, there was no room at that time for programming based on rights or targeted to particular Aboriginal cultural groups.

Friendship Centres represent the earliest formal urban Aboriginal organizations. The first Friendship Centre opened in Winnipeg in April 1959. At present there are 117 Friendship Centres in cities throughout the country. Since the 1950s many other Aboriginal organizations have emerged to address a wide variety of issues. Some recent research in Edmonton, Winnipeg, and Saskatoon found that, in addition to Friendship Centres, non-profit housing organizations have existed for many decades (Peters, 2005). In larger cities, urban Aboriginal organizations are now found in a wide variety of policy sectors, including economic development, child, youth, family, and senior services, education, and justice, as well as in cultural fields (e.g., language, dance, theatre, music, and media).

As the urban Aboriginal populations grew in many cities, the emphasis increasingly shifted away from pan-Aboriginal approaches to particular cultures and histories. Especially in prairie cities, separate First Nations and Métis organizations emerged to provide services for community members and encourage a sense of strong cultural identity. A recent study of First Nations and Métis identities in Saskatoon found that most respondents identified with a particular First Nations origin (Cree, Saulteax, Dene, etc.) and that they also differentiated between western prairie Métis Nation identity and the definition of Métis as 'mixed-blood' (Peters, Maaka, and Laliberte, forthcoming). Because many urban Aboriginal people desire to practise their particular cultures, pan-Aboriginal organizations often attempt to involve elders and representatives from a variety of cultures in their programs and services. Nevertheless, urban Aboriginal people who live in cities where their culture of origin is a minority in the urban Aboriginal population find that, to access culturally specific services, they need to visit or return to their communities of origin (Wilson and Peters, 2005).

The size of the urban Aboriginal population influences the number and diversity of organizations a particular city can support. Table 22.2 provides information from the 2006 census for census metropolitan areas (CMAs) with Aboriginal populations of 10,000 or more. Winnipeg had the largest Aboriginal population (nearly 70,000) in 2006. The next largest urban Aboriginal populations were found in Edmonton and Vancouver, respectively. The composition of the urban Aboriginal population is also an important influence on the structure of urban Aboriginal organizations. In some prairie cities, half or more of this population self identifies as Métis, and in many of these cities Métis organizations have emerged to provide culturally specific services.

Obtaining predictable and adequate funding is a challenge to urban Aboriginal organizations. The federal government has maintained that it is responsible only for registered Indians and that these responsibilities are limited to reserve borders. The federal government has regarded all other Aboriginal people as a provincial responsibility. In turn, the provinces have argued that the federal government has responsibility for all Aboriginal people. The Royal Commission on Aboriginal Peoples (1996a: 538) noted that the result is a 'policy vacuum'. Although some urban programs have been established through federal, provincial, and municipal funding, these initiatives are unevenly distributed, with short-term and often limited funding (Hanselmann, 2001). Annual grants place an enormous administrative burden on Aboriginal organizations and limit their ability to build successful programs over time. Dependence on government funding creates concerns about sustainability and the ability to shape aspects of programming to reflect cultural needs (Graham and Peters, 2002).

The emphasis on culture, rights, and self-government raises some unique challenges for policy-making for urban Aboriginal peoples. The rationale for government funding for most urban Aboriginal programming is the amelioration of poverty and unemployment, not the promotion of Aboriginal rights to self-government. Although the RCAP (1996a) suggested several models through which urban Aboriginal people could have access to self-government, there has been virtually no progress on this front since the release of the report. While urban Aboriginal organizations have some scope for decision-making, it is severely circumscribed by program regulations and underfunding. Moreover, jurisdictional disputes between federal and provincial governments continue to generate complex and fragmented programming. The emphasis on the importance of culture and self-government results in a focus on particular cultural groups and histories, with the result that the configuration of Aboriginal organizations is complex and varies in different cities. Policy-making for urban Aboriginal communities, then, has some distinct elements associated with Aboriginal cultures, rights, and status.

Socio-economic Diversity in Urban Areas

Urban Aboriginal people are most often depicted as socio-economically marginalized populations, and most programs and services in urban areas address this marginalization. A literature beginning around the 1940s suggests that Aboriginal migration to cities creates challenges for the migrants, and that, because of their poverty, their movement into cities also challenges the capacity of municipal

Table 22.2 Cultural Characteristics of the Aboriginal Identity Population in Selected Cities, 2006

	Vancouver	Victoria	Edmonton	Calgary	Regina	Saskatoon	Winnipeg	Thunder Bay	Toronto	Ottawa–Gatineau	Montreal
Total Aboriginal	40,310	10,905	52,100	26,570	17,110	21,535	68,385	10,055	26,575	20,590	17,865
% of CMA	1.9	3.4	5.1	2.5	8.9	9.3	10.0	8.3	0.5	1.8	0.5
% First Nations	60.6	64.4	44.2	42.0	56.8	54.3	38.5	75.4	68.6	55.3	60.6
% Métis	38.9	34.3	54.6	57.0	43.0	45.4	61.0	24.1	30.1	41.0	36.0
% Inuit	0.5	1.3	1.2	1.0	0.1	0.3	0.5	0.5	1.3	3.7	3.4
% Total Aboriginal	100	100	100	100	99.9	100	100	100	100	100	100

Source: Statistics Canada (2008b).

governments to provide for them (Peters, 2000b). The homogenization of urban Aboriginal communities primarily in terms of their socio-economic status hides the socio-economic diversity of these people, both within and between cities.

A comparison of socio-economic indicators for Aboriginal and non-Aboriginal people in Canada's largest cities suggests urban Aboriginal people are, in aggregate, considerably less well-off than non-Aboriginal people (Table 22.3). The unemployment rate among urban Aboriginal people is more than double that of the non-Aboriginal population in most cities. Aboriginal people are under-represented in managerial, supervisory, and professional occupations. Median incomes are substantially lower for Aboriginal than for non-Aboriginal people. The proportion of parents or spouses (including common-law partners) who are lone parents is much higher among urban Aboriginal people than among non-Aboriginal people, and Aboriginal people are more likely to live in dwelling units that need major repairs. Aboriginal people are much less likely than others to have a university degree. In addition to these statistical descriptions, increased vulnerability to homelessness, addictions, gang membership, violence, and incarceration also mark the lives of many urban Aboriginal people (Cullhane, 2003; Grekul and Laboucane-Benson, 2008; Hanselmann, 2001; LaPrairie, 2002).

It is important to provide a context for urban Aboriginal socio-economic marginalization. Interviews by Silver et al. (2006: 11–15) with 26 urban Aboriginal community leaders identified a number of factors affecting Aboriginal people's economic situation in urban areas, including the failure of both residential and non-residential schools to provide them with the skills required in urban employment and the experience of racism, often on a daily basis, and the resulting destruction of self-esteem and identity. The urbanization of Aboriginal people in Canada occurred at a time when urban economies increasingly required education levels and skills that relatively few Aboriginal people received during their schooling. Challenges facing urban Aboriginal people also need to be situated within the larger context of colonization, which dispossessed them of their lands and languages, sent many children to residential schools, and impoverished reserves and rural Métis communities (RCAP, 1996b).

At the same time, there are considerable variations in Aboriginal socio-economic status between cities. Winnipeg, Regina, and Saskatoon have the highest Aboriginal unemployment rates and the largest differentials between Aboriginal and non-Aboriginal populations. Calgary (7.3 per cent) and Toronto (8.7 per cent) have the lowest Aboriginal unemployment rates. Generally, Aboriginal people are most disadvantaged in Thunder Bay, Winnipeg, Regina, and Saskatoon, and differences between Aboriginal and non-Aboriginal populations are smallest in Toronto, Ottawa–Gatineau, and Montreal. These differences illustrate the importance of 'place-based policies' that take into account local situations (Bradford, 2005).

There are also variations within urban Aboriginal populations. Siggner and Costa's (2005) study of urban Aboriginal people in large cities reports that between 1981 and 2001 school attendance among Aboriginal youth improved and rates of post-secondary completion increased. Employment rates improved in most cities, dependence on government transfer payments decreased, and there was a 28 per cent growth of Aboriginal income earners making $40,000 or more in annual income (adjusted for comparison with 1981). Silver et al. (2006: 18) identify three important strategies—adult education, employment in Aboriginal organizations, and involvement in their children's activities—that Aboriginal community leaders in Winnipeg used to gain socio-economic mobility. Focus groups with Aboriginal middle-class Toronto residents highlighted the importance of recognizing the socio-economic diversity of urban Aboriginal people (Urban Aboriginal

Table 22.3 Socio-Economic Characteristics, Aboriginal Identity Population, Selected Cities

		Vancouver	Victoria	Edmonton	Calgary	Regina	Saskatoon	Winnipeg	Thunder Bay	Toronto	Ottawa–Gatineau	Montreal
Unemployment rates	Aboriginal	10.7	8.8	9.8	7.3	13.8	14.6	11.3	9.3	8.7	8.8	8.7
	Non-Aboriginal	5.6	4.3	4.6	4.0	4.8	5.2	5.0	7.4	6.7	5.7	6.9
% Management occupations	Aboriginal	6.9	6.7	6.2	6.9	6.6	6.6	6.0	8.0	10.1	11.5	7.9
	Non-Aboriginal	11.3	10.7	10.5	11.1	9.6	9.0	9.3	8.0	11.6	11.1	10.3
Median individual income ($)	Aboriginal	18,203	18,132	19,735	24,329	17,842	16,480	18,620	16,724	24,138	25,838	20,362
	Non-Aboriginal	27,596	28,541	29,195	30,831	29,308	26,112	26,334	27,546	26,754	32,219	25,161
% of Parents or spouses who are lone parents	Aboriginal	20.7	19.3	23.5	17.5	33.4	29.7	26.9	23.1	17.7	16.5	15.7
	Non-Aboriginal	8.2	8.7	8.8	7.7	10.6	9.8	10.2	10.0	9.2	8.9	10.0
% Units needing repair	Aboriginal	14.0	11.9	11.6	9.0	12.8	11.8	14.2	12.6	11.3	12.7	14.1
	Non-Aboriginal	6.8	5.9	5.8	4.8	7.8	6.0	8.4	7.4	6.0	6.5	7.7
% with University degree or certificate	Aboriginal	8.7	10.1	5.9	9.4	8.4	11.2	8.3	12.8	12.8	15.5	10.3
	Non-Aboriginal	24.6	23.6	18.3	24.7	18.4	19.4	19.0	14.8	26.7	28.7	21.0

Source: Statistics Canada (2008b).

Strategy, 2005). The Toronto study indicated that middle-class urban Aboriginal people did not access Aboriginal organizations because these organizations were mostly service organizations focusing on a variety of social problems. Instead, they emphasized the need for Aboriginal language and cultural programs that addressed their aspirations.

It is important to recognize this socio-economic diversity in the design and delivery of programs and services. Emphasizing only marginalization can perpetuate negative stereotypes that view all urban Aboriginal people as destitute, deflect attention from the success that many urban Aboriginal residents experience, and create the perception that there is no capacity among Aboriginal people to contribute to both the Aboriginal and the non-Aboriginal communities in urban areas. An emphasis on socio-economic marginalization can also homogenize the design and delivery of programs and services within and between cities and fail to respond to the aspirations of Aboriginal peoples who do not fall into this group.

The homogenization of urban Aboriginal peoples as economically marginalized is linked to a history of concern about their settlement patterns by governments, social agencies, and a variety of academic researchers. With increasing numbers in cities, researchers and policy-makers expected that they would create poverty-stricken concentrations in inner-city areas (Peters, 2005). Some more recent government reports show that concerns remain about the possibility of concentration and its implications for Aboriginal people and for cities (Sgro, 2002: 21).

Concern about Aboriginal concentration has roots, explicitly or implicitly, in a literature on the emergence of inner-city ghettoes in large US cities. In the US, the concept of the 'underclass' was developed to describe intense poverty, its concentration over very large areas, and the resulting social isolation from mainstream society and values. Wilson (1987) described how the movement

of employment opportunities to suburban locations drew away working- and middle-class families, leaving behind an increasingly isolated and politically powerless 'underclass'. Inner-city disinvestment, coupled with growing welfare and illicit economies in response to the lack of employment opportunities, resulted in the collapse of public institutions and the development of a set of attitudes and practices of everyday life that isolated populations from the rest of urban society. Other work has explored 'neighbourhood effects', suggesting that concentration itself can have negative impacts, such as the development of antagonistic cultures and isolation from the rest of urban society (Buck, 2001).

Aboriginal people are over-represented among the urban poor in Canadian cities and are more likely than the non-Aboriginal population to live in poor urban neighbourhoods (Heisz and McLeod, 2004: 7). Yet researchers have consistently found relatively low rates of segregation among urban Aboriginal peoples (Walks and Bourne, 2006). A recent study (Peters and Starchenko, 2008) assesses levels of segregation and explores whether the neighbourhood effects identified with levels of segregation in US and ethnic studies are found with urban Aboriginal people in Canadian cities. The study employed the five classical indices identified by Massey and Denton (1988)—evenness, exposure, concentration, clustering, and centralization—and compared the Aboriginal identity population and the white Caucasian population as defined by the Canadian Employment Equity Act of 1986. The study found that Aboriginal settlement patterns were generally characterized by:

- even to moderately even distribution across census tracts (evenness);
- low likelihoods of exposure only to other Aboriginal people (exposure);
- relatively high levels of concentration, in other words, Aboriginal people occupy a relatively small amount of urban space (concentration);

- low likelihood that census tracts inhabited by Aboriginal people adjoin each other (clustering); and
- high tendencies to live close to the city centre in prairie cities and moderate tendencies to live near the city centre in the eastern cities and in Vancouver (centralization).

In combination, these dimensions of the settlement patterns of urban Aboriginal people suggest that characteristics linked to the emergence of **ghettos** in the United States—isolation from the mainstream as the result of large numbers of contiguous census tracts with a high proportion of the population being black or Hispanic (Hughes, 1990)—are not characteristics of the Canadian situation.

Recognizing the difficulty of identifying neighbourhood effects (Buck, 2001), the study used a stepwise regression procedure to assess the significance of the relationship between levels of segregation and neighbourhood outcomes, controlling for the mediating characteristics of the state of urban economies and the characteristics of individual residents. The analysis explored the three main effects that could be assessed using census data: the emergence of oppositional subcultures, decline in access to employment, and decline in the quality of services. The analysis attempted to assess whether different levels of segregation were associated with these effects.

There was no evidence of the emergence of oppositional cultures with higher levels of segregation, as indicated by the measure often used in the US literature as a proxy—the labour force participation rates of adult minority group males (i.e., black or Hispanic in the American context; Aboriginal in the Canadian context). In more segregated cities, fewer employed Aboriginal people lived and worked in the same census tract. This finding is consistent with the trend in many cities for employment opportunities to move out of inner-city areas. Also, no evidence suggested that higher levels of segregation were associated with declining service quality as measured by high

school dropout rates. Clearly, Aboriginal people are disproportionately unemployed and dropout rates are relatively high, but these characteristics do not seem to be associated with the emergence of segregated Aboriginal neighbourhoods.

Table 22.4 provides more information about Aboriginal settlement patterns over time, describing changes in patterns of urban Aboriginal residency in inner-city areas of the prairie cities of Edmonton, Saskatoon, Regina, and Winnipeg between 1996 and 2006. These are cities with relatively large numbers of Aboriginal residents representing a relatively large proportion of the total population. Inner-city areas were defined as those census tracts where the proportion of housing built before 1946 was twice the metropolitan average in 2006. Census tracts that did not meet these criteria but were surrounded on three sides by inner-city tracts were included to incorporate areas that have been redeveloped. Table 22.4 shows that between 1996 and 2006 the proportion of the inner-city Aboriginal population did not increase as much as the total Aboriginal population did. In other words, many of the individuals who contributed to the increase in urban Aboriginal populations over that decade are found in areas outside the core. This is reflected in the drop in the total proportion of the Aboriginal population living in the core in Edmonton, Saskatoon, and Winnipeg, with the proportion remaining the same in Regina.

Table 22.4 also addresses perceptions that urban Aboriginal residents are overwhelmingly represented in high-poverty neighbourhoods. Following Jargowsky's (1997) definition, high-poverty neighbourhoods are those where 40 per cent or more individuals have incomes below the poverty line. Clearly, Aboriginal people are over-represented in high-poverty neighbourhoods, compared to the proportion they represent of the total population of these cities. However, they comprise considerably less than half of the population in those areas; in other words, the majority of residents in high-poverty areas are non-Aboriginal. Moreover, a relatively

Table 22.4 Aboriginal Identity Population in Relation to Prairie Inner-City and High Poverty Areas, 1996–2006

	Edmonton	Saskatoon	Regina	Winnipeg
% increase in Aboriginal identity population, 1996–2006	39.4	33.3	25.7	49.5
% of inner city that is Aboriginal, 1996	6.6	11.2	14.8	15.7
% of inner city that is Aboriginal, 2006	6.5	13.2	18.6	18.5
% of total CMA Aboriginal population in inner city, 1996	22.6	19.5	16.8	51.6
% of total CMA Aboriginal population in inner city, 2006	16.9	16.8	16.9	40.4
% of CMA that is Aboriginal, 2006	5.1	9.3	8.9	10.0
% of high-poverty tract population that is Aboriginal, 2006	12.7	33.5	40.8	28.1
% of total Aboriginal population in high-poverty tracts, 2006	1.5	17.0	15.3	22.5

Source: Statistics Canada 2008b.

small proportion of the total Aboriginal population lives in poor neighbourhoods. Nonetheless, the over-representation of Aboriginal people in high-poverty areas is not trivial, and programs and services need to address this over-representation. At the same time, it is important to recognize that urban Aboriginal people are not homogeneously poor or residing in core areas or in neighbourhoods with a concentration of Aboriginal people.

The fact that Aboriginal people are relatively dispersed in urban areas indicates that Aboriginal issues in cities are not only inner-city issues, and raises questions for targeting programs and services. Located in inner-city areas, programs and services may not be accessible to middle- and upper-class Aboriginal residents living in more suburban areas. If programs and services for Aboriginal people are found only in inner-city areas, they can lead to the inadvertent clustering of Aboriginal people in those areas. Finally, initiatives that target areas of high poverty need to recognize that the majority of people living in those areas are not Aboriginal people.

Conclusion

The analysis in this chapter suggests a number of challenges for public policy concerning Aboriginal urbanization. First, the perception that Aboriginal people are leaving reserves and rural communities and moving to urban areas needs to be carefully examined. Census data show that an increasing proportion of the Aboriginal population lives in cities, but there are variations by Aboriginal group, and under-enumeration, back-and-forth movement between cities and reserves, and changing patterns of self-identification complicate that picture. While the proportion of Aboriginal people living in the city has increased according to census data, rural areas and particularly reserves continue to have importance in many Aboriginal peoples' lives, raising questions about the appropriate geographies for public policy intervention.

Aboriginal cultures can facilitate successful adaptation to urban life for urban Aboriginal residents. These cultures can also contribute to the richness of urban cultural life. Like other cultural groups, urban Aboriginal cultures are dynamic and creative in attempts to adapt values and practices to a new environment. Urban Aboriginal cultures are reflected in, and reinforced by, a variety of urban Aboriginal organizations, and these organizations continue to be the main way urban Aboriginal people experience self-government. While these organizations face challenges related to cultural complexity and unstable government funding, they are an important element of contemporary urban Aboriginal communities. Challenges for public policy emerging from the importance of Aboriginal cultures in urban areas include promoting these cultures as contributors to cultural liveliness in cities, recognizing the diversity of urban Aboriginal cultures, creating transparent, stable, and adequate funding, and finding ways to recognize Aboriginal rights to self-government.

Many depictions of urban Aboriginal people view them primarily as poor and service-dependent. Clearly, poverty is a serious issue in urban Aboriginal communities, and programs and services addressing poverty amelioration are essential. However, a focus only on marginalization can hide the history of colonialism that failed in the economic and cultural development of reserves and rural communities, education systems that did not teach Aboriginal children the skills necessary for economic success in contemporary economies, and racism and discrimination in urban areas. Emphasizing marginalization can also hide the successes of urban Aboriginal people. Finally, despite continuing concerns about Aboriginal peoples' concentration in inner-city areas, their settlement patterns are quite different from those of minority groups in US inner cities. Inner-city areas continue to have Aboriginal and non-Aboriginal residents, and, increasingly, Aboriginal people are found in all areas of cities. These settlement patterns challenge policy-makers to develop programs and services for diverse and dispersed populations.

Aboriginal cultures have the potential to make unique and valuable contributions to the cultural life of cities. Their status as Aboriginal peoples with Aboriginal rights makes them unique from other urban residents. The poor socio-economic conditions of many urban Aboriginal people and the often negative relationships between them and majority populations mean that the challenges associated with Aboriginal urbanization need to be taken very seriously. Relatively little research, however, has focused on urban Aboriginal people. As a result, academics and policy-makers assume that studies of other cultural groups (ethnic groups, immigrants, US inner-city populations, marginalized peoples) are directly applicable to the situation of urban Aboriginal people. It is important to examine the demographics and experiences of urban Aboriginal people in order to evaluate whether these models are appropriate. This work adds to our understanding of the social, economic, and political dynamics of these cities.

Review Questions

1. Why do Aboriginal peoples move to cities?
2. What are the situations Aboriginal peoples face in cities in terms of access to services and cultural identity?
3. What urban problems are specific to Aboriginal peoples? What can be done about them—by Aboriginal peoples themselves or by governments?

Note

1. I use the terms 'Indigenous' and 'Aboriginal' to refer to the original inhabitants of North America and their descendants. I capitalize the terms 'Indigenous' and 'Aboriginal' in the same manner that words such as 'European' and 'American' are capitalized when referring to specific peoples (Johnson et al., 2007).

References

Berkhoffer, R.F. 1979. *The White Man's Indian: Images of the American Indian from Columbus to the Present*. New York: Vintage.

Bradford, N. 2002. *Why Cities Matter: Policy Research Perspectives for Canada*. CPRN Discussion Paper No. F/23. Ottawa: Canadian Policy Research Networks.

———. 2005. *Place-based Public Policy: Towards a New Urban and Community Agenda for Canada*. Ottawa: Canadian Policy Research Network.

Buck, N. 2001. 'Identifying neighbourhood effects on social exclusion', *Urban Studies* 38: 2251–75.

Canada. 1997. *Gathering Strength: Canada's Aboriginal Action Plan*. Ottawa: Minister of Public Works and Government Services Canada.

Cullhane, D. 2003. 'Their spirits live within us: Aboriginal women in Downtown Eastside Vancouver emerging into visibility', *American Indian Quarterly* 27: 593–606.

Del Popolo, F., A.M. Oyarce, B. Ribotta, and J. Rodríguez. 2007. *Indigenous Peoples and Urban Settlements: Spatial Distribution, Internal Migration and Living Conditions*. Santiago de Chile: Latin American and Caribbean Demographic Centre Population Division. At: <www.eclac.org/publicaciones/xml/9/32549/PyD78-final.pdf>. (June 2009)

Graham, K., and E.K. Peters. 2002. 'Aboriginal communities and urban sustainability', in F.L. Seidle, ed., *The Federal Role in Canada's Cities: Four Policy Perspectives*. Ottawa: Canadian Policy Research Networks.

Grekul, J., and P. LaBoucane-Benson. 2008. 'Aboriginal gangs and their (dis)placement: Contextualizing recruitment, membership, and status', *Canadian Journal of Criminology and Criminal Justice* 50: 59–82.

Guimond, E. 2003. 'Fuzzy definitions and population explosion: Changing identities of Aboriginal groups in Canada', in D. Newhouse and E. Peters, eds, *Not Strangers in These Parts: Aboriginal People in Cities*. Ottawa: Policy Research Initiative.

Hanselmann, C. 2001. *Urban Aboriginal People in Western Canada*. Calgary: Canada West Foundation.

Heisz, A., and L. McLeod. 2004. *Low Income in Census Metropolitan Areas, 1980–2000*. Catalogue no. 75–001–XIE. Ottawa: Statistics Canada.

Howard-Bobiwash, H. 2003. 'Women's class strategies as activism in Native community building in Toronto', *American Indian Quarterly* 27: 566–82.

Hughes, M.A. 1990. 'Formation of the impacted ghetto: Evidence from large metropolitan areas, 1970–1980', *Urban Geography* 11: 265–84.

Hylton, J.H. 1999. 'The case for self-government: A social policy perspective', in J.H. Hylton, ed., *Aboriginal Self-Government in Canada*. Saskatoon: Purich Publishing.

Johnson, J.T. 2008. 'Indigeneity's challenges to the white settler-state: Creating a third space for dynamic citizenship', *Alternatives* 33: 29–52.

———, G. Cant, R. Howitt, and E.J. Peters, eds. 2007. *Geographical Research: Journal of the Institute of Australian Geographers* 45, 2: 117–210.

Jargowsky, P.A. 1997. *Poverty and Place: Ghettos, Barrios and the American City*. New York: Russell Sage Foundation.

Kalbach, W.E. 1987. 'Growth and distribution of Canada's ethnic populations, 1871–1981', in L. Dreidger, ed., *Ethnic Canada: Identities and Inequalities*. Toronto: Copp Clark Pitman.

LaPrairie, C.P. 2002. 'Aboriginal over-representation in the criminal justice system: A tale of nine cities', *Canadian Journal of Criminology* 44: 181–208.

Massey, D.S., and N. Denton. 1988. 'The dimensions of residential segregation', *Social Forces* 67: 281–315.

Meredith, P. 2008. 'Urban Māori', *Te Ara—The Encyclopedia of New Zealand*. At: <www.TeAra.govt.nz/NewZealanders/MaoriNewZealanders/UrbanMaori/en>. (June 2009)

Morgan, G. 2006. 'Aboriginal politics, self-determination and the rhetoric of community', in S. Herbrechter and M. Higgins, eds, *Returning (to) Communities: Theory, Culture and Political Practice of the Communal*. New York: Rodopi.

Nagel, J. 1995. 'American Indian ethnic renewal: Politics and the resurgence of identity', *American Sociological Review* 60: 947–65.

National Urban Indian Family Coalition. 2007. *Urban Indian America*. Seattle: National Urban Indian Family Coalition. At: <www.nuifc.net/programs/files/NUIFC_PUBLICATION_FINAL.pdf>. (June 2009)

Newhouse, D.R. 2000. 'From the tribal to the modern: The development of modern Aboriginal societies', in R.F. Laliberté, P. Settee, J.B. Waldram, R. Innes, B. Macdougall, L. McBain, and F.L. Barron, eds, *Expressions in Canadian Native Studies*. Saskatoon: University of Saskatchewan Extension Press.

Norris, M.J., and S. Clatworthy. 2009. 'Urbanization and migration patterns of Aboriginal peoples in Canada: A half century in review (1951–2006)', paper presented at the workshop on Indigenous Urbanization Internationally, Saskatoon, 29–30 Oct.

Ouart, P. 2009. 'The Saskatoon Indian and Métis Friendship Centre and the Community Liaison Committee: Laying the Groundwork for Self-Government, 1968–1982', MA thesis, University of Saskatchewan.

Peters, E.J. 2000a. 'Aboriginal people and Canadian geography: A review of the recent literature', *Canadian Geographer* 44: 44–55.

———. 2000b. 'Aboriginal people in urban areas', in D. Long and O.P. Dickason, eds, *Visions of the Heart: Canadian Aboriginal Issues*. Toronto: Harcourt Canada.

———. 2002. '"Our city Indians": Negotiating the meaning of First Nations urbanisation in Canada, 1945–1975', *Historical Geography* 30: 75–92.

———. 2005. 'Indigeneity and marginalisation: Planning for and with urban Aboriginal communities in Canada', *Progress in Planning* 63: 325–404.

———, R. Maaka, and R. Laliberte. Forthcoming. 'I'm sweating with Cree culture not Saulteaux culture and there goes the beginning of pan-Indianism', in A. Romanic and F. Trovato, *Urban Aboriginal Cultural Identities*. Toronto: University of Toronto Press.

——— and V. Robillard. 2007. 'Urban hidden homelessness and reserve housing', in J.P. White, P. Maxim, and D. Beavon, eds, *Aboriginal Policy Research*. Toronto: Thompson Educational Publishing.

——— and O. Starchenko. 2008. *Neighbourhood Effects and Levels of Segregation of Aboriginal People in Large Cities in Canada*. Ottawa: Canada Mortgage and Housing Corporation.

Royal Commission on Aboriginal Peoples. 1993. *Aboriginal Peoples in Urban Centres*. Ottawa: Minister of Supply and Services.

———. 1996a. *Perspectives and Realities: Report of the Royal Commission on Aboriginal Peoples*, vol. 4. Ottawa: Minister of Supply and Services.

———. 1996b. *Looking Forward, Looking Back. Report of the Royal Commission on Aboriginal Peoples*, vol. 1. Ottawa: Minister of Supply and Services.

Rundstrom, R., D. Deur, K. Berry, and D. Winchell. 2000. 'Recent geographical research on Indian and Inuit in the United States and Canada', *American Indian Culture and Research Journal* 24: 85–110.

Sahlins, M. 2000. *Culture in Practice: Selected Essays*. New York: Zone Books.

Sgro, Judy. 2002. *A Vision for the 21st Century*. Ottawa: Prime Minister's Caucus Task Force on Urban Issues, Final Report.

Siggner, A. 2003. 'The challenge of measuring the demographic and socio-economic condition of the urban aboriginal population', in D. Newhouse and E. Peters, eds, *Not Strangers in These Parts: Aboriginal People in Cities*. Ottawa: Policy Research Initiative.

——— and R. Costa. 2005. *Aboriginal Conditions in Census Metropolitan Areas, 1981–2001*. Catalogue no. 89–613–MIE, No. 008. Ottawa: Statistics Canada.

Silver, J., P. Ghorayshi, J. Hay, and D. Klyne. 2006. *In a Voice of Their Own: Urban Aboriginal Community Development*. Winnipeg: Canadian Centre for Policy Alternatives.

Statistics Canada. 2006. *Aboriginal Identity, Area of Residence, Age Groups and Sex for the Population of Canada, Provinces and Territories, 2006 Census of Population*. Catalogue no. 97–558–XCB2006006. Ottawa: Statistics Canada.

———. 2008a. *Aboriginal Peoples in Canada in 2006: Inuit, Métis and First Nations, 2006 Census*. At: <www12.statcan.ca/english/census06/analysis/aboriginal/index.cfm>. (June 2009)

———. 2008b. *Aboriginal Population Profile, 2006 Census of Population*. Catalogue no. 92–594–XWE. Ottawa: Statistics Canada. At: <www12.statcan.ca/english/census06/data/profiles/aboriginal/index.cfm?Lang=E>. (June 2009)

Straus, A.T., and D. Valentino. 2001. 'Retribalization in urban Indian communities', in L. Susan and K. Peters, eds, *American Indians and the Urban Experience*. New York: Altamira Press.

Taylor, J. 2009. 'Urbanisation and Indigenous Australians', lecture at University of Saskatchewan, 18 Mar.

Tobias, J.L. 1983. 'Protection, civilization, assimilation: An outline history of Canada's Indian policy', in A.L. Getty and A.S. Lussier, eds, *As Long as the Sun Shines and Water Flows: A Reader in Canadian Native Studies*. Vancouver: University of British Columbia Press.

Todd, R. 2000–1. 'Between the land and the city: Aboriginal agency, culture, and governance in urban areas', *London Journal of Canadian Studies* 16: 48–66.

Urban Aboriginal Strategy. 2005. *Urban Aboriginal Task Force Progress Report 1*. At: <www.servicecanada.gc.ca/eng/on/epb/uas/reports/uatfphase1.shtml#intro>. (June 2009)

Walker, R. 2003. 'Engaging the urban Aboriginal population in low-cost housing initiatives: Lessons from Winnipeg', *Canadian Journal of Urban Research* 12: 99–118.

———. 2006. 'Searching for Aboriginal/Indigenous self-determination: Urban citizenship in the Winnipeg low-cost housing sector, Canada', *Environment and Planning A* 38: 2345–63.

———. 2008a. 'Improving the interface between urban municipalities and Aboriginal communities', *Canadian Journal of Urban Research* 17: 20–36.

———. 2008b. 'Aboriginal self-determination and social housing in urban Canada: A story of convergence and divergence', *Urban Studies* 54: 185–205.

Walks, R.A., and L.S. Bourne. 2006. 'Ghettos in Canada's cities? Racial segregation, ethnic enclaves and poverty concentration in Canadian urban areas', *Canadian Geographer* 50: 273–97.

Wilson, K., and E.J. Peters. 2005. '"You can make a place for it": Remapping urban First Nations spaces of identity', *Society and Space* 23: 395–413.

Wilson, W.J. 1987. *The Truly Disadvantaged: The Inner City, the 'Underclass' and Public Policy*. Chicago: University of Chicago Press.

CHAPTER 23

The Built Environment and Obesity: Trimming Waistlines through Neighbourhood Design

Jason Gilliland

Introduction

Obesity is a burgeoning public health problem in Canada. One in four Canadian children aged 2–17 and nearly six out of 10 adults (18 years and older) are overweight or obese (Shields and Tjepkema, 2006). Childhood obesity rates in Canada have nearly tripled over the past 25 years (Shields, 2005). Similar increases in obesity rates have been identified in the United Kingdom, the United States, and other developed nations. The increased prevalence of obesity has been linked to the concurrent rise of many socio-psychological and physical health problems. Not surprisingly, policy-makers are struggling to find effective and cost-efficient methods for obesity prevention.

It is understood that obesity results from an imbalance between energy intake and energy expenditure, in other words, by eating too much and exercising too little. While health researchers have focused traditionally on individual-level factors, such as genetic predisposition, recent work suggests that environmental factors also may influence body weights and overall well-being through the opportunities they provide for promoting or hindering healthy behaviours. Recent literature suggests that the ways in which cities have been designed and constructed over the past half-century or so, with increasing automobile dependence and limited opportunities to walk

for utilitarian purposes, may be at least partly to blame for the rise in obesity rates in North America (see Bunting and Filion, Chapter 2; Filion and Bunting, Chapter 3; Perl and Kenworthy, Chapter 11). Furthermore, since dietary intake and energy expenditure are modifiable behaviours, it is crucial to understand how environments influence behaviour, as environmental interventions may be among the most effective strategies at mitigating the obesity epidemic that now plagues many developed nations. Unfortunately, the paucity of Canadian data on environmental determinants of physical activity and food consumption patterns limits the development of evidence-based recommendations for policy and practice. But this is a very new and exciting area of research and there is every reason to believe that policy-related measures might see implementation within the next few years.

Following a brief overview of the obesity problem in Canada, this chapter examines the evidence on the environment–obesity link and then discusses potential interventions for building healthier communities.

The Problem: Obesity 'Epidemic' in Canada

Obesity is an important risk factor for a number of chronic diseases, and it entails substantial financial burden on Canada's health-care system. The percentage of Canadian adults considered obese increased from 10 per cent in 1970 to 23 per cent in 2004—from 8 to 23 per cent in men and from 13 to 22 per cent in women (Luo et al., 2007).[1]

Obesity has been linked to many physical health problems, including type-2 diabetes, hypertension, heart disease, sleep apnea, pulmonary diseases, and certain cancers (Marcus et al., 1996; Fagot-Campagna, 2000; Figueroa-Colon et al., 1997; Figueroa-Munoz et al., 2001). Obesity has also been linked to increased prevalence of socio-psychological problems such as discrimination, behavioural problems, negative self-esteem, anxiety, and depression (Strauss and Pollack, 2003). Furthermore, recent estimates show that obesity places a significant financial burden on the Canadian health-care system, responsible for about $4.3 billion per year in direct and indirect costs (Katzmarzyk and Janssen, 2004).

Distinct regional variations in obesity exist across Canada. In 2004, 34 per cent of adults in Newfoundland and Labrador were considered obese, compared to only 19 per cent of adults in British Columbia (Shields and Tjepkema, 2006). Perhaps even more dramatic are the variations with respect to degree of urbanization. In general, Canadian adults who reside in census metropolitan areas are much less likely to be obese than those living outside of CMAs (20 per cent versus 29 per cent). Furthermore, in CMAs with 2 million or more people, only 17 per cent of adults were obese. Adult obesity prevalence was 24 per cent in CMAs with a population of 100,000 to 2 million, and 30 per cent in urban centres with a population of 10,000 to 100,000. A closer look at individual cities indicates that St John's has the highest rates of adult obesity at 36.4 per cent, and Vancouver the lowest at 11.7 per cent. If we consider adults who are also overweight, then Hamilton takes the title as Canada's 'fattest city', with 74.3 per cent of adults obese or overweight, and Halifax has the lowest levels of obese/overweight adults at 47.8 per cent (see Table 23.1). A simple analysis of available

data suggests that there may be a negative relationship between city size and levels of obesity (r = −0.53) and overweight (r = −0.42). Such correlations have led some researchers in the US and Canada to argue that differences in obesity prevalence may be due to the respective level of 'urban sprawl' (see also Donald and Hall, Chapter 16).

Exploring the Link between the Built Environment and Obesity

Although it is widely accepted that today's obesity epidemic is due to changing behaviours associated with energy intake and energy expenditure, the underlying determinants of those obesity-related behaviours are not as straightforward. Public health promoters and researchers have long argued that people's abilities to lead healthy lives depend on characteristics of the environments in which they live, work, learn, and play. In one of the earliest studies to propose an ecologic paradigm for understanding obesity, Egger and Swinburn (1997) posited that increasing rates of obesity should be considered as a normal response to an abnormal environment. Following this argument, an increasing amount of research over the past decade has been aimed at exploring if, what, and how obesity-related, or 'obesogenic', behaviours are related to the built environment. Simply defined, the built environment refers to those components of our physical surroundings constructed by humans, such as buildings, parks, and transportation networks; its design can have a significant effect on many lifestyle decisions and behaviours related to obesity, as conceptualized in Figure 23.1.

The Role of the Built Environment on Energy Intake

Eating behaviours have a significant impact on the overall health of an individual. Four of the 10 leading causes of death in the United States are diseases directly related to diet (Zenk et al., 2005).

Table 23.1 Proportion of Adults Who Are Overweight or Obese, Canadian CMAs, 2004

CMA	Population (000s)	Overweight (BMI >25)*	Obese (BMI >30)
Hamilton	452	74.3	34.6
Kingston	81	70.1	28.9
St John's	159	70.0	36.4
St Catharines–Niagara	346	69.3	23.1
Saint John	124	68.9	34.7
Saskatoon	147	64.5	27.0
Gatineau	199	63.6	n/a
Oshawa	208	63.5	29.6
Victoria	251	62.6	19.0
Kitchener	450	62.3	30.7
Edmonton	946	62.2	20.1
Greater Sudbury	72	62.1	26.1
Ottawa	636	62.0	19.7
London	470	61.6	26.6
Thunder Bay	185	60.0	32.6
Abbotsford	110	58.3	25.0
Winnipeg	525	58.2	25.2
Regina	151	58.1	31.8
Quebec	552	56.8	17.3
Trois-Rivières	139	56.6	n/a
Windsor	99	56.5	33.2
Calgary	765	53.8	25.7
Sherbrooke	97	52.4	n/a
Saguenay	141	52.3	18.9
Vancouver	1,720	51.8	11.7
Montreal	2,577	51.6	21.2
Toronto	3,772	50.9	15.6
Halifax	284	47.8	18.4

*Includes obese

Source: Statistics Canada, 2004 Canadian Community Health Survey.

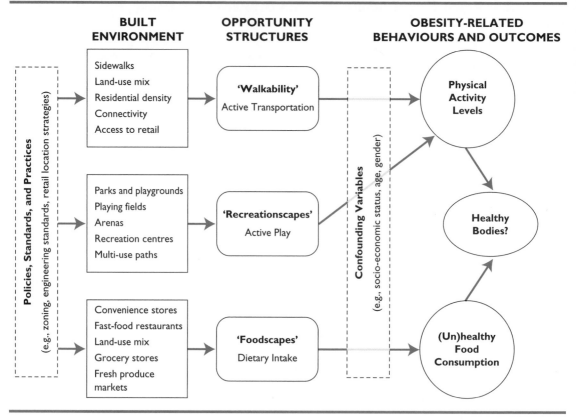

Figure 23.1 Conceptual diagram illustrating the potential influence of the built environment on obesity

The majority of obesity-related health problems are caused by a diet with low fruit and vegetable consumption and high fat and sugar content (WHO, 2003). Dietary habits have changed significantly in recent decades, as a greater proportion of food consumption is taking place outside of the home (Cummins and Macintyre, 2006; Nielsen et al., 2002; Blay-Palmer, Chapter 24). The growing belief among planning and health researchers is that the changing dietary habits of North Americans are associated with major shifts in the food retail environment, which has limited the access to healthy food retailers such as supermarkets and increased the availability of unhealthy foods through the proliferation of 'junk-food' outlets.

The Potential Influence of 'Food Deserts' on Diet and Health

The recent emergence of '**food deserts**', or areas of cities with relatively poor access to healthy and affordable food, has been attributed to the ongoing suburbanization of food retailers in North America and the United Kingdom. In many US cities, low-income, minority-dominated urban neighbourhoods are often considered food deserts, as grocers have vacated these communities in recent decades (Eisenhauer, 2001; Weinberg, 2000; Zenk et al., 2005). Several studies in the UK have also identified food deserts in socially deprived areas, typically local authority housing estates in suburban rather than inner-city locations (Clarke et al., 2002;

Sooman et al., 1993; Wrigley, 2002). Canadian findings are mixed. In London, Ontario, Larsen and Gilliland (2008) identified low-income, inner-city neighbourhoods as 'food deserts' using a variety of GIS-based techniques of network analysis to assess supermarket accessibility within a 10–15-minute walk or public transit ride. Moreover, they showed how spatial inequalities in access to supermarkets in London had increased over time: in 1961, more than 75 per cent of London's inner-city population lived within a kilometre of a supermarket; by 2005, that number was less than 20 per cent (see Figure 23.2). On the other hand, Apparicio and colleagues (2007) found that food deserts are 'missing'

in Montreal and access to healthy food is not a major issue for Montreal residents. Furthermore, Smoyer-Tomic and colleagues (2006) found that, on average, low-income neighbourhoods near Edmonton's city centre actually had the best supermarket access in the city; however, certain high-needs neighbourhoods still had poor access.

Grocery retailing practices in North American cities have changed repeatedly over the past century; small independent food markets gave way to small chain food stores, which were then superseded by larger supermarkets. Since the early 1990s, the trend has been to erect giant superstores on suburban lands that 'offer more land for parking,

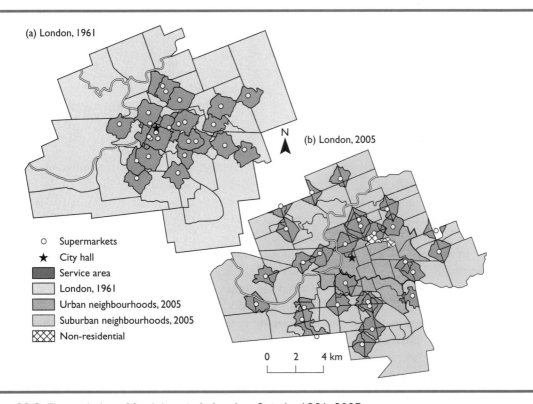

Figure 23.2 The evolution of food deserts in London, Ontario, 1961–2005

The location of supermarkets and 500-metre network service areas reveals retail migration out of the urban core.

Source: Larsen and Gilliland (2008).

easier loading and unloading for trucks, convenient access to highways, and a development context for much larger stores' (Pothukuchi, 2005: 232). These superstores have created new challenges for urban planners and engineers around traffic, parking, public transit, and other issues that occur in the planning review process. The abandonment of smaller inner-city supermarkets has also presented challenges for public health policy-makers due to the uneven distribution of healthy, affordable food opportunities.

Recent empirical studies have identified high rates of obesity and individuals who are overweight among residents living in neighbourhoods without a supermarket (Morland et al., 2006). Since supermarkets are the primary source of healthy and affordable food for most people, poor access to a supermarket makes it harder to maintain a healthy diet. A large US study involving over 10,000 respondents across four states (Maryland, North Carolina, Mississippi, and Minnesota) discovered that white Americans' fruit and vegetable consumption increased by 11 per cent with the presence of one or more supermarkets in their census tract, and black Americans' fruit and vegetable intake increased by 32 per cent for each additional supermarket in the census tract (Morland et al., 2002). Using a 'quasi-experimental' research design, Wrigley and colleagues (2002) reported that improved retail food access in an under-served area of Glasgow had a positive impact on fruit and vegetable consumption among residents who shopped at the new supermarket. Empirical research in the UK, US, and Canada indicates that residents of food deserts pay higher prices for groceries at small food shops and convenience stores where healthy foods are also scarcer (Chung and Myers, 1999; Kayani, 1998; Larsen and Gilliland, 2008). In addition, Bertrand and colleagues (2008) found that 40 per cent of Montreal residents without an automobile had poor access to fresh fruits and vegetables. Such findings remind us that 'access' to healthy food is not only a geographic issue but is fundamentally economic, as food prices and transportation costs also influence healthy food choices.

In North American cities today, the reality is that most new grocery superstores are found, along with other 'big-box' outlets, in expansive retail centres, which are almost always built in areas beyond walking distance from residential land uses, which essentially makes them accessible only to consumers with automobiles (see Figure 23.3).

The Built Environment and 'Junk-Food' Consumption

It is commonly believed that fast food is contributing to an overall increase in childhood obesity. The average American adolescent has been reported to frequent fast-food restaurants twice a week (about 100 times per year), accounting for approximately one-third of his or her away-from-home meals (Guthrie et al., 2002). Children are now consuming more fat in their diets than ever before, and this rate continues to rise as children enter adolescence (Crownfield, 2004).

Changing neighbourhood food environments may be partly responsible for the rise in obesity; easy access to 'junk-food' outlets such as fast-food restaurants has a significant influence on dietary behaviours. Population health studies have suggested that obesity may be more prevalent among the poor than the rich in industrialized countries because the poor are more likely to be exposed to—or constricted in their choice to—fast food (Story et al., 2002). One study in Melbourne, Australia, for example, showed that the density of fast-food restaurants is 2.5 times higher in low- versus high-status neighbourhoods (Reidpath, 2002). Likewise, in Edmonton the odds of being exposed to fast-food outlets are greater in socially disadvantaged areas with more Aboriginal residents (Peters, Chapter 22), lone parents (Townshend and Walker, Chapter 8), renters (Walker and Carter, Chapter 20), and low-income households (Walks, Chapter 10) (Smoyer-Tomic et al., 2008). Positive associations have also been identified between the density

Figure 23.3 A 'club' format food retailer: automobile-oriented development on the rural fringe of London, Ontario. (J. Gilliland)

of fast-food restaurants and adult obesity rates at the state level (Chou et al., 2004; Maddock, 2004).

For children and youth, the food environment around their schools has a significant impact on diet, and in turn, may influence obesity levels (French et al., 2001; see also Rosenberg and Wilson, Chapter 21). Ongoing research in London, Ontario, indicates that fast-food restaurants and convenience stores are significantly clustered around elementary schools, particularly those in low-income neighbourhoods (see Figure 23.4); furthermore, increased proximity and density of junk-food opportunities within walking distance (500 metres) around the school and/or home were linked with increased purchasing of junk food (Gilliland and Tang, 2007). These findings support the call for additional research and policy interventions to address the proliferation of junk food in youth-oriented neighbourhoods.

Role of the Built Environment on Energy Expenditure

It is well known that a physically active lifestyle is supportive of overall health. For older adults, physical activity can reduce the risk of heart disease and high blood pressure, and can improve musculoskeletal health and overall quality of life (Brandon et al., 2009; Greenberg and Renne, 2005). Physical activity can also improve independence and functional ability, which can reduce hospitalization, morbidity, and mortality (Ackermann et al., 2008; Berke et al., 2007; Takano et al., 2002). Higher physical activity levels among children and

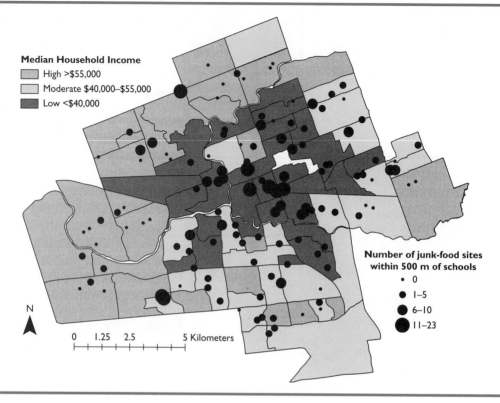

Figure 23.4 Junk food targeted to poor kids?

Elementary schools in low-income areas of London, Ontario, are exposed to three times more junk-food outlets than are high-income schools.

youth have been correlated with academic achievement, healthy body weight promotion, positive self-esteem, and positive attitudes and behaviours (Canadian Paediatric Society, 2002). Unfortunately, the majority of Canadians, both young and old, are not physically active enough to reap the health benefits (Allison et al., 2007).

An increasing number of planning and public health researchers are recognizing that prevailing patterns of land use and development (see also Grant and Filion, Chapter 18) in North American cities negatively impact physical activity levels in that they discourage active modes of travel such as walking and restrict opportunities for physical recreation. Evidence for these arguments is explored below.

Built Environments for Physical Activity: 'Recreationscapes'

Opportunities for recreation within neighbourhood environments obviously are important for facilitating physical activity. Numerous studies in the US, Australia, and Canada have found that convenient access to publicly provided recreational opportunities such as parks, playgrounds, and other recreation facilities is consistently associated with higher rates of physical activity and decreased overweight, especially among children and adolescents (Gordon-Larsen et al., 2006; Maziak et al., 2007; Norman et al., 2006; Sallis et al., 2000). As early as 1990, James Sallis, the pioneer of 'active living research', and his colleagues showed that the frequency of exercise among San Diego residents was related to the distance between their homes and available exercise facilities. In a recent study in London, Ontario, it was discovered that children who had two or more public recreation facilities within 500 metres of their home were active for 16 additional minutes after school per day than were those who had fewer amenities (Tucker et al., 2009). Opportunities located within walking distance of home may be doubly important for stimulating active behaviours as both the route and the destination contribute to overall activity levels.

A few recent studies have adopted an 'environmental justice' approach to exploring parks, as access to affordable, quality recreation facilities and programs may not be equitably distributed across neighbourhoods (Gilliland et al., 2006; Taylor et al., 2006). Children from low-income households may have fewer opportunities for health-promoting activity due to reduced access to high-quality parks, a reduced ability to afford fee-based recreation programs and/or transportation to free facilities outside their own neighbourhood, as well as parental concerns regarding neighbourhood safety (Gilliland et al., 2006, 2007; Tucker et al., 2007; see also Cowen, Siciliano, and Smith, Chapter 17). Numerous US studies have indicated that parks are less likely to be available, and more likely to be of poorer quality, in low-income neighbourhoods and/or those with high proportions of visible minorities (Loukaitou-Sideris, 2002; Wolch, 2005). Contrary to US studies, Smoyer-Tomic and colleagues (2004) found in Edmonton that low-income, high-social-need neighbourhoods had equitable spatial access to playgrounds; however, the condition of playgrounds in these neighbourhoods was often poorer than those in higher-income neighbourhoods. In their evaluation of a small sample of 28 parks selected from six urban neighbourhoods in Montreal, Coen and Ross (2006) discovered that parks in poor neighbourhoods were of significantly lower quality than those in high-income neighbourhoods. Based on detailed field audits of every

public park (n = 208) in London, Ontario, Gilliland and colleagues found no distinct socio-spatial inequity with respect to the availability and quality of public parks in that city, but did identify the existence of 'recreational deserts', or areas where children did not have easy access to a good place to play (Gilliland et al., 2006, 2007).

The perception of poorer-quality facilities has been associated with lower levels of physical activity among adolescents (Romero, 2005). In London, Ontario, it was found that high-quality parks were heavily frequented in all neighbourhoods, whereas poor-quality parks were virtually abandoned in all but the lowest-income neighbourhoods (Gilliland et al., 2007). Recent interviews with parents in London revealed that their perceptions of poor neighbourhood safety and the perceived long distances to activity facilities were significant barriers to physical activity for young children (Tucker et al., 2007). Similarly, speaking with elementary school students in London, Tucker and colleagues (2008) learned that youth *want* to use parks, yet often avoid them because of their undesirable qualities, such as garbage, safety issues, or lack of lighting. These findings highlight the need for city planners to better understand and address both children's and parents' perceptions of features of the built environment that support or hinder physical activity and restorative play.

Built Environments for Physical Activity: Walkability and Active Travel

The most common form of physical activity for people of all ages is walking. Even moderate increases in walking have been linked to numerous health benefits among adults, such as better body weight and overall fitness, as well as protection from high blood pressure, diabetes, cancer, cardiovascular disease, osteoarthritis and hip fractures, stress, and depression (Demers, 2006; Frank et al., 2003). Both walking and biking are common forms of recreational physical activity, but they are also utilitarian forms of physical activity, as they are a both a practical mode of transport to work, school, or shop. For many adults, the journey to work represents the most convenient opportunity to be physically active on a regular basis; however, recent data on the commuting patterns of Canadian adults from the 2006 census indicate that only a small percentage of the employed population take an active mode of travel such as walking or biking to their job sites. Furthermore, as seen in Figure 23.5, the data suggest that cities with the highest proportions of their working population commuting via a private vehicle also have the highest percentage of overweight citizens.

For children and youth, the journey to school represents a significant opportunity to increase daily levels of physical activity. Research suggests that children who walk or bike to school also tend to be more physically active during other parts of the day (Cooper et al., 2005). Nevertheless, the proportion of students who use active transportation modes (e.g., walking, cycling, skating) to and from school has decreased dramatically in recent years; a nationwide survey of children in the US revealed a decline from 41 per cent to 13 per cent over the last 25 years (McDonald, 2007). Clearly, some of this change is related to the growth in

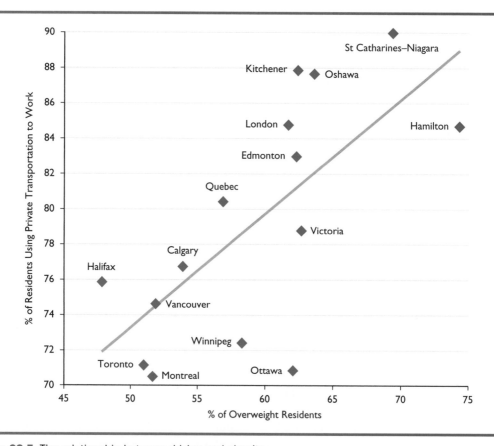

Figure 23.5 The relationship between driving and obesity

Cities with a high proportion of workers commuting by private automobile also tend to have a high proportion of overweight or obese residents ($r^2 = 0.5$).

Source: Data on 15 largest CMAs from 2006 census.

private schools, charter schools, and amalgamations that make it difficult if not impossible for children to transport themselves to school as well as to parental fear of predators, unsafe neighbourhoods, etc. (see Cowen, Siciliano, and Smith, Chapter 17). Further research will be needed to sort out the strength of the effect of the environment when these other factors are controlled for.

Either way, a growing body of research indicates that environmental features such as the presence of sidewalks, multi-use pathways, and retail shops, as well as the configuration of streets and the mix of land uses, have a significant influence on walking behaviours (Frank et al., 2003, 2006; Owen et al., 2004) (see Table 23.2). The majority of this research has focused on the environmental factors that influence utilitarian travel (i.e., to work, to shop, to school) rather than walking for pleasure, exercise, or overall health. Indeed, the earliest studies on this subject were performed mostly by urban planners rather than health scientists, and more specifically by transportation planners, whose pioneers include Susan Handy (Handy et al., 2002), Robert Cervero (Cervero and Duncan, 2003), and Lawrence Frank (Frank and Kavage,

2008). Because much of this work has focused on commuting, subjects have typically been adults of traditional 'working age', between 18 and 65 years old. The current popularity of 'new urbanist' communities (Grant and Filion, Chapter 18) as an alternative to the traditional suburb owes its popularity, at least implicitly if not explicitly, to a dawning societal recognition that the way we produce communities has a great impact on individual health and well-being; indeed on our social construction of ourselves as a society.

Much less is known about environmental factors influencing active travel behaviours among youth, despite the fact that they are typically less mobile and the group most likely to benefit from increasing walkability in local neighbourhoods. Research on active travel among youth, which has largely drawn its variables from studies of adults, has suggested that neighbourhood factors such as distance to school, land-use mix, and various characteristics of urban design such as the presence/absence of sidewalks influence decisions regarding a child's mode of travel to school (Kerr et al., 2007; McMillan, 2005, 2007). Current evidence from US and Canadian studies indicates that distance

Table 23.2 Established Environmental Factors Correlated with Increased Walking

Environmental Support	Selected References
Population density	Cervero, 1996; Frank and Pivo, 1995
Employment/office density	Lee and Moudon, 2006; Frank and Pivo, 1995; Kockelman, 1997
Retail/commercial density	Lee and Moudon, 2006; Cervero and Duncan, 2003; Cervero, 1996
Land-use mix	Larsen et al., 2009; Cervero and Duncan, 2003; Frank and Pivo, 1994
Sidewalks, paths, and trails	Kitamura et al., 1997; Hess et al., 1999; Troped et al., 2001
Intersection density	Cervero and Duncan, 2003
Street lighting	Cervero and Kockelman, 1997
Street trees	Larsen et al., 2009
Shorter trip distance	Larsen et al., 2009; Kockelman, 1997

Source: Gilliland, Jason and Kristian Larsen, 2010. 'Planning Pathways to Health: Assessing the Impact of the Built Environment on Walking Behaviours', in Pathways to the Future: Proceedings of the International Planning Conference (Christchurch, NZ: New Zealand Planning Institute).

between home and school is the most important barrier to choosing to walk to school (Larsen et al., 2009; Schlossberg et al., 2006). A recent study of the journey-to-school among over 800 children aged 11–13 in London, Ontario, revealed that 52 per cent took an active mode of travel to school (48 per cent walked and 4 per cent used bike, scooter, or rollerblades); however, that figure rose to 62 per cent for those who lived within 1.6 km of school and to 93 per cent for those who lived within 500 metres of school (Larsen et al., 2009).

Research also suggests that the layout or design of certain features of the built environment plays an important role in choice of travel mode, although the evidence is somewhat mixed regarding exactly how this occurs. The density of street intersections, for example, is related to route options and 'connectivity' in the local neighbourhood, and it has been shown to have positive associations with rates

of active travel (Braza et al., 2004; Kerr et al., 2007) (see Figure 23.6). Intersection density is also related to increased number of roadway crossings, however, which raises safety concerns that may negatively affect rates of active travel, particularly among school-aged children and the elderly. Likewise, a few studies have indicated that higher residential densities are an important factor towards increasing active travel among adolescents (Frank et al., 2007; Kerr et al., 2007), but other studies have found no relationship between residential density and walking to school (Larsen and Gilliland, 2009). The presence of sidewalks, which arguably can increase pedestrian safety, has been linked to increased walking and bicycling to school in the US (Fulton et al., 2005; Kerr et al., 2006), but the variable has not appeared to be significant in Canadian studies (Larsen et al., 2009). In addition, a recent Canadian study has indicated that increased tree coverage

A. Typical urban school neighbourhood   B. Typical suburban school neighbourhood

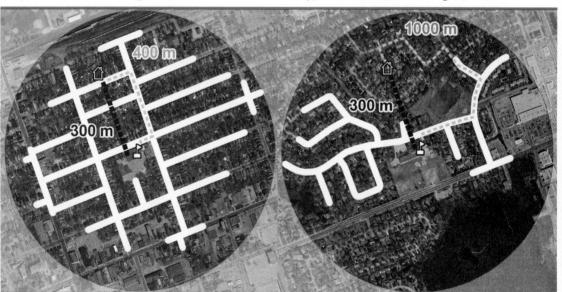

Figure 23.6 Street network patterns and neighbourhood walkability

In school neighbourhood B, the student must travel 2.5 times farther from home to school than in A, even though they are the same distance apart (300 metres) 'as the crow flies'.

Source: Gilliland, J. 2010. 'Healthy by Design: Planning for Children's Well-Being', Designing Auckland: A Mayoral Conversation, City of Auckland, New Zealand.

along streets is positively associated with likelihood of walking to school (Larsen et al., 2009).

Finally, higher density and mix of land uses have been linked to increased rates of walking and physical activity in adults for utilitarian travel, as such factors increase the number of potential nearby walking destinations (Frank et al., 2004; Owen et al., 2004; Saelens et al., 2003;). The relationship between the density or mix of land uses and children's travel is less clear: it may be that neighbourhoods of high density and land-use mix provide a more interesting walk to school than uniformly zoned residential neighbourhoods; on the other hand, highly dense and highly mixed areas may be perceived by parents as being unsafe. Recent work in London, Ontario, by Larsen and colleagues (2009) indicates that rates of walking or cycling to school among adolescents were higher in areas with greater land-use mix, after controlling for other key factors. In a US setting, Kerr and colleagues (2006) also found a positive correlation between land-use mix and non-motorized travel to school, but Ewing and colleagues (2004) found the opposite. Although some results are mixed, research on younger populations is scant but growing, and current findings suggest that both community design and geographical factors influence levels of active transportation among youth, and possible environmental interventions may exist for making neighbourhoods more walkable.

Discussion and Conclusions: Where Do We Go from Here?

Given the growing body of evidence linking the built environment to dietary behaviours, physical activity patterns, and expanding waistlines, it can be argued that the way we have built our communities in Canada in recent decades has been a major cause of the recent obesity epidemic. The remainder of this chapter presents a road map for future work on building healthy communities, beginning with avenues for future research

and concluding with suggestions for (re)designing environments for health.

How Can We Build Healthier Communities?

What can policy-makers, city planners, engineers, developers, builders, and other agents of change in the built environment learn from this growing body of research? How do we modify current and future development to alter critical obesity-related behaviours? What suggestions can be put forward for retrofitting, or should we say 'retrofitnessing', cities for improving the body weights and overall health of Canadians?

Cultivating healthier community food environments is critical for improving dietary habits and promoting healthy body weights. To attract supermarkets back to underserved, inner-city food deserts, municipalities could introduce financial incentives, such as building restoration initiatives or land 'swaps', or change zoning or parking regulations. Any municipal strategy for inner-city redevelopment must recognize the positive correlation between supermarkets and population density. The existence of a local supermarket makes a neighbourhood more attractive to (potential) residents; likewise, the existence of a large number of residents (potential customers) makes a neighbourhood more attractive to a potential supermarket owner.

In locations that cannot attract or support a supermarket, alternative strategies, such as farmers' markets, are available for improving access to fresh, healthy, and affordable food. Using a 'pre-test/post-test' research design, Larsen and Gilliland (2009) identified how the opening of a farmers' market greatly improved the price and availability of fresh and healthy food in one disadvantaged neighbourhood of London, Ontario, that previously had been considered a food desert. Furthermore, they estimated that the annual savings to the household grocery budget due to improved access associated with the market was equivalent to the average monthly rent in that city; thus, they further argued that the particular disadvantaged neighbourhood under surveillance could no longer be considered a food desert. Farmers' markets have long been fixtures in many of Canada's larger cities. The St Lawrence Market in downtown Toronto, for example, has fed local residents and visitors for over two centuries. Likewise, the Atwater and Jean Talon markets in Montreal are embedded in the fabric of their low-income neighbourhoods, but each one also serves a larger clientele, being easily reached via public transit. It is possible that many older, inner-city neighbourhoods in other Canadian cities have a stock of underutilized warehouse buildings that could support a permanent indoor market; where buildings are not available, outdoor markets can occupy sections of under-utilized parking lots (particularly on weekends) or space on retail-oriented streets that are temporarily closed to automobiles during the hours of market operation. Furthermore, outdoor markets can be moved throughout the city, serving different neighbourhoods on different days of the week.

Planning and policy interventions are necessary to limit the concentration of junk-food retailers within a certain distance of schools and to restrict junk-food sales and advertising targeted to children and youth. Some US cities, including Los Angeles and New York City, have introduced zoning restrictions to ban the sale of trans fats; to date, no Canadian city has enacted similar strategies. Given Canada's government-funded universal health-care system, where all taxpayers ultimately bear the burden of the costs associated with obesity, it is reasonable to believe that Canadians would be even more accepting of a legislative approach than US citizens.

Designing, building, and retrofitting neighbourhoods to encourage physical activity is critical. Creating high-quality public places to play for different populations in all neighbourhoods is critical for increasing levels of physical activity. Given the scarcity of large parcels of open space in older (already developed) areas of the city, in order to rectify current inequities city planners need to adopt innovative strategies for providing new recreation spaces in older neighbourhoods, such as the acquisition and redevelopment of vacant lots and underutilized schoolyards, as well as any available '**brownfield' sites** (i.e., former industrial land) and '**greyfield' sites** (e.g., former commercial plazas). In addition, future investments of increasingly scarce municipal resources should be aimed at creating multi-purpose recreation spaces that appeal to multiple demographic groups.

Besides improving neighbourhood 'recreationscapes' for participation in sports, a more effective method to get Canadians moving will be to build or retrofit our cities so that they are more supportive of active modes of travel, such as walking or biking. Increasing reliance on the automobile over the past half-century has led to increasingly sedentary lifestyles. The first step in promoting walkable urban forms is to rework or remove obsolete zoning laws and engineering standards. Numerous studies confirm that children and adults in neighbourhoods with a high mix of land uses walk much more often than their counterparts in more homogeneous residential neighbourhoods; the former have more destinations easily reachable on foot. Therefore, zoning laws that still serve to segregate land uses should be rewritten to promote walking and support public health. It is also time to critically review long-standing engineering standards focused on outdated 'minimums' (e.g., minimum parking requirements or road widths), which have long served to prioritize the automobile. A number of strategies have been proposed for 'taming' city streets to make them more pedestrian-friendly: reduce street widths; build or widen sidewalks; install clearly marked bike paths; redesign crosswalks; make pedestrian lights longer; plant trees along streets; and reduce speed limits on streets of potentially high pedestrian traffic. While the specific impacts of these planning and design strategies on walking and physical activity are still being empirically validated, these suggestions should be relatively easy to introduce and will undoubtedly enhance the pedestrian experience in Canadian cities.

Many researchers also suggest that walking levels can be increased by improving neighbourhood 'connectivity', which essentially refers to how directly or efficiently a person can move from one location to another using existing circulation networks. The majority of previous studies have identified neighbourhoods of high connectivity as those having a high density of street intersections, and by extension they have shorter blocks, more direct routes, and a greater variety of route options; such ideas were first popularized by Jane Jacobs in 1961, in her now classic text, *The Death and Life of Great American Cities*. But connectivity is about more than just street intersections; to lessen dependency on the private automobile, planners must carefully plan the courses of bike trails, multi-use pathways, and bus routes, and also (re)think links between transport modes to ensure direct and efficient routes to significant places (e.g., universities, shopping centres, and workplaces). A new bicycle path, for example, will have little impact on encouraging active commuting if it is constructed in low-density areas far away from residences and/or workplaces. On the other hand, residents in a 'food desert' neighbourhood could potentially experience a great improvement in dietary choices if an existing bus route was redrawn to directly and efficiently connect them with an affordable grocery store in a nearby neighbourhood.

While talking about the weather is a popular Canadian pastime, very few physical activity researchers discuss climate (Tucker and Gilliland, 2007). Physical activity levels of Canadians decrease significantly in winter (Brandon et al., 2007; Stephens et al., 1986). A logical explanation for seasonal fluctuations in physical activity is the fact that most major cities in Canada are covered in snow for up to four months of the year (Environment Canada, 2009). Unfortunately, urban parks tend to be abandoned with the first snow and most municipalities do little to encourage

physical activity during wintertime, besides building outdoor skating rinks or expensive indoor hockey arenas, which appeal to only a small proportion of the population. Unlike Nordic cities (e.g., Copenhagen), few Canadian cities maintain their bicycle and walking paths in winter. Winterizing bike and pedestrian pathways is just one relatively simple solution to promoting active travel and keeping some people out of their cars year-round.

The walkability of a neighbourhood is also greatly impacted by school-siting decisions. Throughout Canada and the United States, inner-city schools are being closed, or threatened with closure, while new ones are being opened up on the rural fringe. Such decisions are typically a response to bureaucratic calculations such as 'floor space per student' ratios, without a holistic vision regarding the potential broader, long-term economic, social, and health impacts of a school closing. This is a short-sighted, narrow-focused approach, as closing neighbourhood schools leads to longer commutes for youth and ultimately more children have to be bused, or chauffeured by parents, between home and school. School location decisions should be made with the goal of encouraging physically active travel.

The weight of evidence clearly indicates that prevailing forms of development in North American cities are 'obesogenic' in that they deter walking and recreation, while promoting the consumption of unhealthy foods. Unlike motivational programs or educational campaigns, which have had limited, short-term impacts, any changes in behaviour due to changes in the built environment should have more lasting effects on the battle against obesity. Breakthroughs in our understanding of how the built environment impacts obesity, and what we can do about it, have come from combining expertise from multiple disciplines and arenas. Future studies and interventions require strong collaborations among researchers of various disciplines, between health and social scientists (e.g., epidemiologists, health promoters, psychologists, economists, human/urban geographers) and those who study, plan, and construct the built environment, such as city planners and engineers, as well as property developers and builders. Furthermore, with growing co-operation and exchange among researchers, planners, policy-makers, health professionals, advocacy groups, and citizens, we will soon see how this exciting field of research will lead to effective policy and interventions aimed at creating healthier built environments for all.

Acknowledgements

I wish to acknowledge the ongoing support of graduate students and collaborators, including: Kristian Larsen, Paul Hess, Sisir Sarma, Janet Loebach, Trish Tucker, Jen Irwin, Meizi He, David Stubbs, Kevin VanLierop, Richard Sadler, Martin Healy, Godwin Arku, and Eric Simard. Thanks to Trudi Bunting and Ryan Walker for comments on a draft of this chapter. I am also grateful for the financial support of the London Community Foundation, the Canadian Institutes of Health Research, the Social Sciences and Humanities Research Council of Canada, and the University of Western Ontario.

Review Questions

1. What reasons are suggested for increasing the walkability of our cities? Are you able to suggest others?
2. How can we build healthier communities?

Note

1. Overweight and obesity are typically assessed using the body mass index (kg/m²). For adults, a BMI of >25 is considered overweight, and a BMI of >30 is considered obese.

References

Ackermann, R.T., B. Williams, H.Q. Nguyen, E.M. Berke, M.L. Maciejewski, and J.P. LoGerfo. 2008. 'Healthcare cost differences with participation in a community-based group physical activity benefit for medicare managed care health plan members', *Journal of American Geriatrics Society* 56, 8: 1459–65.

Allison, K.R., E.M. Adlaf, J.J.M. Dwyer, D.C. Lysy, and H.M. Irving. 2007. 'The decline in physical activity among adolescent students: A cross-national comparison', *Canadian Journal of Public Health* 98: 97–100.

Alter, D., and K. Eny. 2005. 'The relationship between the supply of fast-food chains and cardiovascular outcomes', *Canadian Journal of Public Health* 96, 3: 173–7.

Apparicio, P., M.S. Cloutier, and R. Shearmur. 2007. 'The case of Montreal's missing food deserts: Evaluation of accessibility to food supermarkets', *International Journal of Health Geographics* 6: 4.

Berke, E.M., L.M. Gottlieb, A.V. Moudon, and E.B. Larson. 2007. 'Protective association between neighborhood walkability and depression in older men', *Journal of American Geriatrics Society* 55, 4: 526–33.

Berrigan, D., and R.P. Troiano. 2002. 'The association between urban form and physical activity in U.S. adults', *American Journal of Preventative Medicine* 23, 2 (suppl.): 74–9.

Bertrand, L., F. Therien, and M.S. Cloutier. 2008. 'Measuring and mapping disparities in access to fresh fruits and vegetables in Montréal', *Canadian Journal of Public Health* 99: 6–11.

Brandon, C.A., G.R. Jones, M. Speechley, and J. Gilliland. 2007. 'Influence of winter weather on total, daily physical activity values in older, community-dwelling adults', *Applied Physiology, Nutrition and Metabolism* 32, 1: S14.

———, D.P. Gill, M. Speechley, J. Gilliland, and G.R. Jones. 2009. 'Physical activity levels of older community-dwelling adults are influenced by summer weather variables', *Applied Physiology, Nutrition and Metabolism* 34, 2: 182–90.

Braza, M., W. Shoemaker, and A. Seeley. 2004. 'Neighborhood design and rates of walking and biking to elementary school in 34 California communities', *American Journal of Health Promotion* 19: 128–136.

Canadian Paediatric Society (CPS). 2002. 'Healthy active living for children and youth', *Paediatrics & Child Health* 7: 339–45.

Canadian Population Health Initiative (CPHI). 2006. *Improving the Health of Canadians: Promoting Healthy Weights.* Ottawa: Canadian Institute for Health Information.

Cervero, R. 1996. 'Mixed land-uses and commuting: Evidence from the American Housing Survey', *Transportation Research-A* 30: 361–77.

——— and M. Duncan. 2003. 'Walking, bicycling, and urban landscapes: Evidence from the San Francisco Bay area', *American Journal of Public Health* 93: 1478–83.

——— and K.M. Kockelman. 1997. 'Travel demand and the 3Ds: Density, diversity, and design', *Transportation Research-D* 2: 199–219.

Chou, S.Y., M. Grossman, and H. Saffer. 2004. 'An economic analysis of adult obesity: Results from the behavioral risk factor surveillance system', *Journal of Health Economics* 23: 565–87.

Chung, C., and S.L. Myers. 1999. 'Do the poor pay more for food? An analysis of grocery store availability and food price disparities', *Journal of Consumer Affairs* 33, 2: 276–96.

Clarke, G., H. Eyre, and G. Cliff. 2002. 'Deriving indicators of access to food retail provision in British cities: Studies of Cardiff, Leeds and Bradford', *Urban Studies* 39, 11: 2041–60.

Coen, S., and N.A. Ross. 2006. 'Exploring the material basis for health: Characteristics of parks in Montreal neighborhoods with contrasting health outcomes', *Health and Place* 12: 361–71.

Cooper A.R., L.B. Andersen, N. Wedderkopp, A.S. Paige, and K. Froberg. 2005. 'Physical activity levels of children who walk, cycle, or are driven to school', *American Journal of Preventive Medicine* 29: 179–84.

Crownfield, P. 2004. 'The girth of a nation', *Dynamic Chiropractic* 22: 46.

Cummins, S., and S. Macintyre. 2006. 'Food environments and obesity—Neighbourhood or nation?', *International Journal of Epidemiology* 35, 1: 100–4.

Cummins, S., M. Petticrew, C. Higgins, A. Findlay, and L. Sparks. 2005. 'Large scale food retailing as an intervention for diet and health: Quasi-experimental evaluation of a natural experiment', *Journal of Epidemiological Community Health* 59: 1035–40.

Cunningham, G.O., and Y.L. Michael. 2004. 'Concepts guiding the study of the impact of the built environment on physical activity for older adults: A review of the literature', *American Journal of Health Promotion* 18: 435–43.

Demers, M. 2006. *Walk For Your Life!* Ridgefield, Conn.: Vital Health.

Egger, G., and B. Swinburn. 1997. 'An "ecological" approach to the obesity pandemic', *British Medical Journal* 315: 477–80.

Eisenhauer, E. 2001. 'In poor health: Supermarket redlining and urban nutrition', *GeoJournal* 53: 125–33.

Ewing, R., W. Schroeer, and W. Greene. 2004. 'School location and student travel: Analysis of factors affecting mode choice', *Transportation Research Record* 1895: 55–63.

Fagot-Campagna, A. 2000. 'Emergence of type 2 diabetes mellitus in children: Epidemiological evidence', *Journal of Pediatric Endocrinology and Metabolism* 13: 1395–1402.

Figueroa-Colon, R., F.A. Franklin, J.Y. Lee, R. Aldridge, and L. Alexander. 1997. 'Prevalence of obesity with increased blood pressure in elementary school-aged children', *Southern Medical Journal* 90: 806–13.

Figueroa-Munoz, J.I., S. Chinn, and R.J. Rona. 2001. 'Association between obesity and asthma in 4–11 year old children in the UK', *Thorax* 56: 133–7.

Frank, L.D., M. Andresen, and T. Schmid. 2004. 'Obesity relationships with community design, physical activity, and time spent in cars', *American Journal of Preventative Medicine* 27, 2: 87–96.

———, P. Engelke, and T. Schmid. 2003. *Health and Community Design: The Impact of the Built Environment on Physical Activity.* Washington: Island Press.

——— and S. Kavage. 2008. 'Urban planning and public health: A story of separation and reconnection', *Journal of Public Health Management Practice* 14: 214–20.

———, J. Kerr, J. Chapman, and J. Sallis. 2007. 'Urban form relationships with walk trip frequency and distance among youth', *American Journal of Health Promotion* 21, 4 (suppl): 305–11.

——— and G. Pivo. 1995. 'Impacts of mixed use and density on utilization of three modes of travel: Single-occupant vehicle, transit, and walking', *Transportation Research Record* 1466: 44–52.

———, J.F. Sallis, T.L. Conway, J.E. Chapman, B.E. Saelens, and W. Bachman. 2006. 'Many pathways from land use to health—Associations between neighborhood walkability and active transportation, body mass index, and air quality', *Journal of the American Planning Association* 72: 75–87.

French, S.A., M. Story, D. Neumark-Sztainer, J.A. Fulkerson, and P. Hannan. 2001. 'Fast food restaurant use among adolescents: Associations with nutrient intake, food choices and behavioral and psychosocial variables', *International Journal of Obesity* 25, 12: 1823–33.

Fulton, J.E., J.L. Shisler, M.M. Yore, and C.J. Caspersen. 2005. 'Active transportation to school: Findings from a national survey', *Research Quarterly for Exercise Sport* 76: 352–7.

Gilliland, J., M. Holmes, J.D. Irwin, and P. Tucker. 2006. 'Environmental equity is child's play: Mapping public provision of recreation opportunities in urban neighbourhoods', *Vulnerable Children & Youth Studies* 1: 256–68.

———, ———, J. Loebach, K. Larsen, P. Tucker, J. Irwin, M. He, and P. Hess. 2007. 'Is the grass greener on the other side of town? Examining relationships among the availability, quality, and use of public play spaces', paper presented to the Canadian Public Health Geomatics Conference, Ottawa.

——— and K. Tang. 2007. 'Neighbourhood characteristics associated with the clustering of junk food outlets around schools', paper presented to the annual meeting of the Association of American Geographers, San Francisco.

Gordon-Larsen, P., M.C. Nelson, P. Page, and B.M. Popkin. 2006. 'Inequality in the built environment underlies key health disparities in physical activity and obesity', *Pediatrics* 117: 417–24.

Greenberg, M.R., and J. Renne. 2005. 'Where does walkability matter the most? An environmental justice interpretation of New Jersey data', *Journal of Urban Health-Bulletin of the New York Academy of Medicine* 82, 1: 90–100.

Guthrie, J.F., B.H. Lin, and E. Frazao. 2002. 'Role of food prepared away from home in the American diet, 1977–78 versus 1994–96: Changes and consequences', *Journal of Nutrition Education and Behaviour* 34, 3: 140–50.

Handy, S., M. Boarnet, R. Ewing, and R. Killingsworth. 2002. 'How the built environment affects physical activity: Views from urban planning', *American Journal of Preventative Medicine* 23 (2S): 64–73.

Hess, P.M., A. Vernez Moudon, M.C. Snyder, and K. Stanilov. 1999. 'Site design and pedestrian travel', *Transportation Research Record* 1674: 9–19.

Jacobs, J. 1961. *The Death and Life of Great American Cities.* New York: Random House.

Katzmarzyk, P.T., and I. Janssen. 2004. 'The economic costs associated with physical inactivity and obesity in Canada: An update', *Canadian Journal of Applied Physiology* 29, 1: 90–115.

Kayani, N. 1998. *Food Deserts: A Practical Guide.* London: Chadwick House.

Kerr, J., L. Frank, J.F. Sallis, and J. Chapman. 2007. 'Urban form correlates of pedestrian travel in youth: Difference by gender, race-ethnicity and household attributes', *Transportation Research D* 12: 177–82.

———, D. Rosenberg, J.F. Sallis, B.E. Saelens, L.F. Frank, and T.L. Conway. 2006. 'Active commuting to school: Associations with environment and parental concerns', *Medicine & Science in Sports & Exercise* 38: 787–94.

Kitamura, R., P.L. Mokhtarian, and L. Laidet. 1997. 'A micro-analysis of land use and travel in five neighborhoods in the San Francisco Bay area', *Transportation* 24: 125–58.

Kockelman, K.M. 1997. 'Travel behavior as function of accessibility, land use mixing, and land use balance: Evidence from San Francisco Bay area'. *Transportation Research Record* 1607: 116–25.

Larsen, K., and J. Gilliland. 2008. 'Mapping the evolution of "food deserts" in a Canadian city: Supermarket accessibility in London, Ontario, 1961–2005', *International Journal of Health Geographics* 7.

——— and ———. 2009. 'A farmers' market in a food desert: Evaluating impacts on the price and availability of healthy food', *Health and Place* 15, 3: 1158–62.

———, ———, and P. Hess. 2008. 'Individual and environmental influences on a child's mode of travel to school', paper presented to the annual meeting of the Association of American Geographers, Boston.

———, ———, ———, P. Tucker, J. Irwin, and M. He. 2009. 'The influence of the physical environment and sociodemographic characteristics on children's mode of travel to and from school', *American Journal of Public Health* 99: 520–6.

Lee, C., and A.V. Moudon. 2006. 'Correlates of walking for transportation or recreation purposes', *Journal of Physical Activity and Health* 3: S77–S98.

Lopez, R. 2004. 'Urban sprawl and risk for being overweight or obese', *American Journal of Public Health* 94, 9: 1574–9.

Loukaitou-Sideris, A. 2002. 'Children in Los Angeles parks: A study of equity, quality, and children's satisfaction with neighbourhood parks', *Town Planning Review* 73: 467–88.

Luo, W., H. Morrison, M. Groh, C. Waters, M. DesMeules, E. Jones-McLean, A.M. Ugnat, S. Desjardins, M. Lim, and Y. Mao. 2007. 'The burden of adult obesity in Canada', *Chronic Diseases in Canada* 27, 4: 135–44.

McDonald, N.C. 2007. 'Active transportation to school: Trends among U.S. schoolchildren, 1969–2001', *American Journal of Preventive Medicine* 32: 509–16.

McMillan, T.E. 2005. 'Urban form and a child's trip to school: The current literature and a framework for future research', *Journal of Planning Literature* 19: 440–56.

———. 2007. 'The relative influence of urban form on a child's travel mode to school', *Transportation Research A* 41: 69–79.

Maddock, J. 2004. 'The relationship between obesity and the prevalence of fast food restaurants: State-level analysis', *American Journal of Health Promotion* 19: 137–43.

Marcus, C.L., S. Curtis, C.N. Koerner, A. Joffe, J.R. Serwint, and G.M. Loughlin. 1996. 'Evaluation of pulmonary function and polysomnography in obese children and adolescents', *Pediatric Pulmonology* 21, 3: 176–83.

Maziak, W., K.D. Ward, and M.B. Stockton. 2007. 'Childhood obesity: Are we missing the big picture?', *Obesity Reviews* 9: 35–42.

Morland, K., A.V. Diez Roux, and S. Wing. 2006. 'Supermarkets, other food stores, and obesity: The Atherosclerosis Risk in Communities Study', *American Journal of Preventive Medicine* 30, 4: 333–9.

———, S. Wing, and A. Diez Roux. 2002. 'The contextual effect of the local food environment on residents' diets: The Artherosclerosis Risk in Communities Study', *American Journal of Public Health* 92: 1761–8.

Nielsen, S.J., A.M. Siega-Riz, and B.M. Popkin. 2002. 'Trends in food locations and sources among adolescents and young adults', *Preventive Medicine* 35, 2: 107–13.

Norman, G.J., S.K. Nutter, S. Ryan, J.F. Sallis, K.J. Calfas, and K. Patrick. 2006. 'Community design and access to recreational facilities as correlates of adolescent physical activity and body-mass index', *Journal of Physical Activity and Health* 3: S118–S128.

Openshaw, S. 1984. *The Modifiable Areal Unit Problem*. Norwich, Conn.: Geo Books.

Owen, N., N. Humpel, E. Leslie, A. Bauman, and J. Sallis. 2004. 'Understanding environmental influences on walking', *American Journal of Preventive Medicine* 27: 67–76.

Pothukuchi, K. 2005. 'Attracting supermarkets to inner-city neighbourhoods: Economic development outside the box', *Economic Development Quarterly* 19, 3: 232–44.

Reidpath, D.D., C. Burns, J. Garrard, M. Mahoney, and M. Townsend. 2002. 'An ecological study of the relationship between social and environmental determinants of obesity', *Health and Place* 8: 141–5.

Robinson, W.S. 1950. 'Ecological correlations and the behavior of individuals', *American Sociological Review* 15: 351–7.

Romero, A. 2005. 'Low-income neighborhood barriers and resources for adolescents' physical activity', *Journal of Adolescent Health* 36, 3: 253–9.

Ross, N.A., S. Tremblay, S. Khan, D. Crouse, M. Tremblay, and J.-M. Berthelot. 2007. 'Body mass index in urban Canada: Neighbourhood and metropolitan area effects', *American Journal of Public Health* 97: 500–8.

Saelens, B., J. Sallis, and L. Frank. 2003. 'Environmental correlates of walking and cycling: Findings from the transportation, urban design, and planning literatures', *Annals of Behavioural Medicine* 25: 80–91.

Sallis, J.F., and K. Glanz. 2006. 'The role of built environments in physical activity, eating, and obesity in childhood', *The Future of Children* 16: 89–108.

———, M.F. Hovell, C.R. Hofstetter, J.P. Elder, M. Hackley, C.J. Caspersen, and K.E. Powell. 1990. 'Distance between homes and exercise facilities related to frequency of exercise among San Diego residents', *Public Health Reports* 105: 179–85.

———, J.J. Prochaska, and W.C. Taylor. 2000. 'A review of correlates of physical activity of children and adolescents', *Medicine and Science in Sports and Exercise* 32: 963–75.

Schlossberg, M., J. Greene, P.P. Phillips, B. Johnson, and B. Parker. 2006. 'School trips: Effects of urban form and distance on travel mode', *Journal of the American Planning Association* 72: 337–46.

Shields, M. 2005. 'Measured obesity: Overweight Canadian children and adolescents', *Nutrition: Findings from the Canadian Community Health Survey*. Ottawa: Statistics Canada.

——— and M. Tjepkema. 2006. 'Regional differences in obesity', *Health Reports*. Ottawa: Statistics Canada.

Smoyer-Tomic, K.E., J.N. Hewko, and M.J. Hodgson. 2004. 'Spatial accessibility and equity of playgrounds in Edmonton, Canada', *Canadian Geographer* 48, 3: 287–302.

———, J.C. Spence, and C. Amrhein. 2006. 'Food deserts in the prairies? Supermarket accessibility and neighbourhood need in Edmonton, Canada', *Professional Geographer* 58, 3: 307–26.

———, ———, K.D. Raine, C. Amrhein, N. Cameron, V. Yasenovskiy, N. Cutumisu, E. Hemphill, and J. Healy. 2008. 'The association between neighborhood socioeconomic status and exposure to supermarkets and fast food outlets', *Health and Place* 14: 740–54.

Sooman, A., S. Macintyre, and A. Anderson. 1993. 'Scotland's health: A more difficult challenge for some? The price and availability of healthy foods in socially contrasting localities in the west of Scotland', *Health Bulletin* (Edinburgh) 51, 5: 276–84.

Story, M., D. Neumark-Sztainer, and S. French. 2002. 'Individual and environmental influences on adolescent eating behaviors', *Journal of the American Dietetic Association* 102: S40–S51.

Strauss, R.S., and H.A. Pollack. 2003. 'Social marginalization of overweight children', *Archives of Paediatrics & Adolescent Medicine* 157, 8: 746–52.

Takano, T., K. Nakamura, and M. Watanabe. 2002. 'Urban residential environments and senior citizens' longevity in megacity areas: The importance of walkable green spaces', *Journal of Epidemiology & Community Health* 56, 12: 913–18.

Taylor, W., W. Poston, L. Jones, and M. Kraft. 2006. 'Environmental justice: Obesity, physical activity, and healthy eating', *Journal of Physical Activity and Health* 3: 30–54.

Timperio, A., D. Crawford, A. Telford, and J. Salmon. 2004. 'Perceptions about local neighborhood and walking and cycling among children', *Preventative Medicine* 38: 39–47.

Troped, P.J., R.P. Saunders, and R.R. Pate. 2001. 'Associations between self-reported and objective physical environmental factors and use of a community rail-trail', *Preventive Medicine* 32: 191–200.

Tucker, P., and J. Gilliland. 2007. 'The effect of season and weather on physical activity: A systematic review', *Public Health* 121, 12: 909–22.

———, ———, and J.D. Irwin. 2007. 'Splashpads, swings, and shade: Parents' preferences for neighborhood parks', *Canadian Journal of Public Health* 98: 198–202.

———, J.D. Irwin, J. Gilliland, and M. He. 2008. 'Adolescents' perspectives of home, school and neighborhood environmental influences on physical activity and dietary behaviors', *Children, Youth and Environments* 18, 2: 12–35.

———, ———, ———, ———, K. Larsen, and P. Hess. 2009. 'Environmental influences on physical activity levels in youth', *Health & Place* 15: 357–63.

Weinberg, Y. 2000. 'No place to shop: Food access lacking in the inner city', *Race, Poverty and the Environment*: 22–4.

World Health Organization (WHO). 2003. *Diet, Nutrition and the Prevention of Chronic Diseases*. Geneva: WHO.

Wolch, J., J.P. Wilson, and J. Fehrenbach. 2005. 'Parks and park funding in Los Angeles: An equity mapping analysis', *Urban Geography* 26, 1: 4–35.

Wrigley, N. 2002. '"Food deserts" in British cities: Policy context and research priorities', *Urban Studies* 39, 11: 2029–40.

Zenk, S., A. Schultz, B. Israel, S. James, S. Bao, and M. Wilson. 2005. 'Neighborhood racial composition, neighborhood poverty, and the spatial accessibility of supermarkets in metropolitan Detroit', *American Journal of Public Health* 95: 660–7.

CHAPTER 24

Food Systems and the City

Alison Blay-Palmer

Food is a sustaining and enduring necessity. Yet among the basic essentials for life—air, water, shelter, and food—only food has been absent over the years as a focus of serious professional planning interest. This is a puzzling omission because, as a discipline, planning marks its distinctiveness by being comprehensive in scope and attentive to the temporal dimensions and spatial interconnections among important facets of community life. (APA, 2007: 1)

Introduction

Food offers a prism for understanding more about the challenges faced in cities and regions (Born and Purcell, 2006; Morgan, 2008). Food can help us envision more sustainable, livable urban spaces that are better integrated with their adjacent landscapes. Urban geography, planning, and sociology contribute to rethinking the role of food as we move to more sustainable communities. By placing food at the centre of thinking about quality and the (re)imagining of producer–consumer linkages, we are able to conceive of space and relationships in new ways (Donald, 2008; Hinrichs and Lyson, 2008; Marsden, 2008; Morgan et al., 2006). As planners and others clamour for new approaches and tools, food has surfaced as a missing link in the urban (re)creation project. However, since

agriculture has traditionally been considered a rural preoccupation, separate from urban and consumer issues, engaging with food as an integral part of the urban system is still considered very unusual. This is confirmed by the above statement from the American Planning Association regarding the lack of serious attention given to food in the planning process. Based on the recognition of the potential for a more integrated approach to food, this chapter addresses how our cities could and should be fed; as such, it also delivers a strong rationale for using food as a lens to frame policy and develop even more relevant perspectives on urban growth and development.

The chapter unfolds in three parts. First, to provide a meaningful context, the issue of food and planning in Canada is tackled. Next we explore innovative moves by planning and policy groups that signal a radical change in the role of food as a framing tool. We then turn to case studies that offer examples of different initiatives that can assist urban experts and practitioners interested in putting food on the agenda. In many ways, academic and policy concerns about how a city feeds itself reflect the most profound changes within the nascent transition from modern to coherent postmodern world views. The evolution of land-use practices, community well-being, and the rise of science and technology are all well represented in the Canadian food story. These shifts provide us with a context for the evolution of food as a consideration in Canadian cities.

Food and Urban Spaces in Canada

The history of food in Canada encapsulates many changes in our society since the mid-1800s. The place of rural, urban, and food communities in our planning tradition also reflects these shifts. In the early days, planning and food intersected peripherally through land preservation initiatives as people attempted to protect fertile rural land. In a broader societal context, the association of food with rural landscapes is founded in our farm community roots. At the turn of the twentieth century some 63 per cent of Canadians lived in rural areas (Statistics Canada, 2005). As modern society took hold, however, there was a gradual shift in population from rural to urban communities and, as the economy became much more deeply entrenched in manufacturing, farm size expanded, partially in response to rural depopulation. This was also the time when increased mechanization and the application of nitrogen-based fertilizers and chemical pesticides meant that fewer people were needed to farm bigger tracts of land. By 2006, we had 229,373 farms remaining, with an average of 728 acres per farm, an increase in acreage per farm of 73 per cent from the early twentieth century (Statistics Canada, 2007). With fewer connections to food producers, places of food production and food consumption became increasingly disconnected.

In addition to demographic changes, **food systems** also reflect technological changes and help us understand the challenges that innovation poses to planners. Since the mid-nineteenth century, the transformation of food production from a local, small-scale, manual activity to a predominately technologically complex, industrial, monoculture production system mirrors key changes in our relationship to nature. The 'green revolution' that

began in the 1950s marked a technological shift, with increased reliance on irrigation, chemicals, and sophisticated seed packages for food production, plus distribution via new, long-distance, refrigerated modes of transport. The stated goal was to eliminate famines by mass-producing food. Coincident with this green revolution was the reorientation of large chemical companies away from wartime production towards more domestic markets, a shift dominated by the manufacture and distribution of agricultural chemicals such as pesticides and fertilizers. As food production became more industrialized and corporatized under this food production regime (Goodman and Redclift, 1981; Friedmann and McMichael, 1989), there was a growing separation of urban and rural spaces as people became physically and psychologically distanced from their food sources (Kneen, 1995). This has accelerated so that most of our food now comes from a global industrial food system where the average food molecule travels over 2,400 kilometres from field to fork (Pirog, 2001). As modern society took hold, cars and suburbs became de rigueur.

In the past, planners have been challenged to reconcile conflicting needs across jurisdictions and interest groups. On the one hand, there is the need to preserve valuable farmland often in the shadow of urban development; on the other are urbanites who leave cities for suburban, peri-urban, and rural areas, bringing along their idealized visions of rural landscapes. These idyllic imaginaries must be reconciled with the realities of farming, including odours and access to roadways and the conservation of agriculturally productive lands (Bryant et al., 1982; Joseph et al., 1988; Paquette and Doman, 2003). In the face of pressures from 'city folk' to gentrify the countryside and turn more land from food production to housing development, the primary provincial response has been the preservation of agricultural land. The 1976 Ontario Agricultural Code of Practice (Caldwell, 2000), the Agricultural Land Reserve (ALR) enacted in British Columbia in 1973, and the Greenbelt legislation enacted in Ontario in 2005 (Suzuki Foundation Report, 2006) all reflect the desire to set aside land for food production. The closer one gets to urban centres, the more intense the conflict—as of 2001, half of Canadian urban centres were built on agriculturally viable land (Statistics Canada, 2001, cited in Suzuki Foundation Report, 2006). However, despite provincial initiatives to protect land, urban pressures for outward expansion and development frequently supersede the need to protect quality land for food production. The case of the BC ALR is instructive. Despite the provincial government having set aside land for agriculture in the 1970s, the reality is that poor-quality land in remote parts of the province has been swapped for high-quality land closer to urban centres (Suzuki, 2006) or for second homes in 'cottage' communities such as Invermere (Bell, 2007). In this way, development trumps food production.

Our lifestyles also affect how we eat. In recent times in Canada the pace of life has accelerated—families, often out of economic necessity, moved from one to two incomes and food preparation emerged as more of an efficient assembling of prefabricated ingredients than the daily preoccupation of a stay-at-home caregiver. This has produced tension in Canadian's attitude to food, which is captured through the debate about affordable access to quality food, such as organically grown foods, contrasted against the trade-offs linked to large-scale retail developments, our reliance on a suburban car culture, and the apparent efficiency offered by large retailers that regularly opt for quantity over quality. Locally based food production–consumption linkages, new health and fitness awareness, and environmental concerns all are reflections of individual attempts to redress some of these challenges (see especially Gilliland, Chapter 23).

As part of the move to healthier living, in the last decade we have witnessed an increased desire on the part of consumers to reconnect with and understand what they are putting into their bodies (Whatmore, 2002). This has created opportunities for many innovative relationships between people and their food. The creation of Community Supported Agriculture, farmers' markets, the slow food movement, and the '100-mile diet' (Smith and MacKinnon, 2006) all demand that spaces exist in or near urban centres for consumers and farmers to connect. The recent move by the Canadian Medical Association calling for calorie counts on fast-food meals is another facet of the quest to eat better and regain an understanding of what we are consuming. Other emerging trends are linked to urban food production. These include a suite of garden types, including rooftops, backyards, and shared lands through community gardens. The majority of these new types of food relationships tend to include organic growing methods and help to provide more sustainable urban living environments. For this change to be significant and lasting, a reconceptualization of how people get access to food will ultimately be required. This, in turn, is connected to how we conceive of city spaces.

As the archetypical big-box store, Wal-Mart provides interesting insights into food and retail (Donald, 2008). Wal-Mart's commitment is to 'Save money. Live better.' by being fiercely competitive through its 'lean retailing' strategy that (1) concentrates purchasing power; (2) vertically integrates global production–delivery chains; and (3) maintains low-wage operations (Christopherson, 2007: 453). When Wal-Mart entered the organic food market in 2006, it used its global connections to access fresh produce and encouraged its multinational corporate suppliers such as Kellogg's to develop lines of organic processed food. The move by Wal-Mart into organics provoked widespread controversy about alternative food systems, specifically organics (Donald and Blay-Palmer, 2007). Organic supporters' vision is founded on small-scale, trust-based, locally rooted ecological production. Wal-Mart, on the other hand, sees organic food as a new niche opportunity and revved up its suppliers to produce more of what Julie Guthman has dubbed 'yuppie chow' (Guthman, 2003,

2007). While organic production methods lower the use of chemicals, other core organic/sustainability principles, including ecologically sound production methods, fair labour standards, and animal welfare, are not on Wal-Mart's table (see also Rees, Chapter 5; Connelly and Roseland, Chapter 14). As well, to get access to Wal-Mart's organics, consumers need to be part of the sprawling suburban car culture (Donald and Blay-Palmer, 2006; Dupuis, 2000; Guthman, 2003).

Clearly, then, food represents a complex and nuanced challenge. But nothing hits closer to home than food. Given the growing advocacy on the part of some consumer groups,[1] food could help to define the debate about how we see the cities of the future. In the case of food in the cityscape, there is considerable public stakeholder support for locally produced foods but not a lot of knowledge. Food offers a unique lens for recalibrating city spaces as it integrates land-use planning, transportation, water management, and environmental quality, as well as regulations and regulatory regimes concerned with human and environmental health. While conservation initiatives are important and have resulted in planning regulations to address tensions related to agriculture and food production as well as urban/rural tensions, most policies contain nothing that directly relates to the food system per se (Mackenzie, 2008). As we consider the impact that cities have on the environment and human health, there is a pressing need for a holistic approach in order to develop relevant policy. This is even more urgent given the increasingly complex challenges related to food, such as the environmental impact of food miles and questions of urban density and greening cities, as well as heightened consumer awareness about food safety and food-related health issues. In addition, new citizen-based awareness stimulates change from the ground up but remains disjointed, fragmented, and small-scale. At this stage, public enthusiasm for alternative food systems is important but is no match for heavy-handed and systematic corporate strategies such as those of Wal-Mart. Today more than ever a synthesizing approach is urgently needed to offer expertise, as is a large-scale, integrated model such as 'smart growth' that could empower municipalities and encourage city officials to take steps in new directions (Grant and Filion, Chapter 18). As a group, professional planners may well be in a unique position to offer just such direction.

To understand how food can help make linkages to more sustainable cities, we need to understand how food and cities are connected. If we adopt a food perspective to frame urban challenges, food accounts for a substantial level of a city's **ecological footprint** (see Rees, Chapter 5). For example, food accounts for 40 per cent of the ecological footprint of London, England. Every year its residents and citizens consume 2.5 million tons of food, which in turn generate 883,000 tons of organic waste (Petts, 2001). As we have now crossed the threshold where more people live in urban areas than in rural areas, planning for food as part of the urban environment promises to make cities greener in many ways. In recent years there has been concern about:

- environmental impacts of cities and their growing greenhouse gas footprints;
- the escalating cost of transporting food, particularly in the context of the food-for-fuel debate;
- the need to recycle waste within the urban system.

As experts at the Global Footprint Network explain:

> The global effort for sustainability will be won, or lost, in the world's cities, where urban design may influence over 70 per cent of people's Ecological Footprint. High-Footprint cities can reduce this demand on nature greatly with existing technology. Many of these savings also cut costs and make cities more livable. (Wackernagel et al., 2006)

Research by geographers on food issues clarifies changing relationships between people, food, and the city, particularly in the context of creating alternative food networks through shorter, increasingly trust-based food chains. Central to these discussions are farmers' markets (Connell et al., 2008; Feagan, 2008), alternative supply chains (Goodman, 2003; Maye et al., 2007; Morgan et al., 2006), and the role of retailers (Jackson et al., 2007) as supply chains shift in the existing food system. Food planning and policy can help address all of these challenges.

Food, Planning, and Policy

Various bodies within North America and the European Union (EU) have incorporated food into their planning agendas. The most notable accomplishment at the organizational level is that of the American Planning Association, which in 2007 produced a Food Planning Policy Guide. The inclusion of food in planning is not exclusive to the United States. In 2008, the Association of European Schools of Planning (AESOP) established a special thematic group to study food and planning, marking another watershed. Food joins other critical planning issues such as property rights, new technologies, urban design, ethics, transportation, and risk mitigation. Food is an essential addition to planning, as it offers a lens to see the entire urban landscape as a system. As such, this new focus can facilitate the creation of healthy communities. Thus, the importance of getting planners to respond immediately to food in broad-based policy initiatives cannot be underestimated.

In Canadian planning circles, food is just now creeping onto the municipal agenda. On the policy side, food has been on the radar for somewhat longer. Organizations, most notably the Toronto Food Policy Council (TFPC), have helped pave the way for planners to include food in their toolboxes. Early initiatives opened up opportunities and inspired others within the city bureaucracy to use food as a vehicle to deal with a range of challenges, from community beautification to improving youth self-esteem. As TFPC manager Wayne Roberts explains, 'food security is good for Toronto's integrity, cohesion and reputation as "the city that works" which in turn attracts tourists and business. Everyone benefits when fresh, local food is available at vibrant farmers' markets, lively main street grocery stores and picturesque community gardens' (Roberts, 2008). One recent initiative from the TFPC reflects this multi-faceted policy approach to food: 'healthy' food-vending carts are being introduced as a way to add to the progressive image of the city and contribute to the health of Toronto citizens and visitors. The TFPC's vision has spread to other cities in North America and the EU as many municipalities—including London (UK), Ottawa, Vancouver, Philadelphia, Berkeley, Hartford, Portland, Tacoma, and Kamloops—have adopted food charters and/or created food policy councils of their own. The appeal of this approach is the integrative perspective that comes from having a food policy council.

With this holistic approach in mind, planners at the Region of Waterloo Public Health Unit take a systems approach to food, albeit at a regional scale. As a result they have produced groundbreaking work on food miles as they connect the dots between what the region could produce for local consumption, what food the region exported, and the resulting unnecessary food miles due to a mismatch between the two. Using data about existing production capacity, Xuereb (2005) estimated that if the region ate what it could produce, it would reduce its GHG emissions by the equivalent of over 16,000 cars every year. Waterloo's public health planners have also been leaders in producing maps showing where local food producers are located as part of their broader undertaking to understand the linkages between health, food prices, local sustainability, and rural–urban relationships.

Meanwhile, a handful of municipal governments have raised the profile of food by taking on various aspects of the food cause. For example, in June 2008, the mayor of Markham, Ontario, announced a local food procurement initiative that would 'help support Ontario's farm economy, address climate change, reduce greenhouse gas emissions and pesticide use, curb urban sprawl, protect Ontario farm lands and promote sustainable farming practices' (Scarpitti, 2008). The City of Toronto announced a similar plan in the fall of 2008 for 37 daycares with the potential to expand the program to senior citizen homes and homeless shelters. The directive to purchase up to 50 per cent of food from local sources was unanimously approved by Toronto Council. Local environmental groups claimed that '[i]ncreasing the amount of local produce will support farmers in the Greenbelt and Southern Ontario and the decrease in greenhouse gases emitted from food being transported to Toronto from the other side of the globe will improve air quality' (Hanes, 2008). The EU is far ahead of North America on issues of public procurement—an excellent summary is available in Morgan and Sonnino's *The School Food Revolution* (2008). For example, Rome undertakes to purchase local, organic ingredients where possible for meals made from scratch served in its public schools (Morgan and Sonnino, 2008). The move to ban trans fats in New York, Texas, and California restaurants signals a convergence of food-based public health challenges in response to widely recognized problems of obesity, dwindling health-care resources, and public outcry about food quality.

In this regard, the intersection of policy and food reflects the importance of good **governance**. In the cases of both Toronto and Markham, public input and discussion preceded the creation of the purchasing policies. In the short term, this produced thoughtful policy and is helping to reshape government so that it supports more sustainable procurement in the long term. The hierarchical, silo approach that dominates much of current policy and planning results in disconnected outcomes that serve the needs of a few and often create more problems than solutions. A systemic approach founded on networks and informed by a broad grasp of challenges that include economic, social, and environmental considerations could improve the urban environment. Unfortunately, in Canada we have yet to achieve this broad-based and integrative mandate, but we are making progress on a number of fronts, as illustrated by the case studies in the next section.

Case Studies

Numerous projects in Canada currently capture the reality and/or potential offered by strengthening local food systems. The few examples offered here are from multiple perspectives—production, distribution, and waste—and illustrate how more sustainable food initiatives could fit into a systematic urban **regeneration** context.

Urban Gardens

Community gardens are not new in North America. During World War I and especially World War II, households and neighbourhoods were encouraged to plant Victory Gardens, and communal plots were set aside so that food could be produced locally, thereby saving costs inherent to the system of capitalist market production. The success of community gardens lies in the ability to meet a range of community needs, including the provision of beautiful and safe community spaces, informal meeting places, and spaces for building personal well-being. Community gardens can be found across Canada. In Montreal, the first community garden was created in 1936. Today the city boasts 97 community gardens with over 8,200 plots. The gardens are multi-functional, meeting recreational, food, and community needs of Montreal citizens. As well, some gardens are designed to be accessible for people in wheelchairs. The city provides seeds, soil, flowers, tool storage facilities, and access to water to support these initiatives (Montreal, 2008).

In Winnipeg, community gardens can be rented from the city for a nominal amount every year ($25 for an unserviced plot and $37 for a serviced plot). Plots are managed by the Public Works Department, which stipulates strict conditions of use. Vancouver also offers community gardening opportunities, primarily on its own parkland. In this case the city prepares the garden space—breaking ground, applying compost, and setting out individual plots—and the community is expected to manage the land in an open and inclusive way (Vancouver, 2009).

In Toronto, community gardens are seen as places to build safe community spaces where people can interact, create beauty, and get some physical activity and mental peace (Parker, 2008). As part of its network, Toronto has an eight-acre urban farm that operates in one of the most economically disadvantaged areas of the city. This garden provides youth with leadership opportunities. It also aims to break down barriers between gangs as youth get to know their peers from other parts of the city.[2]

Despite the value of these communal gardens, there are typically two problems. First, the costs associated with infrastructure and maintenance include the need to insure gardens, and second, pressures from land development are unavoidable. In the first case, basic infrastructure must be provided to gardeners, including reliable access to water, compost, tree trimming, and general maintenance. In some instances, gardens are made available to communities at no charge as municipalities absorb operational expenses. The most onerous cost is insurance. This is such a burden it has prevented some gardens from getting off the ground. In Kingston, Ontario, urban agriculture and community gardening activities are being proposed as part of the Official Plan. As part of this initiative, the designation of urban agriculture (UA) as a 'community facility' is being recommended so that gardens could be covered under city insurance policies. If approved, this will mean that UA will be a recognized land use and garden participants will not have to be concerned about liability costs.

With respect to the threat from development, the benefits and success of community gardens can also be their downfall. The textbook case in this regard comes from New York City. Starting in the mid-1990s with the bulldozing of 20 community gardens, the city moved to sell off land that had been used by communities as gardens since 1918 as part of efforts to revitalize communities. The rationale to sell garden land was that the lots had previously been building sites, so they were only temporary community garden sites. Ironically, once 'blighted' neighbourhoods that were revived in part by community gardens had witnessed increasing land prices and gone upscale, the garden lands came under pressure for housing and retail space. A second issue was the need for more land for affordable housing. In 2002, a settlement was reached between the city and the state. It protected over 200 gardens, some of which ended up in land trusts, while the rest were designated as parkland (Elder, 2005).

Positive economic considerations also are emerging from urban agriculture. The most radical of these initiatives is SPIN (small plot intensive) farming, a new farming model that is rapidly taking hold in cities across North America. The first SPIN 'farm' was developed in Saskatoon on half an acre of land pieced together from rented residential plots ranging in size from 500 to 3,000 square feet. SPIN farmers report that urban farming reduces wind and pest damage and that more plantings of fast-growing produce (e.g., lettuce and radishes) can be harvested in one season. Meanwhile, a government-funded pilot project in Philadelphia documented that gross revenue of over $50,000 can be earned from half an acre of urban land using SPIN techniques. SPIN farming has also been reported in Vancouver, Guelph, Ontario, and Iowa (SPIN Farming, 2009). The most-cited benefits of SPIN farming are the low start-up costs, profitability, and the maintenance of urban green spaces.

In Detroit and other cities where the economic meltdown has been felt especially acutely and is layered on top of earlier economic challenges, urban agriculture offers some hope. It is estimated there are 100,000 abandoned properties in Detroit alone (Roberts, 2009). New forms of urban farming such as SPIN offer economic alternatives as well providing a positive and empowering way to move from urban blight to beautification. Supportive planning and policy initiatives, such as accreditation in official plans, are needed for urban agriculture projects to become more than just scattered transitional community-based endeavours and to have a lasting place in cities.

Rooftop Gardens

Rooftop gardens offer another opportunity to cities as they look to change. As Ayalon (2006) reports, green roofs can pay multiple dividends to cities and property owners. Green roofs reduce heating and cooling costs, relieve pressure on storm water management systems by decreasing impervious surfaces, extend the life of a roof surface, remediate poor air quality, and improve aesthetic features in a city. It is also anticipated that green roofs will improve property values (CMHC, 2008).

A city can promote green roofs in many ways. It can charge taxes to make it more expensive for people who do not have a green roof (for example, storm water disposal taxes) or incentives can be tied to green roofs (for example, builders may be allowed to use a different development density if their project incorporates green roofing) (Ayalon, 2006: 45). Regulations can provide minimum expectations for new buildings. In Tokyo, new developments are required to have a minimum of 20 per cent green roof surface. Philadelphia, on the other hand, provided variances for green roofs to allow builders more flexibility to retrofit or to incorporate green roofs into new building projects.

Green roof demonstration projects also can be part of public education about the benefits of greener cities. In 1995, the Toronto Food Policy Council and the City of Toronto sponsored a green roof demonstration project in conjunction with Canadian Mortgage and Housing Corporation and identified substantial decreases to storm water management requirements due to green roofs and found that water leaving green roofs did not have to be cleaned. Research indicated that by installing 8 per cent green roof cover on existing Toronto rooftops, the city would save approximately $40 million annually (Ayalon, 2006). This work resulted in the creation of Green Roofs for Healthy Cities, an innovative NGO that provides expertise and resources about green roofs. In 2006, Toronto committed to install green roofs on existing buildings and to use green roofs on 50–75 per cent of new city buildings. In May 2009, the city passed a bylaw requiring all buildings to incorporate green roofs into their design.

Vancouver, too, has shown leadership in this area (Ayalon, 2006). In the fall of 2008 the city, as part of its new eco-density commitment, altered bylaws to remove barriers to green construction initiatives such as green roofs and renewable technologies (Vancouver, 2008). In the United States, Chicago Mayor Richard Daley has been very supportive of demonstration projects. Green roofs are encouraged through a combination of density bonuses and the city's energy conservation code (Ayalon, 2006). Another way to stimulate green roof construction is through direct subsidies. This approach has been used successfully in Linz, Austria, Basel, Switzerland, and Stuttgart, Germany. These policies subsidize up to 50 per cent of building costs and have resulted in the conversion of up to 15 per cent of existing roofs.

Redistributing Food: Eliminating the Food Desert

Important work has been going on in North America and the UK in the last decade to identify and find solutions to problems associated with '**food deserts**' (for the UK, see Wrigley, 2002; for Chicago, Block and Kouba, 2006; for Alberta,

Smoyer et al., 2006). According to work done in the UK, a food desert exists in 'areas of relative exclusion where people experience physical and economic barriers to accessing healthy food' (Reisig and Hobbiss, 2000: 138). Food deserts are linked with poor diet, and that, in turn, is linked to compromised health. In North America food deserts are related to the advent of big-box stores that require car access and can result in diminished food access in urban areas. Planners address food access challenges through zoning and other development options. The move to walkable communities, including higher-density urban use, is part of the solution. A recent study in Saskatoon highlights useful strategies. Research there points to the potential of zoning to encourage multipurpose neighbourhoods of high population density with incentives for retailers to establish in 'food desert' areas (Peters and MacCreary, 2008).

While radical shifts to more **livable cities** are being established, interim measures can be adopted. One way to provide healthier food is to rethink how spaces are used. Small farmers' markets, for example, can be located where commuters congregate and/or in leftover space in low-income areas. The Region of Waterloo Public Health Department is experimenting with this option. In the summers of 2007 and 2008, weekly markets were set up in parking lots around the City of Waterloo. Community centres, hospitals, and other public venues were included. Disadvantaged communities were targeted as a way to educate and expose people to fresh, healthy food. While the project is still being assessed, preliminary results indicate that people who had access to the markets increased the quantity and variety of food they ate. The markets also became a meeting place and helped to build a stronger sense of community.

Another example involves taking food to people, or people to food. While both solutions are less than ideal, they offer ways to reconnect people with their food; a pilot project funded by the New York State Health Department illustrates

this approach (Figures 24.1 and 24.2). In this case, Veggie Mobiles provide low-income communities with much-needed fresh produce. The concept is simple and allows people such as diabetics, who need fresh, low-sodium food, access to appropriate produce. They also provide food education by offering a 'Taste and Take' event on Wednesdays when people can sample and take home a selected produce item.

Santropol Roulant in Montreal is another example of creative food delivery. This project also targets disadvantaged citizens, in this case people who are elderly and/or have mobility challenges. Affiliated with McGill University, the project provides 'Meals on Wheels' by delivering hot meals using bicycles and bringing together youth and senior citizens. Santropol Roulant grows its own produce as much as possible on rooftops and in small urban spaces such as balconies. By using intense growing techniques that combine hydroponics with biodynamic and permaculture growing strategies, they are able to produce up to six times more food than normally would be expected from the small areas under 'cultivation'.

Other jurisdictions are addressing poverty through food access programs. In this vein, the Toronto FoodShare program is among the most recognized of its kind in Canada. Established in 1985, FoodShare improves access to food through school meal programs, the good food box, incubator kitchen facilities, youth programs, and food education programs (FoodShare, 2009). The Child Hunger Eradication Program (CHEP) in Saskatoon is another Canadian success story. By focusing on food security and child poverty issues, the program mobilizes parents, school groups, volunteers, and the local business community to improve food education and access to children living in poverty (CHEP, 2009). Both FoodShare and CHEP work to build linkages with local farmers as part of their food security strategies. However, it is important to stress that while these programs all are remarkable initiatives that reflect people's ability to work

Figure 24.1 The Veggie Mobile, a pilot project of the New York State Department of Health. (Capital District Community Gardens)

with limited resources, they also are small in scale and exist largely outside the main operations of cities; as such, they reflect fundamental problems in urban environments and highlight the pressing need to make food access an integrated consideration for urban development.

Vertical Farming

From an awareness of the need for cities to produce their own food, Dickson Despommiers and his students at Columbia University have worked to conceptualize vertical farms (Despommiers et al., 2009). While still on the drawing board, they combine principles of urban agriculture and hydroponics in multi-storey indoor farms (Figure 24.3).

Vertical farms, they suggest, would be able to provide food year-round while using space more efficiently. Depending on the crop, an acre could be 4–6 times as productive as an acre of outdoor land, while for crops such as strawberries, indoor cultivation would be 30 times as productive per acre. Growing indoors also addresses issues of drought and other weather-related challenges faced by farmers. As food would be raised in controlled conditions and grown organically, it also would be healthier for consumers. Proposed water recirculation systems would make vertical farms closed-loop food production systems as waste and water could be captured and reflowed into the system. There is also the potential to put energy onto the grid as methane could be

Figure 24.2 Interior of the Veggie Mobile. (Capital District Community Gardens)

produced from composting organic waste. Potentially, this offers the chance to reduce demands for energy inputs (fossil fuel, hydroelectric, or nuclear-generated power) as fresh produce would be grown using low inputs in the places where it would be needed. Vertical farms also offer urban employment opportunities as this type of food production would be labour-intensive. The proposal for vertical farms offers much hope for the future of healthy food production within more sustainable cities.

Conclusion

Clearly, food offers an integrative lens to urban geographers, policy-makers, planners, and other practitioners who seek to move cities in more sustainable directions. This chapter outlined the trajectory Canada has taken in addressing food issues. The overview makes it clear that food is rarely considered explicitly in the urban context. Connections to peri-urban and rural communities are also tenuous at best. While the case studies point to places where food is making its way onto the urban planning and policy agenda, much more is needed.

Through food we can develop progressive dialogues about and address challenges facing urban centres. A food prism allows us to unpack challenges such as production, distribution, and waste management in terms that are relevant to many members in a community. People need food. It provides a language and a space for uniting seemingly disparate interests. Food provides a place from which to appreciate the extent of connectivity within and beyond city systems and to situate urban areas within their foodsheds (Kloppenberg et al., 1996). This connects consumers to producers and urban with rural spaces. Food offers the chance to break down the silos between health, social justice, environmental well-being, and economic development. Recalling the emphasis the American Planning Association has placed on the positive role food can play for planners, we can begin to understand how food can help us to address the most pressing issues for cities—deteriorating urban environments (Rees, Chapter 5; Connelly and Roseland, Chapter 14), crumbling economies (Vinodrai, Chapter 6; Donald and Hall, Chapter 16), and social decay (Walks, Chapter 10; Bourne and Walks, Chapter 25)—to create more robust and resilient living spaces. While it is essential to recognize the pressures on urban resources and capacities, integrated food systems offer solutions to many related and pressing problems.

For this to take place, large-scale integration and incorporation with policies such as sustainability and smart-growth plans are necessary. Specific references to food are needed in official plans to

Figure 24.3 Prototypical model of a vertical farm. (Courtesy of Blake Kurasek)

give food-based initiatives the traction to move forward. Planners and municipal officials can help move food onto the agenda. In a newspaper interview, Paula Jones, a San Francisco food activist, explains,

> All the individual efforts are super important, but we need policymakers and business at the table, too. . . . Government can bring in not only the policy but also the funding and technical expertise that it takes to drive large-scale, systemic changes. (Rich, 2008)

Food, then, can provide the basis for partnerships between public and private interests as we conceive of and make real our cities of the future. Bringing projects such as vertical urban farms

from the drawing board into urban centres will require many partners at the table. To facilitate this process, we will need to legitimize, integrate, and standardize the role of food in urban design and policy. As recognized by the APA, food needs to be accorded a policy home so citizens, policy-makers, and bureaucrats can take ownership and develop a food-centric mandate. As this chapter has demonstrated, this is happening in some jurisdictions.

However, a permanent set of champions could bring food onto the agenda and ensure that hard dollars are committed to food-based initiatives so that we avoid relying largely or entirely on soft money and volunteers. This task calls for professional engagement at all scales—from the local to the international—to take ownership of food and to provide mandates that need to be entrenched in law. With these points in mind, it appears that the realm of planning may offer the best place for food. Planning could offer the integrated perspective now missing for food. While food will continue to be connected to issues of rurality, communities, and health, urban planning may be a way to bring food forward as a focal point for a new understanding of architecture, land-use planning, and environmental quality in urban areas.

Food provides a rallying point to re-imagine cities so they are truly 'just cities' (Fainstein, 2008). By planting our city spaces and growing our own food we have the chance to redefine our cities in sustainable terms and to empower citizens to be engaged in everyday acts of change, lifting common, daily acts to ones of transformation (Lefebvre, 1991). Food can provide a platform in the creation of the 'new city'.

Review Questions

1. Why should professional planners be concerned with urban food supplies and distribution?
2. How have urban food systems changed over the last 100 years?
3. How is the urban food system changing today?

Notes

1. Sales of organic food increased an average of 20 per cent annually from the early 1990s into the twenty-first century. Forecasts set organic sales in the US as increasing by 71 per cent between 2006 and 2011 (Knudson, 2007).
2. See also the award-winning work by Will Allen, *Growing Power in Milwaukee and Chicago*, at: <www.macfound.org/site/c.lkLXJ8MQKrH/b.4537249/k.29CA/Will_Allen.htm>.

References

American Planning Association. 2007. *Policy Guide on Community and Regional Food Planning*. At: <www.planning.org/policy/guides/adopted/food.htm>. (5 Feb. 2010)

Ayalon, O. 2006. 'Making Rooftops Bloom: Strategies for Encouraging Rooftop Greening in Montréal', Masters in Urban Planning thesis, McGill University School of Urban Planning.

Block, D., and J. Kouba. 2006. 'A comparison of the affordability of a market basket in two communities in the Chicago area', *Public Health and Nutrition* 9: 837–45.

Born, B., and M. Purcell. 2006. 'Avoiding the local food trap: Scale and food systems in planning research', *Journal of Planning Education and Research* 26, 2: 195–207.

Boye, S. 2008. *Community Gardens in the City of Toronto: Benefits of Community Gardens*. Parks and Recreation, City of Toronto. At: <www.toronto.ca/parks/programs/community.htm>. (11 Jan. 2009)

Bryant, C.R., L.H. Russwurm, and A.G. McLennan. 1982. *The City's Countryside: Land and Its Management in the Rural–Urban Fringe*. London: Longman.

Caldwell, W. 2000. 'Rural Non-Farm Development: Its Impact on the Viability and Sustainability of Agricultural and Rural Communities', research proposal submitted to Research and Corporate Services Division Research Branch, Ontario Ministry of Agriculture, Food and Rural Affairs.

Canadian Human Rights Commission. n.d. *Human Rights in Canada, Population and People, January 1, 1900*. At: <www.chrc-ccdp.ca/en/getBriefed/1900/population.asp>. (19 Jan. 2009)

Child Hunger Eradication Program (CHEP). 2009. 'Good Food Incorporated'. At: <www.chep.org/index.html>. (15 Apr. 2009)

Connell, D., J. Smithers, and A. Joseph. 2008. 'Farmers' markets and the "good food" value chain: A preliminary study', *Local Environment* 13, 3: 169–85.

Despommiers, D., et al. 2009. 'Vertical farms: Home'. At: <www.verticalfarm.com/>. (15 Apr. 2009)

Donald, B. 2008. 'Food systems planning and sustainable cities and regions: The role of the firm in sustainable food capitalism', *Regional Studies* 42, 9: 1251–62.

Dupuis, M. 2000. Not in my body: BGH and the rise of organic milk', *Agriculture and Human Values* 17, 3: 285–95.

Elder, R.F. 2005. 'Protecting New York City's community gardens', *New York University Environmental Law Journal* 13: 769–800. At: <www3.law.nyu.edu/journals/envtllaw/issues/vol13/3/v13_n3_elder.pdf>. (16 Jan. 2009)

Feagan, Robert. 2008. 'Direct marketing: Towards sustainable local food systems?', *Local Environment* 13, 3: 161–7.

FoodShare. 2009. 'Good, Healthy Food for All'. At: <www.foodshare.net/index.htm>. (15 Apr. 2009)

Friedmann, H., and P. McMichael. 1989. 'Agriculture and the state system: The rise and decline of national agricultures, 1870 to the present', *Sociologia Ruralis* 29, 2: 93–117.

Goodman, D. 2003. 'The quality "turn" and alternative food practices: Reflections and agenda', *Journal of Rural Studies* 19, 1: 1–7.

———— and M. Redclift. 1989. *From Peasant to Proletarian: Capitalist Development and Agrarian Transitions.* New York: Blackwell.

Guthman, J. 2003. 'Fast food/organic food: Reflexive tastes and the making of "yuppie chow"', *Social and Cultural Geography* 4, 1: 45–58.

Hanes, A. 2008. 'Toronto's target: 50% local food', *National Post*, 31 Oct. At: <http://network.nationalpost.com/np/blogs/toronto/archive/tags/Local+food/default.aspx>. (11 Jan. 2009)

Hinrichs, C., and T. Lyson, eds. 2008. *Remaking the North American Food System: Strategies for Sustainability.* Lincoln: University of Nebraska Press.

Jacobs, J. 1961. *The Death and Life of Great American Cities.* Toronto: Random House.

————. 2004. *Dark Age Ahead.* Toronto: Random House.

Jaffe, J., and M. Gertler. 2006. 'Victual vicissitudes: Consumer deskilling and the (gendered) transformation of food systems', *Agriculture and Human Values* 23: 146–62.

Joseph, A.E., P.D. Keddie, and B. Smit. 1988. 'Unravelling the population turnaround in rural Canada', *Canadian Geographer* 32: 17–30.

Kloppenberg, J., et al. 1996. 'Coming into the foodshed', *Agriculture and Human Values* 13, 3: 33–41.

Kneen, B. 1995. *From Land to Mouth: Understanding the Food System*, 2nd edn. Toronto: NC Press.

Knudson, W. 2007. *The Organic Food Market.* East Lansing: Michigan State University, Strategic Marketing Institute Working Paper 01-0407.

Lefebvre, H. 1991. *The Production of Space.* London: Blackwell.

Marsden, T., ed. 2008. *Sustainable Communities: New Spaces for Planning, Participation and Engagement.* London: Elsevier.

Masterman, B., and M. Sturk. 2005. *Paradise Preserved.* Calgary: Bayeux Art.

Montreal. 2008. 'Montreal, living in Montreal, community gardens'. At: <http://ville.montreal.qc.ca/pls/portal/docs/page/plan_urbanisme_en/media/documents/061030_2_1_en.pdf>. (11 Jan. 2009)

Morgan, K. 2008. 'Greening the realm: Sustainable food chains and the public plate', *Regional Studies* 42, 9: 1237–50.

————, T. Marsden, and J. Murdoch. 2006. *Worlds of Food: Place, Power and Provenance in the Food Chain.* London: Oxford University Press.

Nemore, C. 1997. *Rooted in Community: Community Gardens in New York City.* A Report to the New York State Senate. At: <www.cityfarmer.org/NYcomgardens.html>. (16 Jan. 2009)

Parker, E. 2008. Personal communication, City of Toronto.

Peters, E., and T. McCreary. 2008. 'Poor neighbourhoods and the changing geography of food retailing in Saskatoon, Saskatchewan, 1984–2004', *Canadian Journal of Urban Research* 17, 1: 78–106.

Planners for Tomorrow. 2006. 'Planning for tomorrow', University of British Columbia Community and Regional Planning. At: <www.plannersfortomorrow.ca>. (11 Jan. 2009)

Pirog, R. 2001. *Food, Fuel, and Freeways: An Iowa Perspective on How Far Food Travels, Fuel Usage, and Greenhouse Gas Emissions.* Ames: Iowa State University, Leopold Center for Sustainable Agriculture, June.

Rich, D. 2008. 'Farming the city: Planners start thinking of how to feed us', *San Francisco Chronicle*, 22 Mar., F-1.

Reisig, V., and A. Hobbiss. 2000. 'Food deserts and how to tackle item: A study of one city's approach', *Health Education Journal* 59: 137–49.

Santropol Roulant. 2006. At: <www.santropolroulant.org/2006/E-home.htm>. (16 Jan. 2009)

Scarpitti, F. 2008. 'Announcement on Markham's agreement with Local Food Plus to adopt LFP standards and zero waste program for municipal catering and food services', Municipality of Markham, 4 June, 2008 Smog Summit, Toronto. At: <www.organicfoodcouncil.org/files/downloads/Announcement%20on%20Markham.pdf>. (10 Jan, 2009)

Smoyer-Tomic, K., J. Spence, and C. Amrhein. 2006. 'Food deserts in the prairies? Supermarket accessibility and neighbourhood need in Edmonton, Canada', *Professional Geographer* 58, 3: 307–26.

Southern Alberta Land Trust Society (SALTS). 2009. 'What we've accomplished'. At: <www.salts-landtrust.org/who_5_accomplishments.html#5>. (19 Jan. 2009)

SPIN Farming. 2009. 'SPIN makes agriculture accessible to anyone, anywhere'. At: <www.spinfarming.com/whatsSpin/>. (11 Jan. 2009)

Statistics Canada. 2007. *A Statistical Portrait of Agriculture, Canada and Provinces: Census Years 1921 to 2006.* At: <www.statcan.gc.ca/pub/95-632-x/2007000/t/4185571-eng.htm>. (13 Dec. 2008)

Toronto. 2009. 'Living in Toronto: Green Bin Program'. At: <www.toronto.ca/greenbin/index.htm>. (16 Jan. 2009)

Vancouver, City of. 2008. 'City of Vancouver Fact Sheet: Green Building Made Easier'. At: <www.vancouver-ecodensity.ca/webupload/File/Green%20Barriers%20Fact%20Sheet.pdf>. (16 Jan. 2009)

————. 2009. *Vancouver's Community Gardens*. Parks and Recreation. At: <http://vancouver.ca/parks/parks/comgardn.htm>. (11 Jan. 2009)

Wackernagel, M., et al. 2006. The ecological footprint of cities and regions; Comparing resource availability with resource demand', *Environment and Urbanization* 18, 1: 103–12. Excerpts at: <www.footprintnetwork.org/en/index.php/GFN/page/footprint_for_cities/>.

Whatmore, S. 2002. *Hybrid Geographies: Natures, Cultures, Spaces*. London: Sage.

World Hunger Year (WHY). 2007. 'Food Security Learning Centre: Community gardens'. At: <www.whyhunger.org/programs/fslc/topics/community-gardens.html>. (15 Jan. 2009)

Wrigley, N. 2002. '"Food deserts" in British cities: Policy context and research priorities', *Urban Studies* 39, 11: 2029–40. At: <www.informaworld.com/smpp/content~db=all~content=a713707798>.

Xuereb, M. 2005. *Food Miles: Environmental Implications of Food Imports to Waterloo Region*. Waterloo, Ont.: Region of Waterloo Public Health. At: <http://chd.region.waterloo.on.ca/web/health.nsf/0/54ED787F44ACA44C852571410056AEB0/$file/FOOD_MILES_REPORT.pdf>.

Part IV

Towards a Sustainable, Healthy, and 'Smart' Future for Canadian Urban Communities

Part 4, Chapter 25, by two of Canada's leading urban geographers, Larry S. Bourne and R. Alan Walks, synthesizes the lessons learned from the preceding chapters, identifies several themes present throughout *Canadian Cities in Transition*—complexity, contradictory tendencies, and uncertainty—and discusses the challenges and opportunities facing urban Canada today. In considering the challenges that cities must confront, the authors focus on the growing problem of homelessness. In exploring the planning and policy nexus, they especially consider the tension implicit in the goal of the 'compact city' and the reality of urban sprawl.

CHAPTER 25

Conclusion: Challenges and Opportunities in the Twenty-First-Century City

Larry S. Bourne and R. Alan Walks

By most international standards Canadian cities score high on rankings of the quality of life offered to their residents. They exhibit a number of positive features as well as a relative absence of serious negative attributes. Canadian cities, for example, have not suffered the extremes of poverty, ghettoization, and social segregation evident in cities in many other parts of the Americas. Nor do they exhibit the same levels of congestion, pollution, crime, and corruption that inflict other cities, especially those in the developing world. Moreover, the level of local public goods and services provided in Canadian cities, including health care and education, has continued to be of relatively high standard. Nonetheless, there are also indicators of rising stress, for example, widespread recognition of both persistent problems and newly emerging challenges. There is, for instance, growing evidence that living conditions for many residents in our cities are deteriorating, and as a consequence increasingly vocal arguments stress that co-ordinated actions by governments, agencies of civil society, and the private sector are necessary to address these conditions.

The previous chapters in this volume have demonstrated, often in dramatic fashion, the wide range and scale of challenges facing Canadian cities. In this chapter we identify a number of these challenges—summarized for convenience as clusters of related issues—and outline some of the implications of these issues for the future economic viability and living environments of our cities. We then use these clusters to illustrate three cross-cutting themes deduced from previous chapters: first, the complexity of change in today's cities; second, the contradictions inherent in urban growth and in our policy responses to that growth; and third, the destabilizing effects of uncertainty in the economy, in public policy, and in everyday urban life. The second section offers two case studies to illustrate these themes as well as brief speculations on alternative future directions in urban development.

Cross-Cutting Themes

Complexity, contradictory tendencies, and uncertainty, of course, are not unique to cities; indeed, they are common in all social systems. They are, however, particularly relevant to an understanding of the processes of urban change and to the task of devising appropriate strategies for public policy and collective action. In theory, *complex social systems*, unlike mechanical systems (e.g., car engines) or electronic systems (e.g., computers), are characterized by a large number of quasi-independent parts and by individual, corporate, and institutional actors whose decisions and behaviour can transform the system, or at least shift it to a new development trajectory. Those decisions are not made in a vacuum; they are conditional on past events and actions, and are restricted by societal structures and regulations, as well as by inherited norms and rules of behaviour. Thus, the combined outcomes of such decisions are almost impossible to predict. Moreover, another important lesson learned for researchers and policy-makers is that in complex social systems there is no necessary and predetermined connection between a stimulus (e.g., introducing a new technology, or a new policy initiative) and a specific response in terms of how the system changes. Contemporary cities represent special cases of these attributes.

In this context *complexity* refers to the immense scale, diversity, and density of interactions that are the hallmark of the contemporary metropolitan area. The thousands of businesses and commercial activities in the modern metropolis, the high rate of turnover of those activities, the embedded diversity of the city's social geography, the thick overlay of institutions and government agencies, and the myriad regulations on land use and development (e.g., zoning, building bylaws, fire regulations) illustrate this complexity. Today's urban landscapes are thus characterized both by inertia (the 'sunk costs'), which discourages change, and the fluidity of turnover in businesses and social conditions. Further, one should appreciate the almost infinite variety of connections—the linkages and flows of people, goods, and money—among firms, individuals, and public bodies and between places within a large urban region. These interactions, both real and virtual, are the glue that ties the urban area together as a social and spatial system. The daily journey to and from work is but one obvious example; it is, simultaneously, a spatial linkage (between home and workplace), a sectoral linkage (between labour and the sphere of production), an actual physical flow (of people and vehicles), and a major transportation problem.

In sum, this complexity makes it difficult, as Chapters 2 (Bunting and Filion) and 3 (Filion and Bunting) have shown, to conceptualize **urban form**, to define and measure urban structure, to understand how that structure is spatially organized

and how (and why) it is changing over time. At one time fairly simple models of urban form, based largely on the experience of the industrial city of the United States, were accepted as explaining the form and structure of cities almost everywhere. These were followed by more complex models based on late industrial and early post-industrial cities examined in numerous urban contexts. We now have a situation in which the sheer complexity of urban processes makes even a multiplicity of models seem incomplete. This is one key characteristic of our current juncture in terms of urban theory and public policy.

Complexity, in turn, poses a specific challenge for any attempt at comprehensive urban planning. As we know, everything is related to everything else; a planning initiative undertaken in one geographic area of the city or in one sector invariably has impacts—or *externalities*—on other areas and sectors. These consequences may be positive, as the policy intended, but often they are unintended, and sometimes they are negative and inequitable in the sense of an uneven distribution of costs and benefits. As a simple example, adding or expanding green space in cities is generally considered a good idea; but in a market economy it may also raise real estate values in the adjoining neighbourhhoods, which in turn may lead to the displacement of lower-income households or a reduction in levels of affordability. Stressing the complexity of cities is not, however, a justification for inaction; rather, it is an argument for careful thought and analysis before any deliberate action is undertaken.

The second theme, the presence of *contradictory tendencies*, is also a characteristic typical of complex social systems. In most such systems for every trajectory of change, for every force or set of forces moving the system in one direction, there are countervailing actions or tendencies that seek to maintain the status quo or to move the system in a different, if not opposite, direction. In contemporary cities there are numerous examples to illustrate the importance of this concept. One is the simultaneous pressure for *decentralization*, the suburbanization of economic activities and residential populations searching for more and cheaper land and highway accessibility, and *centralization*, the attraction of the revitalized and intensified central core and gentrified neighbourhoods. These two broad structural processes are related but not in any simple or straightforward manner. At the local level, rapid decentralization can lead to uneven development with some households left behind (e.g., with limited access to suburban job opportunities), while residential renaissance in older neighbourhoods, a direction of change strongly encouraged by policy-makers, developers, and local business, invariably leads to less affordable housing, displacement of low-income households into less accessible locations, and higher real estate prices for all (see Skaburskis and Moos, Chapter 13).

Contradictory tendencies, of course, are literally built into the urban process under capitalism, and some would argue that the same is true in socialist and centrally planned economic systems, although in rather different forms. Within capitalism, as David Harvey noted years ago (1989), there is always a tension between the desire to hold onto current capital assets and the felt need to tear down what exists in order to create more valuable assets. On the one hand, people see the need to maintain and nurture investments in both the physical and built environments, and in associated social arrangements, so that these investments can provide sustained economic benefits for urban residents and firms over many years; on the other hand, people identify the need to destroy and remake urban physical and social landscapes so as to compete with rates of return on investment elsewhere in the economy and thus provide for sufficient capital accumulation.

This tension is accentuated by the durability and spatial fixity of urban investments (e.g., in buildings and infrastructure), which are amortized over very long periods of time, while the conditions for maintaining profitability typically undergo much more rapid change. Thus, the capitalist city is often built up during periods of rapid economic growth to reflect the economic and social requirements for capital accumulation at that time, only to have that built environment and the social relations it supports act as a weight on future rounds of innovation and growth. This is one reason for the periodic economic crises that result in widespread urban unemployment and social problems, and why these are often followed by periods of 'creative destruction' of varied parts of the urban landscape and the urban labour market.

The third theme, *uncertainty*, is also a universal feature of complex social systems. This is the case not simply because they are complex but because they are open to external influence and subject to transformations deriving from inside the system. Cities clearly fit this description. Cities depend on imports of resources (e.g., energy, food), human capital, and labour from their hinterlands and beyond, just as they export goods and services (and waste) beyond their borders. A shift in any one of these exports or imports, or in the balance of the two, can change the trajectory of growth and the form of urban development. The boundaries of cities as functional entities are fuzzy and their economies are open to the impacts of policy decisions and changing conditions elsewhere in the world over which they have little or no control. These external factors or drivers of change—with obvious examples being international competition, trade, new technologies, and immigration (see Filion and Bunting, Chapter 1)—have very different impacts in different cities, on different spaces within the city, and on particular firms and social groups. These same drivers may also shift dramatically in a short period of time, as in the case of the recent restructuring of resource sectors in BC and in the financial services and automobile sectors in the cities of southern Ontario.

While economic uncertainty has always been a product of Canadian urbanization, and indeed of the growth and development of all cities, a plausible argument suggests that the level of uncertainty has increased in recent years. This uncertainty is likely to remain high for the foreseeable future in response to the combined effects, among other factors, of global competition, instability in financial markets, the economic weakness of Canada's main trading partner, and political events that cannot be predicted, not to mention climate change and potential worldwide energy and water shortages.

Uncertainty invariably increases with the size of the system and with the level of complexity in that system. For cities it also increases with the degree of dependence on international markets and external sources of growth, as Chapter 4 by Peter Hall demonstrated. With the emergence of a new period of significant global economic restructuring, the level of uncertainty has again increased, posing a further challenge for public policy formation and for those projecting longer-term trends. How does one incorporate uncertainty into planning policy and practice, and in public policy formulation in general? How does one develop forecasts of population and employment within a context of heightened uncertainty in the components of both?

This issue is magnified in times of slow growth. Indeed, many urban scholars are now bracing for the prospect of widespread urban decline, given historically low fertility levels and thus reduced population growth. Research has examined how to plan for population and economic decline in ways that minimize social harm and economic disruption, while maintaining necessary public services (see Donald and Hall, Chapter 16; Simmons and Bourne, 2007). This obviously poses a major challenge for planners and policy-makers whose unquestioned assumption over many years has been the continuation of urban growth.

Urban Challenges and Policy Issues: A Summary

With the above discussion as context, this section summarizes the major challenges facing Canadian cities in the twenty-first century. The purpose is to encourage readers and students of the city to bring together the diverse issues discussed earlier into an integrated framework. For ease of presentation, these are then grouped as *clusters of related policy issues* (see Table 25.1). Although every reader will have his or her own list, and will attach his or her own descriptive labels, the following list covers most of the current policy agenda in Canadian cities and the main subjects currently under debate in academic circles and public forums. Of course, the list could be longer or shorter, and the decision of how long such a list should be is always somewhat arbitrary. For each of 11 clusters we identify a particular *vision* of what constitutes the 'good' city, a vision that encompasses the ideal direction of change and the ultimate goal that could be achieved if the specific issues in question are seriously addressed. The ordering of issues and visions here does not necessarily imply a hierarchy of interest or of priority for public policy. For each set there is also a relevant body (or bodies) of theory, a huge portfolio of literature, both academic and professional, and a proactive 'community of interest' committed to raising the visibility of the issue as well as elevating the priority that politicians, public agencies, and the community at large assign to achieving the stated goal.

Each set of challenges, by definition, suggests a specific set of policy responses. Thus, if urban poverty, and the concentration of poverty within certain neighbourhoods in the city, is viewed as a major problem (or problems), then appropriate policies are necessary to reduce the overall level of poverty, enhance the redistribution of income, and raise the quality and accessibility of public services available to the most disadvantaged groups in those neighbourhoods. The goal, in this case, would be a more inclusive 'equitable' or 'just' city.

The 11 sets of issue areas cover almost every facet of urban development and contemporary urban life in Canada. The visions offer a heuristic device for summarizing the immense diversity of perspectives in the literature on the preferred or ideal futures for Canadian cities. They also represent a short-hand description of the different priorities assigned to sets of issues. Thus, for some observers, the desired future is for a 'green' city (a.k.a. the sustainable city); for others it is the 'prosperous' or economically **competitive city**, the 'efficient' city (in terms of productivity and the movement of people and goods), or the 'equitable' (or socially just) city. Still others argue, under different labels, for the 'affordable' city, the 'healthy' city, the 'learning' city, and for 'safe' cities, and finally, the planners' current delight, the 'compact' city.

The first set of issues assigns priority to the state of the economy, to economic renewal, local economic development, job creation, and improving labour market conditions. Recent public opinion polls, by the Trudeau Foundation, suggest that these are the current priority issues for most residents. They also are the issues most dependent on external economic conditions. The history of economic growth and development and the place of Canadian cities within global networks of production and trade are discussed in Chapters 1 (Filion and Bunting), 4 (Hall), 5 (Rees), and 6 (Vinodrai) of this volume.

The second theme identifies the links between stresses on the urban environment (natural and man-made) and ecological health and sustainability. As Chapter 5 makes clear, there is a contradiction between urban development and environmental sustainability. The linear metabolism of city economies exacerbates the tendencies towards overconsumption of the earth's resources, while the urban concentration of population grounds the problem and the outcomes of overshoot (in which the **ecological footprint** of the humans is larger than then carrying capacity of the earth) directly in cities. Unfortunately, Canadian cities are particularly wasteful and energy-intensive, and these attributes are unlikely to be sustainable over the long term. In turn, they also add to the problems and contradictions of food production and distribution (see Blay-Palmer, Chapter 24).

The third thematic cluster incorporates road congestion, transit needs, and the widening gap between infrastructure needs and their provision and maintenance (e.g., water, sewers). Chapter 11 (Perl and Kenworthy) in this volume examines networks of transportation and communication in the city, and demonstrates many of the contradictions deriving from how Canadian cities have developed, including health effects, and the relatively inefficient urban form that has evolved to serve automobile transportation. Canadian cities are some of the most auto-dependent cities in the world, although less so than US cities, and thus are especially vulnerable to oil and energy shocks.

The fourth theme groups the challenges of maintaining **social cohesion** and harmony while addressing concerns over poverty, social alienation, and disadvantage, which leads to the question of what is the socially just, equitable city. As discussed in detail in Chapter 10 (Walks), Canadian cities have become less equal and more polarized over the last 30 years, particularly since the early 1990s. This is mostly articulated in occupational terms as the Canadian labour market, following that in the United States, with divergence between the executive/managerial class, who have seen their incomes grow, and those working in other occupational sectors, who have seen their incomes either decline or remain flat bifurcates. Thus, while Canadian cities are not beset by US-style racial and ethnic segregation, they exhibit considerable levels of class and neighbourhood inequality, and most cities have become considerably less equitable and less socially just over time.

The fifth theme is also related to the issue of equity, but specifically identifies access to affordable housing and the persistence of homelessness as serious social issues. These problems are discussed in more detail below.

The sixth issue area acknowledges the critical role of maintaining, if not improving, the quality (and financing) of public spaces, and more broadly the *public goods and social services* (e.g., schools, police, fire, waste disposal, hospitals, community centres) delivered to urban residents. Many Canadian cities have experienced severe fiscal pressures, and this has impacted their ability to deliver quality public services. While not covered in detail by any of the chapters in this volume, this is an area of increasing concern. It has also raised the thorny question of the privatization of public services, as has happened, for example, in the UK. Many municipalities have been going deeper into debt to maintain basic public services, yet they face a reduced ability to attain revenue from local own-source revenues (e.g., property taxes, user fees) as well as inadequate transfers from provincial governments. The Canadian Federation of Municipalities estimates that in 2008 the deficit for infrastructures was $123 billion.

The seventh cluster of issues focuses on improving local **governance**, addressing the need to make civic participation more meaningful while encouraging greater political accountability. Participation and accountability are intended to instill an enhanced sense of place and thus to reduce the 'democracy deficit' evident in political apathy and low voter turnout at the polls. Unfortunately, a policy agenda based on financial stringency and neo-liberal principles has been adopted by upper levels of government over the last decade, and this has forced a more entrepreneurial and less democratic approach to governance onto urban municipalities. Chapter 12 (Allahwala, Boudreau, and Keil) provides a detailed account of these issues in the context of the Canadian city.

The eighth theme concerns the vision of the 'learning city'. This encompasses more than merely formal education, important as that is, and emphasizes the challenge of improving education and skills training, thereby augmenting human capital and improving the viability of local labour markets. This, of course, feeds back into the ability of cities to foster technological and social innovation, especially as such innovation relates to economic sustainability and resilience (Wolfe, 2009). Chapter 6 by Vinodrai examines the issues of creativity and innovation within the context of the cultural economy of Canadian cities.

Ninth, a host of interrelated challenges confront cities in their attempts to adapt to rapidly increasing ethnocultural and racial diversity, as well as to the ongoing demographic transition, specifically to a rapidly aging population. These come together in the form of the 'inclusive city'. Chapter 9 (Hoernig and Zhuang) examines the history and geography of immigration in Canada and its effects on Canadian cities, while Chapter 8 (Townshend and Walker) deals with the issue of aging and lifestyle changes. The latter are rapidly becoming a challenge for the continued vitality of cities (see also Rosenberg and Wilson, Chapter 21), and for efforts to encourage inclusive development, as well as for labour markets (either in terms of unemployment or labour shortages, or both). While Canadian cities, unlike their US counterparts, do not contain any '**ghettos**' (see Walks, Chapter 10), the larger metropolitan areas nonetheless face many problems related to high rates of immigration, labour force discrimination, and the racialization of poverty, and to a degree exhibit patterns similar to those found in some European cities.

The tenth grouping brings together issues related to crime, including youth violence, antisocial behaviour, and alienation, and the need to enhance feelings of safety, security, and an overall sense of belonging. While remaining relatively safe, Canadian cities are characterized by increasing levels of fearfulness (see Cowen, Siciliano, and Smith, Chapter 17), and this perception is reflected in the growing number of gated and guarded housing estates within certain Canadian cities (see Grant and Filion, Chapter 18; Lynch and Ley, Chapter 19). Furthermore, the logic of the entrepreneurial and corporate city compels urban administrations to enact legislation that can have negative impacts on the most disadvantaged populations. This may contribute to criminalizing certain populations and to privatizing or otherwise limiting access to public space and services (see Allahwala, Boudreau, and Keil, Chapter 12).

Finally, we acknowledge the challenges of defining the appropriate roles for, and instruments

Table 25.1 Challenges Facing Canadian Cities

Clusters of Policy Issues Reflecting Contradiction and Uncertainty	Urban Vision
1. economic development; unemployment; job quality	prosperous city (competitive city)
2. environment and ecology: air, water, forests, food	green city (sustainable city)
3. congestion; transportation and infrastructure provision	efficient city
4. social cohesion; inequalities; poverty	equitable city (socially just city)
5. access to housing; affordability; homelessness	affordable city
6. quality of public spaces; nutrition; services; social capital	healthy city
7. governance; local democracy; citizenship	democratic city
8. education; skills and knowledge generation	learning city
9. immigration; demographic change; aging	inclusive city
10. crime and safety; alienation; civility	safe city
11. planning practice: smart growth and decline	compact city

of, urban planning and public policy. This issue is elaborated below.

These visions, of course, are not necessarily either compatible or incompatible, nor are they inherently contradictory, although some clearly are. For example, green cities, in theory at least, can be prosperous, equitable, healthy, and compact. Cities with efficient transportation systems and adequate stocks of affordable housing can be economically competitive, prosperous, and socially equitable. Indeed, some would argue that cities of the future need to be all of these things at the same time. In keeping with the emphasis on complexity, it needs to be noted that all of these issue areas are related, and all of them factor into a discussion of just what constitutes the 'good' city. Yet there are areas of potential conflict, and in some instances trade-offs are necessary. What are those trade-offs, and how do we decide on priorities? Is there necessarily a trade-off between economic efficiency and environmental quality, or can they be achieved simultaneously? Can a city be entrepreneurial and inclusive at the same time? Is the goal of more affordable housing at odds with the principles of smart growth? It is also true that otherwise

reasonable policies in one area can have negative effects in other areas. In any case, it is essential for governments, communities, and policy-makers to set priorities, to establish timetables for action and schedules for social investment, since not everything can be done at once.

The importance attached to each cluster of issues, the priority assigned to attaining each vision of the good city, and the nature of the policy response also will differ among cities, as well as among communities within an urban area. And each will vary in importance over time, depending, for example, on the business cycle and on changing political circumstances. Thus, while encouraging economic development is a challenge for all communities in Canada, the assets available for economic renewal will differ widely. Congestion is not likely to be a problem in smaller cities or in cities facing modest or negative population growth. Environmental stresses appear almost everywhere but frequently in different forms in different cities. In some locations air pollution is the dominant concern; in others it is the quality of water supply, a shortage of green space, the presence of toxic soils, or, almost universally, garbage disposal.

Other examples can be used to illustrate the point of an immense geographical variability in urban issues across the country. Access to affordable, reasonably priced housing is more of a challenge in the larger and growing metropolitan areas, whereas in smaller cities the supply is often there but not the income to purchase or rent suitable housing (see Walker and Carter, Chapter 20; Donald and Hall, Chapter 16). Public goods and services, such as policing, schools, hospitals, and parks, tend to be overwhelmed in large and growing urban areas by increasing demand, while in slower-growing cities the challenge is to secure revenue to maintain existing levels of services and infrastructure. In communities with declining populations the explicit challenge is to downsize the infrastructure base and in some instances the public sector (e.g., school closings) in a humane, equitable, and efficient manner. And obviously, the challenges for planning practice will differ widely depending on local history and economic and political circumstances. There is no 'one-size-fits-all' classification of challenges facing Canadian cities, nor can the same strategies be expected to meet these challenges in all circumstances.

Homelessness in the Canadian City

The problem of urban homelessness demonstrates well the themes of complexity, contradiction, and uncertainty that we have highlighted. Furthermore, alternative policy approaches to homelessness reflect the myriad ways those challenges facing cities are conceptualized and addressed.

Homelessness is perhaps the most visible expression of extreme poverty and of growing social inequality in urban Canada. It is very much related to how the poor are taken care of and how income is redistributed within the **welfare state**, and thus this issue is tied directly to the idea of a socially just city. According to the United Nations, homelessness is defined as (1) having no place to

call home and being forced to sleep either outside or in a temporary shelter, or (2) having access to housing that is lacking in one or more of: sanitation, protection from the elements, safe water, security of tenure, affordability, personal safety, and accessibility to daily needs (particularly employment, education, and health care). This definition, then, includes those literally without a home (often termed *absolute homelessness*), but also those whose current housing situation is precarious and insufficient (*relative homelessness*), such as when someone sleeps on a couch at a friend's home for a period of time or when people (usually women and children) are forced to seek temporary shelter outside the home because of abusive relationships.

Indeed, it is possible to consider a continuum from the most extreme form of homelessness—sleeping rough outside (under bridges, over subway grates, in automobiles, etc.)—and homefulness, which provides the most secure form of tenure and the greatest degree of control over one's housing situation (Figure 25.1). Of those sleeping rough, surveys conducted in Halifax and Greater Vancouver inquired where respondents had slept the previous night. The most common response was that they had slept at someone else's place (28 per cent and 56 per cent in Halifax and Vancouver, respectively), followed by sleeping outside, squatting in another building, staying in a shelter, and/or sleeping in a car, garage, or public building (GVRD, 2002; HRM, 2004). It might be noted that lack of shelter has more deleterious and dangerous effects in Canada than in many other countries, due to the extremely cold winter weather. Every winter, a number of those sleeping rough die of exposure in Canadian cities (Layton, 2000).

As a form of poverty, homelessness is directly tied to access to employment for low-skilled workers, overall levels of income and affluence, and forms of economic development. It is thus related to how prosperous and competitive a city is, and in turn to the structure of demand within the labour market. Of course, even if one has a job, one may

Figure 25.1 Continuum of homefulness to homelessness

end up homeless if the wages are not sufficient to cover the costs of housing plus other daily necessities. Those whose housing is sufficient but who are paying so much on housing that it cuts into their ability to meet other daily needs are experiencing *housing affordability stress*, and thus are considered to be *at risk of homelessness*. Stone (1993) calls this situation *shelter poverty* because in such cases housing costs are the cause of households' current poverty situation, and of their risk of falling into homelessness. Homelessness is simultaneously an issue of housing affordability and of how government policies relate to the production of both market-rate and socially subsidized housing.

As Chapter 20 (Walker and Carter) makes clear, since the early 1990s upper levels of government in Canada have drastically cut back funding for new **social housing** construction and, at the same time, have relaxed restrictions governing landlord–tenant relationships, tenure conversions, and the construction of new high-end housing. One result is a relative decline in affordability in both the rental and owner-occupied stock. Growing inequality (see Walks, Chapter 10) fostered in part by neo-liberal policy reforms (see Allahwala, Boudreau, and Keil, Chapter 12) has meant that median incomes have essentially remained flat, while the lowest-income earners have seen their real incomes decline. Declining or flat incomes, coupled with rising housing costs, result in more people whose demand cannot be met in the housing market,

leading to increasing homelessness, particularly if little new social housing is being built. This is an important reason for the growth in homelessness in Canadian cities over the last 30 years.

Of course, a number of factors explain why some poor individuals become homeless while others do not. Traditional debates regarding the causes of poverty and homelessness have fluctuated between two models: (1) an *individual agency* model and (2) a *structural* model. Explanations derived from an individual agency perspective assume that homelessness is the result of personal failing or various individual 'risk factors' (substance abuse, physical disability, mental illness, criminal behaviour, family breakup, poor job skills, etc.). In contrast, the *structural* paradigm explains homelessness as the result of the workings of capitalist housing and labour markets, which typically produce cyclical and structural unemployment, income inequality, and high housing costs, and these markets have worked to limit the construction of affordable rental housing over the last 20 years (Bacher, 1993; see Walker and Carter, Chapter 20). Recent work points to the need to integrate these two perspectives. O'Flaherty (2003) suggests that while individual-level factors may predict *who* is at highest risk of ending up homeless within a particular city, rent levels and housing market conditions determine the *proportions* of such persons left without housing across cities.

According to surveys undertaken in Canadian cities through the early 2000s (Table 25.2),

Table 25.2 Demographic Characteristics of the Homeless, Selected Canadian Cities (%)

	Calgary	Edmonton	Halifax	Montreal	Toronto	Sudbury	Vancouver
Date of Study	**2008**	**1999**	**2003**	**1996–7**	**2006**	**Jan. 2003**	**2002**
Gender							
Women	22.1	27.0	33.0	25.6	26.4	44.0	32.0
Men	77.9	73.0	67.0	74.4	72.7	56.0	68.0
Age							
Children under age 18	10.9	23.0	25.0	4.6		28.2	13.0
under age 21					8.2		
Adults 18–64	81.7	67 (19–54)*	52.0	93.2		69.7 (19–59)*	87.0
Young adults 18–24	9.4		16.0				15.0
Young Adults 18–29				34.3			
Seniors	1.9	9.0 (55+)*	5.0	2.2	1.6	2.1 (60+)*	1.0
Marriage status (adults)				(1991):			
Single/divorced/widowed	na	87.4	89.0	87.1	na	87.0	90.0
Married	na	12.6	11.0	12.9	na	12.8	10.0
Family households %	na	24.0	20.0	na	na		na
Race/ethnicity				(1991):			
White	62.5	53.0	63.0	94.2	na	62.0	69.0
Aboriginal	15.1	42.0	14.0	na	16.2	27.0	17.0
Other visible minority	10.6	5.0	17.0	na	na	2.0	14.0
Black	na	na	9.0	3.1	na	Na	2.0
Asian	na	na	7.0	na	na	Na	3.0
Other	na	na	1.0	2.6	na	Na	
Source of income	(2002):			(1991)**			
Panhandling/begging	4.7	na	5.0	4.7	17.4**	Na	6.0
State benefits, any form	5.7	na	48.0	73.3	94.7**	42.3	51.0
Pension	na	na	7.0	9.7	na	2.8	2.0
Disability	na	na	18.0	na	21.7**	12.2	9.0
Welfare/training	na	na	23.0	63.6	62.2**	28.3	40.0
Charity/street allowance	na	na	na	na	10.8**	Na	na
Employment or EI	52.0	na	12.0	24.1	na	6.5	13.0
Employment	52.0	na	na	16.0	23.2**	3.6	11.0
EI	0.0	na	na	8.1	na	2.9	2.0
Family/friends help	na	na	na	na	17.3**	Na	na
No income	na	na	23.0	8.9	na	48.3	21.0
Other	9.4	na	11.0	6.7	na	1.4	8.0

**Multiple responses were allowed.

*Age range given in parentheses.

Note: Only those studies enumerating the homeless are included here. The 2002 Calgary study only provided sources of income for unemployed respondents, implying that employment income was the main source for the 52 per cent of respondents who were employed in some fashion. The Toronto study only asked source of income for those with income.

Sources: BCMSDES (2001); City of Calgary (2002, 2008); City of Toronto (2006); Fournier (1991); Fournier et al. (1998); GVRD (2002); HRM (2004); Kuappi (2003).

the 'absolute' homeless (those sleeping rough or in homeless shelters) are more likely to be older, single men and those of Aboriginal ancestry, and they are less likely to be visible minorities or immigrants (except in Halifax). As well, a number of the homeless suffer from depression or mental illness, and have experienced problems with substance abuse (Table 25.3). The homeless also disproportionately suffer from physical disabilities and other medical conditions, and are likely to have entered their predicament due to eviction, domestic abuse, or family breakdown. Women and children, and families, are the fastest-growing subset of the homeless (FCM, 2005).

While many Canadians may imagine the homeless as panhandlers or street beggars, this does not fit the reality of homelessness in Canadian cities. Instead, the majority of the homeless depend on state benefits of some form for their income, mainly disability payments, welfare, and, to a lesser extent, pension income. Furthermore, a significant proportion of the homeless are employed full-time or part-time, particularly in Calgary where the housing market became especially tight in the past decade. Notably, very few homeless persons report being involved in any illegal income-producing activities. Surveys from homeless people across Canada indicate a significant level of diversity of the homeless situation, and thus a large degree of both complexity and uncertainty. The multiplicity of pathways to homelessness in turn demands a multiplicity of policy approaches. The degree to which a city can accommodate such diversity and meet the needs of different disadvantaged groups is related to the concept of the inclusive city.

The rise of homelessness in the Canadian city reveals tendencies towards contradiction. Of course, it is difficult to measure homelessness, not least because those without a fixed address are very hard to locate, and the true numbers of the homeless are thought to be far higher than what can be gauged from traditional sources. Most estimates of the number of homeless are based on either counts of unique individuals using emergency shelters, or occasional enumerations ('snapshots') of those sleeping rough on the streets, under bridges, and in cars (Figure 10.2 in Chapter 10 by Walks provides one such example from Calgary). The first street enumeration in Toronto in 2006 found 5,062 homeless people sleeping rough (City of Toronto, 2006), while accounts from other Canadian cities demonstrate that the problem is both acute and growing.[1] Highly troubling is that the growth of homelessness occurred during a time of economic boom, creating anxiety and uncertainty about how the global economic transformation, which unveiled itself with such force in the fall of 2008 with the onset of a deep global recession, might produce new rounds of homelessness.

Part of the contradiction is that growing homelessness is at least partly related to rising affluence among the highest-income earners who then outbid others for space. As Chapter 20 by Walker and Carter makes clear, the problem in Canada is not a lack of housing—indeed, our housing industry is one of the most efficient in the world and there is more than enough housing to go around—but distribution of resources, that is, distribution by tenure type, by price level, and by geographic area. Some regions and cities have a surplus of housing, while in others vacancy levels are near zero. High incomes drive up housing costs and encourage those at the top of the income ladder to consume both more housing units and more commodious accommodation than they otherwise would. This has the unintended effect of holding back investment that could otherwise be put towards improving rental markets. Neo-liberal policies, coupled with the derivatives/credit bubble of the 2000s (see Immergluck, 2009), have induced many high-income earners to speculate with real estate assets, so driving **gentrification** (tenure conversions, etc.) of the inner-city low-income stock (see Walks and Maaranen, 2008a, 2008b), and pushing the housing market to concentrate disproportionately on building high-end condos and monster houses (Bain, Chapter 15).

Table 25.3 Characteristics of the Homeless Situation, Selected Canadian Cities

	Calgary	Halifax	Ottawa	Sudbury	Vancouver
Date of Study	**2002**	**2003**	**2002–3**	**Jan. 2003**	**2002**
Reasons for situation					
Evicted	7.2	7.0	29.0	15.8	12.0
Moving/stranded	3.5	14.0	5.0	11.2	13.0
Can't afford to rent	17.5	19.0	23.0	15.2	10.0
Can't find accommodation	8.1	12.0	na	na	Na
Lost job/unemployed	12.5	na	na	10.3	1.0
Abuse, family breakdown	16.3	19.0	28.0	19.1	26.0
Addiction (incl. gambling)	5.1	11.0	8.0	9.5	5.0
Mental illness/disability	na	na	33.0*	9.7	2.0
Out of jail/treatment	na	3.0	na	7.1	6.0
Other (transient, fire, refugee, not said, etc.)	7.4	14.0	na	2.2	25.0
Home community					
% from local area	27.4 (>15 yrs)	62.0	46.0	na	71.0
% from elsewhere	72.6	38.0	54.0	na	29.0
Health conditions				(year prior):	
Physical disability/injury	12.5	6.0	**	29.1	15.0
Addiction (alcohol, drugs)	18.0	26.0	27.0/39.0	34.8	39.0
Depression, mental illness	48.3	21.0	31.0	51.5	23.0
Other medical condition	**	16.0	**	**	30.0
Length of stay in shelter					
Less than 6 months	na	79.0	68.0	na	68.0
2–7 days	na	31.0	na	na	12.0
1–4 weeks	na	28.0	na	na	26.0
1–6 months	na	20.0	na	na	30.0
6 months–1 year	na	5.0	13.0	na	12.0
More than 1 year	32.4	16.0	18.0	na	20.0

*Overlapping responses.

**A number of health conditions were listed but data were insufficient to determine accurate percentages for these categories for all homeless in the sample.

Sources: Aubry et al. (2003); BCMSDES (2001); City of Calgary (2002); GVRD (2002); HRM (2004); Kauppi (2003).

As the situation of the destitute becomes worse and more distinct the longer they remain homeless, middle-class families become more likely to keep half-empty houses even as they struggle to maintain and afford them. Their response is to press for tax cuts instead of offering their space for rent, fearful of the loss of privacy and the myriad regulations (e.g., building codes) on renting. The result is declining availability of affordable housing, as well as a divided urban fabric where the intersections of class, gender, race/ethnicity, and geography increasingly determine life chances. A major uncertainty remains as to whether this must characterize the future of Canadian cities. If present trends continue, the Canadian city will become more divided socially and spatially, and the extreme poverty of homelessness will become more visible. Growing homelessness does not mean a less safe city or a less competitive city, but instead a less equitable, less inclusive, less sustainable, less democratic, and less healthy city. The situation of the homeless could be improved in the decades ahead, but only if significant changes are made. There is the potential that the growing visibility of poverty and homelessness will foster a more inclusive city, in which the true nature of unregulated markets and capitalist production becomes evident to a greater proportion of the population, spurring demands for the radical changes that will be required.

Planning, Urban Policy, and Sprawl

The second example of a set of interrelated policy challenges selected for elaboration here, and the eleventh on our list in Table 25.1, involves planning practice and policy as it takes place in metropolitan areas, and in relation to the specific and contentious issue of sprawl and the associated vision of the 'compact city'. The cross-cutting themes identified earlier in this chapter—complexity, contradictory tendencies, and uncertainty—are as applicable to discussions of planning and urban policy formulation as they are to analyses of the socio-economic and ecological systems they are attempting to manage and guide. As extensive background on these same issues has been provided in previous chapters, notably Chapter 12 (Allahwala, Boudreau, and Keil) on governance, Chapter 13 (Skaburskis and Moos) on the economics of land use, Chapter 18 (Grant and Filion) on land-use planning, and Chapter 20 (Walker and Carter) on housing, they do not need to be described in detail here.

For present purposes, 'planning' is defined as the deliberate and rational attempt to improve the quality of life in our cities (Grant, 2008; Hodge and Gordon, 2008). As practised in Canada, however, urban planning is more limited in scope; it is essentially concerned with 'physical' planning, focused primarily on (but not limited to) negotiating issues with respect to land use, transportation, and the built environment. Other forms or dimensions of planning, for example, social planning, economic development, and environmental planning, are also in play but these are often second-order activities, at least at the municipal level. In most provinces the planning system operates within, and draws legitimacy from, the legal and legislative frameworks provided by provincial statutes (e.g., the Municipal Act, the Planning Act), and these frameworks vary widely across the country and, in practice, from one municipality to another within the same province. Already there is a source of potential contradiction here between the need for uniform provincial guidelines and standards and the increasingly diverse needs and priorities of local governments and communities.

Complexity also arises from the fact that planning is but one part of a larger envelope or cluster of public policies and governance structures that shape our cities. These other strands of urban policy formulation emanate from all three levels of government, federal, provincial, and municipal or local (including regional), each of which has different objectives, regulatory tools, and political constituencies. More to the point, one might think of planning practice and policies as the architects of urban form and land use at the micro scale (e.g., subdivision design), but in reality other policies may have an equal or greater effect in structuring our urban regions at more macro scales. Examples of the latter include provincial housing and taxation policies, and federal policies on immigration, trade, welfare, and the like. Combined, these latter policies constitute an *implicit urban policy regime*. Even though the stated objectives of such policies seldom include explicit urban objectives, in aggregate they drive the urban development process. These also are policy decisions over which local governments have little or no influence, yet they deal with the consequences.

Of course, in a highly urbanized country such as Canada almost all policies of the federal and provincial governments have urban impacts, but some have very uneven and indeed uncertain impacts on different cities and regions. For example, immigration policy, which has its own objectives and priorities, has become the major driver of the growth of a few large metropolitan areas in the last decades (see Hoernig and Zhuang, Chapter 9). This, in turn, has encouraged a further concentration of population and economic activity, which we assume was not one of the initial objectives. Trade policies also have very specific impacts on different cities because of the country's high level of regional economic specialization (e.g., automobiles in southern Ontario; forestry in BC and Quebec; oil and gas in Alberta and Saskatchewan). What happens to the gateway cities if immigration trends change and the flow of immigrants declines? What happens if certain forms of energy become prohibitively expensive? These possibilities introduce a great deal of uncertainty into public policy.

In this policy context we return to a discussion of the specific issue of urban sprawl and the attached vision of the compact city. As such, sprawl demonstrates several of the conundrums and overarching themes with which we began. Sprawl is typically defined as scattered, discontinuous, and unplanned urban development on the edges of urban areas. By this rigid standard we have relatively little sprawl in Canada. But the term is often expanded to encompass all forms of suburban development, especially low-density residential suburbs, and is applied as a label to highlight the disconnection among suburban environments; notably between exclusively residential subdivisions and the expansive low-density landscapes of employment and other uses—manufacturing, warehousing and storage, and retail malls and **power centres**. By another definition, 'sprawl' occurs when an urban area is expanding in land area at a faster rate than the rate of population (or household) growth. By this definition, virtually all Canadian metropolitan areas are afflicted by some form and level of sprawl.

The debate on sprawl is useful here precisely because it demonstrates the importance of differing perceptions of a problem and of the policies that appear most appropriate based on those perceptions. Without a doubt the peripheries of our cities continue to be highly 'contested grounds', in theory and practice, and the views differ depending on where one stands (Bourne et al., 2004). And, of course, a long history of research on the origins of suburbanization frames these contemporary debates (Beauregard, 2006; Harris, 2004). To some observers, sprawling forms of suburbanization are not necessarily a problem; indeed, they may be viewed as good, or at least as acceptable, given the benefits they offer to some households and producers (Bogart, 2006; Bruegmann, 2006; Richardson and Bae, 2004). Low-density suburbanization is seen, for example, as offering more choice in housing and living environments than do inner-city areas, and at lower cost. For some households, especially those with children, access to ground-level housing is a high priority in their choice of residential location.

For many employers the suburbs offer significant efficiencies, notably lower land and production costs, access to highways for truck movements, and the availability of space and facilities for warehousing, storage, and distribution (see Hutton, Chapter 7). The prevalence of advanced (and labour-saving) production technologies and just-in-time delivery systems in the modern industrial economy make these same space and accessibility benefits even more attractive. These uses are the primary source of declining densities of employment in urban regions in Canada, although few notice. Interestingly, as an example of contradictory discourses, few municipalities are willing to restrict commercial and industrial uses regardless of their density precisely because they generate tax revenues, require few services, and add jobs.

To other observers, in contrast, low-density suburbia, particularly exurban development, has an uncertain future due to many of the problems cited above—e.g., auto dependency, excessive energy consumption, congestion, environmental pollution, and the loss of agricultural land—as well as, in some instances, social exclusion and isolation (Bullard, 2009; Lindstrom and Bartlin, 2003; Sewell, 2009). Critics of sprawl argue its costs may be artificially depressed because suburban households and employers do not pay the full marginal cost of their low-density environments; specifically, they do not pay the full costs of infrastructure or regional services such as fire protection and social welfare. The reality of the modern suburban environment, even with its uneven distribution of costs and benefits, lies somewhere between these divergent interpretations. To be sure, suburbs have become considerably more diverse, and in some instances higher-density in recent years (e.g., Mississauga City Centre), than is depicted in the conventional image of suburbia. Yet they still are expensive to maintain, and in the case of newer suburbs and exurbs they are often disconnected from each other and from the larger urban fabric.

The pluses and minuses of suburban development further illustrate the difficulty of generalizing about processes as complex as urban growth and suburbanization. There are many purposes for, and many actors involved in, the suburbanization process, and these actors have different reasons for being involved. The process also illustrates our earlier thesis on the critical role of contradictions and countervailing tendencies. The initial reasons for suburban development (e.g., lower costs, less congestion) may quickly disappear, often generating their own downstream *contradictions*, which eventually reduce the initial benefits and/or shift the burdens to others. Moreover, suburban development often has negative implications for the rest of the urban fabric if it undermines the viability of older suburban areas or the inner city. When the overall growth rate for an urban region is high all three zones can, in theory, prosper and grow. But in conditions of slow (or negative) growth, continued suburbanization can drain limited resources from older parts of the urban region. Consider the challenges facing a no-growth (or declining) city. Does it impose severe restrictions on further suburban expansion to encourage infill and intensification? And if so, how?

Differences in urban form, coupled with the emerging dominance of suburban voters, may even influence divergences in political preferences that work to channel resources away from needed public infrastructure and welfare-state protections, and into tax reductions, making it difficult to address urban challenges (Walks, 2004a, 2004b, 2008). But are these problems of suburbanization per se and its highly selective nature, or are they instead reflections of the poor planning and short-sighted design of many suburban environments? Or are they the consequence of the lack of integration between transportation systems, housing, and land use? Or, more broadly, do they offer evidence of the larger contradictions of contemporary market-based urban development?

One response to the perception of sprawl as a problem involves efforts to limit low-density suburban development—one of the premises of smart growth (and new urbanism) strategies based on the vision of the compact city (see Grant and Filion, Chapter 18; Bullard, 2007). These efforts, however, face similar kinds of complexity and built-in contradictions. Ontario's Places to Grow (2006) plan for the Greater Golden Horseshoe and the accompanying Greenbelt Act (2005), establishing a green cordon around the urbanized core of the region, represent recent examples of such strategies. These are laudable plans, and represent innovative efforts to establish a vision of the sustainable, compact city and to outline an agenda for regional planning at an unusually extensive spatial scale. But they also illustrate both the potential conundrums of planning interventions and the complexities involved in trying to resolve competing interests and to address the countervailing processes of change. Further, by indicating where growth can occur and where it cannot in the GTA, they demonstrate the potential for a massive redistribution of the costs and benefits of urban and suburban growth among various fractions of the population. Whether this redistribution is in a direction that is in the broader public interest has to be addressed in each case.

Thus, well-intentioned policy initiatives can have counterproductive and contradictory consequences. For example, tighter limits on suburban growth, intended to restrict low-density suburban development and designed to encourage intensification and the development of transit nodes, also invariably raise land and housing prices for all. Are the assumed benefits of compact development greater than the costs of higher prices? And who benefits and who pays? Unless other policies are introduced more or less simultaneously, such as tax penalties for low-density development and tax incentives for intensification, as well as aggressive transit provision and new infrastructure, road congestion and air pollution will likely increase. Without strict planning regulations and incentives targeted at encouraging jobs and housing to 'co-locate', commuting levels and distances also may increase. As a final example of the trade-offs we face, the new greenbelt around the core of the GTA

(much like the earlier greenbelt in Ottawa and agricultural land restrictions in the Lower Mainland of British Columbia), while an excellent idea in theory with wide public support in practice, could lead some households and firms to leapfrog over the belt, thus augmenting wasteful levels of energy consumption and increasing long-distance commuting.

Conclusion: Alternative Futures

One could sketch a number of alternative scenarios for the future of Canadian cities, both collectively as an urban system and for individual places. These scenarios could vary in the extreme from the downright pessimistic and depressing—the image of decaying rust-belt cities with deteriorating infrastructure and growing homelessness, or depopulating resource-based communities—to the Utopian image of prosperous, healthy, comfortable, equitable, safe, and socially inclusive places where everyone walks or bikes to work. Many different urban futures are possible, but some are clearly more probable than others. In the case of Canadian cities there likely will be examples in the future arranged over the entire spectrum of possibilities, excepting the full Utopian option. The task, for policy-makers and students of the future city, is to define the outlines of a preferred future (and vision) for each place and then to make it happen.

Each chapter in this book has identified, explicitly or implicitly, anticipated directions of change in Canadian cities. By doing so, each chapter provides insight into the possible future states of living and working environments that these cities will offer to their residents and to newcomers. However, the issues summarized in Table 25.1 likely will provide the most explicit visions of what constitutes the 'good city'. These visions represent alternative normative scenarios for the future. How do we incorporate competing interests and different visions in our planning process? Whose visions should have priority and how should such priorities be determined? How do we balance the costs and benefits flowing from the adoption of one set of priorities against those of another set of priorities? Is it possible to achieve a realignment of competing interests—a kind of renewed civic engagement—moving us towards a common set of objectives? These questions—for planners, policy-makers, and concerned citizens—will be at the forefront of discussions and actions regarding the future of our cities in the years to come.

Review Questions

1. How might one discuss Table 25.1 in conjunction with the impact of globalization on Canadian cities?
2. Repeat the above question for all substantive chapters in the book.

Note

1. Comparison of the use of services and shelters for the homeless provides an indication of just how quickly this problem has grown. In only two years, between 1999 and 2001, those using shelters and sleeping rough increased by 21.8 per cent in Edmonton (ETFH, 1999; NHHN, 2001) and 9.6 per cent in Sudbury (Kauppi, 2003). For emergency shelters alone, usage increased by 11.2 per cent in the Municipality of Ottawa–Carleton over the 1996–2004 period (RMOC, 1999; ATEH, 2005) and by a whopping 59.1 per cent in Vancouver between 1999 and 2002 (GVRD, 2002).

References

Alliance to End Homelessness (ATEH). 2005. *Experiencing Homelessness: The First Report Card on Homelessness in Ottawa, 2005*. Ottawa: Alliance to End Homelessness/United Way of Ottawa.

Aubry, T., F. Klodawsky, E. Hay, and S. Birnie. 2003. *Panel Study on Persons Who Are Homeless in Ottawa: Phase 1 Results*. Ottawa: City of Ottawa and University of Ottawa. At: <www.socialsciences.uottawa.ca/crecs/eng/documents/PanelStudyonPersonsWhoAreHomelessinOttawa-03-12.pdf>. (Nov. 2009)

Bacher, J. 1993. *Keeping to the Marketplace: The Evolution of Canadian Housing Policy*. Montreal and Kingston: McGill-Queen's University Press.

Beauregard, R. 2006. *When America Became Suburban*. Minneapolis: University of Minnesota Press.

Bogart, W. 2006. *Don't Call It Sprawl: Metropolitan Structure in the 21st Century*. New York: Cambridge University Press.

Bourne, L.S., M. Bunce, N. Luka, and L. Taylor. 2004. 'Contested ground: The dynamics of peri-urban growth in the Toronto region', *Canadian Journal of Regional Science* 27: 251–69.

British Columbia Ministry of Social Development and Economic Security (BCMSDES). 2001. *Homelessness—Causes and Effects*, vols 1–4. Victoria: Government of British Columbia Ministry of Social Development and Economic Security, and British Columbia Housing Management Commission.

Bruegmann, R. 2006. *Sprawl: A Compact History*. Chicago: University of Chicago Press.

Bullard, R.D. 2007. *Growing Smarter: Achieving Livable Communities, Environmental Justice and Regional Equity*. Cambridge, Mass.: MIT Press.

City of Calgary. 2002. *The 2002 Count of Homeless Persons—2002 May 15*. Calgary: City of Calgary Community Vitality and Protection, Community Strategies, Policy and Planning Division.

———. 2008. *Biennial Count of Homeless Persons in Calgary 2008 May 14*. Calgary: Community and Neighbourhood Services Social Research Unit.

City of Toronto. 2006. *2006 Street Needs Assessment: Results and Key Findings*. Toronto: City of Toronto Community Services Committee Staff Report.

Edmonton Task Force on Homelessness (ETFH). 1999. *Homelessness in Edmonton: A Call to Action*. Edmonton: City of Edmonton.

Federation of Canadian Municipalities (FCM). 2005. *Incomes, Shelter, and Necessities: Quality of Life in Canadian Communities, Theme Report #1*. Ottawa: Federation of Canadian Municipalities.

Fournier, L. 1991. *Itinérance et santé mentale à Montréal, étude descriptive de la clientèle des missions et refuges*. Montréal: Unité de Recherché Psychosociale, Centre de Recherché de l'Hôpital Douglas.

———, S. Chevalier, M. Ostoj, R. Caulet, R. Courtemanche, and N. Plante. 1998. *Dénombrement de la clientèle itinérante dans les centres d'hébergement, les soupes populaires*

et les centres de jour des villes de Montréal et de Québec,
1996–97. Québec: Santé Québec.

Grant, J. ed. 2008. *Canadian Planning: Linking Theory and*
Practice. Toronto: Thompson/Nelson.

Greater Vancouver Regional District (GVRD). 2002. *Home-*
lessness in Greater Vancouver. Vancouver: Greater Vancouver Regional District.

Halifax Regional Municipality (HRM). 2004. *Homeless-*
ness in HRM: A Portrait of Streets and Shelters. Halifax:
Planning and Development Services, Halifax Regional
Municipality.

Harris, R. 2004. *Creeping Conformity: How Canada Became*
Suburban, 1900–1960. Toronto: University of Toronto
Press.

Harvey, D. 1989. *The Urban Experience.* Baltimore: Johns
Hopkins University Press.

Hodge, G., and D.L.A. Gordon. 2008. *Planning Canadian*
Communities: An Introduction to the Principles, Practice and
Participants. Toronto: Thompson/Nelson.

Immergluck, D. 2009. 'Core of the crisis: Deregulation,
the global savings glut, and financial innovation in the
subprime debacle', *City and Community* 8: 341–5.

Kauppi, C. 2003. *Report on Homelessness in Sudbury: Com-*
parison of Findings: July 2000 to January 2003. Sudbury:
City of Greater Sudbury and Social Planning Council
of Sudbury.

Layton, J. 2000. *Homelessness: The Making and Unmaking of a*
Crisis. Toronto: Penguin.

Lindstrom, M., and J. Bartling, eds. 2003. *Suburban Sprawl:*
Culture, Theory and Politics. Oxford: Rowman & Littlefield.

National Housing and Homeless Network (NHHN). 2001.
State of the Crisis: A Report on Housing and Homelessness in
Canada. Toronto: National Housing and Homelessness
Network and the Toronto Disaster Relief Committee.
At: <http://tdrc.net/resources/public/Report-01-11-
NHHN.htm>. (May 2005).

O'Flaherty, B. 2003. 'Wrong person and wrong place: For
homelessness, the conjunction is what matters', *Journal*
of Housing Economics 13: 1–15.

Regional Municipality of Ottawa–Carleton (RMOC). 1999.
Homelessness in Ottawa–Carleton. Ottawa: Regional
Municipality of Ottawa–Carleton.

Richardson, H., and C. Bae, eds. 2004. *Urban Sprawl in West-*
ern Europe. Aldershot: Ashgate.

Sewell, J. 2009. *The Shape of the Suburbs: Understanding Toron-*
to's Sprawl. Toronto: University of Toronto Press.

Simmons, J., and L.S. Bourne. 2007. 'Living with population growth and decline', *Plan Canada* 47, 2: 13–21.

Stone, M.E. 1993. *Shelter Poverty: New Ideas on Housing*
Affordability. Philadelphia: Temple University Press.

Walks, R.A. 2004a. 'Place of residence, party preferences,
and political attitudes in Canadian cities and suburbs',
Journal of Urban Affairs 26: 269–95.

———. 2004b. 'Suburbanization, the vote, and changes in
federal and provincial political representation and influence between inner cities and suburbs in large Canadian urban regions, 1945 to 1999', *Urban Affairs Review*
39: 411–40.

———. 2008. 'Urban form, everyday life, and ideology:
Support for privatization in three Toronto neighbourhoods', *Environment and Planning A* 40: 258–82.

——— and R. Maaranen. 2008a. 'Gentrification, social
mix, and social polarization: Testing the linkages in
large Canadian cities', *Urban Geography* 29: 293–326.

——— and ———. 2008b. *The Timing, Patterning, and*
Forms of Gentrification and Neighbourhood Upgrading in
Montreal, Toronto, and Vancouver, 1961–2001. Toronto:
University of Toronto Centre for Urban and Commun-
ity Studies/Cities Centre, Research Paper 211.

Wolfe, D. 2009. *21st Century Cities in Canada: The Geog-*
raphy of Innovation. Toronto: Conference Board of
Canada.

APPENDIX A

Digital Data in Urban Research

Paul Langlois

This book addresses the notion of transition, and just as Canadian cities have been in transition in recent years, so too have the ways in which students and academics carry out research on cities. On one hand, urban research has been transformed by the adoption of theory and paradigms from other disciplines, such as sociology, cultural studies, and communications studies. This has led to research—primarily qualitative—being carried out from a number of novel and fascinating perspectives, including post-colonial, feminist, and queer theory. On the other hand, there has been an equally impressive re-engagement with approaches that employ quantitative methods due to the ever-increasing availability of spatial and statistical data in digital formats. Researchers now have a wealth of data at their disposal, from high-resolution satellite imagery and street network files to demographic information and census boundary maps. The crucial advance has been the ability to link together these disparate data types using GIS and statistical software. This allows researchers to link events and processes to specific areas and to explore the importance of place on human activity by highlighting patterns at different scales.

Academics are not the only ones who have realized the power and benefits of being able to easily map processes and statistics spatially. All levels of governments, for example, have come to use spatial data heavily, most obviously to analyze census data but also for more immediate concerns such as determining the optimal location for facilities such as fire stations and transit stops, or for monitoring the prevalence of low income or mortgage foreclosures or any number of other circumstances. The private sector, too, now uses spatial data for many tasks, including logistical purposes such as route planning by firms that use the street network, and also for tasks such as evaluating the feasibility of locations for retail or manufacturing facilities by analyzing the socio-demographic makeup of the surrounding population or the accessibility of the site.

As a result, a vast amount of spatial and statistical data exists, some of it free, some of it accessible through special agreements, and some of it available only by purchase. Faculty and students of Canadian universities and colleges are particularly fortunate in having access to substantial resources of digital data, including data otherwise available only through purchase. Most large university and college libraries maintain on their websites lists of available data. Many institutions have at least one staff member dedicated to spatial data, as well as student volunteers, all of whom can help locate data and demonstrate how it can be used.

For anyone studying Canadian cities, Statistics Canada (www.statcan.ca) is an invaluable data resource. The United States Census Bureau (www.census.gov) is the American counterpart. Statistics

Canada makes a wealth of information available to the public in digital form, from raw census data and summary tables to time series and boundary files for all census administrative units. More detailed data files are typically available only for purchase by the general public, but can usually be accessed free of charge by faculty and students of Canadian universities and colleges. Statistics Canada also will create custom data for individuals and organizations, but the costs can be prohibitive.

Over the last decade, municipal, regional, and provincial levels of government have become significant producers—and users—of digital data. Unfortunately, not all of these data are generally available to the public, although governments may grant access to students individually or to their institution under some circumstances. In addition, the various levels of governments may not provide spatial data in sufficient detail or scope for certain kinds of urban research. Governments are primarily interested in political boundaries and therefore tend to produce data based on geographic units such as census tracts. Data not directly linked to political boundaries, such as maps that show the locations of things like schools, parks, or various land uses, may only be available for purchase from private-sector vendors.

Spatial data, therefore, can require some effort to track down. Other than Statistics Canada, good starting points include Geobase (www.geobase.ca), GeoConnections (www.geoconnections.org), National Resources Canada (www.nrcan.gc.ca), and GeoGratis (geogratis.cgdi.gc.ca). As mentioned, check with your institution's library for further information about data sources, including special arrangements with other academic institutions, local governments, or private-sector vendors. Special mention must also be made of Google Maps (maps.google.com) and Google Earth (earth.google.com) software, both of which provide free access to high-resolution satellite imagery of most major urban areas on the planet. Because of their ease of use, these programs can be a very useful aid to many forms of urban research, particularly the qualitative aspects of place that are difficult to capture with spatial data alone.

While many profound advantages and opportunities are provided by digital data, there are drawbacks as well. Comparing data from multiple censuses, for example, can be extremely time-consuming when census tract boundaries change. In other cases, accurate comparison may be difficult to achieve if the wording of census questions has changed or the list of possible answers is different from one census to the next. Beyond these kinds of instrumental circumstances, it is widely accepted that the accuracy and reliability of census information vary widely for a number of reasons related to how the census is configured and how it is carried out. Census data, in other words, are invaluable, but no one considers such statistical information to be perfect. A similar caveat applies to more purely spatial data, such as street network files and maps of amenity locations. The creation of these sorts of data requires considerable painstaking human effort and errors inevitably creep in, often requiring many hours of 'cleaning' and verification before these files can be reliably used.

The relatively widespread availability of digital data has transformed how we formulate and carry out research. It is now, for example, the work of literally a few minutes to map out demographic or economic data that would have represented days or weeks of labour only a few years ago. The corollary, which we would do well to remember, is that with the widespread availability of spatial and numeric data, we also now have the ability to make more mistakes more quickly than ever before.

Appendix B Selected Data on Canada's Census Metropolitan Areas[1] (minimum and maximum values in bold)

Rank	CMA	Population (000s) 2006	Population Δ(%)	Foreign Born (% of population) 2006	Foreign Born Δ(%)[3]	Recent Immigrants[2] (% of population) 2006	Recent Immigrants Δ(%)[3]	Single-Person Households (%) 2006	Single-Person Households Δ(%)[3]	Population Aged 65 and Over (%) 2006	Population Aged 65 and Over Δ(%)[3]	Unemployment Rate (%) 2006	Unemployment Rate Δ(%)[3]	Average Personal Income ($) 2006	Average Personal Income Δ(%)[3,4]	Average Household Income 2006	Average Household Income Δ(%)[3,4]	Incidence of Low Income (%) 2006	Incidence of Low Income Δ(%)[3]	University Degree[5] (% of population) 2006	University Degree Δ(%)[3]
1	Toronto	**5,113**	31.1	**46.4**	22.5	**15.8**	26.5	22.9	5.3	11.0	6.2	6.7	-21.2	40,704	7.2	87,820	12.1	18.4	**26.0**	26.7	60.6
2	Montreal	3,636	13.3	20.9	24.9	7.1	28.2	31.6	16.1	12.5	11.0	6.9	-41.0	34,196	8.4	63,038	10.2	**21.1**	-4.1	21.0	57.8
3	Vancouver	2,117	32.1	40.4	34.8	14.2	42.9	28.5	6.0	11.7	-4.1	5.6	-39.1	36,123	4.6	73,258	10.0	20.8	19.5	24.6	71.5
4	Ottawa-Gatineau	1,131	20.1	18.3	25.1	5.8	17.3	27.0	12.7	10.7	11.1	5.7	-21.9	41,765	11.6	80,838	12.8	14.7	1.4	28.7	42.3
5	Calgary	1,079	**43.1**	23.9	18.2	8.7	29.6	24.9	11.7	**8.6**	11.1	**4.0**	**-50.0**	48,878	**37.0**	**98,253**	**41.7**	13.4	-22.1	24.7	51.4
6	Edmonton	1,035	23.0	18.7	2.1	5.0	-11.6	26.5	14.6	10.2	20.4	4.6	-44.6	39,901	23.3	79,163	26.9	14.1	**-25.0**	18.3	39.0
7	Quebec	716	10.8	3.7	77.7	1.8	106.7	32.9	23.8	13.1	22.8	4.6	-49.5	33,866	10.5	60,884	9.6	16.0	-13.5	20.2	45.7
8	Winnipeg	695	5.2	17.8	2.5	5.0	-0.6	30.2	11.2	12.4	-3.8	5.0	-41.9	33,838	13.3	64,533	14.9	18.8	-7.4	19.0	46.3
9	Hamilton	693	15.5	24.6	5.0	5.6	29.5	25.5	13.7	13.8	7.8	6.0	-32.6	38,299	13.4	76,787	15.6	15.7	4.0	17.5	58.2
10	London	458	20.0	19.5	4.3	4.7	-4.3	28.1	12.4	12.6	4.0	6.1	-28.2	36,720	11.0	70,345	12.0	13.7	0.7	18.3	**35.1**
11	Kitchener	451	26.6	23.3	8.8	6.7	12.4	23.1	12.6	10.6	3.5	5.6	-37.8	38,381	16.5	78,223	18.8	10.5	-10.3	18.4	54.5
12	St Catharines-Niagara	390	7.1	18.4	-2.0	3.3	40.9	26.7	21.0	**16.2**	8.3	6.2	-34.7	33,170	8.8	65,053	9.9	12.5	-3.1	13.1	59.3
13	Halifax	373	16.3	7.5	14.8	2.2	29.7	27.7	29.3	11.0	15.9	6.3	-31.5	35,031	9.6	66,325	7.6	14.3	1.4	24.0	42.5
14	Oshawa	331	37.7	16.5	-15.5	2.4	-20.3	**20.1**	19.7	10.4	15.4	6.4	-24.7	39,644	9.7	82,205	12.5	**9.3**	0.0	13.1	57.2
15	Victoria	330	14.7	19.3	-6.9	3.2	18.1	33.3	14.8	15.7	**-15.3**	4.3	-44.2	37,065	13.7	67,838	13.0	13.2	-2.9	23.6	61.0
15	Windsor	323	23.4	23.6	37.8	7.8	62.1	26.9	11.2	12.0	-5.8	8.3	-29.7	37,330	15.1	72,796	18.1	14.1	-4.1	17.8	**72.6**
17	Saskatoon	234	10.9	7.8	-3.8	2.4	27.0	28.8	11.4	11.1	7.0	5.2	-40.2	35,147	18.0	66,059	20.1	16.3	-13.8	19.4	36.7
18	Regina	195	1.7	7.7	-7.2	2.2	14.8	29.4	16.7	11.8	8.3	4.9	-32.9	36,272	12.9	68,280	12.9	13.5	-14.6	18.4	42.2
19	Sherbrooke	187	32.9	5.7	53.4	2.9	85.9	34.0	20.8	13.1	14.1	6.9	-36.7	30,451	10.9	53,301	10.5	16.8	-16.4	17.6	46.6
19	St John's	181	5.4	2.9	5.2	0.9	6.8	22.4	52.2	10.4	11.8	**10.0**	-37.9	32,756	11.2	65,852	4.9	15.5	-4.3	18.8	58.6

Rank	CMA	Population (000s)		Foreign Born (% of population)		Recent Immigrants[2] (% of population)		Single-Person Households (%)		Population Aged 65 and Over (%)		Unemployment Rate (%)		Average Personal Income ($)		Average Household Income		Incidence of Low Income (%)		University Degree[5] (% of population)	
		2006	1991–2006 Δ (%)	2006	1991–2006 Δ (%)[3]	2006	1991–2006 Δ (%)[3]	2006	1991–2006 Δ (%)[3]	2006	1991–2006 Δ (%)[3]	2006	1991–2006 Δ (%)[3]	2006	1991–2006 Δ (%)[3,4]	2006	1991–2006 Δ (%)[3,4]	2006	1991–2006 Δ (%)[3]	2006	1991–2006 Δ (%)[3]
20	Abbotsford	159	40.0	23.9	29.8	6.8	na	22.9	na	12.1	-11.4	5.5	na	31,149	na	66,041	na	14.0	na	11.6	na
20	Greater Sudbury	158	0.4	6.7	-18.3	0.7	6.2	27.0	26.0	13.9	33.0	7.9	**-8.1**	35,941	9.1	68,071	7.0	12.7	-5.9	13.2	47.0
21	Kingston	152	11.7	12.6	-6.3	2.3	-0.8	27.5	15.3	14.1	14.9	6.5	-13.3	36,386	11.6	69,185	11.5	13.4	2.3	21.7	35.7
22	Saguenay	152	**-5.8**	**1.2**	67.7	**0.6**	158.1	28.8	**54.9**	14.1	**55.3**	8.8	-33.3	30,377	**0.0**	55,552	2.2	14.2	-11.8	12.5	50.5
23	Trois-Rivières	142	3.8	2.2	**85.0**	1.1	**206.0**	**34.7**	29.9	15.5	31.8	7.3	-45.1	**29,614**	5.6	**51,683**	5.2	18.5	-9.3	13.6	49.8
26	Thunder Bay	123	-1.6	10.4	**-21.0**	0.9	**-30.2**	30.1	23.0	14.6	9.4	7.4	-22.1	34,245	4.9	64,470	**1.6**	12.8	4.1	14.8	56.0
27	Saint John	**122**	-2.7	4.2	-8.8	1.1	47.3	25.4	17.4	12.5	0.8	8.0	-29.8	31,920	9.0	61,234	10.6	14.7	-13.0	14.1	58.3
	Mean	766	16.2	15.9	15.9	4.5	35.7	27.7	19.4	12.4	10.5	6.3	-33.5	35,895	11.8	69,522	12.8	14.9	-4.7	18.7	37.9
	Weighted Mean[6]	—	22.4	27.1	21.0	8.8	28.9	27.1	12.7	11.7	8.1	5.8	-31.4	33,643	12.1	61,736	12.1	14.7	-4.4	17.0	37.2
	Median	330.6	14.7	17.8	5.2	3.2	26.8	27.5	15.7	12.4	9.4	6.2	-34.0	35,941	11.0	67,838	11.7	14.2	-4.2	18.4	38.4
	Std. Deviation	1149.2	13.8	11.1	28.5	3.9	52.9	3.7	11.9	1.9	14.4	1.4	10.5	4,128	6.9	10,345	8.1	2.8	11.2	4.6	9.9

Notes:

[1] Based on 1991 CMAs, for comparison purposes.

[2] Arrived in Canada 1996–2006.

[3] Values based on proportional, not absolute, change.

[4] 1991 dollars converted to 2006 dollars using CPI.

[5] Bachelor's degree or higher.

[6] Weighted mean uses 2006 population.

Sources: 2006 data obtained from 2006 CMA profiles; 1991 data obtained from 1996 CMA profiles; 1991 EA profiles by region (long form); 1991 2B profile (detailed questionnaire); 1991 BST CT short form. (Compiled by Paul Langlois)

Glossary

age-friendly city/community A concept advanced by public agencies that asks local governments to take into account the particular needs of older people, who make up an increasing proportion of the population. This vision often draws on a three-part framework: (1) participation; (2) health; and (3) security and independence of the older population.

bid-rent curves A modelling concept in economic geography used to understand and depict the trade-offs made by economic agents (e.g., a household, a firm) between rent and distance. At any point along one bid-rent curve the economic agent is equally satisfied with the combination of location (in relation to the urban centre) and the rent cost to occupy that location. Any negative change in the desired distance from the urban centre, along one bid-rent curve, is compensated for by an equally desirable change in rent cost, such that the economic agent remains indifferent.

brownfield sites Former industrial locations that can become the object of redevelopment efforts and may require decontamination; see *greyfield sites*.

business improvement areas Parts of cities, usually primarily retail and older, where business owners have banded together, agreeing to pay costs (usually through an added municipal tax) to support renovations to make the area more attractive and functionally up-to-date and competitive (e.g., street furniture and planting, parking, pedestrian amenities). Usually some level of partnering is provided by one or more higher levels of government.

citizenship Formal legal rights and responsibilities conferred automatically upon the citizens of a state, as well as rights (e.g., attending public meetings, voting) from which minority groups might feel/be 'excluded'.

commodification Making a commodity of some intangible attribute of urban space. Commodification of the core, for example, would entail the notion that one can purchase (or own) some of the ambience that is attributed to a core area; see *milieu effect*.

community gardens Land space provided by a municipality to individuals and/or groups who contract to actively use and maintain vegetation they have planted. Such gardens are believed to be a step towards municipal food self-sufficiency.

competitive city A city that competes, economically and culturally, with other cities on a national, continental, and especially a global scale. Today, competition to gain 'world city' status, or to strengthen a city's position in the global network, is believed to be a primary factor underlying the urban agenda of a city, especially larger, fast-growing cities.

core housing needs A measure of the housing circumstances of Canadians that combines three standards for housing: (1) adequacy, such that it does not require major repair; (2) suitability, as defined by the National Occupancy Standards for number and type of household members per room; and (3) affordability, as defined by the shelter cost-to-income ratio of 30 per cent of gross household income. A household that fails to meet any of the three standards and is unable to access alternative local housing is said to be in core housing need.

dislocation   The exodus of a major occupant or occupants or of a specific use from a distinctive geographic zone within the city, e.g., the dislocation of low-income residents from the centre of the city, in which case gentrification is most often identified as the dislocating force.

ecological footprint   The resource requirements of an urban area measured in terms of the surface of the earth needed to produce these resources. Ecological footprint can also refer to the surface of the planet needed to absorb (neutralize) the pollution generated by an urban area.

ecological modernization   A weak approach to environmental sustainability that focuses on technological solutions to environmental problems, such that symptoms of our ailing biophysical environment are treated but the underlying processes, structures, and values that create environmentally unsustainable communities remain unchallenged. For example, the development of and incentives to purchase hybrid vehicles might be seen as ecological modernization, while designing pedestrian-, bicycle-, and transit-oriented neighbourhoods and trying to change the cultural values underlying transportation mode choices could be seen as sustainable community development.

entrepreneurial municipal regimes   Forms of municipal administration that emphasize the support of private-sector initiatives or that orient municipal policy-making principally around economic development objectives.

food deserts   Areas of a city, usually of low income, without accessible outlets that provide healthy and affordable food for household consumption.

food systems   The areas and agents that constitute the supply end of the food chain along with all the components of food distribution and consumption in cities.

Fordism   A period of economic development that lasted roughly from the 1920s until the late 1970s, when growth rested on a correspondence between rising consumption and increasing mass production. Fordism required ongoing Keynesian-type government interventions to stimulate consumption.

Fordist-Keynesian   Economic development and economic and social policy-making that relied on government intervention in the form of various welfare-state and demand stimulation measures. The period lasted from the end of World War II until the late 1970s; see *Fordism*; *Keynesianism*.

gentrification   The process whereby high-income households purchase and upgrade central-city housing that once was occupied by residents of a significantly lower income. Today, some would consider other kinds of residential upgrading such as condominium development as gentrification.

ghettos   Space in cities that segregate low-income and/or minority households who lack the freedom, as a consequence of income and/or prejudice, to move into residential zones elsewhere in the city. Originally used in the eighteenth, nineteenth, and early twentieth centuries to refer to neighbourhoods that housed segregated Jewish populations.

governance   The work of government institutions, along with all the instances and processes with an impact on government decision-making. Governance thus provides a much broader perspective on the political process than the concept of government does.

greyfield sites   Abandoned retail locations; see *brownfield sites*.

heartland   The part of Canada where the industrial economy is concentrated. The heartland is also the location of the largest metropolitan regions. The Canadian heartland runs from Quebec City to Windsor.

hinterland   Parts of Canada that depend on natural resources. The hinterland includes all the country with the exception of the heartland.

intermediate goods   Products, finished or semi-finished, that represent an input into a final demand product—e.g., fenders or seatbelts to auto-assembly lines—or to another good that will ultimately be input to a final demand good.

Keynesianism   Economic approach formulated by John Maynard Keynes according to which the market economy benefits from countercyclical government spending. Keynesianism has been associated with public-sector economic development and social programs.

knowledge-based economy   Perspective by which economic development increasingly depends on the presence of an educated workforce. The importance of knowledge in the economy is related to deindustrialization, automation, and the growth of the high-order tertiary sector.

knowledge-intensive economic activity   That part of the economy based on ideas and higher-order services, as opposed to manufacturing and primary (resource) production.

land rent   A value derived within a land market for the use of land, affected by site characteristics such as location. An economic agent (e.g., firm or household) is willing to pay a certain rent to the landowner for the use of the owner's property for a period of time. For comparability of land values across an urban area, it is common conceptually to think of landowners who use that land themselves (e.g., for their private home), instead of renting it to others, as effectively paying 'rent' to themselves for use of their property.

life course   A concept recognizing that individuals move through stages in life defined in part by their personal biographies but also converging around transitional events that are roughly in common throughout a population (e.g., leaving school, leaving the parental home, entering a conjugal relationship). Life course transitions can be examined schematically by grouping key transitional events into meaningful life stages.

livable cities   Cities generally agreed to be 'good' places to live. Often, livability is assessed using clearly defined indicators. Canadian cities generally have ranked high in published statistical reports that claim to measure urban quality of life or livability.

micro-spaces of the core   The concept that the urban core is comprised of specialized sub-areas, usually of a pedestrian or walkable scale, and most often identifiable by function—e.g., law courts, hospital/medical complexes, entertainment districts, retail areas—or by district affiliation—e.g., Gastown, Yorkville. In the twenty-first-century city, these spaces also might include distinctive residential areas, historic districts, and spaces with distinctive landscape features.

milieu effect   The positive and/or negative overall sense of a place associated with a distinctive locale.

mixed-use development   Forms of urban development that comprise different types of activities. Mixed-use developments are often proposed as an instrument to reduce the dependence on the automobile.

multiculturalism   The official policy of the Canadian government that minority groups participate fully in Canadian society while also maintaining distinctively different social values, practices, and institutions, provided the latter adhere to the Canadian Charter of Rights and Freedoms and provincial human rights legislation.

neo-liberalism   Tendency for a withdrawal of governments from the economic and social scene, so as to increase reliance on the private sector and market processes. Neo-liberalism was meant to reverse Keynesian policies.

new economy An economy that reflects recent economic changes stemming from de-industrialization, the rise of high-order tertiary activities, and globalization.

NIMBY (not in my back yard) Reactions against changes happening around one's residence. NIMBY movements are usually targeted at intensification of land use, infrastructure developments, and uses and activities that local residents do not want near them, such as strip clubs, halfway houses, group homes, and landfills. These movements can be locally based or consist of federations of local groups.

non-governmental organizations (NGOs) Organizations that provide/deliver goods or services that might normally be delivered by a government agency—e.g., a homeless shelter. During neo-liberal times, Canadian cities have relied more heavily on NGOs to provide important municipal services that would have been provided by an arm of government during the modern era of the welfare state.

path dependence A perspective by which certain tendencies are long-lasting and difficult to alter because they are supported by institutional arrangements and processes.

place-making Planning efforts to insert physical/architectural features and events into the urban environment to help make a city or a particular part of a city more appealing, hence more 'place-ful' and competitive globally as well as at more local scales.

polarization A distribution that is skewed towards the two ends of the attribute that is being measured. Under conditions of the new economy, income is said to be polarized because major segments in the population fall into either relatively high- or low-income groups.

post-Fordism The period succeeding Fordism characterized by a dismantling of Fordist mechanisms and their replacement by more market-oriented (neo-liberal) processes.

power centres Clusterings of specialized stores of different size along with discount department stores in an automobile-oriented environment. In contrast with shopping malls, there is little common space in power centres, notwithstanding large parking areas.

producer services Services contracted out that cater to producers of final demand goods or services—e.g., contracted legal work, accounting, maintenance, and cleaning.

productive diversity The attraction to new enterprise of a city that is economically diversified and boasts a talented labour force.

push and pull factors Circumstances that influence households' decision to migrate or move; 'push' factors are negative attributes of the current place of residence and 'pull' factors represent the attraction of a relocation alternative.

qualitative development An approach to urban development that departs from a fixation on urban expansion and population growth (i.e., quantitative development), focusing instead on the existing built environment, infilling and redeveloping, and conserving or adapting existing buildings for reuse, with attention to preserving and accentuating a sense of place and urban quality, often at a pedestrian scale.

revitalization/regeneration Renewal or regrowth of an obsolete sector of the economy or area of the city, such as the reinvigoration of the core and inner city in large Canadian metropolitan areas in the twenty-first century.

slow-growth cities Cities where population growth over a 10-year period is less than 10 per cent. Given the high proportion of the Canadian urban system on slow-growth trajectories or in decline (losing population), urbanists are calling for more sophisticated and realistic approaches to urban development that are not centrally focused on unrealistic expectations of continuous growth; see *qualitative development*.

social cohesion The strength of social bonds in society between people from different ethno-cultural backgrounds and socio-economic classes. Strengthening social cohesion is a common social policy goal of state bureaucracies and politicians, and a target for social programming, particularly in diverse societies like Canada.

social housing Government-funded housing provided to low-income households whose housing needs are not adequately met by the private real estate industry. Rent is subsidized such that the household does not pay more than 30 per cent of its gross income.

survival curve Depiction of the proportion of a population surviving at a particular age in life. Given the very low infant mortality rate and significantly reduced mortality at older ages, demographers and human health experts discuss the possibility of nearly all humans living to a genetically fixed age limit as mortality at earlier ages becomes less common, creating a rectangular survival curve.

temporary foreign workers (TFWs) Workers allowed into a country for a prescribed period in specified employment. As such, most rights of citizenship are not available to TFWs.

topophilia Love of place, a term coined by geographer Y.-F. Tuan. It pertains to the growing interest for place in planning and an awareness of the importance of place for many people. The opposite term, 'topophobia', denotes fear of place.

Tower in the Park Model of urban development conceived by Le Corbusier, which consists of high-rise buildings set in a park-like environment. The model has been popular all over the world and has been criticized by Jane Jacobs.

transnational A term used in reference to an immigrant who attains citizenship in one country but keeps up ties with his/her place of origin and/or former residence.

transportation demand management (TDM) A recent strategy used by transportation planners. In the past traffic was simply forecast and accommodated, but TDM attempts to change the demand itself rather than simply accommodate demand—e.g., shifting hours of work in one or more large employment sectors in order to reduce congestion during periods of rush hour or peak load.

urban dynamics Human behaviour taking place in cities; also, journey patterns within urban areas.

urban ecosystem How natural systems function within the built environments of cities.

urban form The configuration of urban areas. Urban form can pertain to the distribution and density of activities within metropolitan regions or to design features of specific places within cities.

urban renewal Strategic reuse of an area of the city that is underused and often run down due to forces of change and transition. Urban renewal schemes are usually planned comprehensively under the direction of professional planners and at least partially funded by one or more levels of government.

urban sustainability Conditions required to assure the long-term availability of the natural resources (including pure water and air) required for the existence of urban settlements. Urban sustainability is increasingly perceived in a global context, such as the contribution of cities to planetary environmental degradation, e.g., global warming. Sustainability can also be defined in more narrow economic terms.

vertical farms The use of high-density urban space for purposes of food cultivation. The term spans a spectrum of practices, from roof gardens to 'factory farms'.

walkability Configurations of urban space that are pedestrian-friendly and so promote walking from place to place within walkable sub-areas. A major goal of twenty-first-century land-use planning is to increase the walkability of Canadian cities.

welfare state Strong state/government involvement in the provision of basic needs, such as health care, housing, and old age security, as well as government intervention in matters more typically dealt with by the private sector, such as wage rates.

In Canada the term is most often associated with the Fordist period of urban economic growth.

world city, global city Very large cities that interact as much or more—in terms of the flows of information, finances, goods, and people—with other places globally as with cities in their own country, and where growth is propelled by global rather than local factors. Various typologies rate different cities' position on a global hierarchy.

Index

This completely revised fourth edition of *Canadian Cities in Transition* examines in depth the major transformations taking place in urban Canada—and the transformations that must be set in motion if the society is to survive. Presenting the city in all its facets—historical evolution, economic dynamics, environmental impacts, urban lifestyles, cultural makeup, social structure, infrastructures, governance, planning, appearance—it is designed to help the next generation address the urban problems they are inheriting: traffic congestion, environmental damage, crime, social segregation, 'food deserts', governance. Topics new to this edition include Aboriginal peoples in urban Canada, urban food systems, the need for more 'walkable' cities to stem the growing obesity epidemic, and the startling but accurate concept of cities as human 'feedlots'.

All of the 25 chapters have been written specifically for this edition by experts in the fields of urban geography, planning, governance, transportation, and environmental studies.

FEATURES

- ❖ A **new** glossary facilitates comprehension of key concepts and terms.
- ❖ **New** part openers organize the text into four parts and introduce the major themes.
- ❖ **New** end-of-chapter review questions and in-text cross-chapter references encourage students to explore material in greater depth.
- ❖ A **new** two-colour design makes the text more accessible, and an expanded art program of maps, figures, and photographs helps readers contextualize key locales and important concepts.
- ❖ Updated appendices with the latest census data on metropolitan areas and information on recent trends in urban research ensure a clear understanding of the most current material.
- ❖ **New** online ancillaries include a student study guide to enhance students' comprehension and a test bank to aid instructors' class preparation.
- ❖ Foundational chapters from the third edition are provided online as resource material.

TRUDI BUNTING is a professor in the Department of Geography and School of Planning, University of Waterloo. PIERRE FILION is a professor in the School of Planning, University of Waterloo. RYAN WALKER is an assistant professor in the Department of Geography and Planning, University of Saskatchewan.

 www.oupcanada.com/bunting4e

OXFORD
UNIVERSITY PRESS

ISBN 978-0-19-543125-4

9 780195 431254